W9-ABW-312

For Reference 1989

DEC

Not to be taken from this room

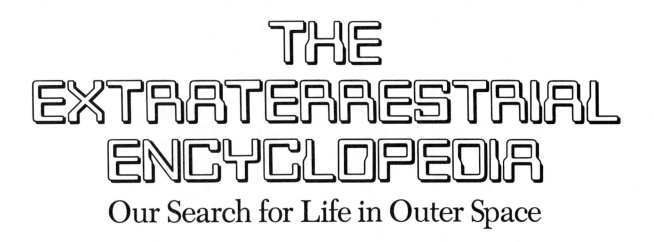

THE EXTRATERRESTRIAL ENCYCLOPEDIA

Our Search for Life in Outer Space

By Joseph A. Angelo, Jr.

Facts On File Publications
New York, New York ● Oxford, England

The Extraterrestrial Encyclopedia
Our Search for Life in Outer Space

Copyright © 1985 by Facts On File, Inc.

Library of Congress Cataloging in Publication Data
  Angelo, Joseph A.
    The extraterrestrial encyclopedia.

    Bibliography: p.
    1. Life on other planets—Dictionaries. 2. Interstellar communication—Dictionaries. I. Title.
  QB54.A523 1984   574.999   83-5599
  ISBN 0-87196-764-2

British Library Cataloguing in Publication Data

Angelo, Joseph A.
  The extraterrestrial encyclopedia : man's search
  for life in outer space.
  1. Space flight—Dictionaries and encyclopedias
  I. Title
  629.4'1'0321   TL790

  ISBN 0-87196-764-2

Printed in the United States of America
10 9 8 7 6 5 4 3 2 1

This book is dedicated to my wife, Joan, on our twenti-
eth wedding anniversary. As we explore and expand out
into the Universe, may we do so with the same love and
understanding that she has given here on Earth to my
family these past two decades.

CONTENTS

ACKNOWLEDGMENTS

This book could not have been prepared without the generous support and assistance of the National Aeronautics and Space Administration. Special acknowledgement must be given to those individuals who were especially helpful in the development of this book. These individuals include: Mr. Les Gaver (Special Assistant, Media Services, NASA Headquarters); the men and women of the Audiovisual Branch, NASA headquarters; Mr. Edward (Larry) Noon (Jet Propulsion Laboratory); Ms. Lisa Vazquez (Media Services Corp., Johnson Space Center); Mr. Terry White (Public Information Office, Johnson Space Center); Ms. Vera Buescher (Ames Research Center); Dr. John Billingham (Ames Research Center); Dr. Bernard Oliver (Ames Research Center); Mrs. Alice Price (Air Force Art Collection); Mr. William A. Rice (Boeing Aerospace Company); Mr. Georg von Tiesenhausen (Marshall Space Flight Center); Mr. Donald Engel (Public Information, Eastern Space & Missile Center); and Dr. Henry Robitaille (The LAND, Walt Disney World's EPCOT Center). The late Dr. Krafft Ehricke provided a special spark of imagination and excitement about the entire concept of our extraterrestrial civilization and his eternally optimistic thoughts form a cornerstone of this book. I would also like to thank my three children, Joseph, James and Jennifer, who spent many weekends helping me sort hundreds of pages of manuscript.

INTRODUCTION

Humanity's Extraterrestrial Civilization

Long before the start of the Space Age in 1957, technological visionaries imagined human settlements on the planets and man-made cities floating majestically in space. Although such large-scale space settlements are still part of future history, they came a giant step closer to realization in January 1984, when President Ronald Reagan (in his State of the Union Message to the U.S. Congress and people) gave the National Aeronautics and Space Administration (NASA) firm directions for a national space program. This dynamic program would be anchored by a strong American commitment to the human presence in space. In the President's own words:

We can follow our dreams to distant stars, living and working in space for peaceful economic and scientific gain. Tonight, I am directing NASA to develop a permanently manned space station and to do it within a decade.

This permanent space station will form the core of many future space activities. In it, some six to eventually 20 astronauts will live, work and even play in low Earth orbit for periods of up to six months at a time. When completed, this station and the space shuttle to get us there will represent a tiny, but permanent, human beachhead in the universe. It will mark the initiation of our extraterrestrial civilization. Even now space technologists and planners are looking beyond low Earth orbit toward the moon and Mars. Lunar bases, human expeditions to Mars and permanent settlements on the Red Planet are all distinct technological possibilities in the next few decades.

But why should people want to live and work in space in the first place? Perhaps the most important reason is because it is part of being human—it is of our very essence to want to explore. People need challenge, variety and adventure. These are the stimuli that help civilizations flourish and grow. Unfortunately, we now live at a time—the first in human history—when there are no great new land or sea frontiers on Earth to explore and conquer. Men and women have explored the depths of the oceans, climbed the peaks of the highest mountains, trekked across the most desolate, sun-scorched deserts, sailed alone around the world, and journeyed to the North and South Poles. Except for the vast icebound reaches of the Arctic and Antarctic, every major spot on our planet has been mapped and charted.

But while there is not much left to explore *outward* across our planet, there is an entire universe to explore *upward*. Space technology has arrived just in time! By its means, outer space, unreachable in all previous ages, now becomes an exciting new frontier. Space provides the human race with our greatest opportunity of all: an essentially limitless environment in which to develop and grow.

Modern advances in space technology will pave the way for the creation of our extraterrestrial civilization—a major development in the evolution of intelligent life on Earth paralleled only by the migration of life from the ancient seas to the land eons ago. Today, we are all witnesses to the start of the migration of intelligent life from the land to the stars. As we watch (and some of us actually participate), human beings will begin to occupy first cislunar and then heliocentric space on our march to neighboring star systems. The exciting news is that we are no longer creatures confined to a limited planetary civilization; through space technology, we have acquired the tools needed to respond to our cosmic destiny.

When we think in terms of life among the stars, we also ask the obvious question: Are we alone in this vast universe? Is the human being the best the universe has been able to do in the evolution of intelligent life over the last 15 to 20 billion years? If we are indeed alone, then it is our destiny to become the first intelligent species to sweep through the Galaxy spreading life, knowledge and love where there is now only emptiness. Perhaps, on the other hand, the universe is teeming with advanced forms of life. Our emergence as a "Solar System Civilization" might provide us with the levels of maturity and technology necessary to participate in any such galactic community. Either alternative is fascinating. In fact, we have now reached a level of technological sophistication and planetary development where it is the universe or nothing! We can choose to seek our destinies among the stars, at peace with one another and with our advanced technologies, or, sadly, we can elect to turn our backs on the stars and perish quietly on one tiny planet in an average star system at the fringes of the Galaxy.

This book has been written for everyone who wonders about things beyond the planet Earth, about things and beings extraterrestrial. In one volume, you will find a discussion of the major space technologies and developments that mark our search for extraterrestrial life and the initiation of our extraterrestrial civilization. One thought has especially stimulated and inspired me during the development of this book: Somewhere out there, possibly reading this very sentence, are the young man and young woman who will be the parents of the first Martians. Regarding the possibility of other intelligent life in the universe, we can now say (at the very least): "We have met the extraterrestrials, and they are US!"

Dr. Joseph A. Angelo, Jr.
Cape Canaveral (April 1985)

abiotic Not involving living things; not produced by living organisms.

abundance of elements (in the Universe) Analyses of solar and stellar spectra have shown that hydrogen and helium are by far the most abundant elements in the Universe. All other elements taken together make up only some 2 percent of the mass of the Universe. The "cosmic abundance" of an element can be expressed as its percentage of the total mass found in the Universe. Using the abundances found in our own Solar System as as "cosmic standard" and incorporating other astrophysical data (such as stellar spectra), the estimated cosmic abundance of the most common materials in the Universe is:

Element	Abundance (percentage of total mass)
Hydrogen (H)	73
Helium (He)	25
Oxygen (O)	0.8
Carbon (C)	0.3
Neon (Ne)	0.1
Nitrogen (N)	0.1

See also: **electromagnetic spectrum**

Advanced X-Ray Astrophysics Facility (AXAF) A planned NASA X-ray astrophysics facility designed to complement visual and radio-frequency observations made from the ground and from space-based observatories such as the Space Telescope. The basic objectives of the AXAF are to determine the positions of X-ray sources; their physical properties, such as composition and structure; and the processes involved in X-ray production. The Advanced X-ray Astrophysics Facility will study stellar structure and evolution, large-scale galactic phenomena, active galaxies, clusters of galaxies, quasars and cosmology. This facility will be a grazing incidence X-ray telescope with nested pairs of mirrors and a 9- to 12-meter focal length. It will have an overall length of approximately 13 to 16 meters and a mass of between 10,000 and 12,000 kilograms. The AXAF will be placed in a 28.5-degree inclination, approximately 500-kilometer altitude orbit by the Space Shuttle, which will revisit it periodically for maintenance, repair or the inclusion of new experiment pack-

ages. X-ray observations are now providing basic clues to the evolution of giant systems of galaxies. However, we have only a few tantalizing clues at present, and the Advanced X-Ray Astrophysics Facility, planned for launch in the late 1980s, should make the picture much clearer.

See also: **astrophysics**

aerobic microbes Microorganisms that are capable of living in the presence of oxygen.

albedo The fraction of incident (that is, falling or striking) light or electromagnetic radiation that is reflected by a surface or an entire object.

alien (In the sense used throughout this book) an extraterrestrial; an inhabitant (presumably intelligent) of another world.

Alpha Centauri The star system nearest to our own Sun, approximately 4.3 light-years away. It is actually a triple star system, with two stars orbiting around each other and the third star, called Proxima Centauri, revolving around the pair at some distance. In various celestial configurations, Proxima Centauri becomes the closest known star to our Solar System—approximately 4.2 light years away.

amino acid An acid containing the amino (NH_2) group, a group of molecules necessary for life. More than 80 amino acids are presently known, but only some 20 occur naturally in living organisms, where they serve as the building blocks of proteins. Amino acids have been synthesized nonbiologically under conditions that simulate those that may have existed on the primitive Earth, followed by the synthesis of most of the biologically important molecules. Amino acids and other biologically significant organic substances have been found to occur naturally in meteorites and are not considered to have been produced by living organisms. Meteorites, chunks of extraterrestrial matter arriving on Earth, represent a very interesting source of information about the occurrence of prebiotic (prelife) chemistry beyond the Earth. In 1969, for example, analyses of meteorite materials provided the first convincing proof of the existence of extraterrestrial amino acids. Since then, a rather large quantity of information has accumulated, showing that many more of the molecules considered necessary for life are also present in meteorites. As a result of these observations, it appears that the chemistry of life is not unique to the planet Earth itself.

See also: **exobiology; life in the Universe**

anaerobic microbes   Microorganisms that are killed by the presence of oxygen.

ancient astronauts   Was the Earth visited in the past by extraterrestrial explorers, by powerful intelligent beings from another planet or even from the stars?

Some people think so. As evidence, they may point to legends that they feel actually describe visits by ancient astronauts. Many of the world's cultures, for instance, have traditional tales of beings with superhuman powers who helped early humans.

Take the story of how prehistoric people first learned to use fire, which comes to us from Greek mythology. Prometheus, whose name means "forethought," and his brother Epimetheus, whose name means "afterthought," were members of a group of giant gods called Titans. It was Epimetheus's task to give the animals of the Earth the powers they needed to survive. But he gave them so many powers that when it came time to help humans, he had no gift left to give.

Prometheus, seeing the helplessness of humans, took pity on them. A far wiser god than his brother, he stole the gift of fire from the other gods and brought it down from the heavens to give humankind. With it, people could light small fires at the entrance of their caves and protect themselves from dangerous prowling animals. And with it, eventually, they could develop a technology that would make them the supreme animals on Earth.

Zeus, the king of the gods, was greatly angered at Prometheus's foresight. He had him chained to a rock, says the legend, and sent an eagle each day to tear at his liver. Prometheus suffered for thousands of years until Zeus's anger finally subsided. Then Hercules killed the eagle and freed Prometheus from his chains.

On one level, a legend like this seems to offer the tantalizing suggestion that gods such as the Titans were really ancient astronauts who helped early humans. Unfortunately, there is no evidence whatsoever to show that such legends are based on historic fact.

There are other theories about ancient astronauts, too. The major claim is that extraterrestrials affected the biological development of primitive apelike creatures and caused them to become intelligent human beings. How else, the reasoning goes, could primitive humans have produced advanced civilizations such as those that arose in ancient Egypt and Central America? This idea is sometimes offered as a complement to the two traditional theories of human origins. The first, the creationist theory, holds that God created humans and all the other animals as unique and separate species. The second, the evolutionist theory, holds that animals evolved, or developed, from earlier species; in the case of humans, from apelike ancestors.

There are several variations on the idea that human development was affected by extraterrestrials. One version has it that a migratory wave of ancient astronauts visited the Earth, planted the seeds of civilization among the primitive creatures they found on the planet and then continued on to the edge of the Galaxy.

Another version of the theory suggests that extraterrestrial visitors performed some form of genetic engineering on apes or even primitive humans. The result, it suggests, was the evolution of human intelligence and the emergence of modern humans.

A third variation on the theme suggests that humans themselves arose on another planet. Traveling from a distant homeland, some of these early humans landed on Earth and established a flourishing civilization. Then suddenly this advanced civilization was devastated by some kind of cosmic catastrophe. Right now, the theory goes, we are rebuilding civilization on the ruins of that ancient society.

Such ideas have given rise to many popular books and movies that seek to link the achievements of ancient human civilizations with visits from extraterrestrials. But what is the proof? Where is the scientific evidence? It rests, insist people who believe in the theory, on the existence of enormous and astounding ancient architectural structures that they believe "primitive" humans could not have built alone.

The main examples offered are Egypt's spectacular pyramids, built beginning around 2,650 B.C. as tombs for the pharaohs, or kings. The largest, the Great Pyramid, was one of the Seven Wonders of the World. Its base alone covers some 13 acres, an area large enough to hold 10 football fields. Rising originally to a height of 481 feet, it is built of more than 2 million stone blocks, each cut to fit snugly with the next, and each weighing an average of 2½ short tons.

How, indeed, did primitive humans achieve such a colossal feat? Well, in fact, they didn't—simply because people of this period were *not* primitive, at least not in the sense of lacking intelligence in the first place. All archaeological evidence shows that their brains, both in form and function, were exactly the same as our own. Geologists have been able to construct a picture of Earth's history that spans billions of years, and humans' part in that history goes back not the mere 5,000 years to the time of ancient Egypt, but back some 3½ million years or more. Ancient Egyptians were modern people: curious, ingenious and intelligent, just like their 20th-century descendants.

It is true, of course, that their technology was crude and primitive by today's standards; they had no machinery or iron tools, for instance. Nevertheless, they were capable of building the pyramids with the technology available at the time. They were able to cut the stone blocks with copper chisels and saws, drag them from the quarries to the site on rope-bound sledges, then haul them into position up ramps made of earth, wood and sun-dried brick.

Then, too, a pharaoh had enormous manpower at his disposal. Herodotus, the Greek historian, said 400,000 workers labored for 20 years to build the Great Pyramid. They worked on the tomb during the annual flood, when the river Nile overflowed its banks onto the nearby fields and made farming impossible.

The apparently sophisticated practices of early astronomers are also cited as a reason for believing in the intervention of ancient astronauts. But again, early astronomers did have tools. Working with a transit (a pole or tube with which to mark the passage of celestial objects across a meridian), they could have become quite practiced in observing the heavens. The astronomers of ancient Egypt, for example, discovered that the seasonal flooding of the Nile River corresponded with the time when the star Sirius first became visible in the morning sky. As a result, they were able to create a very accurate calendar, consisting of 365.25 days.

Perhaps the most solid proof against the ancient-astronaut theory is the sheer lack of evidence. Almost two centuries of serious archaeological and geological exploration of our planet have failed to yield the slightest shred of credible evidence that ancient astronauts once walked with early humans. Archaeologists have, on the other hand, uncovered thousands of purely terrestrial ancient tombs, cities and campsites.

Where, then, one might ask, are the extraterrestrial visitors' former landing sites, installations and exploratory equipment? Any starfaring civilization would surely have left behind at least one or two samples of their sophisticated technology. Did we of Earth not leave our footprints and exploratory equipment on the Moon? Are not two Viking craft even now silently guarding the Martian landscape?

Androcell An innovative and bold concept, proposed by Dr. Krafft A. Ehricke, involving a human-made new world that is an independent self-contained human biosphere not located on any naturally existing celestial object. These human-made miniworlds, or planetellas, would use mass far more effectively than the natural worlds of our Solar System, which formed out of the original nebular material. For example, naturally formed celestial bodies are essentially "solid" spherical objects of great mass. Their surface gravity forces result from the attraction of a large quantity of matter. However, except for the first kilometer or so, the interior of these natural worlds is essentially "useless" from a human exploitation and habitation point of view.

Instead of large quantities of matter, the Androcell would use rotation (centrifugal inertia) to provide variable levels of artificial gravity. The unutilizable solid interior of a natural celestial body is now replaced (through human ingenuity) with many useful, inhabitable layers of cylinders. Therefore, Androcell inhabitants would be able to enjoy a truly variable life-style in a multi-gravitational-level world. There would be a maximum gravity level at the outer edges of the Androcell, tapering off to essentially zero gravity in the inner cylinder levels closest to the central hub.

The Androcell will not be tied to the Earth-Moon system but rather—with its giant space-based factories, farms and fleets of merchant spacecraft—will be free to seek political and economic development throughout heliocentric (Sun-centered) space. Its inhabitants might trade with Earth, the Moon, Mars or other Androcells. These giant space settlements of 10,000 to perhaps 1,000,000 inhabitants are most analogous to the city-state of ancient Greece. The multi-gravity-level life-style would encourage migration to and from other worlds—perhaps a terraformed Mars or subdued Venus, or maybe even one of the moons of the giant outer planets. In essence, the Androcell represents the "cellular division" of humanity—since as autonomous extraterrestrial city-states, their inhabitants could choose to pursue culturally diverse life-styles.

Of course, we already have our initial, natural Androcell—we call it "Spaceship Earth." In time, inhabitants of our parent world will be able to use their technical skills and human intelligence to fashion a series of such Androcells, or large space settlements, throughout the Solar System. As the number of such artificial human habitats grows, a swarm of settlements might eventually encircle the Sun, capturing and using its entire energy output. At that point our extraterrestrial civilization will have created a Dyson sphere, and the next stage of cosmic mytosis, migration to the stars, would occur.

See also: **extraterrestrial civilizations; space settlement.**

android A term from science fiction describing a robot with near-human form or features.

See also: **cyborg**

Andromeda Galaxy The Great Spiral Galaxy in Andromeda is our neighboring galaxy and the most distant object visible to the naked eye of an observer on Earth. It is as large as our own Milky Way, or larger, and is some 2.2 million light-years (670 kiloparsecs) away. The Andromeda Galaxy, also called M31, emits only about one-tenth as much energy in the form of infrared radiation as our own galaxy. X-ray images taken with a telescope on NASA's HEAO-2 (High Energy Astronomy Observatory-2) satellite have revealed more than 70 X-ray sources, which appear to be binary stars and supernova remnants in this neighboring galaxy. Figure C1, color page A, is a photograph of the Andromeda Galaxy as seen by the 200–inch optical telescope at the Palomar Observatory. It reveals faint dust lanes threading a collection of millions of stars, one seemingly indistinguishable from another.

See also: **astrophysics; stars**

androsphere A term, developed by Dr. Krafft A. Ehricke, that describes the synthesis of the terrestrial and extraterrestrial environments. It relates to our productive integration of the Earth's biosphere, which contains the major terrestrial environmental regimes, and the material and energy resources of the Solar System, such as the Sun's radiant energy and the Moon's mineral resources.

angstrom [symbol: Å]   A unit of length commonly used to measure wavelengths of electromagnetic radiation in the visible portion of the spectrum.

1 angstrom (Å) = 10^{-10} meters = 0.1 nanometers

This unit is named after Anders Jonas Ångstrom (1814–74), a Swedish physicist.

annihilation   The conversion of a particle and its corresponding antiparticle into electromagnetic radiation (annihilation radiation) upon collision. This annihilation radiation has a minimum energy equivalent to the resting masses (m_0) of the two colliding particles. For example, when a positron and an electron collide, the minimum annihilation radiation consists of a pair of gamma rays, each of 0.511-million electron volts (MeV) energy. The energy of annihilation is derived from the mass of the disappearing particles according to the famous Einstein mass-energy equivalence formula, $E = m_0c^2$.

See also: **antimatter**

antigalaxy   A galaxy composed of antimatter.

See also: **antimatter; antimatter cosmology**

antimatter   Matter in which the ordinary nuclear particles (such as neutrons, protons and electrons) are conceived of as being replaced by their corresponding antiparticles—that is, antineutrons, antiprotons, positrons and so on. For example, an "antiparticle hydrogen atom" would consist of a negatively charged antiproton with an orbital positron. Normal matter and antimatter mutually annihilate each other upon contact and are converted into pure energy, called annihilation radiation. Although individual antiparticles have been discovered and their behavior observed in laboratories, bulk quantities of antimatter have yet to be found in the Universe.

See also: **annihilation**

antimatter cosmology   A cosmological model proposed by the Swedish scientists Alfvén and Klein as an alternate to the Big Bang model. In their model the early Universe is assumed to consist of a huge, spherical cloud, called a metagalaxy, containing equal amounts of matter and antimatter. As this cloud collapsed under the influence of gravity, its density increased and a condition was reached in which matter and antimatter collided—producing large quantities of annihilation radiation. The radiation pressure from the annihilation process caused the Universe to stop collapsing and to expand. In time, clouds of matter and antimatter formed into equivalent numbers of galaxies and antigalaxies. (An antigalaxy is a galaxy composed of antimatter.)

There are many technical difficulties with the Alfvén-Klein cosmological model. For example, no observational evidence has yet been obtained of large quantities of antimatter existing in the Universe. If these antigalaxies existed, large quantities of annihilation (gamma-ray) radiation would certainly be emitted at the interface points between the matter and antimatter regions of our Universe.

See also: **antimatter; astrophysics; "Big Bang" theory; cosmology**

antiparticle   Every elementary particle has a corresponding real (or hypothetical) antiparticle, which has equal mass but opposite electric charge (or other property, as in the case of the neutron and antineutron). The antiparticle of the electron is the positron; of the proton, the antiproton; and so on. However, the photon is its own antiparticle. When a particle and its corresponding antiparticle collide, they are converted into energy in a process called annihilation.

See also: **annihilation; antimatter**

aperture synthesis   The use of a variable-aperture radio interferometer to mimic the "full dish" of a huge equivalent radio telescope.

See also: **radio astronomy**

arc-minute $1/60$th of a degree of angle. This unit is associated with precise measurements of motions and positions of celestial objects as occurs in the science of astrometry.

See also: **arc-second; astrometry**

arc-second $1/3600$th of a degree of angle. This unit is associated with very precise measurements of stellar motions and positions in the science of astrometry.

See also: **arc-minute; astrometry**

Arecibo Interstellar Message   To help inaugurate the powerful radio/radar telescope of the Arecibo Observatory in the tropical jungles of Puerto Rico, an interstellar message of friendship was beamed to the fringes of the Milky Way Galaxy. On November 16, 1974 this interstellar radio signal was transmitted toward the Great Cluster in Hercules (Messier 13 or M13, for short) which lies about 25,000 light-years away from Earth. The globular cluster M13 contains about 300,000 stars within a radius of approximately 18 light-years.

This transmission, often called the Arecibo Interstellar Message, was made at the 2380 megahertz radio frequency with a 10-hertz bandwidth. The average effective radiated power was 3×10^{12} watts (3 terawatts) in the direction of transmission. The signal is considered to be the strongest radio signal yet beamed out into space by our planetary civilization. Perhaps 25,000 years from now a radio telescope operated by members of an intelligent alien civilization somewhere in the M13 cluster will receive and decode this interesting signal. If they do, they will learn that intelligent life has evolved here on Earth!

The Arecibo Interstellar Message of 1974 consisted of

1679 consecutive characters. It was written in a binary format—that is, only two different characters were used. As shown in figure 1, the characters can be denoted as "0" and "1". In the actual transmission, each character was represented by one of two specific radio frequencies and the message was transmitted by shifting the frequency of the Arecibo Observatory's radio transmitter between these two radio frequencies in accordance with the plan of the message.

The message itself was constructed by the staff of the National Astronomy and Ionosphere Center (NAIC). It can be decoded by breaking up the message into 73 consecutive groups of 23 characters each and then arranging these groups in sequence one under the other. The numbers, 73 and 23, are prime numbers. Their use should facilitate the discovery by any alien civilization receiving the message, that the above format is the right way to interpret the message. Figure 2 shows the decoded message: The first character transmitted (or received) is located in the upper right hand corner.

This message describes some of the characteristics of terrestrial life that the scientific staff at the National Astronomy and Ionosphere Center felt would be of particular interest and technical relevance to an extraterrestrial civilization. The NAIC staff interpretation of the interstellar message is as follows:

The Arecibo message begins with a "lesson" that describes the number system being used. This number system is the binary system, where numbers are written in powers of two (2) rather than of ten (10) as in the decimal system used in everyday life. Staff scientists believe that the binary system is one of the simplest number systems. It is also particularly easy to code in a simple message. Written across the top of the message (from right to left) are the numbers one (1) through ten (10) in binary notation. Each number is marked with a "number label"— that is, a single character which denotes the start of a number. A problem that had to be dealt with carefully was the writing of a number so large that all its digits could not be accommodated in the available space. The NAIC staff's solution to this problem is a special point of the initial "number lesson" in the message and is shown in the numbers 8, 9 and 10. The number sequence was intentionally written so that there was not enough room to write 8, 9, and 10 on a single line. The digits for which there wasn't room written below (here to the left) of the

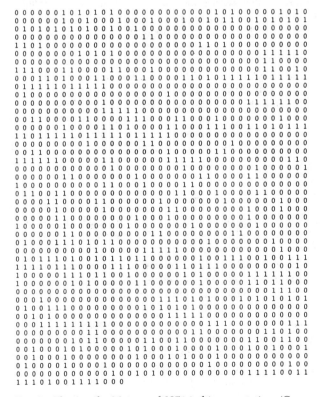

Fig. 1   The Arecibo Message of 1974 in binary notation. (Courtesy of Frank D. Drake and the staff of the National Astronomy & Ionosphere Center, which is operated by Cornell University under contract with the National Science Foundation.)

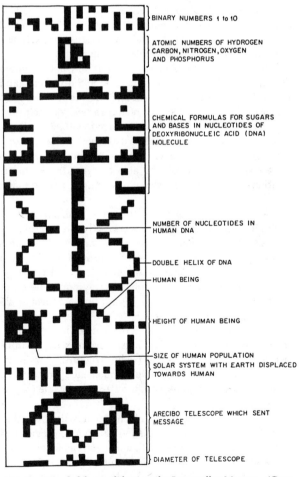

Fig. 2   Decoded form of the Arecibo Interstellar Message. (Courtesy of Frank D. Drake and the staff of the National Astronomy & Ionosphere Center which is operated by Cornell University under contract with the National Science Foundation.)

least significant digits. It is important to realize that the second and successive lines of digits were not written under the number label, but only under characters denoting a numerical value. This was intended to mean that in the message the number label always stands by itself in the upper right hand corner of a number. This procedure then identifies a number, the orientation of a number written in the message, and where that number begins. The NAIC staff felt it necessary to make the interpretation of numbers clearcut and to economize in the use of message characters.

The next block of information sent in the message occurs just below the numbers. It is recognizable as five numbers. From right to left these numbers are: 1, 6, 7, 8, and 15. This otherwise unlikely sequence of numbers should eventually be interpreted as the atomic numbers of the elements hydrogen, carbon, nitrogen, oxygen and phosphorus.

Next in the message are twelve groups on lines 12 through 30 that are similar groups of five numbers. Each of these groups represents the chemical formula of a molecule or radical. The numbers from right to left in each case provide the number of atoms of hydrogen, carbon, nitrogen, oxygen, and phosphorus, respectively, that are present in the molecule or radical.

Since the limitations of the message did not permit a description of the physical structure of the radicals and molecules, the simple chemical formulas do not define in all cases the precise identity of the radical or molecule. However, these structures are arranged as they are organized within the macromolecule described in the message. Intelligent alien organic chemists somewhere in the M13 cluster should eventually be able to arrive at a unique solution for the molecular structures being described in the message.

The most specific of these structures, and perhaps the one that should point the way to correctly interpreting the others, is the molecular structure that appears four times on lines 17 through 20 and lines 27 through 30. This is a structure containing one phosphorus atom and four oxygen atoms, the well–known phosphate group. The outer structures on lines 12 through 15 and lines 22 through 25 give the formula for a sugar molecule, deoxyribose. The two sugar molecules on lines 12 through 15 have between them two structures: the chemical formulas for thymine (left structure) and adenine (right structure). Similarly, the molecules between the sugar molecules on lines 22 through 25 are: guanine (on the left) and cytosine (on the right).

The macromolecule or overall chemical structure is that of deoxyribonucleic acid (DNA). The DNA molecule contains the genetic information that controls the form, living processes, and behavior of all terrestrial life. This structure is actually wound as a double helix, as depicted in lines 32 through 46 of the message. The complexity and degree of development of intelligent life on Earth is described by the number of characters in the genetic code,

that is, by the number of adenine-thymine and guanine-cytosine combinations in the DNA molecule. The fact that there are some four billion such pairs in human DNA is illustrated in the message by the number given in the center of the double helix between lines 27 and 43. Please note that the "number label" is used here to establish this portion of the message as a number and to show where the number begins.

The double helix leads to the "head" in a crude sketch of a human being. The scientists who composed the message hoped that this would establish a connection between the DNA molecule, the size of the helix, and the presence of an "intelligent" creature. To the right of the sketch of a human being is a line that extends from the head to the feet of our "message human." This line is accompanied by the number 14. This portion of the message is intended to convey the fact that the "creature" drawn is 14 units of length in size. The only possible unit of length associated with the message is the wavelength of the transmission, namely 12.6 centimeters. This makes the creature in the message 176 centimeters or about five feet nine inches tall. To the left of the human being is a number, approximately four billion. This number represents the current human population on the planet Earth.

Below the sketch of the human being is a representation of our Solar System. The Sun is at the right, followed by nine planets with some coarse representation of relative sizes. The third planet, Earth, is displaced to indicate that there is something special about it. In fact, it is displaced towards the drawing of the human being, who is centered on it. Hopefully, an extraterrestrial scientist in pondering this message will recognize that Earth is the home of the intelligent creatures that sent it.

Below the Solar System and centered on the third planet is an image of a telescope. The concept of "telescope" is described by showing a device that directs rays to a point. The mathematical curve leading to such a diversion of paths is crudely indicated. The telescope is not upside down, but rather "up" with respect to the symbol for the planet Earth.

At the very end of the message (bottom of figure 2), the size of the telescope is indicated. Here, it is both the size of the largest radio telescope on Earth and also the size of the telescope that sent the message (namely, the Arecibo telescope). It is shown as 2,430 wavelengths across, or roughly 305 meters. (No one, of course, expects an alien civilization to have the same unit system we use here on Earth— but physical quantities, such as the wavelength of transmission, provide a common reference frame.)

This interstellar message was transmitted at a rate ten characters per second and it took 169 seconds to transmit the entire information package. It is interesting to realize, that just one minute after completion of transmission, our interstellar greetings passed the orbit of Mars. After 35 minutes, the message passed the orbit of Jupiter; and after 71 minutes it silently crossed the orbit of Saturn. Some five hours and 20 minutes after transmission, the message passed the orbit of Pluto, leaving the Solar System and entering interstellar space. It will be detectable by telescopes of approximately the same size and capability as the Arecibo facility that sent it anywhere in our Galaxy!

If you had to prepare a message to the stars, what type of information would you decide to transmit?

See also: **Arecibo Observatory; interstellar communication; what do you say to a little green man?**

Arecibo Observatory The Arecibo Observatory, the world's largest radio/radar telescope, is located in the lush tropical jungles of Puerto Rico. (See fig. 1.) It is the main observing instrument of the National Astronomy and Ionosphere Center, a national center for radio and radar astronomy and ionospheric physics operated by Cornell University under contract with the National Science Foundation. The 305-meter (1,000-foot) dish of the observatory's giant telescope fills a spherical bowl that was naturally formed by the collapse of huge limestone caves.

The enormity of the giant concrete, steel and aluminum structure provides an interesting contrast to the surrounding green carpet of lush tropical jungle. Yet, a certain harmony or synthesis is also present. The giant radio telescope actually uses nature to help explore nature. This observatory was located in the tropics because the Moon and the planets pass nearly overhead at the lower latitudes. Astronomers depend on the remote setting and the surrounding hills to reduce radio-wave interference. The enormous telescope itself (three football fields across) sits, as mentioned, in a natural limestone sinkhole, the use of which avoided very expensive excavation costs during

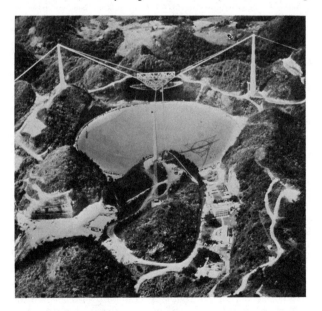

Fig. 1   The Arecibo radio/radar observatory is the largest facility of this kind on the planet Earth. It is located in the tropical jungles of Puerto Rico. (Courtesy of National Astronomy and Ionosphere Center.)

construction. Wild tropical vegetation under the bowl prevents erosion of the underlying terrain. The minimal annual temperature variations enhance structural stability.

When it operates as a radio receiver, the large Arecibo telescope listens for signals from celestial objects at the farthest reaches of the Universe. As a radar transmitter/receiver, it assists astronomers and planetary scientists by bouncing signals off the Moon, off nearby planets and their satellites, or even off the layers in the Earth's ionosphere.

The white steel triangle suspended over the giant dish (see fig. 1) is a support structure for the equipment that receives and amplifies the radio waves collected from space. Once amplified, these signals are then sent to a control building on the ground below for data processing and evaluation.

Most radio telescopes have a steerable dish or reflector that collects the incoming radio signals. At the Arecibo Observatory, the giant collector dish lies immobile in the Earth, while the receiving equipment (all 544 metric tons of it) hangs suspended some 170 meters above the air. This receiving equipment is steered and pointed with the assistance of remote-control devices. In fact, the entire giant observatory has been so designed that a single experimenter can operate it alone.

There is a circular railroad track, winding around the inside of the white steel triangle, for steering the feed arm. The feed arm is the large section that hangs down directly below the steel triangle. (See fig. 1.) Approximately 100 meters long, the feed arm directs the observatory's transmitting and receiving equipment to various parts of the sky. The two carriage houses (containing the transmitting and receiving equipment) can be moved to selected positions along the curved underside of the feed arm by means of a second railroad track. In this way, astronomers can "point" the Arecibo telescope through a combination of feed-arm and carriage-house movements.

The long structures that project from the carriage houses are the telescope's antennas. Their shapes are determined by the reflecting dish's spherical configuration; that is, radio signals hitting the dish are reflected into a "focal line" rather than a focal point. The largest of these "line feeds" is 29 meters.

In addition to receiving radio signals collected by the giant reflector dish, this feed can also direct radio energy down on the reflector for transmitting signals into space. The Arecibo Observatory has four radar transmitters: the 5–12-megahertz transmitter for ionospheric investigations; the 50-megahertz transmitter for ionospheric and lunar studies; the 430-megahertz for ionospheric studies and lunar and planetary radar studies; and the 2,380-megahertz (S-band) for planetary radar studies.

The giant Arecibo telescope can also play a major role in our search for extraterrestrial intelligence (SETI). It can not only listen for interstellar radio signals from intelligent alien civilizations; it can also be used as a transmitter to send our own messages to the stars. In fact, the 450,000-watt radar output of the observatory's S-band transmitter, when concentrated into a narrow beam by the giant reflector, has an effective power 100 times greater than the total electric power production of all the world's generating plants. In other words, the Arecibo telescope can transmit the strongest signal now leaving the Earth. This human-made radio-frequency "beacon" is actually powerful enough to be detected by similar radio observatories located anywhere in the Galaxy! Just think, alien radio astronomers may someday be searching our region of the Milky Way when suddenly they detect an Arecibo transmission that "shines" with a (radio-frequency) brilliance 10 billion times stronger than the Sun. In 1974 a very special message was sent to the stars from this facility.

See also: **Arecibo Interstellar Message; interstellar communication; radar astronomy; radio astronomy**

artificial intelligence (AI) A term commonly taken to mean the study of thinking and perceiving as general information-processing functions—or, if you will, the science of machine intelligence. In the past few decades, computer systems have been programmed to diagnose diseases; prove theorems; analyze electronic circuits; play complex games such as chess, poker and backgammon; solve differential equations; assemble mechanical equipment using robotic manipulator arms and end effectors (the "hands" at the end of the manipulator arms); pilot unmanned vehicles across complex terrestrial terrain, as well as through the vast reaches of interplanetary space; analyze the structure of complex organic molecules; understand human speech patterns; and even write other computer programs. All of these computer-accomplished functions require a degree of "intelligence" similar to mental activities performed by the human brain. Someday, a general theory of intelligence may emerge from the current efforts of scientists and engineers who are now engaged in the field of artificial intelligence. This general theory would help guide the design and development of even "smarter" robot spacecraft and exploratory probes, allowing us to more fully explore and use the resources that await us throughout the Solar System.

Artificial intelligence generally includes a number of elements or subdisciplines. Some of these are: planning and problem solving; perception; natural language; expert systems; automation, teleoperation and robotics; distributed data management; and cognition and learning. Each of these AI subdisciplines will now be discussed briefly.

All artificial intelligence involves elements of planning and problem solving. The problem-solving function implies a wide range of tasks, including decision making, optimization, dynamic resource allocation and many other calculations or logical operations.

Perception is the process of obtaining data from one or more sensors and processing or analyzing these data to assist in making some subsequent decision or taking some subsequent action. The basic problem in perception is to extract from a large amount of (remotely) sensed data some feature or characteristic that then permits object identification.

One of the most challenging problems in the evolution of the digital computer has been the communication that must occur between the human operator and the machine. The human operator would like to use an everyday, or natural, language to gain access to the computer system. The process of communication between machines and people is very complex and frequently requires sophisticated computer hardware and software.

An expert system permits the scientific or technical expertise of a particular human being to be stored in a computer for subsequent use by other human beings who have not had the equivalent professional or technical experience. These expert systems have been developed for use in such diverse fields as medical diagnosis, mineral exploration and mathematical problem solving. To create such an expert system, a team of software specialists will collaborate with a scientific expert to construct a computer-based interactive dialogue system that is capable, at least to some extent, of making the expert's professional knowledge and experience available to other individuals. In this case, the computer, or "thinking machine," not only stores the scientific or professional expertise of one human being but also permits ready access to this valuable knowledge base because of its artificial intelligence, which guides other human users.

Automatic devices are those that operate without direct human control. NASA has used many such automated smart machines to explore alien worlds. For example, the two Viking Landers placed on the Martian surface in 1976 represent one of the great triumphs of robotic space exploration. After separation from the Viking Orbiter spacecraft, the lander (protected by an aeroshell) descended into the thin Martian atmosphere at speeds of approximately 16,000 kilometers per hour (10,000 miles per hour). It was slowed down by aerodynamic drag until its aeroshell was discarded. The lander slowed down further by releasing parachutes and finally achieved a gentle landing by automatically firing retro-rockets. This entire sequence was successfully accomplished automatically by both Viking Landers.

Teleoperation implies that a human operator is in remote control of a mechanical system. Control signals can be sent by means of "hardwire" (if the device under control is nearby) or via electromagnetic signals (for example, laser or radio frequency) if the robot system is some distance away. Of course, in dealing with great distances in interplanetary exploration, a situation is eventually reached when even electromagnetic-wave transmission cannot accommodate real-time control (that is, when the mechanical system carries out an order immediately after it is issued). When the device to be controlled on an alien world is light-minutes or even light-hours away, teleoperation must yield to increasing levels of autonomous, machine intelligence–dependent robotic operation.

Robot devices are computer-controlled mechanical systems that are capable of manipulating or controlling other machine devices, such as end effectors. Robots may be mobile or fixed in place and either fully automatic or teleoperated.

Large quantities of data are frequently involved in the operation of automatic robotic devices. The field of distributed data management is concerned with ways of organizing cooperation among independent but mutually interacting data bases.

In the field of artificial intelligence, the concept of cognition and learning refers to the development of a machine intelligence (MI) that can deal with new facts, unexpected events and even contradictory information. Today's smart machines handle new data by means of preprogrammed methods or logical steps. Tomorrow's "smarter" machines will need the ability to learn, perhaps even to understand, as they encounter new situations and are forced to change their mode of operation.

Perhaps late in the next century, as the field of artificial intelligence matures, we will send fully automatic robot probes on interstellar voyages. Each very smart interstellar probe must be capable of independently examining a new star system for suitable extrasolar planets and, if successful in locating one within a candidate ecosphere, beginning the search for extraterrestrial life. Meanwhile, back on Earth, scientists will wait for its electromagnetic signals to travel light-years through the interstellar void, eventually informing its human builders that the extraterrestrial exploration plan has been successfully accomplished.

See also: **robotics in space; Viking Project**

asteroid A small, solid object orbiting the Sun independently of a planet. The majority of the asteroids—which are also sometimes called minor planets or planetoids—have orbits that lie in the "asteroid belt" between Mars and Jupiter (most travel around the Sun at distances of 2.2 to 3.3 astronomical units), and they possess orbital periods between three and six years. There are more than 500 main-belt asteroids over 50 kilometers in diameter and thousands that are smaller. Approximately 2,000 of these have fairly well known orbits and have been assigned names. Their diameters range from approximately 1,000 kilometers (Ceres) down to 20 kilometers and smaller. Beginning in this century, more than 60 asteroids that cross the Earth's orbit have also been discovered, the majority of these within the last decade. Such asteroids are popularly known as "Earth-crossers," or "Earth-crossing asteroids." Table 1 provides selected information about some of these minor planets.

The surfaces of most asteroids appear dark, suggesting significant carbon content. Chemically bound water also appears to have been detected on some surfaces. The mass of the entire asteroid belt is estimated to be approximately 3×10^{21} kilograms—a value that is about $1/20$th the mass of our Moon.

Asteroids are usually classified on the basis of their spectral characteristics—that is, by how much light of different colors they reflect. The C-type (carbonaceous) asteroids appear to be covered with a dark, carbon-based material and are assumed to be similar to carbonaceous chondrite meteorites. They may contain up to 10 percent water, 6 percent carbon, significant amounts of sulfur and useful amounts of nitrogen. The S-type asteroids are thought to be similar to silicaceous or stony-iron meteorites. They are more common near the inner edge of the main belt. Of particular interest in space industrialization is the fact that S-class Earth-crossing asteroids may contain up to 30 percent free metals—that is, alloys of iron, nickel and cobalt, along with high concentrations of precious metals. The M-type (for metallic) asteroids are thought to be the remaining metallic cores of very small, differentiated planetoids that have been stripped of their crusts by collisions with other asteroids. Three minor asteroid designations—R-type, E-type and U-type—are also encountered. The R-type stands for asteroids that have red features, while the E-type asteroids (estatine achondrites) appear to possess various stony compositions. Finally, U-type asteroids display unusual or unknown characteristics.

To date, there is no general consensus among space scientists on the origin and nature of the various types of asteroids or their assumed relationships to different types of meteorites. The earliest theory about the origin of the asteroids was that they are remnants of the explosion of a "missing planet" between Mars and Jupiter. At present, the tendency is to support other theories, ones suggesting that the material that formed the asteroids gradually accumulated in the same way as the larger planets, with the asteroids eventually suffering many collisions among themselves along with experiencing large orbital disturbances from close approaches to massive celestial objects, such as Jupiter. Space scientists have also hypothesized that some of the dark (that is, low-albedo) asteroids whose perihelia (orbital points nearest to the Sun) fall within the orbit of Mars may actually be extinct cometary nuclei.

Earth-crossing asteroids are of special interest in the development of our extraterrestrial civilization, since they would be the easiest to visit, return samples from and eventually exploit for their material content. By convention, these "Earth-crossers" are divided into three groups: Aten, Apollo and Amor. The Aten group of asteroids possess orbits that overlap the Earth's orbit at aphelion (the point at which Earth is farthest from the Sun), while the Apollo group have orbits that overlap the Earth's orbit at perihelion. The Amor group of asteroids, on the other hand, are those with perihelia lying between 1.017 astronomical units (Earth's aphelion) and 1.38 astronomical units (an arbitrarily chosen figure).

Some scientists have postulated that the impact of an Earth-crossing asteroid may have been responsible for the disappearance of the dinosaurs and many other animal species on Earth some 65 million years ago.

See also: **albedo; asteroid mining; comet; extraterrestrial catastrophe theory**

Table 1   Data for Selected Asteroids

Achilles: This asteroid was the first member of the Trojan group to be discovered.

Diameter	~70 km[a]
Perihelion	4.44 AU[b]
Aphelion	5.98 AU
Orbital period	11.77 yr
Inclination	10.3°
Year discovered	1906

Adonis: A member of the Apollo group of minor planets.

Diameter	0.3 km
Perihelion	0.44 AU
Aphelion	3.30 AU
Orbital period	2.56 yr
Inclination	1.4°
Year discovered	1936

Amor: The major member of the Amor group of asteroids.

Diameter	0.5 km
Perihelion	1.08 AU
Aphelion	2.76 AU
Orbital period	2.77 yr
Inclination	11.9°
Year discovered	1932

Aten: An "Earth-crossing" asteroid with an orbital period of less than one year.

Diameter	~1.1 km
Perihelion	0.79 AU
Aphelion	1.14 AU
Orbital period	0.95 yr
Inclination	17.9°
Year discovered	1976
Features	Type S; probably silicate or metal-rich; albedo = 0.17

Apollo: The asteroid for which the Apollo group of Earth-crossing asteroids is named. This group of asteroids have perihelion values (that is, their distances from the Sun at their point of closest approach) falling inside the orbit of the Earth. This group is named after the asteroid Apollo, which approached within 0.07 astronomical units (AU) of the Earth in 1932—a "near miss" by celestial-encounter standards. Because they cross the Earth's orbit, the Apollo group is of particular interest in asteroid mining concepts.

Diameter	~1 km
Perihelion	0.65 AU
Aphelion	2.29 AU
Orbital period	1.78 yr
Inclination	6.36°
Year discovered	1932

[a] ~70 km = approximately 70 kilometers.
[b] AU = astronomical unit(s).

Ceres: In 1801 the Italian astronomer Giuseppe Piazzi discovered it—the first asteroid to be found and the largest.

Diameter	1,002 km
Perihelion	2.55 AU
Aphelion	2.98 AU
Orbital period	4.60 yr
Inclination	10.6°
Year discovered	1801
Features	Type C

Chiron: A minor planet with an orbit lying almost entirely between Saturn and Uranus, making it the most distant asteroid yet found. It is conceivable that Chiron is one of the brighter members of a distant, as yet to be discovered asteroid belt, or else it may possibly be a dormant comet or even an escaped satellite of Saturn or Uranus. Chiron is known to be in a highly perturbed, "chaotic" orbit that will eventually lead either to collision with a planet or to ejection from the Solar System.

Diameter	~300–400 km (uncertain)
Perihelion	8.5 AU
Aphelion	18.9 AU
Orbital period	50.68 yr
Inclination	6.93°
Year discovered	1977

Eros: The largest member of the Amor group of Earth-crossing asteroids. Under favorable circumstances it passes only 0.13 AU (20 million km) from the Earth, at which time it is one of the brightest asteroids. Scientists believe that Eros's surface is stony or stony-iron in composition. It is highly elongated, with possible dimensions of 18 km × 36 km. Some scientists believe that Eros might actually be two celestial objects orbiting in contact.

Diameter	~20 km (irregular shape)
Perihelion	1.13 AU
Aphelion	1.78 AU
Orbital period	1.76 yr
Inclination	10.83°
Year discovered	1898
Features	Type S

Hermes: A tiny member of the Apollo group of asteroids. It was discovered in 1937 when it approached within 0.006 AU (780,000 km) of Earth, a very close call by astronautical standards.

Diameter	0.5 km
Perihelion	0.62 AU
Aphelion	2.66 AU
Orbital period	2.10 yr
Inclination	6.2°
Year discovered	1937

Icarus: A member of the Apollo group. It has the smallest perihelion of any known asteroid.

Diameter	~1.4 km
Perihelion	0.19 AU
Aphelion	1.97 AU
Orbital period	1.12 yr
Inclination	20.2°
Year discovered	1949

Juno: Discovered in 1804 by the German astronomer Karl Harding. This asteroid was the third to be found.

Diameter	249 km
Perihelion	1.98 AU

Aphelion	3.35 AU
Orbital period	4.36 yr
Inclination	13°
Year discovered	1804
Features	Type S

Pallas: The second minor planet to be discovered—a feat accomplished by the German astronomer Wilhelm Olbers in 1802.

Diameter	583 km
Perihelion	2.11 AU
Aphelion	3.42 AU
Orbital period	4.61 yr
Inclination	34.8°
Year discovered	1802
Features	Type U

Trojan Group: A group of asteroids that lie near the two Lagrangian points (that is, the two "gravity wells," or points of stable equilibrium) in Jupiter's orbit around the Sun. Achilles was the first asteroid in this group to be identified; many subsequently discovered members of this group have been named in honor of the heroes, both Greek and Trojan, of the Trojan War.

Vesta: The brightest minor planet and the third largest, with an approximate diameter of 555 km. Vesta was discovered in 1807 by the German astronomer Wilhelm Olbers. Under favorable conditions it can just be seen with the naked (unaided) eye. Vesta is probably among the best studied of all asteroids. It is apparently unique among the larger minor planets in that it appears to have a surface of basaltic lava, indicating a complex geological history of heating and volcanism. Many space scientists believe that Vesta may be the parent body of a class of meteorites called the eucrite class (indicating a type of stony composition). Although this hypothesis is not yet proven, the theory appears to represent the most probable connection between specific meteorites and an asteroid parent body. Telescopic studies show that Vesta is nearly spherical, with several types of igneous rock on its surface. With its low inclination and 2.36 AU (average) distance from the Sun, Vesta is among the most easily reached of the large main-belt asteroids.

Diameter	555 km
Perihelion	2.15 AU
Aphelion	2.57 AU
Orbital period	3.63 yr
Inclination	7.14°
Year discovered	1807

SOURCE: Developed by the author from the latest NASA data.

asteroid mining   The asteroids, especially Earth-crossing asteroids, can serve as "extraterrestrial warehouses" from which future space workers may extract certain critical materials needed in space-based industries and in the construction of large space habitats. On some asteroids scientists expect to find water (trapped), organic compounds and metals. In time, it may be far more efficient to mine an Earth-crossing asteroid than to lift these very same raw materials from the Earth's surface. Thus, the asteroids could play a major role in the evolution of our extraterrestrial civilization.

See also: **asteroid**

astro-   A prefix meaning "star" or "stars" and (by extension) sometimes used as the equivalent of "celestial," as in *astroengineering.*

astrobiology   The study of living organisms on celestial bodies other than the Earth.

See also: **exobiology**

astrochimps   The nickname frequently given to the primates used during the early U.S. space program. A monkey called Albert was the first primate in space. He was launched on board a V-2 rocket (captured from Germany at the end of World War II) at the White Sands Missile Range in New Mexico on June 11, 1948. Unfortunately for this simian space explorer, the trip proved fatal. During Project Mercury (the first U.S. manned space program), astrochimps—also called "chimponauts"—were used extensively to test space-capsule and launch-system hardware prior to their commitment for use in an actual manned space flight. For example, in January 1961 a 17-kilogram (37-pound) chimpanzee named Ham was launched from Cape Canaveral by a Redstone rocket on a suborbital flight test of the Mercury spacecraft. During this mission the propulsion system developed more thrust than planned, and Ham experienced overacceleration. After the capsule was recovered, however, Ham appeared to be in good physical condition; but when shown another space capsule, his reactions made it clear to all that the astrochimp wanted no further part of the space-flight program! In November 1961 another astrochimp, a 17-kilogram (37.5-pound) chimpanzee named Enos, was successfully launched into orbit from Cape Canaveral by an Atlas rocket. This flight was the final orbital qualification test of the Mercury spacecraft. While in space, Enos performed psychomotor duties and upon recovery was found to be in excellent physical condition.

astroengineering   Incredible feats of engineering and technology involving the energy and material resources of an entire star system or several star systems. The detection of such astroengineering projects would be a positive indication of the presence of a Type II, or even Type III, extraterrestrial civilization in our Galaxy. One example of an astroengineering project would be the creation of a "Dyson sphere"—a cluster of structures and habitats made by an intelligent extraterrestrial species to encircle their native star and effectively intercept all of its radiant energy output.

See also: **Dyson sphere; extraterrestrial civilizations**

astrometry   The science that is concerned with the very precise measurement of the motion and position of celestial objects; a subset or branch of astronomy.

astronomical unit (AU)   An extraterrestrial unit of distance defined as the "mean" distance between the center of the Earth and the center of the Sun—that is, the semimajor axis of the Earth's orbit. One AU is equal to 149.6×10^6 kilometers (approximately 92.9 million miles), or 499.01 light-seconds.

astrophysics Astronomy addresses fundamental questions that have occupied man since his primitive beginnings. What is the nature of the Universe? How did it begin, how is it evolving and what will be its eventual fate? As important as these questions are, there is another motive for astronomical studies. Since the 17th century when Newton's studies of celestial mechanics helped him formulate the three basic laws of motion and the universal law of gravitation, the sciences of astronomy and physics have become intertwined. "Astrophysics" can be defined as the study of the nature and physics of stars and star systems. It provides the theoretical framework for understanding astronomical observations. At times astrophysics can be used to predict phenomena before they have even been .observed by astronomers, such as black holes. The vast laboratory of outer space makes it possible to investigate large–scale physical processes that cannot be duplicated in a terrestrial laboratory. Although the immediate, tangible benefits to mankind from progress in astrophysics cannot always be measured or predicted, the opportunity to extend our understanding of the workings of the Universe is really an integral part of the rise of our extraterrestrial civilization.

Today, astrophysics has within its reach the ability to bring about one of the greatest scientific achievements ever—a unified understanding of the total evolutionary scheme of the Universe. This remarkable revolution in astrophysics is happening now due to the confluence of two streams of technical development: remote sensing and spaceflight. Through the science of remote sensing, we have acquired sensitive instruments capable of detecting and analyzing radiation across the whole range of the electromagnetic (EM) spectrum. Spaceflight, on the other hand, enables astrophysicists to place these remote sensing instruments outside the Earth's atmosphere. The wavelengths transmitted through the interstellar medium and arriving in the vicinity of near–Earth space are spread over approximately 24 decades of the spectrum. (A decade is a group, series or power of 10.) However, most of this electromagnetic radiation never reaches the surface of the Earth, because the terrestrial atmosphere effectively blocks such radiation across most of the spectrum. It should be remembered that the visible and infrared "atmospheric windows" occupy a spectral slice whose width is roughly one decade. Ground–based radio observatories can detect stellar radiation over a spectral range that adds about five more decades to the range of observable frequencies; but the remaining 18 decades of the spectrum are still blocked and are effectively "invisible" to astrophysicists on the Earth's surface. Consequently, information that can be gathered by observers at the bottom of the Earth's atmosphere represents only a small fraction of the total amount of information available concerning extraterrestrial objects. Sophisticated remote sensing instruments placed above the Earth's atmosphere are now capable of sensing electromagnetic radiation over nearly the entire spectrum—and these instruments are rapidly changing our picture of the cosmos.

For example, we previously thought that the interstellar medium was a fairly uniform collection of gas and dust; but spaceborne ultraviolet telescopes are now showing us that its structure is very inhomogeneous and complex. There are newly discovered components of the interstellar medium, such as extremely hot gas that is probably heated by shock waves from exploding stars. There is a great deal of interstellar pushing and shoving going on. Matter gathers and cools in some places because matter elsewhere is heated and dispersed. Besides discovering the existence of the very hot gas, the orbiting telescopes have discovered two potential sources of the gas: the intense stellar winds that boil off hot stars; and the rarer, but more violent, blasts of matter from exploding supernovae.

In addition, X-ray and gamma ray astronomy have contributed substantially to the discovery that the Universe is not relatively serene and unchanging as previously imagined, but is actually dominated by the routine occurrence of incredibly violent events.

And this series of remarkable new discoveries is just beginning. Future astrophysics missions will provide access to the full range of the electromagnetic spectrum at increased angular resolution. They will support experimentation in key areas of physics, especially relativity and gravitational physics. Out of these exciting discoveries, perhaps, will emerge the scientific pillars for constructing an extraterrestrial civilization based on technologies unimaginable in the framework of contemporary physics.

Virtually all the information we receive about celestial objects comes to us through observation of electromagnetic radiation. Cosmic ray particles are an obvious and important exception, as are extraterrestrial material samples that have been returned to Earth (for example, lunar rocks). Each portion of the electromagnetic spectrum carries unique information about the physical conditions and processes in the Universe. Infrared radiation reveals the presence of thermal emission from relatively cool objects while ultraviolet and extreme ultraviolet radiation may indicate thermal emission from very hot objects. Various types of violent events can lead to the production of X-rays and gamma rays.

Although EM radiation varies over many decades of energy and wavelength, the basic principles of measurement are quite common to all regions of the spectrum. The fundamental techniques used in astrophysics can be classified as: imaging, spectrometry, photometry and polarimetry. Imaging provides basic information about the distribution of material in a celestial object, its overall structure and, in some cases, its physical nature. Spectrometry is a measure of radiation intensity as a function of wavelength. It provides information on nuclear, atomic and molecular phenomena occurring in and around the extraterrestrial object under observation. Photometry involves measuring radiation intensity as a function of time. It provides information about the time variations of physical processes within and around celestial objects, as well as their absolute intensities. Finally, polarimetry is a measurement of radiation intensity as a function of polarization angle. It provides information on ionized particles rotating in strong magnetic fields.

High–energy astrophysics encompasses the study of extraterrestrial X-rays, gamma rays and energetic cosmic ray particles. Prior to space–based high energy astrophysics, scientists believed that violent processes involving high energy emissions were rare in stellar and galactic evolution. Now, because of studies of extraterrestrial X-rays and gamma rays, we know that such processes are quite common, rather than exceptional. The observation of X-ray emissions has been very valuable in the study of high energy events, such as mass transfer in binary star systems, interaction of supernovae remnants with interstellar gas, and quasars (whose energy source is presently unknown). It is thought that gamma rays might be the missing link in understanding the physics of high–energy objects such as pulsars and black holes. The study of cosmic ray particles provides important information about the physics of nucleosynthesis and about the interactions of particles and strong magnetic fields. High–energy phenomena that are suspected sources of cosmic rays include supernovae, pulsars, radio galaxies and quasars.

X-ray astronomy is the most advanced of the three high–energy astrophysics disciplines. Space–based X-ray observatories increase our understanding in the following areas: (1) stellar structure and evolution, including binary star systems, supernovae remnants, pulsar and plasma effects, and relativity effects in intense gravitational fields; (2) large–scale galactic phenomena, including interstellar media and soft X-ray mapping of local galaxies; (3) the nature of active galaxies, including spectral characteristics and the time variation of X-ray emissions from the nuclear or central regions of such galaxies; and (4) rich clusters of galaxies, including X-ray background radiation and cosmology modeling.

Gamma rays consist of extremely energetic photons (that is, energies greater than 10^5 electron volts) and result from physical processes different than those associated with X-rays. The processes associated with gamma ray emissions in astrophysics include: (1) the decay of radioactive nuclei; (2) cosmic ray interactions; (3) curvature radiation in extremely strong magnetic fields; and (4) matter–antimatter annihilation. Gamma ray astronomy reveals the explosive, high energy processes associated with such celestial phenomena as supernovae, exploding galaxies and quasars, pulsars and black hole candidates.

Gamma ray astronomy is especially significant because the gamma rays being observed can travel across our entire Galaxy, and even across most of the Universe, without suffering appreciable alteration or absorption.

Therefore, these energetic gamma rays reach our Solar System with the same characteristics, including directional and temporal features, as they started with at their sources, possibly many light years distant and deep within regions or celestial objects opaque to other wavelengths. Consequently, gamma ray astronomy provides information on extraterrestrial phenomena not observable at any other wavelength in the electromagnetic spectrum and on spectacularly energetic events that may have occurred far back in the evolutionary history of the Universe.

Cosmic rays are extremely energetic particles that extend in energy from one million (10^6) electron volts to over 10^{20} eV and range in composition from hydrogen (atomic number Z = 1) to a predicted atomic number of Z = 114. This composition also includes small percentages of electrons, positrons and possibly antiprotons. Cosmic–ray astronomy provides information on the origin of the elements (nucleosynthetic processes) and the physics of particles at ultrahigh energy levels. Such information addresses astrophysical questions concerning the nature of stellar explosions and the effects of cosmic rays on star formation and galactic structure and stability.

Astronomical work in a number of areas will greatly benefit from large, high resolution optical systems that operate outside the Earth's atmosphere. Some of these areas include investigation of the interstellar medium, detailed study of quasars and black holes, observation of binary X-ray sources and accretion disks, extragalactic astronomy, and observational cosmology. The Space Telescope (ST) constitutes the very heart of NASA's space-borne ultraviolet/optical astronomy program through the end of this century. Its ability to cover a wide range of wavelengths, to provide fine angular resolution and to detect faint sources will make it the most powerful astronomical telescope ever built.

In many ways the Space Telescope will function like a major ground–based observatory. It will be operated by an ST Operations Control Center through an independent Space Telescope Science Institute. It will have a permanent staff of astronomers and enough facilities to support a large number of visiting scientists. The Space Telescope will be a long–lived orbiting observatory that accommodates a variety of scientific instruments. These instruments can be changed and updated using the Space Shuttle, as both scientific priorities and sensor technologies evolve.

While the Space Telescope has been designed as a general–purpose instrument with application across a very broad range of astrophysical investigations, there will be a number of problems for which it will not be well suited. Specialized space–based instruments, such as Starlab, will be required to handle these situaions.

Another interesting area of astrophysics involves the extreme ultraviolet (EUV) region of the electromagnetic spectrum. The interstellar medium is highly absorbent at EUV wavelengths (100 to 1,000 angstroms [Å]). EUV data gathered from space–based instruments will be used to confirm and refine contemporary theories of the late stages of stellar evolution, to analyze the effects of EUV radiation on the interstellar medium and to map the distribution of matter in our "solar neighborhood."

Infrared (IR) and radio astronomy encompass studies of the electromagnetic spectrum from 1 to 10,000 micrometers wavelength (infrared), between 1000 and 10^6 micrometers (microwave) and greater than 10^5 micrometers (radio wave). (A micrometer is one millionth of a meter.) Infrared radiation is emitted by all classes of "cool" objects (stars, planets, ionized gas and dust regions, and galaxies) and cosmic background radiation. Most emissions from objects with temperatures ranging from 3 to 2000 degrees Kelvin are in the infrared region of the spectrum. In order of decreasing wavelength, the sources of infrared and microwave radiation are: (1) galactic synchrotron radiation; (2) galactic thermal bremsstrahlung radiation in regions of ionized hydrogen; (3) the cosmic background radiation; (4) 15 degrees Kelvin cool galactic dust and 100 degrees Kelvin stellar-heated galactic dust; (5) infrared galaxies and primeval galaxies; (6) 300 degrees Kelvin interplanetary dust; and (7) 3000 degrees Kelvin starlight.

Gravitation is the dominant long-range force in the Universe. It governs the large-scale evolution of the Universe and plays a major role in the violent events associated with star formation and collapse. Einstein's General Theory of Relativity (presented in 1915) is currently regarded as the best description of gravitational interaction. Although other relativistic theories have been developed, experimental evidence is needed to verify them. Outer space provides the low–acceleration and low–noise environment needed for the careful measurement of relativistic gravitational effects. A number of interesting experiments have been identified for a space–based experimental program in relativity and gravitational physics.

One such space–based experiment is a planned NASA mission called Gravity Probe-B. The objective of the Gravity Probe-B mission is to measure the geodetic precession due to the motion of a gyroscope through a gravitational field (this is called: relativistic spin-orbit coupling); and the precession produced by the twisting of space due to the rotation of the Earth itself (this is called: relativistic spin-spin coupling). High precision measurement of the magnitude of these two effects will represent a fundamental step in our ability to choose between competing theories of gravitation. Measurements with the necessary precision can only be performed in the environment of outer space.

The ultimate aim of astrophysics is to understand the origin, nature and evolution of the Universe. It has been said that the Universe is not only stranger than we imagine, it is stranger than we *can* imagine! Through the creative use of modern space technology and the Space Shuttle, we will witness many major discoveries in astrophysics in the exciting decades ahead—each discovery helping us understand a little better the magnificent Universe in which we live and the place in which we will build our extraterrestrial civilization.

See also: **Advanced X-Ray Astrophysics Facility;**

Cosmic Background Explorer; Extreme Ultraviolet Explorer; Gamma Ray Observatory; Orbiting Very Long Baseline Interferometry Observatory; Shuttle Infrared Telescope Facility; Space Telescope; Starlab; Starprobe; X-ray Timing Explorer

Astropolis An urban extraterrestrial facility in near-Earth space proposed by Dr. Krafft A. Ehricke. The modular design of Astropolis was selected to make maximum use of the different levels of gravity available in a large, rotating space facility. Astropolis represents a logical growth step beyond the space station. It would contain several thousand inhabitants who would live and work in an unusual multi-gravity-level world. This proposed facility would contain residential sections, a dynarium, space industrial zones, space agricultural facilities, research laboratories and even other world enclosures (OWEs).

Astropolis would have the ability to completely recycle air, water and waste materials. Energy would be supplied by either nuclear power plants or solar arrays. The research section of Astropolis would be dedicated to the long-term use of the space environment for basic and applied research, as well as for eventual industrial exploitation. The other world enclosures would be located at various distances from the hub of Astropolis. Using these special OWE facilities, exobiologists, space scientists, planetary engineers and interplanetary explorers will be able to simulate the gravitational environment of all major celestial objects in the Solar System of interest from the perspective of human visitation and possible settlement. These include the Moon, Mars, Venus, the asteroids and certain moons of the giant outer planets. Pioneering work in the OWE facilities of Astropolis could pave the way for the opening up of both cislunar and heliocentric space (that is, the space between the Earth and the Moon's orbit, as well as all other space surrounding the Sun) to human occupancy.

Astropolis is envisioned as a 4,000- to 15,000-ton-class space complex that would be rotated very slowly, at about 925 revolutions per Earth day (24 hours). Because of its low angular velocity, Coriolis forces (sideward force felt by an astronaut moving radially in a rotating system, such as a space station) would cause little disturbance and discomfort—even at the greatly reduced artificial gravity levels occurring closer to the hub. Therefore, research and industrial projects conducted on an orbiting facility like Astropolis would be able to enjoy excellent variable-gravity-level simulations with minimum Coriolis-force disturbances—in contrast to smaller space stations, which would be spinning more rapidly.

See also: **Androcell; dynarium; extraterrestrial civilizations; other world enclosures; space settlements; space station.**

aurora australis A spectacular display of lights occurring in the night skies at high latitudes in the Southern Hemisphere; caused by atomic particles precipitating into the Earth's atmosphere from the magnetosphere. This beautiful phenomenon is also called the "southern lights."

See also: **aurora borealis; Earth's trapped radiation belts**

aurora borealis   A spectacular optical display occurring in the night skies at high northern latitudes, caused by the precipitation of energetic particles from the magnetosphere into the Earth's atmosphere. This beautiful, extra-terrestrially induced atmospheric phenomenon is also called the "northern lights."

See also: **aurora australis; Earth's trapped radiation belts**

bacteria (singular: bacterium)   Any of an extremely flexible group of microorganisms (of the class Schizomycetes) whose members come in a variety of structures, occur singly or in colonies, and can exist as free-living organisms or as parasites. Bacteria can live just about anywhere and possess a wide variety of biochemical properties.

See also: **exobiology; life in the Universe**

Barnard's star   A red dwarf star approximately six light-years from the Sun, making it the fourth nearest star to our Solar System. The absolute magnitude of Barnard's star is 13.2, and its spectral class is M5V. It has the largest known proper motion, some 10.3 arc-seconds per year. Discovered in 1916 by E.E. Barnard, this star is of popular interest because recent investigations of its wobbling motion have hinted at the possible existence of two (extrasolar) planets orbiting around it. One of these planets is thought to have a mass slightly larger than that of Jupiter, and the other, a mass slightly less.

See also: **extrasolar planets; search for extraterrestrial intelligence; stars**

Bernal sphere   Long before the Space Age began, the British physicist and writer J. Desmond Bernal predicted that the majority of humanity would someday live in "artificial globes" orbiting around the Sun. In his 1929 work *The World, the Flesh, and the Devil*, Bernal boldly speculated about the colonization of outer space.

Bernal's concept of spherical space habitats has influenced both early space-station designs (see fig. 1) and very recent space-settlement designs.

See also: **space settlement; space station**

Fig. 1   An early space-station design—obviously influenced by Bernal's concept of spherical space habitats. (Courtesy of NASA.)

"Big Bang" theory   A theory in cosmology concerning the origin of the Universe. According to the Big Bang cosmological model, there was a very large explosion, also called the "initial singularity," that started the space and time of our Universe. The Universe itself has been expanding ever since. It is currently thought that the Big Bang event occurred between 7 billion and 20 billion (10^9) years ago. Recent astrophysical observations and discoveries lend support to the Big Bang model, especially the discovery of the cosmic microwave background radiation in 1965.

Actually, there are two general variations of the Big Bang model. In one, it is assumed that the Universe will expand forever. This is called the "open-Universe model." The other variation, called the "closed-Universe model," assumes that the Universe will eventually recollapse—possibly back into another "singularity."

The Big Bang theory assumes that cosmic expansion was started by a primordial fireball, which emitted radiation and pushed matter apart. Then, as the matter in the Universe expanded, it cooled, condensing into galaxies, which, in turn, gave rise to stars and eventually other celestial objects such as planets, asteroids, comets and so on. The radiation released from this ancient explosion has now cooled and can be observed as the cosmic microwave background radiation.

Although there are quite probably very many distant objects far out in space that we haven't been able to detect yet, space scientists have discovered radiation from a source quite literally at the edge of the observable Universe. As we look deep into space, we are really looking back into time—that is, we are viewing the most distant regions of the Universe as they were very long ago. In the Big Bang model, before the galaxies and stars formed, the Universe was filled with hot, glowing gas that was opaque. Then, at some point in time during the first million years or so after the Big Bang event, this expanding gas cooled and became transparent. Today, we can look out into space and back into time only until we reach that very distant region where we are observing the early Universe when it transitioned from an opaque to a transparent gas. Beyond that point, space is opaque, so that light waves simply cannot reach us. Instead, we see the glow of that primordial hot gas as it cooled to about 10,000 degrees Kelvin and then cleared. This glow was originally emitted as ultraviolet radiation but has now been Doppler-shifted to longer wavelengths by the expansion of the Universe and currently resembles emission from a dense gas at a temperature of only 2.7 degrees Kelvin. Today, scientists observe it as a diffuse background of microwave and thermal radiation. In fact, you can think of this microwave radiation as a distant spherical wall that surrounds us and delimits the edges of the observable Universe.

See also: **astrophysics; Cosmic Background Explorer; cosmology**

binary star   A pair of stars that orbit about their common center of mass. By convention, the star that is nearest the center of mass in a binary star system is called the "primary," while the other (smaller) star of the system is called the "companion." Binary star systems can be further classified as visual binaries, eclipsing binaries, spectroscopic binaries and astrometric binaries. Visual binaries are those systems that can be resolved into two stars by an optical telescope. Eclipsing binaries occur when each star of the system alternately passes in front of the other, obscuring or eclipsing it and thereby causing their combined brightness to diminish periodically. Spectroscopic binaries are resolved by the Doppler shift of their spectral lines as the stars approach and then recede from the Earth while revolving about their common center of mass. It is interesting to note that eclipsing binaries are generally also spectroscopic binaries. In an astrometric binary, one star cannot be visually observed, and its existence is inferred from the irregularities in the motion of the visible star of the system.

Binary star systems are more common in our Galaxy than generally realized. Perhaps 50 percent of all stars are contained in binary systems. The typical mean separation distance between members of a binary star system is on the order of 10 to 20 astronomical units.

See also: **stars**

biological evolution   A term suggesting that all living things on Earth have a common (biological) ancestry from some initial large and complex organic molecules that formed in the primordial oceans of an ancient Earth. These organic molecules developed into the first living cells that were capable of metabolism, respiration, reproduction and the transfer of genetic information.

See also: **life in the Universe**

biome   The complex community of living organisms that are found in a major ecological region, such as the tundra, a desert, woodlands or a coral reef. A biome is characterized by the distinctive life-forms of important "climax" species of plants and animals. (In ecology the word *climax* refers to that stage in evolution or development in which the community of living organisms becomes stable and starts to perpetuate itself.)

See also: **ecosphere; life in the Universe**

biosphere   The life zone of a planetary body; for example, the part of the Earth inhabited by living organisms.

See also: **ecosphere**

blackbody   A perfect emitter and perfect absorber of electromagnetic radiation. It is capable of radiating at all frequencies as well as absorbing incident radiant energy (that is, radiant energy that strikes it) at all frequencies. The radiant energy emitted by a blackbody, called blackbody radiation, is a function only of its temperature. The physical laws describing the behavior and properties of a blackbody come from equilibrium thermodynamics.

See also: **radiation laws**

black holes   Theorized gravitationally collapsed masses from which nothing—light, matter or any other kind of signal—can escape. Scientists today generally speculate that a black hole is the natural end product when a giant star dies and collapses. If the star has three or more solar masses left after exhausting its nuclear fuels (a solar mass is a unit of measure equivalent to the mass of our own Sun), then it can become a black hole. As with the formation of a white dwarf or a neutron star, the collapsing giant star's density and gravity increase with contraction. However, in this case, because of the large mass involved, the gravity of the collapsing star becomes too strong for even neutrons to resist, and an incredibly dense point mass, or "singularity," is formed. Therefore, a black hole is essentially a singularity surrounded by an event region in which the gravity is so strong that absolutely nothing can escape.

Remember, as the massive star collapses, its gravitational escape velocity (the speed an object needs to reach in order to escape from the star) also increases. When a giant star has collapsed to a dimension called the Schwarzschild radius, its gravitational escape velocity is equal to the speed of light. At this point, not even light itself can escape from the black hole!

The Schwarzschild radius is simply the "event horizon," or boundary of no return, for a black hole. This dimension bears the name of the German astronomer Karl Schwarzschild, who wrote the fundamental equations describing a black hole in 1916. Anything crossing this boundary can never leave the black hole. In fact, the event horizon represents the start of a region disconnected from normal space and time. We cannot see beyond this event horizon into a black hole, and time itself is considered to stop there.

As shown in table 1, the event horizon, or Schwarzschild radius, of a black hole is proportional to the mass of the collapsed star. For a star of 10 solar masses, for example, the Schwarzschild radius is 30 kilometers.

Once a black hole has been formed, it crushes anything crossing its event horizon into its incredibly dense singularity. As the black hole devours matter, its event horizon expands. This expansion is limited only by the availability of mass. Gigantic black holes, which contain the crushed remains of billions of stars, are considered theoretically possible. In fact, some astrophysicists speculate that rotating black holes (called Kerr black holes) containing the remains of millions or billions of dead stars may lie at the centers of galaxies. These enormous rotating black holes may be the powerhouses of quasars and active galaxies. Quasars are believed to be galaxies in an early, violent evolutionary stage, while active galaxies are characterized by their extraordinary energy outputs, which occur mostly from their cores, or "galactic nuclei." Scientists think that "normal" galaxies like our own Milky Way are only "quiet" because the black holes at their centers have no more material upon which to feed.

Today, evidence that super-dense stars like white dwarfs and neutron stars really exist has supported the idea that black holes themselves (representing what may be the ultimate in density) must also exist. But how can we detect an object from which nothing, not even light, can escape?

Astrophysicists think they may have found indirect ways of detecting black holes. Their techniques depend upon certain black holes being members of binary star systems. (A binary star system consists of two stars comparatively near to and revolving about each other. Astronomers would say the two stars are "gravitationally bound" to each other.) Unlike our Sun, many stars in the Galaxy belong to binary systems.

If one of the stars in a particular binary system has become a black hole (although invisible), it would betray its existence by the gravitational effects it produces upon the companion star, which is observable. These gravitational effects would actually be in accordance with Newton's universal law of gravitation; that is, the mutual gravitational attraction of the two celestial objects is directly proportional to their masses and inversely propor-

tional to the square of the distance between them. Once beyond the black hole's event horizon, its gravitational influences are the same as exerted by other objects (of equivalent mass) in the "normal" Universe.

Astrophysicists have also speculated that a substantial part of the energy of matter spiraling into a black hole is converted by collision, compression and heating into X rays and gamma rays, which display certain spectral characteristics. This X and gamma radiation emanates from the material as it is pulled toward the black hole. However, once the captured material has been pulled across the black hole's event horizon, this radiation cannot escape.

Black-hole candidates are celestial phenomena that exhibit such black-hole "capture effects" in a binary star system. Several have now been discovered and studied using space-based astronomical observatories (especially X-ray observatories). One very promising candidate is called Cygnus X-1, an invisible object in the constellation Cygnus the Swan. Cygnus X-1 means that it is the first X-ray source discovered in Cygnus. X rays from the invisible object have characteristics like those expected from materials spiraling toward a black hole. This material is apparently being pulled from the candidate black hole's binary companion—a large star of about 30 solar masses. Based upon the candidate black hole's gravitational effects on its visible companion, the hole's mass has been estimated to be about six solar masses. In time, the giant visible companion might itself collapse into a neutron star or a black hole; or else it might be devoured piece by piece by its black-hole companion. This form of stellar cannibalism would significantly enlarge the existing black hole's event horizon.

Big Bang cosmology states that our Universe began with a violent explosion that sent pieces of matter flying outward in all directions. To date, cosmologists and astrono-

Table 1 Schwarzschild Radius as a Function of Mass

Mass of Collapsed Star (solar masses)[a]	Schwarzschild Radius (km)
5	15
10	30
20	60
50	150

[a](One solar mass = mass of our Sun.)

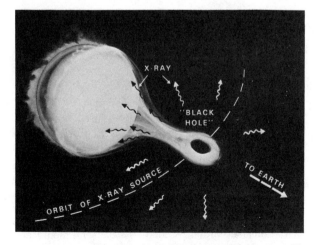

Fig. 1 Artist's concept of a black hole devouring its giant visible binary companion. (This represents what astrophysicists think is taking place between the binary giant star system HDE226868 and the X-ray source Cygnus X-1.) (Courtesy of NASA.)

mers have not detected enough mass in the Universe to reverse this expansion process. The possibility remains, however, that the missing mass may be locked up in undetectable black holes that are more prevalent than anyone currently anticipates.

Do enough black holes exist within the Universe to reverse this expansion? If so, what will then happen to the Universe? Will all of the stars, galaxies and other matter in the Universe eventually collapse inward, just like a massive star that has exhausted its nuclear fuels? Will there ultimately be created one very large black hole, within which the Universe will again collapse back to a singularity?

Extrapolating back more than 10 billion to 15 billion years, some cosmologists trace the present Universe to an initial singularity. Thus, is a singularity both the beginning and the end of our Universe? Is our Universe but a pause or a phase between such singularities?

Other scientists have put forward even more intriguing questions and speculations. If the Universe itself is closed and nothing can escape, perhaps we may already be in a megasize black hole!

As we learn more about our extraterrestrial environment and gather further supporting evidence about the existence and properties of black holes, scientists may someday begin to answer these puzzling questions.

See also: **Cygnus X-1; stars** (white dwarfs, neutron stars and black holes)

bow shock   The outermost boundary of the Earth's magnetosphere. It is created by the magnetogasdynamic interaction of the solar wind and the geomagnetic field.

See also: **Earth's trapped radiation belts; magnetosphere**

C

carbon chauvinism   A strongly prejudiced opinion that extraterrestrial life in accordance with the physical laws of chemistry is based on carbon chemistry. For any chemical element to serve as the basis for organic systems, it must satisfy two distinctive requirements. First, the chemical element must be sufficiently abundant in the Universe, thereby supporting the probabilistic reactions currently thought to be associated with the prebiotic (prelife) stages of chemical evolution. Second, the element should also have a chemistry that supports the formation of long molecular chains, since living systems must be capable of transmitting important information from one generation to another. For example, carbon-based organic systems on Earth pass information from one generation to the next by means of the very long and complex DNA molecule.

If we search the periodic table for candidate chemical elements to support living systems, using the two requirements just stated, only two really obvious candidates emerge: carbon and silicon. Since all life on Earth is carbon-based, we know that this chemical will support the emergence of life, including intelligent life. Silicon-based organic systems are presently unknown. In fact, if we require that silicon-based living systems be capable of passing information from generation to generation via complex molecules like DNA, we encounter a significant problem. For only at low temperatures—say, less than 73 degrees Kelvin (minus 200 degrees Celsius)—will silicon properly form such long molecular chains. Unfortunately, at such low temperatures the reaction rates for the chemical synthesis of prebiotic silicon-based molecules are greatly reduced. Some scientists have estimated that the length of time required for the evolution of an eventually life-bearing primeval chemical "soup" would be on the order of tens to hundreds of billions of years. Such a large span of time is obviously inconsistent with the period normally associated with the appearance of life in the Universe.

Consequently, many contemporary scientists and exobiologists maintain a carbon chauvinism when it comes to a discussion of the chemistry of alien life. This position, perhaps very valid technically, then imposes some severe restrictions on the type of planetary environment that might support the chemical evolution of "life as we know it" (carbon-based).

See also: **extraterrestrial life chauvinisms**

carbon cycle   The chain of thermonuclear fusion reactions thought to be the main energy-generating mechanism in stars much hotter than our Sun. In this cycle hydrogen is converted to helium with large quantities of energy being released and with carbon–12, an isotope of carbon, serving as a reaction catalyst. Interior stellar temperatures greater than about 16 million degrees Kelvin are associated with the carbon cycle.

See also: **fusion**

cargo orbital transfer vehicle (COTV)   An orbital transfer vehicle that would be used during the construction of Satellite Power System (SPS) units to transport up to 4,000 metric tons of cargo between low Earth orbit (LEO) and geosynchronous Earth orbit (GEO). Several classes of COTV have been considered to satisfy cargo-transport requirements: (1) a conventional, chemically fueled, high-thrust COTV with a short trip time and a high degree of reusability (2) a nuclear-rocket COTV featuring a nuclear-reactor propulsion system capable of high thrust and high specific impulse, and (3) an electric-propulsion COTV that provides continuous low-level thrusting to slowly move large payloads between LEO and GEO. The electric power needed to run this type of vehicle's ion engines can be derived from large solar arrays or from compact, well-shielded nuclear-reactor power systems.

See also: **Satellite Power System; space nuclear power; space nuclear propulsion**

Cepheid variable   One of a group of very bright, supergiant stars that pulsate periodically in brightness. Type I, or classical, Cepheid variable stars have characteristic pulsing periods ranging from 5 to 10 days, while Type II Cepheids are characterized by pulsing periods of 10 to 30 days.

See also: **stars** (red giants and supergiants)

CETI   An acronym that stands for "communication with extraterrestrial intelligence." Compare with *SETI*.

See also: **extraterrestrial civilizations; interstellar communication; search for extraterrestrial intelligence**

circadian   Pertaining to events that occur at approximately 24-hour intervals, such as certain biological rhythms.

circumsolar space   A term found in the Soviet astronautical literature describing heliocentric (Sun-centered) space.

circumstellar   Around a star—as opposed to *interstellar*, between the stars.

cislunar   Generally, in or pertaining to the region of outer space between the Earth and the Moon. Cislunar space has also been more rigorously defined as the region between the Earth and the Moon extending from 65,000 kilometers to 344,400 kilometers. By convention, the outer limit of cislunar space is considered to be the outer limit (from the Moon) of its sphere of (gravitational) influence.

closed universe   A model in cosmology in which it is assumed that the Universe will one day stop expanding and eventually recollapse. The geometry of this type of spatially finite, though unbounded, Universe is like the surface of a sphere, and space is taken to have positive curvature. (If parallel lines eventually meet, as happens on the surface of a sphere, we say that the space has a positive curvature.) One of the great debates in modern cosmology centers around the question of whether the Universe is really open or closed.

See also: **astrophysics; "Big Bang" theory; cosmology; open universe**

close encounter (CE)   An interaction with an unidentified flying object (UFO).

See also: **unidentified flying object**

cluster of galaxies   An accumulation of galaxies. These galactic clusters can occur with just a few member galaxies (say, 10 to 100)—such as the Local Group, of which our Milky Way is a part—or they can occur in great groupings involving thousands of galaxies.

comet   A dirty ice "rock" orbiting the Solar System that the Sun causes to vaporize, glow visibly and stream out a long, luminous tail. Comets are generally regarded as samples of primordial material from which the planets were formed billions of years ago. The comet's nucleus is believed to be some type of dirty ice ball, consisting of frozen gases and dust. (See fig. 1.) As the comet approaches the Sun from the frigid regions of deep space, the Sun's radiation causes these ices to sublime (vaporize) and the resultant vapors form an atmosphere or "coma" with a diameter that may reach 100,000 kilometers (60,000 miles). However, cometary nuclei are estimated to have diameters of only a few tens of kilometers. Ions produced in the coma are effected by the charged particles in the solar wind; while dust particles liberated from the comet's nucleus are impelled in a direction away from the Sun by the pressure of the solar wind. The results are the formation of the plasma (Type I) and dust (Type II) cometary tails which can extend for up to 100 million kilometers (60

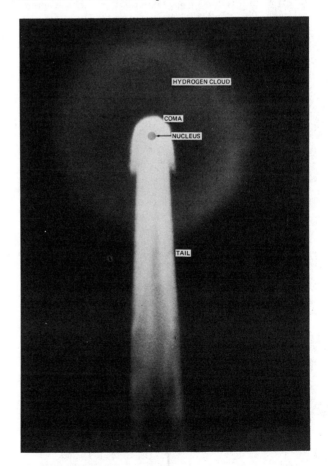

Fig. 1   The major features of a comet. (Courtesy of NASA.)

million miles). The Type I tail, composed of ionized gas molecules, is straight and extends radially outward from the Sun as far as 100 million kilometers (10^8 km). The Type II tail, consisting of dust particles, is shorter, generally not exceeding 10 million kilometers in length. It curves in the opposite direction to the orbital movement of the comet around the Sun. It appears that an enormous cloud of hydrogen atoms also surrounds the visible coma. This hydrogen was first detected in comets in the 1960s.

No astronomical object, other than perhaps the Sun or the Moon, has attracted more attention or interest. Since ancient times, comets have been characterized as harbingers of momentous human events. William Shakespeare wrote in the play *Julius Caesar*:

"When beggars die, there are no comets seen; but the heavens themselves blaze forth the death of princes."

The most spectacular of all periodic comets is Halley's Comet which is due to return in the 1985–86 time period. Table 1 provides some historical and physical data on this famous "long-haired star."

All comets are believed to originate far from the Sun in the Oort Cloud, which is thought to extend out to the limits of the Sun's gravitational attraction, creating a sphere with a radius of approximately 50,000 astronomical units. This cloud, first described by the Dutch astronomer, Jan Hendrik Oort (1900–) is thought to contain billions of comets, whose total mass is estimated to be roughly equal to the mass of the Earth. Small numbers of these comets continually enter the planetary regions of our Solar System, possibly through the gravitational perturbations caused by neighboring stars. Once a comet approaches the Solar System, it is also subject to the gravitational influences of the major planets, especially Jupiter, and the comet may eventually achieve a quasi stable orbit within the Solar System.

By convention, comet orbital periods are divided into two classes: long period comets (which have orbital periods in excess of 200 years), and short period comets (which have periods less than 200 years).

Comets may represent interesting sources of extraterrestrial materials. Table 2 lists the atomic and molecular species that have been observed in cometary coma. A rendezvous with or even an automated sample return mission to a short period comet would answer many of the questions currently puzzling cometary physicists. A highly automated sampling mission, for example, would permit the collection of atomized dust grains and gases directly from the comet's coma. The simplest way to accomplish this exciting space mission would be to use a high velocity flyby technique, with the automated spacecraft passing as close to the comet's nucleus as possible. This probe would be launched on an Earth-return trajectory. Terrestrial recovery of its cosmic cargo could be accomplished by on-orbit rendezvous or by the ejection of the sample into a small atmospheric entry capsule.

Since they are only very brief visitors to our Solar System, you might wonder what happens to these comets after they blaze a trail across the night sky? Well, the chance of any particular "new" comet being captured in a

Table 1   Selected Data for Halley's Comet

Historical Data

Earliest probable recorded apparition	240 B.C.
Number of recorded apparitions From 240 B.C. to 1910 A.D., only the 164 B.C. apparition was not recorded.	28
Shortest period between returns to perihelion	74.42 years (1835–1910)
Longest period between returns to perihelion	79.25 years (451–530)
Closest approach to the Earth	0.04 AU (April 11, 837)
Longest angular tail length recorded	93° (mid-April 837)
Brightest apparent magnitude recorded (approximate)	−3.5 (April 11, 837)

Physical Characteristics

Estimated diameter of nucleus	5 km
Estimated density of nucleus	1 g/cm³
Estimated rotation period	10.3 hours, direct
Observed spectra in 1910	CH, CN, C_2, C_3, Na D, CO⁺, N₂⁺
Observed tails	Type I ion and Type II dust
Associated meteor streams	η Aquarid (Early May) and Orionid (late October)

SOURCE: NASA.

Table 2   Observed Atomic and Molecular Compositions of Comets (Coma)

H	(atomic hydrogen)
C, C_2, C_3	(carbon: atomic and molecular)
CH	(carbon-hydrogen free radical)
CN	(cyanogen radical)
HCN	(hydrogen cyanide)
CH_3CN	(acetonitrile)
NH	(imine radical)
NH_2	(nitrogen hydride molecule)
NH_3	(ammonia)
O	(atomic oxygen)
OH	(hydroxyl radical)
H_2O	(water)
Na	(sodium)
K	(potassium)
Ca	(calcium)
V	(vanadium)
Cr	(chromium)
Mn	(manganese)
Fe	(iron)
Co	(cobalt)
Ni	(nickel)
Cu	(copper)

SOURCE: NASA.

short-period orbit is quite small. Therefore, most of these cosmic wanderers simply return to the Oort Cloud, presumably to loop back into the Solar System eons later. Or else they are ejected into interstellar space along hyperbolic orbits. Sometimes, however, a comet falls into the Sun. One such event was recently observed by instruments on a spacecraft in August 1979. Other comets simply break up because of gravitational (tidal) forces or possible outbursts of gases from within. When a comet's volatile materials are exhausted or when its nucleus is totally covered with nonvolatile substances, we call the comet "inactive." Some space scientists believe that an inactive comet may become a "dark asteroid," such as those in the Apollo group, or else disintegrate into meteoroids. Finally, on very rare occasions, a comet may even collide with an object in the Solar System, an event with the potential of causing a "cosmic catastrophe."

See also: **asteroid;** extraterrestrial catastrophe theory; meteoroids

Comet Atomized Sample Return Mission   Space scientists believe that comets are the most primitive bodies in our Solar System. In fact, they represent the best source of obtainable samples of the original material from which our Solar System was formed. Comets may provide a cosmochemical record of conditions in the interstellar medium and the primordial solar nebula.

Sample-return missions can provide uniquely detailed information about Solar System bodies because of the sensitivity, precision and flexibility of modern terrestrial laboratory techniques. The very productive analyses of lunar rocks and meteorites demonstrate the value of returning extraterrestrial materials to Earth for detailed laboratory study. Of course, special precautions must also be observed to avoid contaminating the Earth's biosphere with extraterrestrial organisms, which could prove disastrously harmful to life-forms that exist today.

Due to the difficulties in obtaining samples of extraterrestrial materials from most Solar System bodies, space technologists generally assume that a sample-return mission should be attempted only after a sequence of earlier reconnaissance and exploration missions. However, for comets, one type of sample return can actually be considered as an early mission. Highly simplified techniques permit particles to be collected directly from the coma (head) of a comet. The spacecraft would be launched in an Earth-return trajectory, and terrestrial recovery could be accomplished by placing the collected sample into a small atmospheric-entry capsule.

This type of comet flyby sample return is called an atomized or plasmatized sample return because the sampled particles are vaporized as they impact and penetrate a special collector on the spacecraft at velocities of 10 to 70 kilometers per second.

There are two main objectives of the Comet Atomized Sample Return Mission. First, it will obtain samples of the volatile and nonvolatile constituents of the comet's coma during a fast fly-through and then return these samples to Earth for analysis. Second, it will determine the densities of coma materials along the flight path.

The comet sample-return spacecraft can be launched by a Space Shuttle/Inertial Upper Stage (IUS) configuration and will travel on a ballistic trajectory that reaches the comet after about two years. Remote sensing instruments on board the spacecraft will be turned on some 60 days before encounter with the comet, which will then be very close to perihelion (its point of closest approach to the Sun). Unless a comet-rendezvous spacecraft is already on station, imagery data will be used for terminal navigation and will provide an accurate determination of the comet's location. The high-speed (10–15 kilometers per second) fly-through of the comet's coma lasts only minutes, but this is still enough time for the collection of atomized dust grains and gases. After the spacecraft encounters the comet, the sample-collector panels are stored in an onboard capsule. Then, a relatively small propulsive maneuver places the spacecraft on an impact trajectory with the Earth. As the sample-return spacecraft reaches the Earth, the sample-containing capsule is ejected, directly entering the atmosphere, and descends on parachutes to the surface for recovery and analysis.

See also: **comet; Comet Rendezvous Mission; extraterrestrial expeditions; Mars Sample Return Mission; orbital transfer vehicle; orbits of objects in space; planetary exploration; space launch vehicles (Space Shuttle); Space Transportation System**

Comet Rendezvous Mission   Is a comet really a dirty snowball? What does the nucleus of a comet look like? The planned NASA Comet Rendezvous Mission (probable launch 1990s) will help answer these and many other intriguing questions about comets. Space scientists consider comets to be among the most primitive bodies in our Solar System. The Comet Rendezvous spacecraft, a Mariner Mark II configuration, will achieve rendezvous with one of several short-period comets and then perform close-up studies of the comet as it passes and achieves perihelion. (Perihelion is the point in an orbit around the Sun when the object is closest to the Sun.) Space scientists refer to "rendezvous missions" when they discuss encounters with comets and Earth-approaching asteroids that have gravitational spheres of influence no more than a few tens of kilometers in extent, while they say "orbiter missions" when discussing encounters involving larger (for example, 50 kilometers in diameter or more) targets among the main-belt asteroids.

Following launch by the Space Shuttle-Centaur, the Comet Rendezvous spacecraft cruises for about five years toward its target—a short-period comet such as Encke or Tempel 2. (Astronomers frequently define a short-period comet as one having an orbital period around the Sun of less than 200 years.) During flight to the comet, propulsive maneuvers are performed to match the trajectory of the spacecraft and the comet several months before the comet experiences perihelion passage. From then on, only small propulsive maneuvers are needed to maintain station keeping and to examine the comet from any desired distance and direction. Remote sensing measurements will start at the initial moment of rendezvous, and as the comet approaches the Sun, the *in situ* (a Latin phrase for "in place") spacecraft instrumentation will measure the expulsion of gases and dust that occurs with progressively greater intensity as the comet nears its closest approach to the Sun. Of course, if the expelled gas and dust environment gets too severe, the rendezvous spacecraft can retreat to a safe distance and then resume a close-encounter position when the solar-heating-driven expulsion of cometary material dies down to acceptable levels. After perihelion passage the rendezvous spacecraft will be inserted into a close (perhaps 10-kilometer) circular polar mapping orbit around the comet. This mission will end some six months after the initial rendezvous.

One option being considered for this mission is to conduct flybys of one or more of the main-belt asteroids on the way to comet rendezvous.

The basic Comet Rendezvous Mission will satisfy the following scientific objectives: (1) a study of the comet's nucleus through a complete perihelion passage; (2) a determination of the chemical and isotopic composition of the comet's nucleus and coma (head); (3) a detailed study and characterization of how gases and dust flow out of the comet when it approaches the Sun and its frozen (volatile) materials are activated by solar energy; and finally, (4) an investigation of how the solar wind interacts with the comet's coma.

Comets are extremely interesting Solar System bodies and represent (especially in the case of extinct comets) a potential source of the extraterrestrial materials needed to expand human civilization into heliocentric (Sun-centered) space. The Comet Rendezvous Mission will produce very important data about these wandering members of our Solar System, which the ancient Greek astronomers called "kometes," or "long-haired stars."

See also: **asteroid; comet; Comet Atomized Sample Return Mission; extraterrestrial resources; orbits of objects in space; planetary exploration**

consequences of extraterrestrial contact   Just what will happen if we make contact with an extraterrestrial civilization? No one on Earth can really say for sure. However, this contact will very probably be one of the most momentous events in all human history! The contact can be direct or indirect. Direct contact might involve a visit to Earth by a starship from another stellar civilization or perhaps take the form of the discovery of an alien probe, artifact or derelict spaceship in the outer regions of our Solar System. Some scientists have speculated, for example, that the hydrogen- and helium-rich giant outer planets might serve or could have served as "fueling stations" for interstellar spaceships from other worlds. Indi-

rect contact, via radio-wave communication, appears to represent the more probable contact pathway (at least from a contemporary terrestrial viewpoint). The consequences of our successful search for extraterrestrial intelligence (SETI) would be nothing short of extraordinary. For example, as part of our own SETI effort, were we to locate and identify but a single extraterrestrial signal, humankind would know immediately one great truth: We are not alone, and it is indeed possible for a civilization to create and maintain an advanced technological society without destroying itself. We might even learn that life, especially intelligent life, is prevalent in the Universe!

The overall impact of this contact will depend on the circumstances surrounding the initial discovery. If it happens by accident or after only a few years of searching, this news, once verified, would surely startle the world. If, however, intelligent alien signals were detected only after an extended effort, lasting generations and involving extensive search facilities, the terrestrial impact of the discovery might be less overwhelming.

The reception and decoding of a radio signal from an extraterrestrial civilization in the depths of space offers the promise of practical and philosophical benefits for all humanity. Response to that signal, however, also involves a potential planetary risk. If we do intercept an alien signal, we can decide (as a planet) to respond—or we may choose not to respond. If we are suspicious of the motives of the alien culture that sent the message, we are under no obligation to respond. There would be no practical way for them to realize that their signal was in fact intercepted, decoded and understood by the intelligent inhabitants of a tiny world called Earth.

Optimistic speculators emphasize the friendly nature of such an extraterrestrial contact and anticipate large technical gains for our planet, including the reception of information and knowledge of extraordinary value. They imagine that there will be numerous scientific and technological benefits from such contacts. However, because of the long round-trip times associated with such radio contacts (perhaps decades or centuries, even with the messages traveling at the speed of light), any information exchange will most likely be in the form of semi-independent transmissions, each containing significant facts about the sending society (such as planetary data, its life-forms, its age, its history, its philosophies and beliefs, and whether it has successfully contacted other alien cultures), rather than an interstellar dialogue with questions asked and answered in rapid succession. Consequently, over the period of a century or more, we Terrans might receive a wealth of information at a gradual enough rate to construct a comprehensive picture of the alien civilization without inducing severe culture shock on Earth.

Some scientists feel that if we are successful in establishing interstellar contact, we would probably not be the first planetary civilization to have accomplished this feat. In fact, they speculate that interstellar communications may have been going on since the first intelligent civilizations evolved in our Galaxy—some four or five billion years ago. One of the most exciting consequences of this type of celestial conversation would be the accumulation by all participants of an enormous body of information and knowledge that has been passed down from alien race to alien race since the beginning of the Galaxy's communicative phase. Included in this vast body of knowledge, something we might call the "galactic heritage," could be the entire natural and social histories of numerous species and planets. Also included, perhaps, would be extensive astrophysical data that extend back countless millennia, providing accurate insights into the origin and destiny of the Universe.

It is felt, however, that these extraterrestrial contacts would lead to far more than merely an exchange of scientific knowledge. Humanity would discover other social forms and structures, probably better capable of self-preservation and genetic evolution. We would also discover new forms of beauty and become aware of different endeavors that promote richer, more rewarding lives. Such contacts might also lead to the development of branches of art and science that simply cannot be undertaken by just one planetary civilization but rather require joint, multiple-civilization participation across interstellar distances. Most significant, perhaps, is the fact that interstellar contact and communication would represent the end of the cultural isolation of the human race. The optimists speculate that we would be invited to enter a sophisticated "cosmic community" as mature, planetary-civilization "adults" proud of our own human heritage—rather than remaining isolated with a destructive tendency to annihilate ourselves in childish planetary rivalries. Indeed, perhaps the very survival and salvation of the human race depends on finding ourselves cast in a larger cosmic role—a role far greater in significance than any human can now imagine.

If a cosmic community of extraterrestrial civilizations really does exist, it is probably composed of individual cultures and races that have learned to live with themselves and their technologies. Cultural life expectancies might be measured in aeons rather than millennia or even centuries. Identifying with these "super" interstellar civilizations and making contributions to their long-term objectives would definitely provide exciting new dimensions to our own lives and would create an interesting sense of purpose for our planetary civilization.

In considering contact with an extraterrestrial civilization, it is not totally unreasonable to think as well about the possible risks that could accompany exposing our existence to an alien culture—most likely far more advanced and powerful than our own. These risks range from planetary annihilation to humiliation of the human race. For discussion, these risks can be divided into four general categories: (1) invasion, (2) exploitation, (3) subversion and (4) culture shock.

The invasion of the Earth is a very recurrent theme in science fiction. By actively sending out signals into the cosmic void or responding to intelligent signals we've detected and decoded, we would be revealing our existence and announcing the fact that the Earth is a habitable planet. Soon thereafter (this risk scenario speculates) our planet might be invaded by an armada of spaceships carrying vastly superior beings who are set on conquering the Galaxy. After a valiant, but futile, fight, humankind is annihilated or enslaved. While such scenarios make interesting motion pictures and science-fiction novels, a logical review of the overall situation does not appear to support the extraterrestrial-invasion hypothesis or its grim (for Earth) outcome. If, for example, as we currently speculate, direct contact via interstellar travel is enormously expensive and technically very difficult—even for an advanced extraterrestrial civilization—then perhaps only the most extreme crisis would justify mass interstellar travel. Any alien race capable of interstellar travel would most certainly possess the technical skills needed to solve planetary-level population and pollution problems. Hence, the quest for more "living space" as a dominant motive for interstellar migration by an alien civilization does not appear to be a logical premise. It is not altogether inconceivable, of course, that members of an advanced civilization might seek to avoid extinction through mass interstellar migration before their native star leaves the main sequence and threatens to supernova. Again, we can logically conjecture that such a powerful, migrant extraterrestrial race would probably not want to compound the problems of a complex, difficult interstellar journey with the additional problems of interstellar warfare. They would most likely seek habitable, but currently uninhabited, worlds upon which to settle and rebuild their civilization. Such habitable worlds could have been located and identified long in advance, perhaps through the use of sophisticated robot probes.

Of course, interstellar travel might also prove much easier than we now predict. If this is the case, then the Galaxy could be teeming with waves of interstellar expeditions launched by expanding civilizations. Maintaining "radio silence" on a planetary scale is consequently no real protection from such waves of extraterrestrial explorers. We would inevitably be discovered by one of numerous bands of wandering extraterrestrials—without the aid of our electromagnetic-wave "homing beacons." In this situation the real question to be raised is the Fermi paradox: Where are they?

Physical contact between Earth and a benign advanced extraterrestrial civilization might also give rise to a silent, unintentional "invasion." Because of vast differences in biochemistry, alien microorganisms introduced into the terrestrial biosphere during physical contact with an alien civilization could trigger devastating plagues that would annihilate major life-forms on Earth.

Another major contact hazard category is exploitation. Some individuals have speculated that to an advanced alien civilization, human beings might appear to be primitive life-forms—ones that represent interesting experimental animals, unusual pets or even gourmet delicacies. Fortunately, differences in biochemistry might also make us very poisonous to eat. Again, the arguments against the "invasion scenario" apply equally well here. It is very difficult to imagine an advanced civilization expending great resources to cross the interstellar void just to bring home exotic pets, unusual lab animals or—perish the thought—"imported snacks." Perhaps it is more logical to assume that when an alien civilization matures to the level of star travel, such cultural qualities as compassion, empathy and a respect for life (in any form) become dominant.

Another major alien contact hazard category that is frequently voiced is subversion. This appears to be a more plausible and subtle form of contact risk. In this case, an alien race—under the guise of teaching and helping us join a cosmic community—might actually trick us into building devices that allow "Them" to conquer "Us." The alien civilization doesn't even have to make direct contact; the extraterrestrial "Trojan horse" might arrive on radio waves. For example, the alien race could transmit the details of a computer-controlled biochemistry experiment that would then secretly create their own life-forms here. The subversion and conquest of Earth would occur from within! There appears to be no limit to such threats, if we assume terrestrial gullibility and alien treachery. Our only real protection would be to take adequate security precautions during a contact and to maintain a healthy degree of suspicion. A form of extraterrestrial xenophobia may not be totally inappropriate. Perhaps special facilities on the Moon or Mars could serve as extraterrestrial contact and communications "ports of entry" into the Solar System. Any alien attempts at subversion could then be rapidly isolated and, if necessary, terminated long before the terrestrial biosphere itself was endangered.

The fourth major risk category involving extraterrestrial contact is massive culture shock. Some individuals have expressed concerns that even mere contact with a vastly superior extraterrestrial race could prove extremely damaging to human psyches, despite the best intentions of the alien race. Terrestrial cultures, philosophies and religions that now place humans at the very center of creation would have to be "expanded" to compensate for the existence of other, far superior intelligent beings. We would now have to "share the Universe" with someone or something better and more powerful than we are. As the dominant species on this planet, could humans accept this new role? While many scientists currently believe in the existence of intelligent species elsewhere in the Universe, we must keep asking ourselves a more fundamental question: Is humankind in general prepared for the positive identification of such a fact? Will contact with intelligent aliens open up a golden age on Earth or initiate cultural regression?

Historians and sociologists, in studying past contacts between two terrestrial cultures, observe that generally (but not always) the stronger, more advanced culture has dominated the weaker one. This domination, however, has always involved physical contact and usually territorial expansion by the stronger culture. If contact has occurred without aggression, the lesser culture has often survived and even prospered. In the case of extraterrestrial contact by means of interstellar communication, the long delays while messages span the cosmic void at the speed of light should enable our planetary civilization to adapt to the changing cosmic condition. There are no terrestrial examples of cultural domination by radio signals alone, and round-trip exchanges of information would require years or even human generations.

Of course, we cannot assume that contact with an alien civilization is without risk. Four general risk categories have just been discussed. Many individuals now feel that the potential benefits (also previously described) far outweigh any possible concerns. Simply to listen for the signals radiated by other intelligent life does not appear to pose any great danger to our planetary civilization. The real hazard issue occurs if we decide to respond to such signals. Perhaps a planetary consensus will be necessary before we answer an "interstellar phone call." It is also interesting to note here, however, that our ultrahigh-frequency (UHF) television signals are already propagating far out into interstellar space and will possibly be detectable out to some 25 to 50 light-years' distance. Are aliens tonight examining a decades-old episode of "Gunsmoke" as a message from Earth?

It is interesting to recognize that the choice of initiating extraterrestrial contact is no longer really ours. In addition to the radio and television broadcasts that are leaking out into the Galaxy at the speed of light, the powerful radio/radar telescope at the Arecibo Observatory was used to beam an interstellar message of friendship to the fringes of the Milky Way Galaxy on November 16, 1974. We have, therefore, already announced our presence to the Galaxy and should not be too surprised if someone or something answers!

See also: **Arecibo Interstellar Message; extraterrestrial civilizations; Fermi paradox; interstellar communication; interstellar spaceship; search for extraterrestrial intelligence; What do you say to a little green man?**

constellation   An easily identifiable configuration of the brightest visible stars in a moderately small region of the night sky. Originally, these constellations, such as Orion the Hunter, were named by ancient astronomers after the heroes and creatures from various mythologies. Today, each constellation has a Latin name; for example, "Ursa Major" (the Great Bear).

Cosmic Background Explorer (COBE)   A proposed NASA explorer-class mission to measure the diffuse infrared and microwave emission of the Universe, from a wavelength of approximately 1 micrometer to 9.6 millimeters. This radiation comes from the "Big Bang" event, the earliest galaxies, interplanetary and interstellar dust, and infrared galaxies. The COBE will be launched in 1988 by the Space Shuttle and ultimately placed into a 900-kilometer altitude, 99°-inclination operational orbit around the Earth. A one-year mission lifetime is planned.

See also: **astrophysics**; **"Big Bang" theory**

cosmic dust Fine microscopic particles drifting in outer space; sometimes called micrometeorites.

cosmic rays Atomic particles (mostly bare atomic nuclei) that have been accelerated to very high velocities and carry great amounts of energy. Cosmic rays move through space at speeds just below the speed of light. They carry an electric charge and spiral along the weak lines of magnetic force that permeate the Galaxy. Although discovered over half a century ago, their origin still remains a mystery. Scientists currently believe they are galactic in origin, but some cosmic rays (especially the most energetic) may actually originate outside the Milky Way Galaxy, making them tiny pieces of extragalactic material.

Cosmic particles represent a unique sample of material from outside the Solar System. Although hydrogen nuclei (that is, protons) make up the highest proportion of the cosmic-ray population, these particles also range over the entire periodic table of elements, from hydrogen through uranium, and include electrons and positrons. Cosmic rays bring astrophysicists direct evidence of processes like nucleosynthesis and particle acceleration that occur as a result of explosive processes in stars throughout the Galaxy.

Cosmic rays also represent a constant ionizing radiation hazard to both human beings and delicate machines traveling through space.

See also: **astrophysics**

cosmological principle Any theory of cosmology must take into consideration the observation that galaxies and clusters of galaxies appear to be receding from one another with a velocity proportional to their separation. Neither our Galaxy (the Milky Way) nor any other galaxy is at the center of this expansion. Rather, scientists now hypothesize that all observers anywhere in the Universe would see the same recession of distant galaxies. This assumption is called the "cosmological principle." It implies space curvature—that is, there is no center of the Universe and therefore no outer limit or surface.

See also: **"Big Bang" theory; cosmology; Hubble's law**

cosmology May be defined as the study of the origin, evolution and structure of the Universe. Figure 1, for example, is a schematic drawing of "our corner of the Universe." It shows the Milky Way Galaxy (of which our Sun is part) moving at a velocity of 1.6 million (10^6) kilometers per hour toward the Hydra group of galaxies. Astrophysical discoveries in recent years tend to support

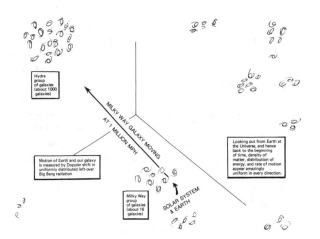

Fig. 1 A schematic view of our sector of the Universe. (Courtesy of NASA.)

the "Big Bang" theory of cosmology—a theory stating that the Universe began in a great explosion (sometimes called the "initial singularity") and has been expanding ever since. The 1965 discovery of the cosmic microwave background with a radio telescope provided observational evidence that there was a very hot phase early in the history of the Universe.

See also: **astrophysics; "Big Bang" theory**

cosmonaut The name used by the Soviet Union for its space travelers, or "astronauts."

See also: **extraterrestrial careers** (astronaut)

culture A colony of cells grown on a nutrient medium under controlled conditions in an incubator. The culture medium provides food for the cells. It is typically a thin layer of jelly made from agar, sugar, water and other materials.

CW laser A continuous-wave (CW) laser, as distinguished from a pulsed laser. A CW laser emits optical radiation for a period in excess of 0.25 second.

cyborg A term originating in the science-fiction literature meaning an artificially produced human being. Very sophisticated robots (that is, those robots with near-human qualities) have been called cyborgs, as have human beings who were artificially created.

See also: **android**

Cygnus X-1 Astrophysicists currently suspect that a few of the brightest X-ray sources in our Galaxy are most likely black holes orbiting closely with relatively ordinary stars. If one of the stars in such a binary system is indeed a black hole, the black hole (although in fact invisible) would reveal its presence by its gravitational effects on the companion star. For once beyond its event horizon, a black hole's gravitational force would appear the same as that exerted by ordinary celestial objects—that is, in accordance with Newton's law of gravitation. According to this universal law, the gravitational attractions of two celestial objects to each other are directly proportional to their masses and inversely proportional to the square of the distance between them. Space scientists think that a substantial portion of the kinetic energy of the matter spiraling into a black hole is converted by collision, compression and heating into X rays and gamma rays, which exhibit certain spectral characteristics. Of course, once this stellar material is pulled across the black hole's event horizon, such radiation cannot escape. Figure 1 on page 17 is an artist's rendering of a binary star system in which a black hole is literally devouring its stellar companion.

The X-ray source called Cygnus X-1 is a most likely candidate for a black hole in just such a binary star system. The term *Cygnus X-1* means that it is the first X-ray source discovered in the constellation Cygnus the Swan. X rays from this invisible object have the characteristics of those predicted when stellar material is pulled toward a black hole. Observations indicate that material is apparently being pulled from the unseen X-ray source's visible binary companion—a very large star of about 30 solar masses in size (that is, it has a mass approximately 30 times that of our Sun). Based upon the invisible X-ray source's gravitational effects on its gigantic, visible companion, the mass of the unseen object appears to be more than 6 times the mass of our Sun. Comparing this deduced mass with theoretical predictions about the properties of black holes, some scientists have postulated that this unseen X-ray source in the Cygnus X-1 binary star system might indeed be a black hole. However, since we really cannot "see" a black hole, current proof is limited to such indirect observations.

See also: **astrophysics; binary star; black holes; stars; X ray**

dark nebula A cloud of interstellar dust and gas sufficiently dense and thick so that the light from more distant stars and celestial objects is obscured. The Horsehead Nebula (NGC 2024) in Orion is an example of a dark nebula.

See also: **nebula**

Deep Space Network (DSN) The worldwide NASA Deep Space Network provides the terrestrial radio communications link for unmanned interplanetary spacecraft, probes and landers. Since 1961 this network of 64-meter diameter radio antennas has provided telecommunications and data-gathering support for a variety of deep-space exploration projects. These space-exploration projects include Ranger (1961–65), Surveyor (1966–68) and Lunar Orbiter (1966–67) missions to the Moon; the Mariner missions to Venus (1962 and 1967), Mars (1964, 1969 and 1971) and Venus-Mercury (1973); the Pioneer inward and outward heliocentric (Sun-centered) orbiters (1965–68); and the Pioneer missions to Jupiter and Saturn (1972 and 1973). The DSN has also provided support for the Pioneer-Venus orbiter and multiprobe (1978); the joint U.S.–West German Helios spacecraft in orbit around the Sun (1974 and 1976); the Viking orbiter-lander missions to Mars (1975); and the Voyager missions to Jupiter, Saturn, Uranus and Neptune (1977). (The dates shown are spacecraft launch dates.)

The extraterrestrial communications record presently belongs to the *Pioneer 10* spacecraft. Launched in 1972 on a mission to Jupiter, this spacecraft successfully encountered the giant planet in 1973 and was then placed on a trajectory that has made it the first terrestrial object to leave the known boundaries of the Solar System. On June 13, 1983, *Pioneer 10* was still sending back data as it passed the orbit of Neptune (Pluto's orbit now has it inside Neptune's orbit), a communication distance of more than 4.5 billion kilometers (about 250 light-minutes).

To maintain contact with spacecraft in deep space, the DSN antennas are located approximately 120 degrees apart around the Earth at Goldstone, California; Madrid, Spain; and Canberra, Australia. Thus, as one antenna dish loses contact due to the Earth's rotation, another one takes over the task of receiving data from the interplanetary spacecraft. Although the primary activity of the DSN is to conduct telecommunications with unmanned planetary spacecraft, these stations are also used as scientific radio telescopes in support of radio astronomy and in experiments involving the search for extraterrestrial intelligence (SETI) through the detection and identification of coherent radio messages from alien civilizations.

See also: **images from space; search for extraterrestrial intelligence**

DNA An acronym standing for "deoxyribonucleic acid." This molecule forms the basis of heredity in many living organisms.

See also: **life in the Universe**

Doppler shift The change in (apparent) frequency and wavelength of a source due to the relative motion of the source and an observer. If the source is approaching the observer, the observed frequency is higher and the observed wavelength is shorter. This change to shorter wavelengths is often called the "blueshift." If, on the other hand, the source is moving away from the observer, the observed frequency will be lower and the wavelength will be longer. This change to longer wavelengths is called the "redshift," since for a visible light source this would mean a shift to the longer-wavelength, or red, portion of the visible spectrum.

Drake equation Just where do we look among the billions of stars in our Galaxy for possible interstellar radio messages or signals from extraterrestrial civilizations? That was one of the main questions addressed by the attendees of the Green Bank Conference on Extraterrestrial Intelligent Life held in November 1961 at the National Radio Astronomy Observatory (NRAO), Green Bank, West Virginia. One of the most significant and widely used results from this conference is the Drake equation (named after Dr. Frank Drake), which represents the first attempt to quantify the search for extraterrestrial intelligence (SETI). This "equation" has also been called the Sagan-Drake equation and the Green Bank equation in the SETI literature.

While more nearly a subjective statement of probabilities than a true scientific equality, the Drake equation attempts to express the number (N) of advanced intelligent civilizations that might be communicating across interstellar distances at this time. A basic assumption inherent in this formulation is the principle of mediocrity—namely, that things in our Solar System (and especially on Earth) are nothing particularly special and represent common conditions found elsewhere in the Galaxy. The Drake equation is generally expressed as:

$$N = R^* f_p n_e f_l f_i f_c L \tag{1}$$

where N is the number of intelligent communicating civilizations in the Galaxy at present

R^* is the average rate of star formation in our Galaxy (stars/year)

f_p is the fraction of stars that have planetary companions

n_e is the number of planets per planet-bearing star that have suitable ecospheres (that is, the environmental conditions necessary to support the chemical evolution of life)

f_l is the fraction of planets with suitable ecospheres on which life actually starts

f_i is the fraction of planetary life starts that eventually evolve to intelligent life-forms

f_c is the fraction of intelligent civilizations that attempt interstellar communication

and L is the lifetime (in years) of technically advanced civilizations.

An inspection of the Drake equation quickly reveals that the major terms cover many disciplines and vary in technical content from numbers that are somewhat quantifiable (such as R^*) to those that are completely subjective (such as L).

For example, astrophysics can provide us a reasonably approximate value for R^*. Namely, if we define R^* as the average rate of star formation over the lifetime of the Galaxy, we obtain

$$R^* = \frac{\text{number of stars in the Galaxy}}{\text{age of the Galaxy}} \tag{2}$$

We can then insert some typically accepted numbers for our Galaxy to arrive at R^*. Namely,

$$R^* = \frac{100 \text{ billion stars}}{10 \text{ billion years}}$$

$$R^* = 10 \text{ stars/year (approximately)}$$

Generally, the estimate for R^* used in SETI discussions is taken to fall between 1 and 20.

The rate of planet formation in conjunction with stellar evolution is currently the subject of much discussion in modern astrophysics. Do most stars have planets? If so, then the term f_p would have a value approaching unity. On the other hand, if planet formation is rare, than f_p approaches zero. Astronomers and astrophysicists currently think that planets should be a common occurrence in stellar-evolution processes. Therefore, f_p is frequently taken to fall in the range between 0.4 to 1.0 in SETI discussions. The value $f_p = 0.4$ represents a more pessimistic view, while $f_p = 1.0$ is taken as very optimistic.

Similarly, if planet formation is a normal part of stellar evolution, we must next ask: How many of these planets are actually suitable for the evolution and maintenance of life? By taking $n_e = 1.0$, we are suggesting that for each planet-bearing star system, there is at least one planet located in a suitable habitable zone, or ecosphere. This is, of course, what we see here in our own Solar System.

We must then ask: Given conditions suitable for life, how frequently does it start? One major assumption usually made (again based on the principle of mediocrity) is that wherever life can start, it will. If we invoke this assumption, then f_l equals unity. Similarly, we can also assume that once life starts, it always strives toward the evolution of intelligence, making f_i equal to 1.

This brings us to an even more challenging question: What fraction of intelligent extraterrestrial civilizations want to communicate with other alien civilizations? All we can do here is make a very subjective guess, based on human history. The pessimists take f_c to be 0.1 or less, while the optimists insist that all advanced civilizations desire to communicate and make $f_c = 1.0$.

Finally, we must also speculate on how long an advanced-technology civilization lasts. If we use the Earth as a model here, all we can say is that (at a minimum) L is somewhere between 50 and 100 years. The tools of high technology have emerged on this planet only in the 20th century. Space travel, nuclear energy, electronic communications and so on have arisen on a planet that daily oscillates between the prospects of total destruction and a "golden age" of maturity. Do most other evolving extraterrestrial civilizations follow a similar perilous pattern in which cultural maturity has to desperately race against new technologies that always threaten oblivion if they are unwisely used? Does the development of a technology base

Table 1   Drake Equation Calculations

THE BASIC EQUATION: $N = R^* f_p n_e f_l f_i f_c L$									
	R^*	f_p	n_e	f_l	f_i	f_c	L	N	*Conclusion*
Very optimistic values	20	1.0	1.0	1.0	1.0	0.5	10^6	$\sim 10^7$	The Galaxy is full of intelligent life!
Your own values									
Very pessimistic values	1	0.2	1.0	1.0	0.5	0.1	100	~ 1	We are alone!

SOURCE: Developed by the author.

necessary for interstellar communication or even interstellar travel also stimulate a self-destructive impulse in advanced planetary civilizations, such that few (if any) survive? Or have most extraterrestrial civilizations learned to live with their evolving technologies, and do they now enjoy peaceful and prosperous "golden ages" that last for millennia to millions of years? In dealing with the Drake equation, the pessimists place very low values on L (perhaps a hundred or so years), while the optimists insist that L is several thousand to several million years. The ultimate limit of a civilization's lifetime for a sunlike star is established by the main-sequence lifetime of the star itself—namely, several billion years. At that point, even a "super interplanetary" society must develop interstellar travel or perish when its sunlike star leaves the main sequence.

Let's go back now to the Drake equation and put in some "representative" numbers. If we take

$R^* = 10$ stars/year
$f_p = 0.5$ (thereby excluding multiple-star systems)
$n_e = 1$ (based on our Solar System as a common model)
$f_l = 1$ (invoking principle of mediocrity)
$f_i = 1$ (invoking principle of mediocrity)
and $f_c = 0.2$ (assuming that most advanced civilizations are introverts or have no desire for space travel)

then the Drake equation yields

$$N \approx L$$

This particular result implies that the number of communicative extraterrestrial civilizations in the Galaxy at present is approximately equal to the lifetime (in Earth years) of such alien civilizations.

Let's now take these "results" one step further. If N is about 10 million (a very optimistic Drake equation output), then the average distance between intelligent, communicating civilizations in our Galaxy is approximately 100 light-years. If N is 100,000, then these extraterrestrial civilizations on the average would be about 1,000 light-years apart. But if there were only 1,000 such civilizations

existing today, then they would typically be some 10,000 light-years apart. Consequently, even if the Galaxy does contain a few such civilizations, they may be just too far apart to achieve communication within the lifetimes of their respective civilizations. For example, at a distance of 10,000 light-years, it would take 20,000 years just to start an interstellar dialogue!

By now you might like to try your own hand at estimating the number of intelligent alien civilizations that could be trying to signal us today. If so, table 1 has been set up just for you. Simply select (and justify to yourself) typical numbers to be used in the equation, multiply all these terms together and obtain a value for N. Very optimistic and very pessimistic values that have been used in other SETI discussions are included in table 1 to help guide your own SETI efforts.

See also: **extraterrestrial civilizations; Fermi paradox; principle of mediocrity; search for extraterrestrial intelligence**

dwarf galaxy   A small, often irregularly shaped galaxy containing a million (10^6) to perhaps a billion (10^9) stars. The Magellanic Clouds, our nearest galactic neighbors, are examples of dwarf galaxies.

See also: **galaxy; Magellanic Clouds**

dynarium   A large, enclosed microgravity facility. Found as part of a large orbiting urban complex, the dynarium would be used by the space-city inhabitants for both business and recreational purposes.

See also: **Astropolis; space settlement**

Dyson sphere   The Dyson sphere is a huge, artificial biosphere created around a star by an intelligent species as part of its technological growth and expansion within a solar system. This giant structure would most likely be formed by a swarm of artificial habitats and mini-planets capable of intercepting essentially all the radiant energy from the parent star. The captured radiant energy would be converted for use through a variety of techniques such as living plants, direct thermal-to-electric conversion devices, photovoltaic cells and perhaps other (as yet undiscovered) energy conversion techniques. In response to the Second Law of Thermodynamics, waste heat and unusable radiant energy would be rejected from the "cold" side of the Dyson sphere to outer space. From our present knowledge of engineering heat transfer, the heat rejection surfaces of the Dyson sphere might be at temperatures of 200 to 300 Kelvin.

This astroengineering project is an idea of the theoretical physicist, Freeman Dyson. In essence, what Dyson has proposed is that advanced extraterrestrial societies, responding to Malthusian pressures, would eventually expand into their local solar system, ultimately harnessing the full extent of its energy and materials resources. Just how much growth does this type of expansion represent?

Well, we must invoke the principle of mediocrity (i.e. things are pretty much the same throughout the Universe) and use our own Solar System as a model. The energy output from our Sun—a G-spectral class star—is approximately 4×10^{26} joules per second. For all practical purposes, our Sun can be treated as a blackbody radiator at approximately 5,800 Kelvin temperature. The vast majority of its energy output occurs as electromagnetic radiation, predominantly in the wavelength range 0.3 to 0.7 micrometers. The available mass in the Solar System for such astroengineering construction projects may be taken as the mass of the planet Jupiter, some 2×10^{27} kilograms. Contemporary energy consumption now amounts to about 10^{13} joules per second, which is about 10 terawatts. Let's now project just a one percent growth in terrestrial energy consumption per year. Within a mere three millennia, mankind's energy consumption needs would reach the energy output of the Sun itself! Today, several billion human beings live in a single biosphere, the planet Earth—with a total mass of some 5×10^{24} kilograms. A few thousand years from now, our Sun could be surrounded by a swarm of habitats, containing trillions of human beings.

The Dyson sphere may therefore be taken as representing an upper limit for growth within our Solar System. It is basically "the best we can do" from an energy and materials point of view in our particular corner of the Universe. The vast majority of these human-made habitats would most probably be located in the "ecosphere" around our Sun—that is, about a one astronomical unit (AU) distance from our parent star. This does not preclude the possibility that other habitats, powered by nuclear fusion energy, might also be found scattered throughout the outer regions of a somewhat dismantled Solar System. (These fusion-powered habitats might also become interstellar space arks.)

Therefore, if we use our own Solar System and planetary civilization as a model, we can anticipate that within a few millennia after the start of industrial development, an intelligent species might rise from the level of planetary civilization (Kardashev TYPE I Civilization) and eventually occupy a swarm of artificial habitats that completely surround their parent star, creating a Kardashev TYPE II civilization. Of course, these intelligent creatures might also elect to pursue interstellar travel and galactic migration, as opposed to completing the Dyson sphere within their home star system (initiating a Kardashev TYPE III civilization).

It was further postulated by Freeman Dyson that such advanced civilizations could be detected by the presence of thermal infrared emission (typically 8.0 to 14.0 micrometer wavelength) from objects in space that had dimensions of one to two astronomical units in diameter.

The Dyson sphere is certainly a grand, far reaching concept. It is also quite interesting for us to realize that the initial permanent space stations and space bases we construct at the close of the 20th century are, in a sense, the first habitats in the swarm of artificial structures that mankind could eventually build as part of our extraterrestrial civilization. No other generation in human history has had the unique opportunity of constructing the first artificial habitat in our own Dyson sphere!

See also: **extraterrestrial civilizations; principle of mediocrity; space settlement; space station**

E

Earth   The Earth is the third planet from the Sun and the fifth largest in the Solar System. Our planet circles its parent star at an average distance of 149.6 million kilometers (93 million miles). Earth is the only celestial body in the Solar System presently known to harbor life. Some of the physical and dynamic properties of the Earth as a planet in the Solar System are presented in table 1.

From space, our planet is characterized by its blue waters and white clouds, which cover a major portion of it (see fig. C2 on color page A). The Earth is surrounded by an ocean of air, consisting of 78 percent nitrogen, 21 percent oxygen and the remainder argon, neon and other gases. The standard atmospheric pressure at sea level is 101,325 newtons per square meter (14.7 pounds per square inch). Surface temperatures range from a maximum of about 60 degrees Celsius (140 degrees Fahrenheit) in desert regions along the equator to a minimum of minus 90 degrees Celsius (minus 130 degrees Fahrenheit) in the frigid polar regions. In between, however, surface temperatures are generally much more benign.

The Earth's rapid spin and molten nickel-iron core (see fig. 1) give rise to an extensive magnetic field. This magnetic field, together with the atmosphere, shields us from nearly all of the harmful radiation coming from the Sun and other stars. Furthermore, most meteors burn up in the Earth's protective atmosphere before they can strike the surface.

Table 1   Dynamic and Physical Properties of the Planet Earth

Radius	
Equatorial	6,378 km
Polar	6,357 km
Mass	5.98×10^{24} kg
Density (average)	5.52 g/cm³
Surface area	5.1×10^{14} m²
Volume	1.08×10^{21} m³
Distance from the Sun (average)	1.496×10^8 km (1 AU)
Eccentricity	0.01673
Orbital period (sidereal)	365.256 days
Period of rotation (sidereal)	23.934 hours
Inclination of equator	23.45 degrees
Mean orbital velocity	29.78 km/sec
Acceleration of gravity, g (sea level)	9.807 m/sec²
Solar flux at the Earth (above atmosphere)	$1,371 \pm 5$ watts/m²
Planetary energy fluxes (approximate)	
Solar	10^{17} watts
Geothermal	2.5×10^{13} watts
Tidal friction	3.4×10^{12} watts
Human-made:	
Coal-burning	2×10^{12} watts
Natural gas–burning	1.4×10^{12} watts
Oil-burning	3×10^{12} watts
Nuclear power	0.3×10^{12} watts
TOTAL HUMAN-MADE	6.7×10^{12} watts
Number of natural satellites	1 (the Moon)

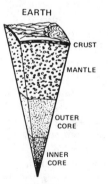

Fig. 1   The interior of the Earth. (Drawing courtesy of NASA.)

In the Space Age manned and unmanned spacecraft have enjoyed a unique vantage point from which to observe the Earth, survey its bountiful resources and monitor its delicate biosphere. The Earth's nearest celestial neighbor, the Moon, is also its only natural satellite (see fig. C17 on color page H).

See also: **Earth's trapped radiation belts; Moon**

Earth-Approaching Asteroid Rendezvous   Unlike main-belt asteroids, which generally remain near the places where they originated, Earth-approaching or Earth-crossing asteroids are in unstable orbits. Some may be fragments of disrupted main-belt asteroids, while others are most likely the nuclei of extinct comets. Space scientists believe that these minor planets are almost certainly related to some meteorites. The Earth-approaching asteroids are also being considered as a potentially valuable source of extraterrestrial materials to support expanded space industrialization and habitation efforts in cislunar space. Consequently, NASA is considering a rendezvous mission to an Earth-approaching asteroid. This mission would help determine such features of the target asteroid as size, shape, density, spin (if any), albedo and surface composition. These data would then help prepare for utilization activities, such as asteroid mining.

The rendezvous spacecraft is proposed to be launched in the 1990's by a Space Shuttle/Inertial Upper Stage configuration and reach the target asteroid in about one or two years. There would typically be as many as five launch opportunities available in any one year. Upon arrival at the asteroid, a propulsive maneuver would be performed to match orbits. Then the spacecraft would remain with the target asteroid for several months, using its remote sensing instruments to examine the "Earth-crosser" from every angle.

See also: **asteroid; asteroid mining; extraterrestrial resources**

Earthlike planet   An extrasolar planet that is located in an ecosphere and has planetary environmental conditions that resemble the terrestrial biosphere—especially a suitable atmosphere, a temperature range that permits the retention of large quantities of liquid water on the planet's surface and a sufficient quantity of energy striking the planet's surface from the parent star. These suitable environmental conditions could then permit the chemical evolution and development of carbon-based life as we know it on Earth. The planet should also have a mass somewhat greater than 0.4 Earth masses (to permit the production and retention of a breathable atmosphere) but less than about 2.4 Earth masses (to avoid excessive surface gravity conditions—that is, to avoid having a gravitational force [g] greater than 1.5).

See also: **ecosphere**

earthshine   A spacecraft or space vehicle in orbit around the Earth is illuminated by sunlight and "earthshine." Earthshine consists of sunlight (0.4- to 0.7-micrometer wavelength radiation, or visible light) reflected by the Earth and thermal radiation (typically 10.6-micrometer wavelength infrared radiation) emitted by the Earth's surface and atmosphere.

See also: **blackbody**

Earth's trapped radiation belts   The magnetosphere is

a region around the Earth through which the solar wind cannot penetrate because of the terrestrial magnetic field. Inside the magnetosphere are two belts or zones of very energetic atomic particles (mainly electrons and protons) that are trapped in the Earth's magnetic field hundreds of kilometers above the atmosphere (see fig. 1). These belts were discovered by Professor James Van Allen of the University of Iowa and his colleagues in 1958. Van Allen made the discovery using simple atomic radiation detectors placed onboard *Explorer 1*, the first American satellite. In his honor, these radiation belts are also called the "Van Allen Belts." Their discovery represents one of the first major discoveries about our extraterrestrial environment to occur in the Space Age. The existence of these radiation belts was an unexpected finding, since knowledge about the Earth's magnetosphere was very limited up to the late 1950s.

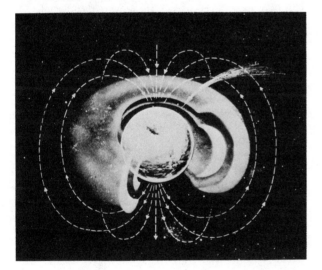

Fig. 1   An artist's concept of the Earth's magnetic field and the Van Allen radiation belts with incoming cosmic ray particles. (Courtesy of NASA.)

The two major radiation belts form a doughnut-shaped region around the Earth from about 320 to 32,400 kilometers (200 to 20,000 miles) above the equator. Energetic protons and electrons are trapped in these belts. The inner Van Allen belt contains both energetic protons and electrons which were captured from the solar wind or were created in nuclear collision reactions between energetic cosmic ray particles and atoms in the Earth's upper atmosphere. The outer Van Allen belt contains mostly energetic electrons that have been captured from the solar wind.

Spacecraft and space stations operating in the Earth's trapped radiation belts are subject to the damaging effects of ionizing radiation from charged atomic particles. These particles include protons, electrons, alpha particles (helium nuclei) and heavier atomic nuclei. Their damaging effects include degradation of material properties and component performance, often resulting in reduced capabilities or even failure of spacecraft systems and experi-

ments. For example, solar cells used to provide electric power for spacecraft are often severely damaged by passage through the Van Allen belts. The Earth's trapped radiation belts also represent a very hazardous environment for human beings traveling in space.

Radiation damage from the Earth's trapped radiation belts can be reduced significantly if the spacecraft or space station is designed with proper radiation shielding. Frequently, crew compartments and sensitive equipment can be located in regions shielded by other spacecraft equipment which is less sensitive to the influence of ionizing radiation. Radiation damage can also be limited by selecting mission orbits and trajectories that avoid long periods of operation where the radiation belts have their highest charged particle populations. For example, for a space station or satellite in low Earth orbit, this would mean avoiding the South Atlantic Anomaly and, of course, the Van Allen Belts themselves.

See also: **cosmic rays; hazards to space workers; magnetosphere; solar wind**

eccentricity (e)   A measure of the ovalness of an orbit. When $e = 0$, the orbit is a circle; when $e = 0.9$, the orbit is a long, thin ellipse.
See also: **orbits of objects in space**

ecliptic   The circle formed by the apparent yearly path of the Sun through the heavens; it is inclined by approximately 23.5 degrees to the celestial equator.

ecology   The field of study that examines the relationship between living organisms and their environment.
See also: **ecosphere; life in the Universe**

ecosphere   That habitable zone or region around a main-sequence star of a particular luminosity in which a planet can maintain the conditions necessary for the evolution and continued existence of life. For life to occur as we know it on Earth (that is, chemical evolution of carbon-based living organisms), global temperature and atmospheric pressure conditions must permit the retention of a significant amount of liquid water on the planet's surface. Conditions that would prevent a habitable Earthlike planet include circumstances in which all the surface water has been completely evaporated (the runaway greenhouse effect) or in which the liquid water on the planet's surface has become completely frozen or glaciated (the ice catastrophe).

For a star like the Sun, an effective ecosphere would typically extend from about 0.7 astronomical units (AU) to about 1.3 AU. In our Solar System, for example, the inner edge of an ecosphere suitable for human life would reach the orbit of Venus, while the outer edge reaches approximately halfway to the orbit of Mars. Because ecospheres appear to be extremely narrow, a planetary system around an alien star will most likely have only one, or perhaps two, planets that are located in a region suitable for the chemical evolution of carbon-based life.

See also: **ice catastrophe**

electromagnetic spectrum When sunlight passes through a prism, it throws a rainbowlike array of colors onto a surface (see fig. 1). This display of colors is called the visible spectrum. It represents an arrangement in order of wavelength of the narrow band of electromagnetic (EM) radiation to which the human eye is sensitive.

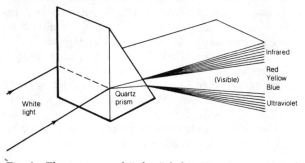

Fig. 1 The spectrum of "white" light. (Drawing courtesy of NASA.)

The electromagnetic spectrum comprises the entire range of wavelengths of electromagnetic radiation, from the shortest-wavelength gamma rays to the longest-wavelength radio waves (see fig. 2). The entire EM spectrum includes much more than meets the eye!.

As shown in figure 2, the names applied to the various regions of the EM spectrum are (going from shortest to longest wavelength): gamma ray, X ray, ultraviolet (UV), visible, infrared (IR) and radio. EM radiation travels at the speed of light (that is, about 300,000 kilometers per second) and is the basic mechanism for energy transfer through the vacuum of outer space.

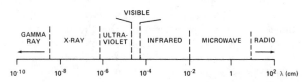

Fig. 2 The electromagnetic spectrum. (Drawing courtesy of NASA.)

One of the most striking discoveries of 20th-century physics is the dual nature of electromagnetic radiation. Under some conditions electromagnetic radiation behaves like a wave, while under other conditions it behaves like a stream of particles, called photons. The tiny amount of energy carried by a photon is called a quantum of energy (plural: quanta). The word *quantum* comes to us from the Latin and means "little bundle."

The shorter the wavelength, the more energy is carried by a particular form of EM radiation. All things in the Universe emit, reflect and absorb electromagnetic radiation in their own distinctive ways. The way an object does this provides scientists with special characteristics, or a

signature, that can be detected by remote sensing instruments. For example, the spectrogram shows bright lines for emission or reflection and dark lines for absorption at selected EM wavelengths. Analyses of the positions and line patterns found in a spectrogram can provide information about the object's composition, surface temperature, density, age, motion and distance.

For centuries, astronomers have used spectral analyses to learn about distant extraterrestrial phenomena. But up until the Space Age, they were limited in their view of the Universe by the Earth's atmosphere, which filters out most of the EM radiation from the rest of the cosmos. In fact, ground-based astronomers are limited to just the visible portion of the EM spectrum and tiny portions of the infrared, radio and ultraviolet regions. Space-based observatories now allow us to examine the Universe in all portions of the EM spectrum. In the Space Age we have also examined the cosmos in the infrared, ultraviolet, X-ray and gamma ray portions of the EM spectrum and have made startling discoveries. We have also developed sophisticated remote sensing instruments to look back on our own planet in many regions of the EM spectrum. Data from these environmental-monitoring and resource-detection spacecraft are providing the tools for a more careful management of our own Spaceship Earth.

See also: **astrophysics; remote sensing; Space Telescope**

electron volt (eV) A unit of energy equivalent to the energy gained by an electron when it passes through a potential difference of one volt. Larger multiple units of the electron volt are frequently used—as, for example; *keV* for thousand, or kilo, electron volts (10^3 eV); *MeV* for million, or mega, electron volts (10^6 eV); and *GeV* for billion, or giga, electron volts (10^9 eV).

$$1 \text{ electron volt} = 1.602 \times 10^{-19} \text{ joules}$$

See also: **Appendix A** (table A-1)

embryo A living organism before birth or hatching; an embryo gradually develops all the structures of the full-grown organism.

enzyme Any one of many proteins that are produced by living organisms and serve as biochemical catalysts in organic processes.

escape velocity The velocity needed to climb out of the gravity well (overcome the gravitational attraction) of a celestial body.

ET An abbreviation for "extraterrestrial."

E.T. In science fiction, written "E.T.," the name given to the affectionate little alien visitor in Steven Spielberg's delightful movie *E.T.—The Extra-Terrestrial.*

ETI   An acronym for *extraterrestrial intelligence.*
See also: **extraterrestrial civilizations; search for extraterrestrial intelligence**

exobiology   The multidisciplinary field that involves the study of extraterrestrial environments for living organisms, the recognition of evidence of the possible existence of life in these environments and the study of any nonterrestrial life-forms that may be encountered. The challenges of exobiology are being approached from several different directions. First, material samples from alien worlds in our Solar System can be obtained for study—as was accomplished during the Apollo lunar expeditions (1969–72); or such samples can be studied *in situ* (Latin for "in their natural or original location") by robot explorers—as was accomplished by the Viking Landers (1976). Lunar rock and soil samples have not revealed any traces of life, while the biological results of the Viking Lander experiments involving Martian soils are still unclear. In particular, the Viking Landers have given us some chemical information about the Martian soil, but we still do not know enough about its nature to predict what reactions will occur when water and nutrients are added to it, as was done in the Viking biological laboratories. Even if the Martian soil is completely sterile and devoid of all life, it is possible that some of the reactions with the added water and nutrients that were observed were just imitations of biological activity. Because of these uncertainties, exobiologists must remain cautious in their final interpretations of the data arising from the Viking Lander biological experiments.

The features of Mars both hostile and conducive to life are summarized in table 1. As just mentioned, the Martian soil samples examined by the Viking Landers exhibited chemical reactions that hinted at, but did not necessarily prove, the existence of life on the Red Planet. These reactions could easily have been the result of nonbiological processes! Most disappointing to exobiologists was the absence of evidence of organic molecules, a key indicator of past and present life or simply that the chemical evolution toward living things had at least begun. The Viking Landers were not capable of cracking open Martian rocks in their search for extraterrestrial life. However, in analog experiments (experiments designed to closely resemble, or be analogous to, conditions existing in a [usually inaccessible] place under study) back on Earth, scientists studying the bleak Antarctic continent broke open rocks from Antarctica's Dry Valleys, a desolate region often considered by exobiologists as the region on Earth most closely resembling the Martian environment. Terrestrial biologists had long regarded the Dry Valleys as lifeless. No animal life is visible on the bare cliffs of the Dry Valleys; and even microscopic investigation of the Valleys' soil revealed no native microbial life. However, when certain rocks from Antarctica's Dry Valleys were carefully broken open, they were found to contain microorganisms!

So the question about life on Mars still remains open. It is possible that native Martian life (most likely microscopic in form) now exists in some crevice of the Red Planet. Or perhaps life began there eons ago but then died out. Scientists today are puzzled about how the Martian climatic conditions could have changed so rapidly. There were once great rivers and floods of water raging over the plains of the Red Planet. This water has now mysteriously vanished, leaving behind a dry, barren and apparently lifeless desert.

A second major approach in exobiology involves conducting experiments in terrestrial laboratories that attempt either to simulate the primeval conditions that led to the formation of life on Earth and extrapolate these results to other planetary environments or to study the response of terrestrial organisms under environmental conditions found on alien worlds.

In 1938 Oparin, a Russian biochemist, proposed a theory of chemical evolution that suggested that organic compounds could be produced from simple inorganic molecules and that life on Earth probably originated by this process. Then, in 1953, at the University of Chicago, American Nobel laureate Harold C. Urey and his former student Stanley L. Miller performed what can be considered the first modern experiments in exobiology. Investigating the chemical origin of life, Urey and Miller demon-

Table 1   Conditions on Mars Hostile to and Supporting Life

Conditions Supporting Life

• Martian atmosphere contains all the chemicals needed for life: nitrogen, carbon, oxygen and water vapor.

• The polar ice caps contain substantial quantities of water.

Conditions Hostile to Life

• The atmospheric pressure on Mars is only 1 percent of that found on Earth (at best). The Martian atmosphere is so thin it resembles air on Earth at an altitude of 32 kilometers. Air this thin would make an animal's or a person's blood boil.

• There is not enough ozone in the thin Martian atmosphere to shield lethal ultraviolet radiation.

• Because of the thin Martian atmosphere and its low pressure, liquid water would vaporize instantly.

• The equatorial-region temperatures on Mars fall as low as minus 100 degrees Celsius (minus 150 degrees Fahrenheit) at night.

SOURCE: NASA.

strated that organic molecules could indeed be produced by irradiating a mixture of inorganic molecules. The historic Urey-Miller experiment simulated the Earth's assumed primitive atmosphere by using a gaseous mixture of methane (CH_4), ammonia (NH_3), water vapor (H_2O) and hydrogen (H_2) in a glass flask. A pool of water was kept gently boiling to promote circulation within the mixture, and an electrical discharge (simulating lightning) provided the energy needed to promote chemical reactions. Within days, the mixture changed colors, indicating that more complex, organic molecules had been synthesized out of this primordial "soup" of inorganic materials.

Other exobiologists and chemists have repeated the overall techniques employed by Urey and Miller. They have experimented with many different forms of energy thought to be present in the Earth's early history, including ultraviolet radiation, high-energy particles and meteorite crashes. These experimenters have subsequently produced in mixtures of inorganic materials many significant organic molecules—organic molecules and compounds found in the complex biochemical structures of terrestrial organisms.

The first compounds synthesized in the laboratory in the classic Urey-Miller experiment were amino acids, the building blocks of proteins. Later experiments have produced sugar molecules, including ribose and deoxyribose —essential components of the deoxyribonucleic acid (DNA) and ribonucleic acid (RNA) molecules. The DNA and RNA molecules carry the genetic code of all terrestrial life.

In the field of exobiology, we call molecules such as proteins and nucleic acids "organic" because they are the building blocks of life—and *not* because they are uniquely formed by living things. As the very important Urey-Miller and other experiments have clearly demonstrated, organic molecules can be formed by nonorganic or nonbiological *natural* processes. So when an exobiologist today talks about a molecule or mixture as being "organic," usually he or she simply means that it contains carbon atoms.

A third general approach in exobiology involves an attempt to communicate with, or at least listen for signals from, other intelligent life-forms within our Galaxy. This effort is often called the search for extraterrestrial intelligence (or SETI, for short). At present, the principal aim of SETI activities throughout the world is to listen for evidence of extraterrestrial radio signals generated by intelligent alien civilizations.

Current theories concerning stellar-formation processes now lead many scientists to believe that planets are normal and frequent companions of most stars. If we consider that the Milky Way Galaxy contains some 100 billion to 200 billion stars, present theories on the chemical evolution of life indicate that it is probably not unique to Earth but may in fact be widespread throughout the Galaxy. Some scientists also speculate that life elsewhere may have

Table 2   Significant Developments in Exobiology

- Nonbiological synthesis of amino acids under conditions simulating those postulated for primitive Earth, followed by the synthesis of most of the biologically important molecules.

- Ammonia (NH_3) and water (H_2O) molecules detected in interstellar space using radio telescopes, followed by the discovery of many more important organic molecules and life precursors.

- Amino acids and other biologically important organic materials have been found to occur indigenously in meteorites and to be of nonbiogenic origin.

- Detailed analysis of lunar rock and soil samples provided no evidence of life, past or present; only traces of amino-acid precursors were found.

- Simulations of the chemistry of the atmospheres of the outer planets and selected moons (for example, Titan) showed that these atmospheres may be sites where extensive abiotic (not biological) synthesis of organic molecules is presently taking place.

- Spectroscopic observation of comets has revealed the presence of biologically important ions, molecules and fragments.

- Viking Lander 1 and 2 experiments have revealed no definite evidence of existing life or organic molecules in Martian soil at two locations on the Red Planet.

- Viking Lander experiments indicate that Martian soil possesses intriguing chemical properties that mimic (in certain respects) some reactions of biological systems.

- Algae, bacteria and fungi were discovered living inside rocks from the coldest and driest regions of Antarctica, regions that are the closest Earthly counterpart of the Martian environment.

- Microfossils have been discovered inside rocks some 3.5 billion years old. This discovery pushes back the estimate of the time when life originated on Earth to within the first billion years after the Earth was formed.

SOURCE: NASA.

evolved to intelligence, curiosity and the ability to build the technical tools required for interstellar transmission and reception of intelligent signals.

Table 2 summarizes some of the significant developments that have occurred in exobiology since the early 1950s. To date, however, none of these efforts has yielded any distinctly positive evidence that life (simple or intelligent) exists beyond our own terrestrial biosphere. Nevertheless, in view of the tremendous impact that such a discovery would have on our cosmic perspective, exobiology is a truly fascinating field of study, very worthy of continued support, participation and growth as we enter the next century.

See also: **extraterrestrial contamination; life in the Universe; Mars** (the search for extraterrestrial life on Mars); **Moon; search for extraterrestrial intelligence; Viking Project**

exogeographer   A person who investigates the surface features of an alien world. Exogeographers use sophisticated spacecraft and advanced remote sensing instruments to study the detailed characteristics of other celestial bodies in our Solar System.

See also: **extraterrestrial careers** (geographer)

exogeologist   A person who studies the origin, history, structure and mineral resources of alien worlds. The Apollo astronauts who walked on the Moon and gathered lunar rock and soil samples conducted our first extraterrestrial field expeditions. In the future, exogeologists will use both manned exploration and sophisticated robot spacecraft to gather geologic specimens from other worlds.

See also: **extraterrestrial careers** (geologist)

Explorer I   The first successful satellite placed in orbit by the United States. It was launched on January 31, 1958, from Complex 26 at Cape Canaveral, Florida, by a Jupiter-C rocket. Instruments on board *Explorer I* detected the Earth's trapped radiation belts. These instruments were developed by Dr. James Van Allen of the State University of Iowa. In honor of his discovery, these trapped radiation belts are now called the Van Allen Belts.

See also: **Earth's trapped radiation belts**

extragalactic   Occurring, located or originating beyond our Galaxy (the Milky Way); typically, farther than 100,000 light-years distant.

extramartian   Occurring, located or originating outside of the planet Mars and its atmosphere. The Martian term equivalent to *extraterrestrial*. The Viking Landers, for example, represent extramartian robotic visitors to the Red Planet.

extrasolar   Occurring, located or originating outside of our Solar System; as, for example, extrasolar planets.

extrasolar planets   Planets that belong to a star other than the Sun. There are two general methods that can be used to detect extrasolar planets: direct (involving a search for telltale signs of planet radiation) and indirect (involving observation of the motion of the parent star).

Why should we be interested in the discovery of extrasolar planets? Well, evidence of planets around other stars would help astronomers validate their current hypothesis that planet formation is a normal part of stellar evolution. Evidence of extrasolar planets, especially if we can also determine their frequency of occurrence as a function of the type of star, would greatly assist scientists in their estimations of the cosmic prevalence of life. If life originates on "suitable" planets whenever it can (as many exobiologists currently hold), then knowing how abundant such suitable planets are in our Galaxy would allow us to make more credible guesses about where to search for extraterrestrial intelligence and what our chances are of finding intelligent life beyond our own Solar System.

See also: **Barnard's star; Drake equation; Project Orion; search for extraterrestrial intelligence**

extraterrestrial   Refers to something that occurs, is located or originates outside of the planet Earth and its atmosphere.

extraterrestrial art   Did you ever wonder what it would be like to explore the frigid oceans of Europa, hike across the red sands of Mars or travel to another star system? In response to these and many other intriguing questions, artists who specialize in cosmic or extraterrestrial subjects are already giving us interpretive answers—many with scientifically realistic accuracy that pushes the limits of our present knowledge. Extraterrestrial or space art can be defined as the specialized art form that deals with aerospace, astronomical, celestial or extraterrestrial themes.

The cosmic artist usually possesses a widely diverse knowledge of such sciences as physics, (exo)biology, astronomy, geology and meteorology—just to name a few. Frequently, the space artist is called upon to translate highly abstract engineering or scientific concepts into visual images that can then be quickly grasped and understood. One critical talent is the artist's ability to "extrapolate"—that is, his or her ability to infer an unknown from something that is actually known. Simply stated, the extraterrestrial artist takes the dreams, ideas and concepts of space scientists, astronomers and engineers and converts them into "realistic," or at least inspiring, images. Because they are not necessarily constrained by the limits of scientific knowledge, space artists can help bring tomorrow to the present and bring the otherwise unreachable within our grasp.

The earliest versions of factual extraterrestrial art appeared in 1865 as the illustrations made by A. de Neuvill and Emile Bayard that accompanied the Jules Verne novel *From the Earth to the Moon*. Jules Verne, of course, was a

Fig. 1   An illustration from the first edition of Jules Verne's science fiction novel *From the Earth to the Moon*, published in 1865. Here, the passengers in Verne's spaceship enjoy their first taste of weightlessness. (Courtesy of NASA.)

precise writer and careful researcher of facts who personally supervised the illustrations used in his novels. (See fig. 1.)

The reportorial artist Tracy Sugarman painted the watercolor entitled *Rollout of the Columbia* (see fig. 2). This aerospace painting depicts the Space Shuttle *Columbia* as it leaves the Vehicle Assembly Building (VAB) at the Kennedy Space Center on December 29, 1980.

Artist Pierre Mion's large painting (3.05 meters [10 feet] tall by 1.02 meters [40 inches] wide) entitled *Astronauts Explore the Moon* dramatizes the immense size of the lunar craters and mountains on the lunar surface (see fig. C3 on color page B).

Fig. 2   *Rollout of the Columbia*—a watercolor by Tracy Sugarman. (Courtesy of the NASA Art Program.)

The *View from Mimas*, painted by artist Ron Miller in March 1981 (see fig 3.), is the first extraterrestrial painting to illustrate known features on a Saturnian satellite and also the first space painting to interpret the features shown in the Voyager images of Mimas, a Saturnian moon. On November 12, 1980, the Voyager spacecraft photographed one of the wonders of our Solar System on Mimas. This natural wonder is an enormous impact crater, more than a quarter the diameter of the small moon and just short of being large enough to have split Mimas in half. The crater is some 130 kilometers across, making it the largest crater (proportional to the size of the celestial body) yet found in the Solar System. In the painting we are looking across this giant crater at Saturn, which is

Fig. 3   *View from Mimas*, by Ron Miller. (Courtesy of the NASA Art Program.)

some 185,000 kilometers (115,000 miles) in the distance. The central, lobed mountain is approximately 5 to 6 kilometers high. Like a fist through a sheet of paper, a giant block of ice has welled up through the dark crater floor, its upper half bright and clear.

Other examples of extraterrestrial art appear throughout this book. These fine artists' renderings represent a wonderful synthesis of technical knowledge and fertile imagination. In many cases, such paintings and drawings are the only way we presently have of exploring some of the most interesting regions of our extraterrestrial domain.

extraterrestrial careers   The first few decades of the next millennium promise to be some of the most exciting times ever experienced in human history. Building on space technologies developed in the 1960s through 1990s, people will start living in permanent space stations, in lunar bases and settlements and in semi-permanent bases on the planet Mars! Extraterrestrial careers in the next century could include such interesting job titles as: space station superintendent; chief engineer, on-orbit manufacturing facility; head, exobiology research module (space station); exobiologist-in-charge, Mars expedition; lunar miner; Martian settler; space farmer (lunar base agricultural facility); extraterrestrial geologist; spaceship captain (Earth-to-Moon run); space construction worker; and maybe even "governor" of the lunar settlement. Graduate students interested in advanced degrees in space technology or planetary science might perform their doctoral research on the space station or as part of the scientific and engineering team at the lunar base. We might even witness the start of a "University of Space"—with the space station serving as our planet's first extraterrestrial campus. Perhaps some of the most exciting breakthroughs in the materials and life sciences will be made by dedicated young investigators exploring the secrets of nature in the microgravity world of the space station. It is not even too unreasonable to speculate that many of tomorrow's Nobel prize-winning concepts in physics, chemistry and medicine will be born out of research conducted in our extraterrestrial environment.

With these exciting ideas in mind, you may ask yourself: "How do I get started in an extraterrestrial career?" Of course, no one today can predict precisely what career pathways and job titles will emerge as we start to build our extraterrestrial civilization. However, it is a pretty safe bet to assume that until the mid-1990s the traditional "way into space" will be through a career as an astronaut. However, even the concept of what an astronaut is and does, has changed dramatically since the early manned space flight programs. In the Shuttle Era, for example, a much larger variety of people have flown into space— broadening the concept of "astronaut" well beyond that of a daring test pilot with nerves of steel who had "the right stuff." Today, guest astronauts, school teachers, scientists and even a politician or two are enjoying the unique experience of space flight. This broadening of the meaning

of astronaut will continue even further as the space station becomes operational in the mid-1990s and an even wider segment of the population experiences the Universe face to face. To help accommodate this expansion of the meaning of astronaut, we will define an astronaut here (for career discussion purposes) as any person who travels above the Earth's sensible atmosphere (150 kilometers). Institutions, such as NASA, may wish to retain the title "astronaut" for selected members of a particularly trained group, such as a Shuttle commander, pilot or mission specialist. However, this often "astropolitical" distinction will eventually give way to a more general 21st-century concept of a space traveler or worker who engages in activities in the extraterrestrial environment. In fact, with the full-scale operation of a permanent space station and lunar base, we can anticipate the creation of a host of exciting and romantic new job titles.

Unfortunately, not all of us will be able to fly in space and view the Earth gliding silently below while we earn a living meeting the Universe face to face. But don't despair! There will be many other interesting extraterrestrial careers in addition to that of astronaut. These include aerospace craftsperson, aerospace engineer, aerospace engineering technician, astronomer, electronics engineer, geographer, mathematician, meteorologist, physicist, science writer or systems analyst, just to name a few. These career areas support astronauts and their activities in space.

We will now explore a few of these extraterrestrial career areas, starting with perhaps the most widely recognized ET career area—that of astronaut.

ASTRONAUT

Through the middle of the 20th century, no one could really describe an astronaut (which literally means "star sailor," from the ancient Greek). The only space travelers existed in fiction, such as Flash Gordon, Buck Rogers or Jules Verne's Moon travelers. However, in 1959 the National Aeronautics and Space Administration (NASA) requested that the military services of the United States identify their members who met very specific qualifications. The search was started to find pilots who would qualify for the exciting new manned space program.

In seeking its first astronauts, NASA emphasized jet aircraft experience and engineering training. The small size of the Mercury capsule then under development helped establish physical stature requirements of the first astronauts. Therefore, the requirements for becoming an astronaut in 1959 included: being less than 40 years of age, being less than 180 centimeters (5 feet 11 inches) tall, being in excellent physical condition, holding a bachelor's degree or its equivalent in engineering, being a qualified jet pilot, having graduated from test-pilot school and having accumulated at least 1,500 hours of flying time.

Despite these very rigorous requirements, more than 500 men qualified. Military and medical records were examined; psychological and technical tests were given;

and personal interviews were conducted by medical and psychological specialists. As a result of this first rather gruelling screening, many astronaut candidates were eliminated, and others simply decided they did not want to be considered further. Not every applicant in 1959 had "the right stuff."

Even more stringent physical and psychological examinations followed in the selection of America's first astronauts. Finally, in April 1959 NASA announced the selection of the first seven astronauts. They were: Navy Lieutenant M. Scott Carpenter; Air Force captains L. Gordon Cooper Jr., Virgil I. "Gus" Grissom and Donald K. "Deke" Slayton; Marine Lieutenant Colonel John H. Glenn Jr.; and Navy lieutenant commanders Walter M. Schirra Jr. and Alan B. Shepard Jr. With the exception of astronaut Slayton, each of these men flew as part of Project Mercury. Slayton was grounded with a previously undiscovered heart condition. However, after medical certification that this condition had cleared up, Deke Slayton realized his ambition to fly in space when he served as a member of the American crew in the Apollo-Soyuz Test Project (ASTP), which occurred in July 1975.

Three years after the selection of the original "Mercury Seven" astronauts, NASA issued another call for astronaut trainees—this time for the Gemini and Apollo programs. Experience in flying high-performance jet aircraft was still stressed, as well as education. The age limit was lowered to 35 years and the maximum height raised to 182 centimeters (6 feet). In addition, the astronaut program was opened to qualified civilians. This second call for astronauts resulted in more than 200 applications. The original list was screened down to 32, and then to 9 in September 1962.

In October 1963, 14 more astronaut trainees were chosen from nearly 300 applicants. By then, the prime selection emphasis had shifted away from jet flight experience to superior academic qualifications. For example, in October 1964 astronaut applications were invited on the basis of educational background alone. These were the "scientist-astronauts"—a name given to the group because the 400-plus applicants who met minimum requirements had a doctorate or equivalent experience in the natural sciences, medicine or engineering.

Applications from such scientist-astronaut candidates were turned over to the National Academy of Sciences in Washington, D.C. for evaluation. Sixteen were recommended to NASA, and six were selected in June 1965.

Another 19 pilot-astronauts were brought into the program in April 1966, and 11 scientist-astronauts were added in mid-1967. Then, when the Air Force Manned Orbiting Laboratory program was cancelled in mid-1969, 7 other astronaut trainees transferred to NASA.

The first group of astronaut candidates selected since the Space Shuttle program was initiated were chosen in January 1978. In July of that year, they started intensive training and evaluation at the Johnson Space Center (JSC). This group of 20 mission specialists and 15 pilots completed training and went from astronaut candidates to active-status astronauts on August 31, 1979. Of this group, 6 were women, and 4 were members of minority groups.

Nineteen new astronaut candidates were selected by NASA in May 1980. These individuals started training in July of that year and achieved astronaut status in August 1981. Eleven astronauts from this group became mission specialists and eight, (Shuttle) pilots. Two are women, and two are members of ethnic minority groups.

Under an agreement between NASA and the European Space Agency (ESA), two European payload specialists for the first Spacelab mission (see fig. 1) also underwent joint training with this group of Shuttle astronaut candidates.

Fig. 1   European Space Agency astronaut Ulf Merbold, *Spacelab 1* payload specialist, conducts one of several experiments in orbit, November 1983. (Courtesy of NASA.)

Each year NASA reviews the current size of the astronaut corps and the anticipated extraterrestrial workload, to determine whether a new recruitment program is required. For example, in 1983 NASA decided to select 12 new astronaut candidates, 6 pilots and 6 mission specialists, with the selected candidates reporting for training at the Johnson Space Center in July 1984.

Current regulations require that citizens of the United States be given preference for appointment to astronaut candidate positions when qualified individuals are available.

Shuttle Crew Positions   The Space Shuttle Orbiter crew positions consist of: commander, pilot, mission specialist and payload specialist. The commander and pilot fly the Orbiter. The mission specialist is normally a spacecraft-proficient crew member who is also skilled in payload operations. With his or her assistance, the payload specialist(s) can function more effectively. One or more (up to four) payload specialists, as noncareer and frequently non-NASA astronauts, may be assigned to a particular Shuttle flight, depending on the payload and the mission.

Shuttle pilot astronauts serve in either the commander or pilot position. Pilot candidates are required to have a bachelor's degree in engineering, the biological or physical sciences, or mathematics from an accredited academic institution. An advanced academic degree—such as a master's or Ph.D. degree—is, however, desirable. The quality of an individual's academic preparation is very important. To meet the minimum qualifications, an applicant must have at least 1,000 hours pilot-in-command time in high-performance jet aircraft. Flight-test experience is also highly desirable. The pilot candidate must be able to pass a NASA Class I spaceflight physical. This Class I physical is similar to military and civilian flight physicals.

Mission-specialist astronaut candidates are required to possess a bachelor's degree in engineering, the biological or physical sciences, or mathematics from an accredited academic institution. This degree must be supplemented by at least three years of related experience. An advanced degree is desirable and can substitute for the experience requirement as follows: a master's degree = one year of experience; a Ph.D. degree = three years. The quality of the applicant's academic preparation is very important! Candidates must also be able to pass a NASA Class II spaceflight physical. This examination is similar to military and civilian flight physicals.

Payload Specialist The payload specialist is a noncareer astronaut who flies as a Space Shuttle passenger and who is responsible for achieving the objectives of a particular experiment or payload. He or she has a great deal of knowledge about the payload, its operations, objectives and supporting equipment. Of course, the payload specialist must also know certain Space Shuttle Orbiter systems, such as hygiene facilities, life-support systems, hatches, tunnels, and caution and warning indicators.

A payload specialist selected by a payload sponsor or nominated by NASA must be certified in writing in the following three-step process. First, he or she must pass a NASA Class III physical examination (or military services equivalent) at a qualified aerospace medicine facility. Next, the prospective payload specialist must obtain a statement of competence from the payload sponsor. This statement indicates that he or she is qualified to handle the particular experiment or payload. Finally, the candidate payload specialist must obtain a statement of successful flight-independent training completion and must also be accepted for flight readiness by the commander of the Shuttle flight. Flight-independent training is the flight-familiarization training that is required for every Shuttle mission. It includes the operation of certain Orbiter systems, such as food and waste-management systems, and normal and emergency procedures that are required for safe crew operations.

Astronaut Training Once you've been lucky enough to be selected as an astronaut, it's back to school to study such basic science and technology subjects as meteorology, mathematics, navigation and guidance, astronomy, physics and computers.

To become familiar with the microgravity or "weightlessness" encountered in spaceflight, astronauts spend time in a special KC-135 aircraft that is flown "over the top" of a parabolic path to simulate microgravity for perhaps 30 seconds (see fig. 2). The effect experienced by the astronauts is similar to that which you experience in a rapidly descending elevator. During periods of microgravity-simulation training, the astronauts practice activities such as drinking, eating, putting on and removing space suits, and testing various types of equipment. Longer periods of simulated microgravity are provided under conditions of "neutral buoyancy" in specially designed water tanks that are large enough to hold full-scale mock-ups of components and equipment that will be used in orbit.

Fig. 2 Astronaut Robert L. Crippen floats in simulated microgravity on board a KC-135 "zero-gravity" aircraft. (Courtesy of NASA/Johnson Space Center.)

The Space Shuttle Orbiter vehicle makes a glider's "dead-stick" approach and landing when it returns to Earth from orbit. In addition to flight simulators, conventional and modified aircraft are used to help Shuttle pilots practice these approaches and landings.

Astronauts are also responsible for knowing about spacecraft and payloads and for launch vehicle design, development and modification activities. Therefore, they must attend many engineering conferences at the Johnson Space Center, where the astronauts are assigned, or at other NASA centers or contractor plants. Of course, because of the great complexity of the space program, no one individual astronaut is expected to know all such changes, so more than one astronaut is usually assigned to specific areas of the program. Frequent technical reports to the rest of the astronaut corps keep everyone up to date.

Flight Assignments When an astronaut is assigned to a Shuttle flight crew, his or her schedule really gets busy. Crews are named for specific flights well in advance of the launch date. Several crews will be in training at the same time. Each crew receives crosstraining so that at least one crewperson can handle the most critical duties of each associate astronaut.

Astronauts participate in spacecraft and payload reviews and test programs. This enables each crew member to become familiar with the payload or spacecraft being carried on a particular flight.

The training pace quickens when the astronauts start working with various simulators. First, they learn the individual tasks necessary for a particular payload or spacecraft. Then they put all these tasks together in a sequence that follows the actual mission profile.

The simulators provide very realistic working conditions. For example, the spacecraft interiors are duplicated, and appropriate instruments (such as guidance and navigation displays) are programmed to give the same readings that they would display during the actual spaceflight. Even out-of-window views of Earth, the stars, the cargo bay and the landing strip are projected on screens where the Orbiter's windows would be.

Astronaut training reaches a peak several weeks before a scheduled Shuttle flight. At this time, the mission simulator is linked with the Mission Control Center and with a simulated version of the tracking station network. Astronaut crews and ground crews thus practice the entire mission in a joint training exercise that proves that everyone and everything is ready for the actual spaceflight.

In between their simulator training sessions, the astronauts will also continuously update their knowledge of the payloads and spacecraft for their particular mission. They will also practice activities related to the mission, such as deploying and retrieving payloads, performing experiments and conducting extravehicular activities.

The job of an astronaut is not done when the flight is over. The crew members will spend several days in debriefing activities—recounting their flight experiences. These postflight activities identify areas where spacecraft systems, payload-handling techniques or training procedures can be improved to benefit future astronaut crews. Then, after a brief vacation, the astronaut resumes his or her training and study schedule in preparation for still another flight into space.

AEROSPACE CRAFTSPERSON

An aerospace craftsperson is among the most skilled of all machinist workers. Persons in this career area are also called instrument makers, modelmakers or simply machinists. The aerospace craftsperson works closely with engineers and scientists, translating ideas and designs into experimental models, custom instruments and one-of-a-kind pieces of equipment. For example, the parts and models they make can range from simple gears to very intricate components for a spacecraft navigation system.

Should you become an aerospace craftsperson? Perhaps, if you have a strong interest in "things mechanical" and a better-than-average ability to work with your hands. You should also possess initiative and resourcefulness, since the aerospace craftsperson must often work with little or no supervision. In this extraterrestrial career area, you will also face new problems and will be required to develop original solutions. Frequently, the aerospace craftsperson must be able to visualize the relationship between individual components and the complete piece of aerospace equipment. He or she must also be capable of understanding the operation of intricate pieces of aerospace instrumentation and equipment. Accuracy is very important, and finished parts must frequently meet very exacting specifications. To accomplish this, the aerospace craftsperson must be able to use precision measuring equipment, including micrometers, verniers, calipers and dial indicators.

Generally, an aerospace craftsperson will learn his or her trade through an apprenticeship that lasts about four years. A typical four-year program includes 8,000 hours of shop training and more than 500 hours of related classroom instruction. Shop training emphasizes the use of hand tools, machine tools, measuring instruments and the working properties of various materials. Classroom instruction, on the other hand, provides related technical subjects such as mathematics, physics, chemistry, metallurgy, electronics, blueprint reading and the fundamentals of instrument design. Apprentices must learn enough shop mathematics so that they can plan their work and use formulas. A basic knowledge of how mechanical things operate is also needed in solving linkage and gear problems.

Employers in the aerospace industry generally prefer high school graduates who have taken algebra, geometry, trigonometry, science and machine shop for their apprenticeship programs. Additional technical education—involving electronics, physics and machine design—is desirable and will often assist future promotions to technician-level positions.

Aerospace craftspersons are usually employed by firms that make spacecraft, aerospace instruments and models of space equipment. Some of these people work for research and development laboratories, while others are employed by the federal government. Because of the delicate nature of the mechanisms upon which they work, aerospace craftspersons sometimes work in controlled environments called "clean rooms" or "white rooms." These enclosures are well lighted, temperature-controlled and dust-free.

Aerospace craftspersons combine their own efforts with those of other technicians to contribute to scientific, technical and industrial advancement. Co-workers can include an aerospace engineer, engineering technician, scientist, quality-control inspector, setup worker, safety engineer, model designer and other skilled workers in the space field.

AEROSPACE ENGINEER

Aerospace engineers must be versatile and creative problem solvers. If you are considering this career area, you should be able to visualize objects in three dimensions and make clear sketches of complex objects and operations. People in this extraterrestrial field work with rockets, reusable space vehicles like the Space Shuttle, spacecraft and space stations. They design and test the structures, power plants, frames, engines and components of spacecraft and space vehicles, giving careful attention to safety, reliability, stability and cost. Aerospace engineers supervise the assembly of spacecraft and space vehicles and the installation of their delicate, often unique, equipment.

Tools and sophisticated equipment are important to aerospace engineers. You must be able to use calculators, technical sketching equipment, multiview drawings and photographs. You must also be able to operate testing and analyzing equipment. Digital and analog computers, graphs, statistics and technical reports assist the aerospace engineer in interpreting and coordinating appropriate design information. As an aerospace engineer, you must also be able to think logically and use the tools of mathematics. In their constant search for a better way of doing things, aerospace engineers are always challenging the established techniques. As technical pioneers and pathfinders, these professionals make valuable contributions to space technology and advance its use to better the quality of life for all.

Aerospace engineers know that their efforts have advanced space-system design and technology and that they are actually shaping the future progress of the planet Earth. Being part of a profession that is helping create humanity's extraterrestrial civilization provides many tangible and intangible rewards. As an aerospace engineer, you belong to a widely diverse career field that offers excellent chances for advancement and recognition. However, many years of preparation and study are required before you can be recognized as a qualified aerospace engineer. In addition, long working hours (frequently under the pressure of deadlines and tight schedules) and many relocations are common as you move from space project to space project.

As an aerospace engineer, you will be expected to continue your education to keep up with the latest advances in space technology and in your specific field of interest. You should be accurate and systematic in finding solutions to challenging problems. In addition, your temperament should be such that it enables you to enjoy finding engineering answers to difficulties with existing designs and operating hardware. Finally, cooperation with others in arriving at such engineering solutions also requires that you have an ability to express your ideas clearly, both in writing and through speaking. (see fig. 3.)

A bachelor's degree in engineering is the minimum education required for most entry positions in the aerospace engineering career field. However, positions in research and in certain aerospace engineering specialties, such as space nuclear power, require advanced study. Interested secondary-school students should follow a college-preparatory curriculum. Courses in mathematics (including algebra, geometry and trigonometry), physics

Fig. 3   Aerospace engineers and technicians test a prototype of the Space Shuttle's remote manipulator system at the Johnson Space Center. (Courtesy of NASA.)

and chemistry are very important!

The large majority of aerospace engineers are employed by industrial firms engaged in the business of designing and manufacturing space vehicles and spacecraft. However, many other aerospace engineers work for the federal government in such organizations as the National Aeronautics and Space Administration (NASA), the Department of Defense (DOD) and the Department of Energy (DOE).

Aerospace engineers work both indoors and outdoors, in manufacturing plants and test facilities and at launch sites and ground tracking stations. In the not too distant future, a few aerospace engineers will also find employment aboard a space station and at lunar bases. And it is distinctly possible that one of the students reading this book will be the first aerospace engineer on Mars!

As an aerospace engineer, you will usually specialize in one aspect of design, testing and production of a modern space system. Thus, you might choose to work in specialties such as structural design, power plants, propulsion systems, communications systems, guidance, navigation and control systems, and so forth.

Career opportunities in this field are expected to grow moderately because both government and private industry are heavily investing in aerospace research and development programs. Space industrialization is frequently cited as a major economic growth area of the next few decades. However, the entire aerospace field is particularly sensitive to changes in the national economy, and most space projects depend on government funding—at least initially.

As an aerospace engineer, you will combine your professional efforts with those of other specialists to contribute to scientific, technical and material advancement. Therefore, the efforts of many specialties must be closely coordinated to provide an effective and efficient aerospace project team. Your co-workers may include drafting technicians, aerospace engineering technicians, geologists, physicists, quality-control inspectors, model designers. reliability engineers, safety engineers and a variety of other skilled individuals.

AEROSPACE ENGINEERING TECHNICIANS

Aerospace engineering technicians assist engineers and scientists in converting scientific theories into Space-Age realities. They work as part of a highly qualified team in all phases of the aerospace industry, from fundamental theory through the construction, testing and actual operation of space systems. In general, these individuals prepare and check engineering drawings, diagrams, specifications, reports or manuals. They also set up and perform tests on materials, components and entire aerospace systems to measure performance and to determine reliability.

As part of their job, aerospace engineering technicians might use wind tunnels, materials-fatigue test machines, acoustical laboratory equipment, vacuum chambers and a great variety of other interesting and sophisticated aerospace instrumentation and testing apparatus.

Is this extraterrestrial career area for you? Well, aerospace engineering technicians should enjoy being part of a team effort, but they should also be capable of working well with little or no supervision. Their assignments frequently involve meticulous attention to detail—in response to the directions of aerospace engineers or in order to determine the precise causes of equipment malfunctions or failures.

To be a successful aerospace engineering technician, you should possess mechanical ability and have good vision (correctable to 20/20). You must also be able to visualize objects in three dimensions as you interpret the requirements specified in engineering drawings and blueprints. Aerospace engineering technicians soon learn that there really are no shortcuts in this aspect of their work. Therefore, they cultivate both patience and perseverance in the meticulous interpretation of engineering drawings and in the careful manufacture of precise aerospace-system components.

Most aerospace engineering technicians take great pride in being part of America's aerospace industry—an industry that is forging exciting new pathways for the development of both our planetary and extraterrestrial civilizations. Successful specialists in this field are constantly aware that the lives of astronauts, fellow space workers and others may ultimately depend on the quality of their efforts. Yet they willingly accept the day-to-day pressure that this awareness imposes on their work.

Working conditions for aerospace engineering technicians are generally good, and salaries are competitive at the technician level. In addition, constant association with design and production engineers provides these technicians with insights into the requirements for advancement to the professional levels of aerospace engineering. Many

technicians are able to pursue technical courses toward engineering degrees while continuing to work at their specialties.

A high-school diploma is the minimum educational requirement for entry into this career field, but the more skilled positions go to those individuals who have earned an associate degree from a vocational-technical institute or community college. A variety of mathematics and science courses are the best preparation for becoming an aerospace engineering technician, along with studies involving mechanical drawing, computer science and English composition. Many industrial and technical firms offer on-the-job experience to complement the formal training offered by technical schools or community colleges. Trainees interested in working with nuclear power systems or radioactive sources must obtain a certificate from the Institute for the Certification of Engineering Technicians (ICET). This institute establishes the minimum standards of education and ability for the field.

Aerospace engineering technicians work in a variety of settings, determined by the space project and their particular job assignments. For example, they can work outdoors at test sites and ground tracking stations or indoors in aerospace manufacturing plants and research laboratories. Employment and job opportunities are expected to increase as the space industry grows. However, the entire aerospace industry is quite sensitive to such factors as the national economy and federal support of the space program. For example, cutbacks in federal funding could cause future job opportunities to be reduced or eliminated.

Aerospace engineering technicians generally work closely with other aerospace-industry specialists in all phases of the space program. Their co-workers include aerospace engineers, development technicians, industrial engineers, draftspersons and mechanical engineers.

ASTRONOMER

Astronomers investigate the nature and properties of the Universe and study its celestial bodies. By studying cosmic forces and processes, they attempt to understand the origin of the Universe. Some astronomers (called planetary scientists) investigate planetary environments and compare them with the terrestrial environment. Astronomers use optical telescopes, spectrometers, photometers, radio telescopes, computers and a wide variety of instruments on board spacecraft, such as X-ray and gamma-ray telescopes, to carry out their work. For example, spectrometers are used to determine the chemical nature of stars and their temperatures and relative motions by measuring the radiant energy they emit, while photometers are used to measure the intensity of stellar light. Space exploration has provided modern astronomers with a wealth of data, gathered above the Earth's intervening atmosphere. In fact, Space-Age discoveries, especially in the area of X-ray astronomy, have literally revolutionized our concepts and models of the Universe in which we live. Astronomers may also teach astronomy, mathematics, physics and space science; manage planetariums; and often act as consult-ants, technical administrators and science writers.

Should you become an astronomer and dedicate yourself to the study of things extraterrestrial? Well, astronomers need very active imaginations and strong powers of obser-vation and concentration. They must be able to see rela-tionships between complex data and form their own hy-potheses from these data. Because they must share the data they obtain and the hypotheses they form with other astronomers and scientists, they must be able to communi-cate their own ideas clearly, both in writing and through speaking. Since advanced education is required in this career field, astronomers need persistence and genuine interest in intellectual and academic pursuits, as well as a real dedication to their work.

Most astronomers take delight in testing major theories concerning the space and time relationships of the Uni-verse. Many are stimulated by the possibility that some of their work will produce major changes in our understand-ing of the cosmos. Competent astronomers can frequently choose a position from the many offered to well-trained individuals and usually find employment in pleasant (sometimes remote) working environments. Some astrono-mers work in facilities on mountaintops in remote areas of the country, while others earn a living teaching at univer-sities, working in space-science laboratories or employed by the federal government.

If you want to be an astronomer, you must have at least a bachelor's degree in astronomy, space science, physics or mathematics to begin work in the field. However, a doc-torate (Ph.D.) is usually necessary if you wish to specialize and advance. The number of colleges and universities in the United States offering advanced or graduate degrees in astronomy is quite small (approximately 50), and students will encounter keen competition for admission to these programs. Interested secondary-school students should prepare for such advanced education by following a college-preparatory curriculum that includes a combina-tion of courses in advanced mathematics, physics, chemis-try, Earth and space sciences, and communicative skills. An undergraduate student should enroll in beginning and advanced physics courses, mathematics courses and courses in astronomy and space science. Undergraduate courses in computer science, statistics, chemistry and elec-tronics are also quite useful for a career in astronomy.

Despite the fact that astronomy is the oldest organized science, it is actually one of the smallest of the extraterres-trial career fields. In 1980, for example, there were only approximately 2,000 professional astronomers at work in the United States. (Of course, there are many thousands of amateur astronomers—that is, persons who enjoy astron-omy as a hobby.) Employment levels in professional as-tronomy may vary as changing national priorities affect the amount of funding for basic research and for the space program in general.

Astronomers are very important members of America's aerospace team. The data they collect and interpret play a significant role in the planning and conduct of the na-tional space program. Some of the people with whom astronomers interact include space photographers, science writers, data systems analysts, computer technicians, as-tronauts, geologists, chemists, physicists, research mathe-maticians, meteorologists and remote sensing specialists. Space Shuttle era advances—including the deployment of the Space Telescope, the construction of a permanent space station and the establishment of lunar bases—will make astronomy an even more exciting field in the decades to come.

ELECTRONICS ENGINEER

Electronics engineers are employed in many areas of electrical and electronic equipment manufacture and use, including the aerospace industry. These engineers design and test new products, such as spacecraft control systems, computers and the electronic portions of sophisticated remote sensing systems used in space exploration and terrestrial resource monitoring. Electronics engineers also design, operate and maintain communication and electri-cal power systems. Some engineers in this career field work in administrative, managerial and teaching posi-tions, while others are involved in the development and sale of technical products, such as communication satel-lites and ground-tracking-station equipment.

Many electronics engineers are employed by aerospace-industry firms that manufacture the electrical and elec-tronic equipment used to support space projects. For some engineers, the work is largely theoretical, such as the research and development of new satellite communication techniques. Most electronics engineers, however, are in-volved in component and hardware development, guiding pieces of aerospace equipment through all stages of manu-facture. They also determine performance standards for new equipment and write maintenance schedules to en-sure that such performance standards will be met.

In the information services and communications indus-try, electronics engineers design and oversee the develop-ment and production of radio and TV broadcasting equip-ment, including communication satellite systems. They are also involved in the operation of these systems.

Is the electronics engineering career field for you? Elec-tronics engineers need the skill and ability to visualize objects in a variety of dimensions. They must be able to study complex objects and operations, as well as possess the ability to think logically and solve problems with ease. Persons in this career field must also be willing to chal-lenge established techniques, in order to advance current knowledge and develop even more exciting aerospace equipment.

Electronics engineers enjoy the challenge and variety of the many specialties from which they can choose. (We will emphasize only the aerospace specialties here—but there are obviously many, many other specialties in today's electronic age.) They work in a field that provides them a good chance for professional advancement and recognition. However, long preparation and intense study are necessary before you can be recognized as a qualified engineer. Electronics engineers frequently experience long working hours, short deadlines and a considerable amount of job relocation as they move from project to project.

The work of electronics engineers is important to the success of the nation's space program, since so many aspects of the electronics industry (computers, communications equipment and so on) are directly involved in the conduct of space missions. Electronics engineers should be willing to study continuously in order to remain technically current in a very rapidly changing industry. These engineers must also display good judgment in decision making and problem solving. They must have the ability to approach solutions from a variety of directions. For example, do we make a spacecraft antenna larger, increase the power for the transmitter or reduce the amount of data to be sent back to scientists on Earth by clever onboard data-processing techniques? Because they are frequently part of an aerospace design team, electronics engineers should be able to express their ideas clearly, both in writing and through the spoken word.

A bachelor's degree in engineering is needed for most entry positions in electronics engineering. Since an engineer's work affects the life, health or property of others, state licensing is often desirable or necessary (depending on the particular job requirements). In order to be licensed, engineering graduates must have a designated amount of relevant work experience and must have passed an appropriate state engineering examination.

If you are interested in becoming an electronics engineer, you should follow a college-preparatory curriculum in high school. Courses in mathematics (including geometry, algebra and trigonometry), physics and chemistry are very important. Language and communications courses, as well as mechanical drawing courses, will also prove quite helpful.

As part of a continuing, on-the-job education program, many engineers take night courses or travel to technical conferences and symposia to exchange new ideas and keep current on the latest developments in a very rapidly changing field.

Most electronics engineers are employed in industrial and manufacturing firms, although others work for the federal government (especially NASA and the Department of Defense). In addition, consulting firms and the communications industry employ a considerable number of electronics engineers.

Electronics engineers in the aerospace industry combine their talents and specialties with those of many other individuals, including physicists, engineering technicians, electricians, quality-control inspectors, machinists and various craftspersons from the communications industry. Some of the most exciting images delivered to Earth by planetary-exploration spacecraft have come to us through the electronic wonders brought about by these engineers.

GEOGRAPHER

Geographers add to our knowledge by investigating the physical, social, economic, political and cultural characteristics of the Earth and its people. They observe the activities of a region relative to such factors as world trade, natural boundaries, population distribution and distribution of soil, vegetation, mineral and water resources. Persons in this career field construct and interpret maps, graphs and diagrams and may travel to remote areas (including space) to gather appropriate data. They use the tools of cartography (mapmaking) and Earth observations from aircraft and spacecraft, as well as computers, to help gather and analyze geographic data. The advent of the Space Age has helped revolutionize modern geography. From space, humans and remote sensing spacecraft have viewed the Earth as never before, and robot spacecraft have visited over a dozen new worlds that now fall into the realm of contemporary geography (or should we say "exogeography"?)

Many geographers teach, write and perform consulting work to help share the vital information they have gathered.

Should you become a geographer, or even an "exogeographer"? To be a successful geographer, you must enjoy working with both people and ideas. You should understand how to interpret maps, charts and graphs and be able to analyze data and solve problems. It is also essential that you be able to express yourself well, both orally and in writing.

One aspect of this field that appeals to many geographers is that they can often customize and tailor their job to suit their individual needs and interests. This includes finding an outlet for particular talents through specialization and meeting fascinating people through teaching and research. They can also find adventure in the many places that their jobs can take them. It is quite possible that a secondary-school student reading this passage might become the first in-residence lunar "exogeographer" and perhaps his or her child the first in-residence Martian "exogeographer." For as the human expansion into space occurs, scientific interests will be followed by industrialization and the rise of extraterrestrial settlements—first on the Moon and then on Mars. Extraterrestrial settlers will want to accurately know where things are on these new worlds. The exogeographers will be there to help them characterize the physical features and resource distributions of these extraterrestrial regions.

Today, entry positions as a professional geographer require at least a bachelor's degree, while a doctorate is generally required if you want to teach at a university or engage in independent research. The usual courses of study include weather, climate, meteorology, cartography (mapmaking) and physical, urban, human and regional geography. Field study is often included in the college curriculum.

Geographers can frequently arrange their work schedules according to their own needs. For example, they may spend a regular workweek in a classroom or in a government facility. Then they might join a research project that takes a few days or perhaps venture off on a field expedition lasting months or even years. Such expeditions on Earth involve living in remote, often primitive, regions; while for our future "exogeographer," the ultimate expedition would be a personal visit to the Moon or Mars. Perhaps a bit closer to home, geographers will also enjoy an interesting view of Earth from either the Space Shuttle or a space station.

According to the U.S. Department of Labor, about 8,000 persons are currently employed as professional geographers in the United States. Even though the field is small and highly competitive, career opportunities are expected to increase in the next decade—especially for individuals with advanced degrees. A person working as a professional geographer may anticipate steady employment and working conditions that are relatively unaffected by changes in social conditions or the national economy.

Geographers work in a variety of fields, including aerospace, urban and community planning, market research and international trade. Perhaps one of the most interesting extraterrestrial career areas that will emerge in the next few years is that of the geographer who is also a remote sensing specialist. This individual will take advantage of the multispectral remote sensing data of the Earth gathered by sophisticated spacecraft and utilize these data in a variety of social, economic and physical geography applications. The ability to accurately locate terrestrial resources from space and to model and predict global markets using spacecraft-derived data will form a very interesting part of the work of tomorrow's geographers.

A geographer frequently works with persons in many other career fields, including cartographers, geologists, photographers, illustrators, engineers, scientists, astronauts and technical writers.

GEOLOGIST

Geologists study the physical aspects of the Earth, including its origin, history, composition and structure. First, they obtain data by drilling, collecting and examining rocks and mineral samples. They may live in the field during this sample-collection time, often for extended periods under primitive conditions. Then the geologists return to their laboratories, where they can perform sophisticated analyses on their rock and mineral specimens and test their hypotheses about the Earth under controlled conditions. Finally, they write technical reports to explain and announce their findings. The Space Age has added an exciting new twist to the study of rocks and minerals. During the Apollo program, the world's first "exogeologists" walked on the lunar surface gathering extraterrestrial rock samples that were returned to Earth for examination. Robot spacecraft have landed on Mars and automatically examined the soil of the Red Planet. And sophisticated multispectral sensors aboard Earth-orbiting spacecraft have returned valuable data of use in analyzing the Earth's surface geology and mineral wealth. All of these exciting additions to the field of geology (or should we now also say "exogeology"?) have come about in the last two decades. The future promises even more interesting developments, including a sample-return mission to Mars and perhaps human expeditions to the Red Planet within the next few decades.

The tools of a traditional, terrestrial geologist include hand lenses, X-ray diffractometers (devices that determine the structure of minerals), petrograph microscopes (devices that help analyze rock formations and assess modifications due to natural processes), geological picks and other complex technical equipment. Modern geologists also use sophisticated remote sensing spacecraft to help locate, characterize and interpret geological phenomena on a global scale. Tomorrow's exogeologist, perhaps accompanying the first contingent of lunar miners, will also use a space suit and surface rover vehicles and be assisted by robotic explorers.

Is the field of geology, or even exogeology, for you? Well, geologists are essentially research scientists and are therefore individuals who possess an inquiring mind and who exhibit an interest in things scientific. You must display an aptitude for science and an above-average intelligence. As a geologist, you must also have the ability to visualize objects in three dimensions. Since a large portion of a geologist's time is spent gathering data in the field, you should also like working outdoors and have the physical stamina and good health necessary to lead a camper's life in the field.

Most geologists enjoy the variety that they experience in their jobs. They have many different tasks that they perform and the opportunity to travel to many interesting places, terrestrial and now even extraterrestrial. However, pursuit of this career field can also mean spending time away from their families and long, physically demanding hours in the field.

A bachelor of science degree is the minimum educational requirement for any position as a geologist. However, beginning positions in research, teaching and exploration usually require a master's degree, while a doctorate is necessary for significant advancement or to specialize. Geologists must study a variety of fields, including geology and related subjects (such as physics, geography, minerals and natural history), the natural sciences and mathematics, and the arts and humanities (especially English composition and economics). Consequently, a secondary-school student should prepare for a career in geology by following a diversified college-preparatory curriculum that includes a combination of courses in English literature and composition, social studies, mathematics and science.

All geologists spend some time in the field, in laboratories and in offices—but the amount of time spent in each of these general work environments will vary with the demands of a particular geological project. The energy crisis of the early 1970s and the resulting emphasis on the location of new energy resources have made the work of contemporary geologists even more valuable. For example, at present approximately 60 percent of all geologists work in the oil and natural gas industries. Employment opportunities in this field should continue to increase with the growing global demands for energy (especially coal and petroleum). Geological specialists are also used extensively in the aerospace industry, especially in the design and operation of planetary-exploration spacecraft, in the evaluation of extraterrestrial data (imagery and actual samples) and in the planning of future misions, such as a manned return flight to the Moon. Geologists work for private industry, for state and federal agencies, for universities and colleges, and even for museums as full- or part-time consultants.

There are many interesting specialties within the overall field of geology. These include careers as an engineering geologist, a geomorphologist (a person who studies the contour and evolution of landforms), a mineralogist, a petrographer (a person who describes and classifies rocks), a petroleum geologist, a petrologist (a person who studies the origin, composition, structure and changes in rocks) or a sedimentologist (a person who studies rocks formed from sediment or those deposited underwater).

Supported by sophisticated Space-Age instruments, geologists are helping to increase the quality of modern life (through resource location, identification and exploitation). They are also supporting the establishment of humanity's extraterrestrial civilization by increasing our understanding of celestial objects and their resource potential. Geologists work as part of a team that often includes data systems analysts, research technicians, environmental engineers, metallurgists, biologists, astronomers, chemists, physicists, geographers and science writers.

MATHEMATICIAN

Modern mathematicians are engaged in a wide variety of activities, ranging from the creation of new theories to the translation of scientific and managerial problems into mathematical terms. Such analytical modeling of physical problems allows us to solve many complicated problems. Modern high-speed computers have supported the extensive use of such modeling efforts by mathematicians and other scientists and engineers. Mathematical work falls into two broad classes: theoretical, or "pure," mathematics and applied mathematics. In practice, these two general classes are not sharply defined and frequently overlap.

Theoretical mathematicians advance mathematical science through the development of new mathematical principles and relationships. Although theoretical mathematicians often seek to increase basic knowledge without necessarily being driven by the immediate, practical use of such new knowledge, this pure and abstract pursuit of knowledge has been very instrumental in producing many scientific and engineering achievements.

Applied mathematicians, on the other hand, use mathematics to develop theories, techniques and methods to solve practical problems in business, government, engineering and the natural and social sciences. In aerospace their work ranges from the precise analysis of the trajectory of an interplanetary probe to the modeling of the origins of life in the Universe and the probability of contact between advanced, intelligent extraterrestrial civilizations. One area of particular excitement in modern aerospace applications is the entire field of artificial intelligence. As we push out into heliocentric (Sun-centered) space with advanced scientific probes and eventually manned spacecraft, very smart machines must be available to make these missions a success. Applied mathematicians working with engineers and scientists are now developing the foundations of such very smart machines and the mathematical framework for their "artificial intelligences."

Much work in applied mathematics is performed by individuals other than professional mathematicians. In fact, the number of scientific and technical workers who depend on expertise in mathematics is many times greater than the number of individuals actually designated as mathematicians.

Should you be a mathematician? Mathematicians need good reasoning ability, persistence and the knack for applying basic principles and solutions to new and unusual problems. As a mathematician, you should also be able to communicate well with others, because a great deal of your work will often involve solving problems for non-mathematicians. You should also have the ability to work independently or as part of a team.

A bachelor's degree is the minimum educational requirement for work as a mathematician in industry, research or business. Undergraduate course work should include analytical geometry, calculus, differential equations, probability and statistics, and mathematical analysis. Courses in computer science are also extremely valuable. However, the entry requirements for many positions in the aerospace industry and in the government demand an advanced degree. A doctorate is essential if you wish to teach at a college or university. Secondary students interested in a career in mathematics should take all available mathematics and science courses, as well as introductory courses in computer science.

The U.S. Department of Labor reported that in 1980 approximately 35,000 persons were employed as mathematicians. Of these, nearly 75 percent were teachers or independent researchers at colleges and universities. Other mathematicians are employed by private industry and the federal government. In the private sector, major employers include the aerospace industry, the electronics industry and the computer industry. NASA and the Department of Defense employ the most mathematicians working for the federal government.

Mathematicians work in all parts of the United States but can be found concentrated in large industrial areas and in regions with large institutions of higher learning.

Mathematicians are very important members of our nation's aerospace team. They frequently work on teams with technicians, statisticians, engineers, computer analysts, systems engineers, operations research analysts and space scientists.

METEOROLOGIST

Meteorologists study the atmosphere or envelope of gases that surrounds the Earth or other celestial bodies in our Solar System. People are most familiar with the work of synoptic meteorologists, who chart and interpret weather data reported by observers and who use these data to predict conditions and make weather forecasts. There are, however, many other interesting specialties within the field of meteorology. Climatologists, for example, study temperature and weather records and deduce climate patterns or trends from these data; dynamic meteorologists study the relationship between physical laws, planetary energy balances and the motion of large masses of air; industrial meteorologists study the relationships between weather conditions and human activities and business pursuits; physical meteorologists try to determine the true nature of the atmosphere; and "exometeorologists," along with planetary scientists, are concerned with weather and climate patterns on alien worlds such as Mars, Venus, Jupiter, Saturn and the Saturnian moon Titan.

Is the field of meteorology for you? Generally, meteorologists are responsible individuals who are curious about the world around them. To be a successful meteorologist, you should have a logical mind; be interested in science and mathematics; be able to combine seemingly unrelated facts and form useful hypotheses from these data; and be capable of thinking abstractly.

Most meteorologists take delight in knowing that people, business enterprises and government agencies depend on the timeliness and accuracy of the valuable information they provide. Meteorologists often work indoors at weather stations, which may be located near large cities or in remote or isolated areas. As a meteorologist, you might have to work irregular hours, since weather stations are open 24 hours a day. However, night and weekend duties are often rotated among all the members of a meteorology team staff.

The minimum educational requirement for entry into this field is normally a bachelor of science degree with at least 20 hours of study in meteorology. Advanced degrees support careers in teaching and the pursuit of independent research.

According to the U.S. Department of Labor, approximately 5,000 persons are currently employed in the United States as meteorologists. About 70 percent of all meteorologists, in turn, work for the federal government, and more than half of these government workers are military personnel. Other meteorologists are employed by the commercial airlines, public utility companies, colleges and universities (mainly as teachers), radio and TV stations, private weather bureaus, consulting firms and insurance companies.

Space-Age advances, particularly the advent of the meteorological satellite, have provided contemporary meteorologists with large amounts of useful environmental and atmospheric data in a timely and economic manner. Employment opportunities in this career field appear good. Meteorologists provide valuable data to aerospace mission planners and are an integral part of this nation's aerospace efforts. It is not too speculative, in fact, to assume that a young man or woman reading this paragraph might even become the first "exometeorologist" in-residence on the Red Planet, helping the early Martian settlers cope with the uncertainties of weather on an alien world.

PHYSICIST

Physicists study and analyze the structure of matter, the different forms of energy and the relationship between matter and energy. Their interest ranges from the tiny world of subnuclear particles to the vast expanses of the galaxies. Their work leads to greater scientific knowledge and to the development of useful technological tools and new materials.

Physicists are engaged in research and development. Some perform basic research, increasing the general body of scientific knowledge. Theoretical physicists are primarily involved with mathematical concepts and formulas, while experimental physicists use systematic observations and measurements to unlock nature's secrets. Experimental physicists often design and develop new instruments to help them conduct their pioneering scientific investigations. Such instruments include, of course, sensor packages placed on board Earth-orbiting spacecraft and interplanetary probes. Engineering physicists are involved in applied research and frequently design new or improved scientific apparatus and products.

Should you be a physicist? As a physicist, like all research scientists, you must have an interest in science and mathematics and possess an inquiring mind. Physicists are generally persevering, can concentrate on detail and possess a disciplined creativity that allows them to sort through a myriad of physical facts and data and create new models of the forces and phenomena that shape our Universe. Because a large portion of a physicist's time is spent working with others in a research laboratory, you must also be able to work as a leader or a member of a talented team. Some physicist projects involve independent research, while others involve the closely coordinated participation of many skilled scientists. As a physicist, you will enjoy the satisfaction of knowing that the world might benefit greatly from the things that you and your colleagues discover.

A bachelor's degree with a major in physics is the minimum educational requirement for work in this career area. However, many entry-level positions require a graduate degree. A doctorate is necessary if you want to engage in independent research, specialize or teach at a university. Prospective physicists must study a variety of scientific and mathematical subjects that then form the foundation for their specialization in physics. Secondary-school students can prepare for a career in physics by taking a college-preparatory curriculum that includes a combination of courses in mathematics, science, the language arts and social studies.

The U.S. Department of Labor indicates that about 40,000 physicists were employed in the United States in 1980. Some worked in applied research for private industry or for the federal government, while others worked as independent researchers, teachers or research assistants in colleges and universities. Experienced physicists often enter the private sector, such as the aerospace industry, where they frequently advance to positions of responsibility in management, research or technical sales. In most academic institutions, promotion to fully tenured professor depends on both the quality of your teaching and the quantity and quality of your published research.

Physicists are very important members of this nation's aerospace team. For example, astrophysicists have revised our model of the Universe (from tranquil to violent) as a result of data collected by instruments carried into space on satellites and probes. Finally, physicists work closely with other scientists such as astronomers, chemists, geologists, geophysicists and biologists. As we expand out into space, future physicists will be able to meet and explore the Universe as never before—in microgravity research facilities or investigating the atmospheric processes of one of the giant outer planets.

SCIENCE WRITER

What does a science writer do? The science writer, sometimes called an engineering writer, organizes, interprets, writes and edits scientific and technical material. A person in this profession must establish a communication link between the scientists or engineers creating the technical information and persons needing the data, such as managers, government officials and other scientists and engineers. Science writers must be able to write in a concise and clear way for general consumer publications, as well as in a highly specialized technical style for communication between experts. These writers are responsible for providing information to managers for decision making and to technicians for the construction, operation and maintenance of equipment.

Science writers prepare rough drafts of a report, paper or publication for review by the project staff. Several revisions are often needed before the final form is considered acceptable. Typical writing assignments may include articles for company newsletters and trade journals in addition to sales literature, proposals and reports, publicity releases, catalogs and brochures. Science writers may also be responsible for preparing manuals, scripts and related technical materials used in training and staff-development programs.

As a science writer, you might also be asked to coordinate writing projects and to arrange illustrations and photographs used in a particular publication.

If you are interested in scientific and technical matters and also have a creative flair with the written language, the science writer career field may be just right for you. In addition to your writing skills and technical expertise, you should be intellectually curious and capable of logical thought. Of equal importance, you must be able to organize a large quantity of detailed information and be very accurate in your work. For example, aerospace audiences don't want to read that "Mars has a few moons"; they demand to know that "Mars has two tiny moons, named Phobos and Deimos." To be a good science writer, you must also have persistence and patience, since gathering and assembling large quantities of technical information can be difficult and time-consuming. Science writing, of course, is not done in a vacuum. You must interact constantly with the scientists, engineers and technicians who created the information. There are other times when you will be required to work alone for long periods with little or no supervision. Writing and editing can be a very lonely profession—just you and the typewriter or word processor. In the end, however, there is a real feeling of satisfaction as the final product leaves your desk and becomes part of the printed literature. As a science writer, you are directly responsible for the timely and accurate dissemination of new discoveries and ideas.

To be successful in this career field, you must be very familiar with the subject of your writing assignment. In the aerospace area you must fully understand all the technologies that go together to form the space program. You must also be able to effectively communicate with scientists, engineers and technicians.

While there is no fixed educational requirement for entry into this career area, a college degree is frequently regarded as essential. In fact, many employers prefer candidates with a degree in science or engineering plus a minor in English, journalism or technical communications. Other employers will emphasize writing ability and look for candidates whose degrees are in journalism, English or technical communication.

Secondary students interested in science writing as a career should prepare by following a college-preparatory curriculum that includes a combination of courses in the language arts, mathematics and the sciences. Special creative writing, journalism and graphic-arts electives will also prove very helpful.

Modern science writers work for large firms in the electronics, aviation, aerospace, chemical and computer industries. Energy technology, computer software and communications firms also employ many science writers, along with research laboratories. Federal agencies, such as NASA, employ many hundreds of science writers in such diverse areas as the physical sciences, agriculture, energy and health.

In aerospace, as in other technical fields, the science writers are often the key to the successful communication of new ideas and discoveries. These writers depend upon many specialists to gather the information that they use in their writing, while scientists, engineers and program managers depend on the science writers to clearly and accurately report new ideas. Other occupational specialties related to the science writer career field include translator, information specialist, illustrator, writer of training materials, editor, specifications writer and media specialist.

In a most general sense, the aerospace science writer opens the Universe up to us. Through his or her creative writing skills and ability to select illustrations, the wonders of the planets, the discoveries made by space probes, the latest findings of Earth-observation satellites are transformed from clusters of numbers and tables of data into a clear and accurate record of scientific discovery. As we push out into space and create humanity's extraterrestrial civilization, tomorrow's science writers will be there to report these magnificent accomplishments to the millions who remain behind on the surface of the third planet from the Sun.

SYSTEMS ANALYST

Many important technical functions and scientific research projects depend on systems analysts to plan efficient methods of processing data and handling the results. Systems analysts begin an assignment by discussing the information-processing problem with project managers or technical specialists. This communication enables them to determine the exact nature of the problem and to break it down into its component parts. For example, if a new remote sensing system is being considered to help monitor terrestrial resources from space, the systems analysts must first determine what new data should be collected, the equipment needed for such data acquisition and the steps to be followed in the processing and distribution of the information. Are all the data collected by the satellite to be transmitted to Earth, then processed and distributed? Or are perhaps only selected data to be processed and transmitted? Finally, is data transmission to be limited to a few selected users on the ground? These are the types of information-processing options or trade-offs that the systems analyst must consider.

Persons in this career field use various techniques, such as mathematical model building, to fully analyze a problem and devise an optimum new information system or data-processing strategy. Once the system has been developed, the systems analyst prepares charts and diagrams that describe its operation in terms that project managers and engineers can understand. They will also help decision makers by preparing cost-benefit analyses.

If the system is accepted, the systems analyst then translates the logical requirements of the system into the capabilities of the computer machinery, or "hardware." Systems analysts also prepare specifications for computer programmers to follow and work with these programmers as they "debug," or eliminate errors in, the data-processing system. Thus, the systems analyst helps create and evaluate a new information system, establishes the optimum flow of data through that system and then assists in the creation of the computer hardware and software necessary to make the information system a practical reality.

Should you become a systems analyst? As a systems analyst, you should be capable of thinking logically and like working with ideas and concepts. You will often have to deal with a number of tasks simultaneously, so the ability to concentrate and pay close attention to detail is quite important. Systems analysts may work independently, or as part of a team on a large aerospace project. You must also be able to communicate effectively with the scientists, engineers, support technicians and computer programmers assigned to a particular project.

There is no one universal formula that describes the requirements needed to become a systems analyst. The requirements are often determined by the nature of the project and the employers' preferences. In general, college graduates are sought for these jobs, and persons with advanced degrees are frequently preferred. In the aerospace industry, candidates with an engineering or scientific background also possessing an extensive amount of computer programming experience are often eagerly sought. Since a great deal of the systems analyst's work involves the use of computer models to simulate system performance and to generate cost-benefit analyses, persons with a computer-science degree are also in demand.

Prior work experience (for example, as an aerospace engineer or scientist) is also important in this career field. Nearly half of the persons entering this occupation have transferred from other fields. In many research laboratory or industrial situations, systems analysts began their employment as computer programmers and then were promoted into analyst positions as they gained experience.

Secondary-school students should study mathematics (including algebra, calculus and trigonometry) and take available introductory computer-science and data-processing courses in preparation for college work in this career area.

In 1980 there were approximately 200,000 individuals employed as systems analysts. Systems analysts working for research and industrial firms supporting the aerospace industry were often required to live in the northeastern and western regions of the United States. Employment prospects for systems analysts are expected to grow faster than the average for most occupations through the 1980s, as computer usage expands.

The demand for systems analysts should continue to rise as computer capabilities increase and as new applications are found for computer technology. Telecommunication networks, global or national information systems and increased use of remote sensing systems to monitor terrestrial resources represent anticipated growth areas for aerospace systems analysts. Over the next decade, systems analysts will also help develop methods for using the fifth-generation computers to solve problems in ways not yet even recognized.

A systems analyst works closely with many other specialists, including technical writers, engineers, scientists, computer programmers, operations research analysts and technicians. A future aerospace systems analyst reading this passage might be the person in charge of modeling the communications network for the first Mars base, developing artificial-intelligence strategies for a robot asteroid survey spacecraft or creating the data-processing algorithms and software supporting a very smart Earth-orbiting crop-survey satellite. All of these exciting challenges will fall within the professional interests of tomorrow's aerospace systems analysts.

In this section we have examined just a few of the many exciting "extraterrestrial" careers that are available. No one can say for sure what new career areas will open up as we begin to build our extraterrestrial civilization. Imagine, for example, someone predicting a hundred years ago that today people would be employed to run electronic computers, would take courses on electronic computers in colleges and universities and would even have computers in the home. At present, we face quite the same challenge in predicting what future extraterrestrial career areas might come into being several decades from now. However, based on a logical extrapolation of the human presence in space from a low-Earth-orbit space station to the Moon to Mars and then beyond, let's hazard a few "educated guesses." Sometime in the next century, a person might earn a living as a lunar miner, a Martian farmer, an asteroid prospector, a space-base magistrate, the owner and operator of an interplanetary shipping fleet or even dean of the University of Outer Space, with its far-flung telecommunication-linked campuses on the Moon, Mars and throughout the space cities in cislunar space. What other future extraterrestrial careers do you predict?

extraterrestrial catastrophe theory For millions of years giant, thundering reptiles roamed the lands, dominated the skies and swam in the oceans of a prehistoric Earth. Dinosaurs reigned supreme. Then, quite suddenly, some 65 million years ago, they vanished. What happened to these giant creatures and to thousands of other ancient animal species?

From archaeological and geological records, we do know that some tremendous catastrophe occurred about 65 million years ago on this planet. It affected life more extensively than any war, famine or plague in human history. For in that cataclysm approximately 75 percent of all animal species were annihilated—including, of course, the dinosaurs. It is most interesting to observe, however, that as long as those enormous reptiles roamed and dominated the Earth, mammals, including humans themselves, would have had little chance of evolving.

Several years ago the scientists Luis and Walter Alvarez and their colleagues at the University of California at Berkeley discovered that a pronounced increase in the amount of the element iridium in the Earth's surface had occurred at precisely the time of the disappearance of the dinosaurs. Since iridium is quite rare in the Earth's crust and much more abundant in the rest of the Solar System, they postulated that a large asteroid (or possibly a comet) might have struck the ancient Earth. This cosmic collision would have promoted an environmental catastrophe throughout the planet. The scientists reasoned that such an asteroid would largely vaporize while passing through the Earth's atmosphere, spreading a dense cloud of dust particles—including quantities of extraterrestrial iridium atoms—uniformly around the globe. Recently, an experimentally observed enhanced abundance of iridium has been found in a thin layer of the Earth's lithosphere (crust) lying between the final geologic formations of the Cretaceous period (which are dinosaur fossil-rich) and the formations of the early Tertiary period (whose rocks are notably lacking in dinosaur fossils). The Alvarez hypothesis further speculated that following this asteroid impact, a dense cloud of dust covered the Earth for many years, obscuring the Sun, blocking photosynthesis and destroying the very food chains upon which many ancient life-forms depended. Of course, this hypothesis also raises many questions. For example, if a giant asteroid did trigger this ancient environmental catastrophe, where is the residual crater? An answer perhaps is that the catastrophe was caused by an encounter with a large comet instead. Since a comet is believed composed largely of ice and dirt, then this type of prehistoric extraterrestrial encounter could still have plunged the world into darkness without leaving a single large crater on the surface of the planet.

There are, of course, many other scientific opinions as to why the dinosaurs vanished. A popular one is that there was a gradual but relentless change in the Earth's climate to which these giant reptiles and many other prehistoric animals simply could not adapt.

Nevertheless, the possibility of an asteroid's or comet's striking the Earth and triggering a catastrophe is quite real. Just look at a recent photograph of Mars or the Moon and ask yourself how those large impact craters were formed. Fortunately, the probability of a large asteroid's striking the Earth is quite low. For example, space scientists now estimate that the Earth will experience one collision with an "Earth-crossing" asteroid (of one kilometer diameter size or greater) every 300,000 years.

See also: **asteroid; comet; Nemesis**

extraterrestrial civilizations How might we categorize levels or types of extraterrestrial civilizations? According to some scientists, intelligent life in the Universe might be thought of as experiencing three basic levels of civilization, when considered on an astronomical scale. The Soviet astronomer N.S. Kardashev, in examining the issue of information transmission by extraterrestrial civilizations in 1964, first postulated three types of technologically developed civilizations on the basis of their energy use. A Type I civilization would represent a planetary civilization similar to the technology level on Earth today. It would command the use of somewhere between 10^{12} and

10^{16} watts of energy—the upper limit being the amount of solar energy being intercepted by a "suitable" planet in its orbit about the parent star. For example, the solar energy flux at the Earth (the solar constant outside the atmosphere) is $1,371 \pm 5$ watts per meter squared, and the spectral distribution of sunlight is that of an approximate blackbody radiator at a temperature of 5,760 degrees Kelvin, which peaks in the visible part of the electromagnetic (EM) spectrum—namely, a wavelength of 0.4 to 0.7 micrometer.

A Type II extraterrestrial civilization would engage in feats of planetary engineering, emerging from its native planet through advances in space technology and extending its resource base throughout the local star system. The eventual upper limit of a Type II civilization could be taken as the creation of a "Dyson sphere." A Dyson sphere is a shell or cluster of habitats and structures placed entirely around a star by an advanced civilization to intercept and use basically all the radiant energy from that parent star. What the American physicist Freeman J. Dyson proposed in 1960 was that an advanced extraterrestrial civilization would eventually develop the space technologies necessary to rearrange the raw materials of all the planets in its solar system, creating a more efficient composite ecosphere around the parent star. Dyson further postulated that such advanced civilizations might be detected by the presence of thermal infrared emissions from such an "enclosed star system," in contrast to the normally anticipated visible radiation. Once this level of extraterrestrial civilization is achieved, the search for additional resources and the pressures of continued growth could encourage interstellar migrations. This would mark the start of a Type III extraterrestrial civilization.

At maturity a Type III civilization would be capable of harnessing the material and energy resources of an entire galaxy (typically containing some 10^{11} to 10^{12} stars). Energy levels of 10^{37} to 10^{38} watts would be involved!

Command of energy resources might therefore represent a key factor in the evolution of extraterrestrial civilizations. It should be noted that a Type II civilization controls about 10^{12} times the energy resources of a Type I civilization; and a Type III civilization approximately 10^{12} times as much energy as a Type II civilization.

What can we speculate about such civilizations? Well, starting with the Earth as a model (our one and only "scientific data point"), we can presently postulate that a Type I civilization could exhibit the following characteristics: (1) an understanding of the laws of physics; (2) a planetary society (for example, global communication network, interwoven food and materials resource networks); (3) intentional or unintentional emission of electromagnetic radiations (especially radio frequency); (4) the development of space technology and spaceflight—the tools necessary to leave the home planet; (5) (possibly) the development of nuclear energy technology, both power supplies and weapons; and (6) (possibly) a desire to search for and communicate with other intelligent life-forms in the Universe. Many uncertainties, of course, are present. Given the development of the technology for spaceflight, will the planetary civilization opt to create a solar-system civilization? Do the planet's inhabitants develop a long-range planning perspective that supports the eventual creation of artificial habitats and structures throughout their star system? Or do the majority of Type I civilizations unfortunately destroy themselves with their own advanced technologies before they can emerge from a planetary civilization into a more stable Type II civilization? Does the exploration imperative encourage such creatures to go out from their comfortable, planetary niche into an initially hostile, but resource-rich star system? If this "cosmic birthing" does not occur frequently, perhaps our Galaxy is indeed populated with intelligent life, but at a level of stagnant planetary (Type I) civilizations that have neither the technology nor the motivation to create an extraterrestrial civilization or even to try to communicate with any other intelligent life-forms across interstellar distances.

Assuming that an extraterrestrial civilization does, however, emerge from its native planet and create an interplanetary society, several additional characteristics would become evident. The construction of space habitats and structures, leading ultimately to a Dyson sphere around the native star would reflect feats of planetary engineering and could possibly be detected by thermal infrared emissions as incident starlight in the visible spectrum was intercepted, converted to other more useful forms of energy and the residual energy (determined by the universal laws of thermodynamics) rejected to space as heat at perhaps 300 degrees Kelvin. Type II civilizations might also decide to search in earnest for other forms of intelligent life beyond their star system. They would probably use portions of the electromagnetic spectrum (radio frequency and perhaps X rays or gamma rays) as information carriers between the stars. Remembering that Type II civilizations would control 10^{12} times as much energy as Type I civilizations, such techniques as electromagnetic beacons or feats of astroengineering that yield characteristic X-ray or gamma-ray signatures may lie well within their technical capabilities. Assuming their understanding of the physical Universe is far more sophisticated than ours, Type II civilizations might also use gravity waves or other physical phenomena (perhaps unknown to us now but being sent through our Solar System at this very instant) in their effort to communicate across vast interstellar distances. Type II civilizations could also decide to make initial attempts at interstellar matter transfer. Fully automated robotic explorers would be sent forth on one-way scouting missions to nearby stars. Even if the mode of propulsion involved devices that achieved only a small fraction of the speed of light, Type II societies should have developed the much longer-term planning perspective and thinking horizon necessary to support such sophisticated, expensive and lengthy missions. The Type II civilization might also utilize a form of panspermia (the diffusion of spores or molecular precursors through space) or even ship microscopically encoded viruses through the interstellar void, hoping that if such "seeds of life" found a suitable ecosphere in some neighboring or distant star system, they would initiate the chain of life—perhaps leading ultimately to the replication (suitably tempered by local ecological conditions) of intelligent life itself. Finally, as the Dyson sphere was eventually completed, some of the inhabitants of this Type II civilization might respond to a cosmic wanderlust and initiate the first "peopled" interstellar missions. Complex space habitats would become "space arks" and carry portions of this civilization to neighboring star systems. Again, however, we must ask what is the lifetime of a Type II civilization? It would appear from an extrapolation of contemporary terrestrial engineering practices that perhaps a minimum of 500 to 1,000 years would be required for even an advanced interplanetary civilization to complete a Dyson sphere.

Throughout the entire Galaxy, however, if just one Type II civilization embarks on a successful interstellar migration program, then—at least in principle—it would eventually (in perhaps 10^8 to 10^9 years) sweep through the Galaxy in a "leapfrogging" wave of colonization, estab-

lishing a Type III civilization in its wake. This Type III civilization would eventually control the energy and material resources of up to 10^{12} stars—or the entire Galaxy! Communication or matter transfer would be accomplished by techniques that can now only politely be called "exotic." Perhaps directed beams of neutrinos or even (hypothesized) faster-than-light particles (tachyons) would serve as information carriers for this galactic society. Or, as Carl Sagan proposed in 1973, they might use tunneling through black holes as their transportation network. Perhaps again, they would have developed some kind of thought-transference telepathy that could form a basic communication network over the vast regions of interstellar space. In any event, a Type III civilization should be readily evident, since it would be galactic in extent and easily recognizable by its incredible feats of astroengineering.

In all likelihood, our Galaxy at present does not contain a Type III civilization. Or else the Solar System is being ignored—intentionally kept isolated—perhaps as as game preserve or "zoo," as some have speculated; or maybe it is one of the very last regions to be "filled in."

Table 1   Characteristics of Extraterrestrial Civilizations

Level	Characteristics	Energy Consumption	Manifestations

TYPE I	• Planetary society
	
	• Developed technology
	  – understanding of the laws of physics
	  – space technology
	  – nuclear technology
	  – electromagnetic communications
	
	• Initiation of spaceflight, interplanetary travel; settlement of space
	
	• Early attempts at interstellar communication
		10^{12}–10^{16} watts
		
		• Starting to push planetary resource limits
			• Intentional or unintentional electromagnetic emissions (especially radio wave)

TYPE II	• Solar-system society
	
	• Construction of space habitats
	
	• "Dyson sphere" as an ultimate limit
	
	• Search for intelligent life in space
	
	• Possible interstellar communication between Type II civilizations
	
	• Reasonably long societal lifetimes (10^3 to 10^5 years)
	
	• Far-term planning perspectives
	
	• Initiation of interstellar travel/colonization
		10^{26}–10^{27} watts
		
		• Ultimately all radiant energy output of native star is utilized
			• Electromagnetic
			  – radio waves
			  – X-rays
			  – gamma rays
			
			• Gravity waves
			
			• Mass transfer
			  – probes
			  – panspermia
			  – stellar ark

TYPE III	• Galactic civilization
	
	• Interstellar communication/travel
	
	• Fantastic feats of astroengineering
	
	• Very long societal lifetimes (10^8–10^9 years)
	
	• Effectively "the immortals," for planning purposes
		10^{37}–10^{38} watts
		
		• Energy resources of entire Galaxy (10^{11}–10^{12} stars) are commanded
			• Feats of astroengineering
			
			• Exotic
			  Communication
			  – neutrinos
			  – tachyons
			  Travel
			  – tunneling through black holes
			  – telepathy
			  – ? ? ? ?

The other perspective is that if we are indeed alone or the most advanced civilization in our Galaxy, we now stand at the technological threshold of creating the first Type II civilization in the Galaxy, and if successful, we then have the potential of becoming the wave of interstellar civilization that sweeps across the Galaxy and establishes a Type III civilization in the Milky Way!

Table 1 summarizes these speculations on the levels of extraterrestrial civilizations and their potential characteristics.

See also: **blackbody; Dyson sphere; Zoo hypothesis**

extraterrestrial contamination In general, the contamination of one world by life-forms, especially microorganisms, from another world. Using the Earth and its biosphere as a reference, this planetary-contamination process is called *forward contamination* if an extraterrestrial sample or the alien world itself is contaminated by contact with terrestrial organisms and *back contamination* if alien organisms are released into the Earth's biosphere.

An alien species will usually not survive when introduced into a new ecological system, because it is unable to compete with native species that are better adapted to the environment. Once in a while, however, alien species actually thrive, because the new environment is very suitable and indigenous life-forms are unable to successfully defend themselves against these alien invaders. When this "war of the biological worlds" occurs, the result might very well be a permanent disruption of the host ecosphere, with severe biological, environmental and possibly economic consequences.

Of course, the introduction of an alien species into an ecosystem is not always undesirable. Many European and Asian vegetables and fruits, for example, have been successfully and profitably introduced into the North American environment. However, any time a new organism is released in an existing ecosystem, a finite amount of risk is also taken.

Frequently, alien organisms that destroy resident species are microbiological life-forms. Such microorganisms may have been nonfatal in their native habitat, but once released in the new ecosystem, they become unrelenting killers of native life-forms that are not resistant to them. In past centuries on Earth, entire human societies fell victim to alien organisms against which they were defenseless; as for example, the rapid spread of diseases that were transmitted to native Polynesians and American Indians by European explorers.

But an alien organism does not have to directly infect humans to be devastating. Who can easily ignore the consequences of the potato blight fungus that swept through Europe and the British Isles in the 19th century, causing a million people to starve to death in Ireland alone?

In the Space Age it is obviously of extreme importance to recognize the potential hazard of extraterrestrial contamination (forward or back). Before any species is intentionally introduced into another planet's environment, we must carefully determine not only whether the organism is pathogenic (disease-causing) to any indigenous species but also whether the new organism will be able to force out native species—with destructive impact on the original ecosystem. The introduction of rabbits into the Australian continent is a classic terrestrial example of a nonpathogenic life-form creating immense problems when introduced into a new ecosystem. The rabbit population in Australia simply exploded in size because of their high reproduction rate, which was essentially unchecked by native predators.

At the start of the Space Age, scientists were already aware of the potential extraterrestrial-contamination problem—in either direction. Quarantine protocols (procedures) were established to avoid the forward contamination of alien worlds by outbound unmanned spacecraft, as well as the back contamination of the terrestrial biosphere when lunar samples were returned to Earth as part of the Apollo program.

A quarantine is basically a forced isolation to prevent the movement or spread of a contagious disease. Historically, quarantine was the period during which ships suspected of carrying persons or cargo (for example, produce or livestock) with contagious diseases were detained at their port of arrival. The length of the quarantine, generally 40 days, was considered sufficient to cover the incubation period of most highly infectious terrestrial diseases. If no symptoms appeared at the end of the quarantine, then the travelers were permitted to disembark. In modern times, the term *quarantine* has obtained a new meaning: namely, that of holding a suspect organism or infected person in strict isolation until it is no longer capable of transmitting the disease. With the Apollo program and the advent of the lunar quarantine, the term now has elements of both meanings. Of special interest in future space missions to the planets and their moons is how we avoid the potential hazard of back contamination of the Earth's environment when robot spacecraft and human explorers bring back samples for more detailed examination in laboratories on Earth.

A Planetary Quarantine Program was started by NASA in the late 1950s at the beginning of the U.S. space program. This quarantine program, conducted with international cooperation, was intended to prevent, or at least minimize, the possibility of contamination of alien worlds by early space probes. At that time, scientists were concerned with forward contamination. In this type of extraterrestrial contamination, terrestrial microorganisms, "hitchhiking" on initial planetary probes and landers, would spread throughout another world, destroying any native life-forms, life precursors or even remnants of past life-forms. If forward contamination occurred, it would compromise future attempts to search for and identify extraterrestrial life-forms that had arisen independently of the Earth's biosphere.

A planetary quarantine protocol was therefore established. This protocol required that outbound unmanned planetary missions be designed and configured to minimize the probability of alien-world contamination by terrestrial life-forms. As a design goal, these spacecraft and probes had a probability of 1 in 1,000 (1×10^{-3}) or less that they could contaminate the target celestial body with terrestrial microorganisms. Decontamination, physical isolation (for example, prelaunch quarantine) and spacecraft design techniques have all been employed to support adherence to this protocol.

One simplified formula for describing the probability of planetary contamination is:

$$P(c) = m \cdot P(r) \cdot P(g)$$

where

$P(c)$ is the probability of contamination of the target celestial body by terrestrial microorganisms

m is the microorganism burden

$P(r)$ is the probability of release of the terrestrial microorganisms from the spacecraft hardware

$P(g)$ is the probability of microorganism growth after release on a particular planet or celestial object

As previously stated, $P(c)$ had a design goal value of less than or equal to 1 in 1,000. A value for the microorganism burden, m, was established by sampling an assembled spacecraft or probe. Then, through laboratory experiments, scientists determined how much this microorganism burden was reduced by subsequent sterilization and decontamination treatments. A value for $P(r)$ was obtained by placing duplicate spacecraft components in simulated planetary environments. Unfortunately, establishing a numerical value for $P(g)$ was a bit more tricky. The technical intuition of knowledgeable exobiologists and some educated "guessing" were blended together to create an estimate for how well terrestrial microorganisms might thrive on alien worlds that had not yet been visited. Of course, today, as we keep learning more about the environments on other worlds of our Solar System, we can keep refining our estimates for $P(g)$. Just how well terrestrial life-forms grow on Mars, Venus, Titan and a variety of other interesting celestial bodies will be the subject of on-site laboratory experiments performed by 21st-century exobiologists.

As a point of history, the early U.S. Mars flyby missions (for example, *Mariner 4*, launched on November 28, 1964, and *Mariner 6*, launched on February 24, 1969) had $P(c)$ values ranging from 4.5×10^{-5} to 3.0×10^{-5}. These missions achieved successful flybys of the Red Planet on July 14, 1965, and July 31, 1969, respectively. Postflight calculations indicated that there was no probability of planetary contamination as a result of these successful precursor missions.

The manned U.S. Apollo missions to the Moon (1969–72) also stimulated a great deal of debate about forward and back contamination. Early in the 1960s, scientists began speculating in earnest, "Is there life on the Moon?" Some of the most bitter technical exchanges during the Apollo program took place over this particular question. If there was life, no matter how primitive or microscopic, we would want to examine it carefully and compare it with life-forms of terrestrial origin. This careful search for microscopic lunar life would, however, be very difficult and expensive because of the forward-contamination problem. For example, all equipment and materials landed on the Moon would need rigorous sterilization and decontamination procedures. There was also the glaring uncertainty about back contamination. If microscopic life did indeed exist on the Moon, did it represent a serious hazard to the terrestrial biosphere? Because of the potential extraterrestrial-contamination problem, time-consuming and expensive quarantine procedures were urged by some members of the scientific community.

On the other side of this early 1960s contamination argument were those exobiologists who emphasized the suspected extremely harsh lunar conditions: virtually no atmosphere; probably no water; extremes of temperature ranging from 120 degrees Celsius at lunar noon to minus 150 degrees Celsius during the frigid lunar night; and unrelenting exposure to lethal doses of ultraviolet, charged-particle and X-ray radiations from the Sun. No life-form, it was argued, could possibly exist under such extremely hostile conditions.

This line of reasoning was countered by other exobiologists who hypothesized that trapped water and moderate temperatures below the lunar surface could sustain very primitive life-forms. And so the great extraterrestrial-contamination debate raged back and forth, until finally the *Apollo 11* expedition departed on the first lunar-landing mission. As a compromise, the *Apollo 11* mission flew to the Moon with careful precautions against back contamination but with only a very limited effort to protect the Moon from forward contamination by terrestrial organisms.

The Lunar Receiving Laboratory (LRL) at the Johnson Space Center in Houston provided quarantine facilities for two years after the first lunar landing. What we learned during its operation serves as a useful starting point for planning new quarantine facilities, Earth-based or space-based. In the future, these quarantine facilities will be needed to accept, handle and test extraterrestrial materials from Mars and other Solar-System bodies of interest in our search for alien life-forms (present or past).

During the Apollo program, no evidence was discovered that native alien life was then present or had ever existed on the Moon. A careful search for carbon was performed by scientists at the Lunar Receiving Laboratory, since terrestrial life is carbon-based. One hundred to 200 parts per million of carbon were found in the lunar samples. Of this amount, only a few tens of parts per million are considered indigenous to the lunar material, while the

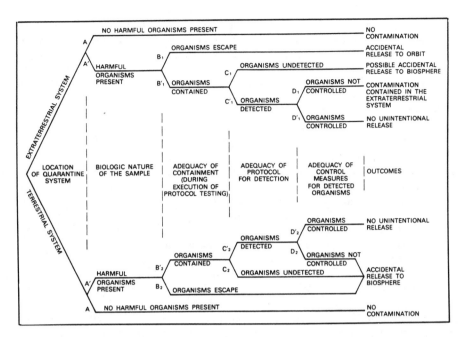

bulk amount of carbon has been deposited by the solar wind. Exobiologists and lunar scientists have concluded that none of this carbon appears derived from biological activity. In fact, after the first few Apollo expeditions to the Moon, even back-contamination quarantine procedures were dropped.

There are three fundamental approaches toward handling extraterrestrial samples to avoid back contamination. First, we could sterilize a sample while it is en route to Earth from its native world. Second, we could place it in quarantine in a remotely located, maximum-confinement facility on Earth while scientists examine it closely. Finally, we could also perform a preliminary hazard analysis (called the extraterrestrial protocol tests) on the alien sample in an orbiting quarantine facility before we allow the sample to enter the terrestrial biosphere. Figure 1 describes the event sequences and possible outcomes for extraterrestrial sample analyses performed in quarantine facilities on Earth and in space. As shown in this figure, these quarantine facilities are characterized by different outcomes, depending on when in the chain of events a containment failure occurs. To be adequate, a quarantine facility must be capable of: (1) containing all alien organisms present in a sample of extraterrestrial material, (2) detecting these alien organisms during protocol testing and (3) controlling these organisms after detection until scientists could dispose of them in a safe manner.

One way to bring back an extraterrestrial sample that is free of potentially harmful alien microorganisms is to sterilize the material during its flight to Earth. However, the sterilization treatment used must be intense enough to guarantee that no life-forms as we currently know them could survive. An important concern here is also the impact the sterilization treatment might have on the scien-

tific value of the alien sample. For example, use of chemical sterilants would most likely result in contamination of the sample, preventing the measurement of certain soil properties. Heat could trigger violent chemical reactions within the soil sample, resulting in significant changes and the loss of important exogeological data. Finally, sterilization would also greatly reduce the biochemical information content of the sample. It is even questionable as to whether any significant exobiology data can be obtained by analyzing a heat-sterilized alien material sample. To put it simply—in their search for extraterrestrial life-forms, exobiologists want "virgin alien samples."

If we do not sterilize the alien sample en route to Earth, we have only two general ways of avoiding possible back-contamination problems. We can place the unsterilized sample of alien material in a maximum quarantine facility on Earth and then conduct detailed scientific investigations, or we can intercept and inspect the sample at an orbiting quarantine facility before allowing the material to enter the Earth's biosphere.

The technology and procedures for hazardous-material containment have been employed on Earth in the development of highly toxic chemical- and germ-warfare agents and in conducting research involving highly infectious diseases. A critical question for any quarantine system is whether the containment measures are adequate to hold known or suspected pathogens while experimentation is in progress. Since the characteristics of potential alien organisms are not presently known, we must assume that the hazard they could represent is at least equal to that of terrestrial Class IV pathogens. (A terrestrial Class IV pathogen is an organism capable of being spread very rapidly among humans; no vaccine exists to check its spread; no cure has been developed for it; and the organism produces high mortality rates in infected persons.)

Judging from the large uncertainties associated with potential extraterrestrial life-forms, it is not obvious that any terrestrial quarantine facility will gain very wide acceptance by the scientific community or the general public. For example, locating such a facility and all its workers in an isolated area on Earth actually provides only a small additional measure of protection. Consider, if you will, the planetary environmental impact controversies that could rage as individuals speculated about possible ecocatastrophes. What would happen to life on Earth if alien organisms did escape and went on a deadly rampage throughout the Earth's biosphere? The alternative to this potentially explosive controversy is quite obvious: locate the quarantine facility in outer space. A space-based facility provides several distinct advantages: (1) it eliminates the possibility of a sample-return spacecraft's crashing and accidentally releasing its deadly cargo of alien microorganisms; (2) it guarantees that any alien organisms that might escape from confinement facilities within the orbiting complex cannot immediately enter the Earth's biosphere; and (3) it ensures that all quarantine workers remain in total isolation during protocol testing (that is, during the testing procedure). (See fig. 2.)

As we expand the human sphere of influence into heliocentric (Sun-centered) space, we must also remain conscious of the potential hazards of extraterrestrial contamination. Scientists, space explorers and extraterrestrial entrepreneurs must be aware of the ecocatastrophes that might occur when "alien worlds collide"—especially on the microorganism level.

With a properly designed and operated orbiting quarantine facility, alien-world materials can be tested for potential hazards. Three hypothetical results of such protocol testing are: (1) no replicating alien organisms are discovered; (2) replicating alien organisms are discovered, but they are also found not to be a threat to terrestrial life-forms; or (3) hazardous replicating alien life-forms are discovered. If potentially harmful replicating alien orga-

nisms were discovered during these protocol tests, then quarantine workers would either render the sample harmless (for example, through heat- and chemical-sterilization procedures); retain it under very carefully controlled conditions in the orbiting complex and perform more detailed analyses on the alien life-forms; or properly dispose of the sample before the alien life-forms could enter the Earth's biosphere and infect terrestrial life-forms.

See also: **Mars Sample Return Mission**

extraterrestrial expeditions Organized voyages by human explorers to other worlds in our Solar System. When future historians look back on the 20th century, they will immediately recognize one of the greatest accomplishments in all human history—the first extraterrestrial expeditions to the Moon conducted by the United States through NASA's Apollo Program.

After countless eons of biological and social evolution, followed by a few millennia of recorded history, in a literal "blink of an eye" we quite suddenly left our native planet for the first time in 1969 to set foot on another world (see fig. C4 on color page B). When Astronaut Neil Armstrong stepped on the lunar surface at 10:56 PM EDT on July 20, 1969, our extraterrestrial civilization was born! While over 500 million of his fellow Earthlings anxiously watched and listened a short 1.3 light-seconds away, Armstrong extended our reach in the cosmos.

How did this magnificent undertaking get started? On May 25, 1961 President John F. Kennedy in his historic address to a joint session of Congress and to the Nation declared:

"I believe this nation should commit itself to achieving the goal, before this decade is out, of landing a man on the Moon and returning him safely to Earth. No single space project in this period will be more impressive to mankind, or more important in the long-range exploration of space; and none will be so difficult or expensive to accomplish."

Less than ten years later, this goal was achieved when Astronaut Armstrong became the first person to walk on the Moon!

Several statements made during this historic expedition should remain in the hearts of people everywhere.
The first radio message from an alien world: (transmitted from the Apollo 11 Lunar Module at touchdown)
"Houston, Tranquility Base here. The Eagle has landed!"
Astronaut Neil Armstrong: (as he became the first man to walk on the Moon)
"That's one small step for man . . . one giant leap for mankind."
Astronaut Buzz Aldrin: (The second man on the Moon, as he first viewed the lunar landscape)
"Beautiful, beautiful. Magnificent desolation."
Former President Richard Nixon: (as he made the first telephone call to another world)
"Neil and Buzz, I am talking to you by telephone

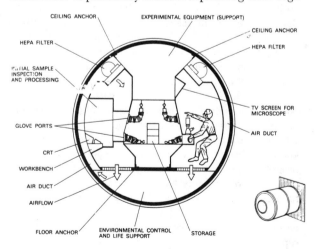

Fig. 2   The laboratory module for an extraterrestrial quarantine facility. (Drawings courtesy of NASA.)

from the Oval Office at the White House, and this certainly has to be the most historic telephone call ever made . . . . Because of what you have done, the heavens have become a part of man's world."

Table 2 on page 125 provides a brief summary of all the Apollo Program missions, including the six expeditions to the Moon's surface. These missions, humankind's first extraterrestrial expeditions, represent a genuine triumph for all humanity. For a comparatively brief moment in history, the skills were established, the people, materials and equipment organized, and the vehicles and facilities built to accomplish a difficult, objective that gave everyone a new feeling of pride, togetherness and confidence. The magnificent success of the Apollo Program provides strong testimony to future generations that human beings can achieve worthwhile, though difficult, objectives. This is perhaps one of the greatest legacies any civilization can bestow on future generations. Through the footsteps of the Apollo astronauts, we all have taken our very first steps on the pathway to the stars!

See also: **Moon; Why humans explore**

extraterrestrial life chauvinisms What characteristics and properties should extraterrestrial life-forms exhibit? Do they resemble terrestrial organisms, or are they entirely different from any living system found on Earth? What properties of terrestrial life are really basic to life found elsewhere in the Universe, and what characteristics of living systems are purely accidents of evolution? These are some of the puzzling questions facing modern scientists and exobiologists as they seek to unravel the mystery of life in the Universe. As you can see, it is not easy to describe what alien life-forms will be like—should they in fact exist elsewhere in the vast reaches of our Galaxy and beyond. At present, we have only one data point (source of information) on the emergence of life in a planetary environment—our own planet Earth. All terrestrial organisms are carbon-based and have descended from a common, single occurrence of the origin of life in the primeval "chemical soup" of an ancient Earth. How can we project this singular fact to the billions of unvisited worlds in the cosmos? We can only do so with great technical caution, realizing full well that our models of extraterrestrial life-forms and our estimates concerning the cosmic prevalence of life can easily become prejudiced, or chauvinistic, in their findings.

Chauvinism can be defined as a strongly prejudiced belief in the superiority of one's group. Applied to speculations about extraterrestrial life, this word can take on several distinctive meanings, each heavily influencing any subsequent thought on the subject. Some of the more common forms of extraterrestrial life chauvinisms are: G-star chauvinism, planetary chauvinism, terrestrial chauvinism, chemical chauvinism, oxygen chauvinism and carbon chauvinism. While such heavily steeped thinking may not actually be wrong, it is important to realize that it also sets limits, intentionally or unintentionally, on

contemporary speculations about life in the Universe.

G-star, or Solar-System, chauvinism implies that life can only originate in a star system like our own—namely, a system containing a single G-spectral-class star. Planetary chauvinism assumes that extraterrestrial life has to develop independently on a particular planet, while terrestrial chauvinism stipulates that only "life as we know it on Earth" can originate elsewhere in the Universe. Chemical chauvinism demands that extraterrestrial life be based on chemical processes, while oxygen chauvinism states that alien worlds must be considered uninhabitable if their atmospheres do not contain oxygen. Finally, carbon chauvinism asserts that extraterrestrial life-forms must be based on carbon chemistry.

These chauvinisms, singularly and collectively, impose tight restrictions on the type of planetary system that might support the rise of living systems, possibly to the level of intelligence, elsewhere in the Universe. If they are indeed correct, then our search for extraterrestrial intelligence is now being properly focused on Earthlike worlds around Sunlike stars. If, on the the other hand, life is actually quite prevalent and capable of arising in a variety of independent biological scenarios (for example, silicon-based or sulfur-based chemistry), then our contemporary efforts in modeling the cosmic prevalence of life and in trying to describe what "little green men" (LGMs) really look like is somewhat analogous to using the atomic theory of Democritus, the ancient Greek, to help describe the inner workings of a modern nuclear-fission reactor. As we continue to explore our own Solar System—especially Mars and the moons of Jupiter and Saturn—we will be able to better assess how valid these chauvinisms really are.

See also: **carbon chauvinism; oxygen chauvinism; planetary chauvinism**

extraterrestrial museum At ceremonies on May 18, 1984, NASA transferred ownership of the *Viking 1* Lander, which is now on the planet Mars, to the Smithsonian's National Air and Space Museum in Washington, D.C. As a result, the museum has an exhibit located on another world!

This transfer included the loan of the official Viking Lander plaque to the museum. This plaque renames the *Viking 1* Lander, the Thomas A. Mutch Memorial Station, in memory of the Viking Lander Imaging Team leader and NASA associate administrator for space science, who died in a tragic climbing accident in the Himalayas in 1980. The plaque will be placed on the original *Viking 1* Lander by American astronauts when they travel to the Red Planet at some time in the future.

The *Viking Landers* (*1* and *2*) set down on the surface of Mars in 1976 and were the first American spacecraft to provide close-up views from the surface of another planet.

See also: **Mars; National Air and Space Museum; Viking Project**

extraterrestrial resources When people think about outer space, visions of vast emptiness, devoid of anything useful, frequently come to their minds. However, when forward-thinking space technologists gaze into the extraterrestrial environment, they see a new frontier rich with resources, including unlimited solar energy, a full range of raw materials and an environment that is both special (for example, high vacuum, microgravity, physical isolation from the terrestrial biosphere) and reasonably predictable.

Since the start of the Space Age, investigations of the Moon, Mars, the asteroids and meteorites have provided tantalizing hints about the rich mineral potential of our extraterrestrial environment. For example, the Apollo missions established that the average lunar soil contains more than 90 percent of the materials needed to construct a complicated space industrial facility. (See table 1.) The soil in the lunar highlands is rich in anorthosite, a mineral suitable for the extraction of aluminum, silicon and oxygen. Other lunar soils have been found to contain orebearing granules of ferrous metals like iron, nickel, titanium and chromium. Iron can be concentrated from the lunar soil (called regolith) before the raw material is even refined by simply sweeping magnets over the regolith to gather the iron granules scattered within.

Some scientists have even suggested that water ice and other frozen gases (or volatiles) may be trapped on the lunar surface in perpetually shaded polar regions. If future exploration of the Moon indicates that this is true, then "ice mines" on the Moon could provide both oxygen and hydrogen—vital resources for our extraterrestrial settlements and space industrial facilities. The Moon would be able both to export chemical propellants for propulsion

systems and to resupply materials for life-support systems. (See fig. C5 on color page B.)

Its vast mineral-resource potential, frozen volatile reservoirs and strategic location will make Mars a critical "extraterrestrial supply depot" for human expansion into the mineral-rich asteroid belts and to the giant outer planets and their fascinating collection of resource-laden moons. Smart explorer robots will assist the first human settlers on the Red Planet, enabling them to quickly and efficiently assess the full resource potential of their new world. As these early settlements mature, they will become economically self-sufficient by exporting propellants, life-support-system consumables, food, raw materials and manufactured products to feed the next wave of human expansion to the outer regions of the Solar System. Trading vessels will also travel between cislunar space and Mars, carrying specialty items to eager consumer markets in both civilizations.

The asteroids, especially Earth-crossing asteroids, represent another interesting source of extraterrestrial materials. Current Earth-based spectroscopic evidence and analysis of meteorites (which scientists believe originate from broken-up asteroids) indicate that carbonaceous (C-class) asteroids may contain up to 10 percent water, 6 percent carbon, significant amounts of sulfur and useful amounts of nitrogen. S-class asteroids, which are common near the inner edge of the main asteroid belt and among the Earth-crossing asteroids, may contain up to 30 percent free metals (alloys of iron, nickel and cobalt, along with high concentrations of precious metals). E-class asteroids may be rich sources of titanium, magnesium, manganese and other metals. Finally, chondritic asteroids, which are found among the Earth-crossing population, are believed

Table 1   Materials Available on the Surface of the Moon

Elements	Percentage by Weight		
	Mare[a]	Highlands[b]	Basin Ejecta[c]
Oxygen	39.7–42.3	44.6	42.2–43.8
Silicon	18.6–21.6	21.0	21.1–22.5
Aluminum	5.5–8.2	12.2–14.4	9.2–10.9
Iron	12.0–15.4	4.0–5.7	6.7–10.4
Calcium	7.0–8.7	10.1–11.3	6.3–9.2
Magnesium	5.0–6.8	3.5–5.6	5.7–6.3
Titanium	1.3–5.7	0.3	0.8–1.0
Chromium	0.2–0.4	0.1	0.2
Sodium	0.2–0.4	0.3–0.4	0.3–0.5
Manganese	0.2	0.1	0.1
Potassium	0.06–0.22	0.07–0.09	0.13–0.46
Hydrogen, carbon, nitrogen, fluorine, zirconium, nickel		100 ppm[d]	
Zinc, lead, chlorine, sulfur, other volatiles		5–100 ppm	

[a]Mare = relatively smooth, dark areas of lunar surface.
[b]Highlands = densely cratered, rugged uplands.
[c]Basin Ejecta = materials ejected out of impact crater basins.
[d]ppm = parts per million.

to contain accessible amounts of nickel, perhaps more concentrated than the richest deposits found on Earth.

Using smart machines, including self-replicating systems, space settlers in the next century will be able to manipulate large quantities of extraterrestrial matter and move it about to wherever it is needed in the Solar System. Some of these materials might even be refined en route, with the waste slag being used as a reaction mass in some advanced propulsion system. Many of these extraterrestrial resources will be used as the feedstock for space industries that will form the basis of interplanetary trade and commerce. For example, atmospheric ("aerostat") mining stations could be set up around Jupiter and Saturn, extracting such materials as hydrogen and helium—especially helium-3, an isotope of great value in nuclear-fusion research. Similarly, Venus could be mined for carbon dioxide in its atmosphere; Europa for water; and Titan for hydrocarbons. Large fleets of robot spacecraft could mine the Saturnian ring system for water ice while a sister fleet of robot vehicles extracts metals from the main asteroid belt. Even comets might be intercepted and mined for their frozen volatiles.

So the next time you gaze up at the night sky, don't think of space as a desolate, empty place; think of it instead as a cosmic frontier, rich in energy and material resources—ready to be harvested through human ingenuity and advances in space technology.

See also: **asteroid; comet; Jupiter; Mars; Moon; Saturn; space industrialization; Venus**

extravehicular activity (EVA) A unique role that people play in the U.S. space program began on June 3, 1965, when Astronaut Edward H. White II left the protective environment of his *Gemini IV* spacecraft cabin and ventured into deep space. (See fig. C6 on color page C.) His mission—to perform a special set of procedures in a new and hostile environment—marked the start of the unique form of space technology called "extravehicular activity" or "EVA" for short. EVA may be defined as the activities conducted by an astronaut outside the protective environment of his or her space capsule, aerospace vehicle or space station. With respect to the Space Transportation System or Space Shuttle, EVA is identified as the activities performed by the Shuttle astronauts outside the pressure hull or within the Orbiter payload bay when the payload bay doors are open. EVA is an optional, payload-related STS service.

Gemini mission proved that EVA was a viable technique for performing orbital mission operations outside the spacecraft crew compartment. Then, as Gemini evolved into the Apollo Program, and Apollo into Skylab, EVA mission objectives pushed the science and art of extravehicular activity to their limit. New, more sophisticated concepts and methods were perfected, extending the capability to obtain scientific, technical and economic return from the space environment. Skylab also demonstrated the application of EVA techniques to unscheduled mainte-

nance and repair operations—salvaging the program and inspiring its participants to new heights of aerospace accomplishment. Because of this success and usefulness, EVA capability has been incorporated into the Space Shuttle program and will also be included as an integral part of future space stations and space construction activities.

The term EVA, as applied to the Space Shuttle, includes all activities for which crewmembers don their space suits and life support systems, and then exit the Orbiter cabin into the vacuum of space to perform operations internal or external to the payload bay volume.

Shuttle-generic EVA can be divided into three basic categories: (1) Planned—that is, the EVA was planned prior to launch in order to complete a mission objective; (2) Unscheduled—that is, an EVA was not planned, but is required to achieve successful payload operation or to support overall mission accomplishments; and (3) Contingency—that is, EVA is required to effect the safe return of all crewmembers.

Each Orbiter mission provides the equipment and consumables needed for three two-person EVA operations, each lasting six hours nominally. Two of the EVAs are available for payload operations and the third is retained for Orbiter contingency EVA. Additional EVAs may be added with the provision of more consumables and equipment.

The extravehicular mobility unit (EMU) consists of a self-contained (that is, having no umbilical cords) life-support system and an anthropomorphic pressure garment with thermal and micrometeoroid protection. It provides a breathing environment and incorporates provisions for internal cooling, communications equipment, special EVA helmet visor protection, crew comfort devices and external restraint and tethering fittings. The unit and the associated life-supporting consumables provide for a six-hour nominal EVA with a subsequent recharge capability for additional EVAs.

Handrails and handholds are provided to assist astronaut "microgravity" movement within the Orbiter's flight and mid-decks, in and on the airlock and around the periphery of the cargo bay.

The Space Shuttle Orbiter also provides ultrahigh-frequency (UHF) duplex communications from the crewmembers in the flight deck to the crewmembers performing EVA operations and between the latter. S-band and Ku-band channels are used for space-to-ground communications between all Orbiter crewmembers and the appropriate ground control centers and for payload data transmission to the ground. Both voice and data can be relayed to the ground by the Tracking and Data Relay Satellite System (TDRSS).

To ensure maximum EVA capabilities, the remote manipulator system (RMS) and tools, restraints and ancillary equipment are available as support equipment for each Shuttle mission. The RMS consists of a large external arm that, in conjunction with certain payload supporting equipment, is capable of deploying, retrieving and operating on payloads weighing as much as 29,500 kilograms (65,000 pounds). The system is an electromechanical device, 15.2 meters (50 feet) long, having shoulder, elbow and wrist joints along with an end effector. The manipulator is controlled from the aft crew station on the flight deck through direct vision complemented by a closed circuit television system with cameras mounted on the manipulator arm and in the payload bay.

When used to assist the EVA crewmembers, the remote manipulator system may perform one or more of the following useful functions:

(1) it can perform multiple transfers of equipment between the EVA work area and the replacement equipment stowage area.

(2) using a handrail on the RMS end effector, an EVA astronaut can use the RMS as a translation path to remote areas on the payload or the Orbiter vehicle itself

(3) the attached lights can be used to supply additional lighting at the work area, and the attached closed circuit TV can aid in payload inspection tasks and in task coordination with the other Orbiter crewmembers and with personnel on the ground.

To perform equipment maintenance, repair and replacement, or assembly operations on orbit, the EVA crewmember requires certain tools, tethers, restraints and portable workstations. The portable workstation serves as the crewmember's restraining platform while he or she performs EVA tasks and provides foot restraints, stowage for tools, tethers, a portable light and other support equipment. The workstation is universal in design and may attach directly to the payload, to the Orbiter structure or even to the RMS in supporting various EVA tasks.

The manned maneuvering unit (MMU) will be flown on rescue missions and on other flights where appropriate. The MMU is a propulsive backpack device that gives the EVA crewmember the capability of reaching areas outside the Orbiter's payload bay not otherwise readily accessible. The unit, a modular device readily attached to the extravehicular mobility unit (EMU) and stowed in the cargo bay, may be donned, doffed and serviced by one EVA crewmember for use as required during the nominal six hour EVA period. A six-degree-of-freedom control authority and automatic attitude-hold capability enable the device to contribute significantly to the EVA objectives of a particular Shuttle flight. The MMU may also be used to effect crew transfer in the event of a rescue operation.

Given adequate restraints, working volume and compatible man-machine interfaces, the EVA astronauts can duplicate almost any task designed for human operation on the ground.

The following typical EVA tasks demonstrate the range of EVA opportunities that are available to space technology planners and payload designers in the Shuttle era—which is also the dawn of humanity's extraterrestrial civilization:

(1) inspection, photography, and possible manual over-ride of vehicle and payload systems, mechanisms and components

(2) installation, removal or transfer of film cassettes, material samples, protective covers, instrumentation and launch or entry tie-downs

(3) operation of equipment, including tools, cameras and cleaning devices

(4) cleaning optical surfaces

(5) connection, disconnection and storage of fluid and electrical umbilicals

(6) repair, replacement, calibration and inspection of modular equipment and instrumentation on the spacecraft or payloads

(7) deployment, retraction and repositioning of antennas, booms and solar panels

(8) attachment and release of crew and equipment restraints

(9) performance of experiments

(10) cargo transfer

These EVA applications can demechanize an operational task and thereby reduce design complexity (automation), simplify testing and quality assurance programs, lower manufacturing costs and improve the probability of task success.

Orbiter rescue operations can be performed from the airlock, the docking module or side hatch. Entrance to the rescue vehicle will be through the docking module or airlock hatch. Several techniques for transferring personnel from a disabled Orbiter or damaged space station to a rescuing Orbiter are under study. The primary exit from a disabled Orbiter is through the cabin airlock into the payload bay. One EMU-equipped crewperson would remain inside the cabin of the disabled Orbiter to transfer the personnel rescue systems (PRSs) into the airlock, while the second EMU-equipped crewperson would be stationed in the payload bay to remove the PRSs from the airlock and attach them to the device for transporting the crewpersons of the disabled space vehicle back to the rescue ship. If the airlock path is blocked, rescue will be performed through the disabled Orbiter's side hatch on the cabin mid-deck level. In this particular situation, the disabled Orbiter's cabin would be depressurized for rescue operations. The two EMU-equipped crewpersons from the disabled Orbiter would then assist in preparation and transfer of the personnel rescue systems. For this side hatch rescue scenario, both EMU-equipped crewpersons would be located in the mid-deck area to assist transfer operations. The transfer techniques under consideration use either the remote manipulator system, the RMS with a transfer line or the MMU to accomplish the transfer of stranded astronauts from the disabled Orbiter or space station to the rescue Shuttle vehicle

See also: **space construction; space station; space suit; Space Transportation System**

extravenusian Occurring, located or originating outside of the planet Venus and its atmosphere. The Venusian equivalent to the term *extraterrestrial*. The Soviet Venera Landers, for example, represent extravenusian visitors.

Extreme Ultraviolet Explorer (EUVE) A NASA astrophysics mission planned for 1987–90. Its main objective will be to make the first all-sky map of all detectable extreme-ultraviolet (EUV) sources. These EUV data will then be used by scientists to confirm current theories about stellar evolution. Astrophysicists will also use data from this mission to study the effects of extreme-ultraviolet radiation on the interstellar medium. The EUVE spacecraft will have a total mass of about 1,150 kilograms and will be placed in a 550-kilometer, 28.5-degree-inclination orbit around the Earth by the Space Shuttle.

See also: **astrophysics**

extreme-ultraviolet (EUV) radiation The short-wavelength portion of the ultraviolet (UV) region of the electromagnetic spectrum; usually defined as wavelengths between 10 and 100 nanometers. The EUV portion of the electromagnetic spectrum is considered the transition region, or "bridge," between ultraviolet radiation and X rays. This transition region between X rays and UV radiation is sometimes called the XUV band.

See also: **electromagnetic spectrum; Extreme Ultraviolet Explorer**

Far Ultraviolet Spectroscopy Explorer (FUSE) The Far Ultraviolet Spectroscopy Explorer will be a one-meter-class orbiting telescope, equipped for very-high-resolution spectroscopy in the 90- to 120-nanometer wavelength region of the electromagnetic (EM) spectrum. The objectives of this mission are to conduct far-ultraviolet (FUV) spectroscopy of distant stars, galaxies and interstellar matter. Information on many spectral features that lie in the 90- to 120-nanometer band is essential to an understanding of interstellar gas, extended stellar atmospheres, galactic nuclei, supernova remnants and processes in the upper atmospheres of planets within our Solar System. For example, the structure of the interstellar medium depends on the medium's temperature, velocity and location in space. The FUSE will be sensitive to all conditions of the interstellar medium. The mission has a planned launch date of 1995–2000.

See also: **astrophysics**

"faster-than-light" travel The ability to travel faster than the known physical laws of the Universe will permit.

In accordance with Einstein's theory of relativity, the speed of light is the ultimate speed that can be reached in the space-time continuum. The speed of light in free space is 299,793 kilometers per second.

Concepts like "hyperspace" have been introduced in science fiction to sneak around this "speed-of-light barrier." Unfortunately, despite popular science-fiction stories to the contrary, most scientists today feel that the speed-of-light limit is a real physical law that isn't likely to change.

See also: **hyperspace; interstellar travel; relativity; tachyon**

Fermi paradox—"Where are they?" The dictionary defines the word *paradox* as an apparently contradictory statement that may nevertheless be true. According to the lore of physics, the famous Fermi paradox arose one evening in 1943 during a social gathering at Los Alamos when the brilliant Italian-American physicist Enrico Fermi asked the penetrating question: "Where are they?" "Where are who?" his startled companions replied. "Why, the extraterrestrials," responded the Nobel prize-winning physicist, who was at the time one of the lead scientists on the top-secret Manhattan Project.

Fermi's line of reasoning that led to this famous inquiry has helped form the basis of much modern thinking and strategy concerning the search for extraterrestrial intelligence (SETI). It can perhaps be summarized best as follows. Our Galaxy is some 10 billion to 15 billion years old and contains perhaps 100 billion stars. If just *one* advanced civilization had arisen in this period of time and attained the technology necessary to travel between the stars, within 50 million to 100 million years, that advanced civilization could have diffused through or swept across the entire Galaxy—leaping from star to star, starting up other civilizations and spreading intelligent life everywhere. But as we look around, we don't see a Universe teeming with intelligent life, nor do we have any technically credible evidence of visitations or contact with alien civilizations, so we must perhaps conclude that no such civilization has ever arisen in the 15-billion-year history of the Galaxy. Therefore, the paradox: While we might expect to see signs of a Universe filled with intelligent life (on the basis of statistics and the number of possible "life sites," given the existence of 100 billion stars in just this Galaxy alone), we have seen no evidence of such. Are we, then, really alone? If we're not alone—where are they?

Many attempts have been made to respond to Fermi's very profound question. The "pessimists" reply that the reason we haven't seen any signs of intelligent extraterrestrial civilizations is that we really are alone. Maybe we are the first technically intelligent beings to rise to the level of space travel. Perhaps it is our cosmic destiny to be the first species to sweep through the Galaxy spreading intelligent life!

The "optimists," on the other hand, hypothesize that intelligent life exists out there somewhere and offer a variety of possible reasons why we haven't "seen" these civilizations yet. We'll discuss just a few of these proposed reasons here. First, perhaps intelligent alien civilizations really do not want anything to do with us. As a planet we may be too belligerent, too intellectually backward or simply below their communications horizon. Other optimists suggest that not every intelligent civilization has the desire to travel between the stars, or maybe they do not even desire to communicate by means of electromagnetic signals. Yet another response to the intriguing Fermi paradox is that *we* are actually *they*—the descendants of ancient astronauts who visited the Earth millions of years ago when a wave of galactic expansion passed through this part of the Galaxy.

Still another group responds to Fermi's question by declaring that intelligent aliens are out there right now but that they are keeping a safe distance, watching us either mature as a planetary civilization or destroy ourselves. A subset of this response is the extraterrestrial zoo hypothesis, which speculates that we are being kept as a "zoo" or wildlife preserve by advanced alien zookeepers who have elected to monitor our activities but not be detected themselves.

Finally, other people respond to the Fermi paradox by saying that the wave of cosmic expansion has not yet reached our section of the Galaxy—so we should keep looking! Within this response group are those who declare that the alien visitors are just now arriving among us!

If you were asked "Where are they?" by Enrico Fermi, just how would you respond?

See also: **ancient astronauts; Drake equation; extraterrestrial civilizations; search for extraterrestrial intelligence; unidentified flying object; Zoo hypothesis**

fission (nuclear)   In nuclear fission, the nucleus of a heavy element, such as uranium or plutonium, is bombarded by a neutron, which it absorbs. The resulting compound nucleus is unstable and soon breaks apart, or fissions, forming two lighter nuclei (called fission products) and releasing additional neutrons. In a properly designed nuclear reactor, these fission neutrons are used to sustain the fission process in a controlled chain reaction. The nuclear-fission process is accompanied by the release of a large amount of energy, typically 200 million electron volts per reaction. Much of this energy appears as the kinetic (or motion) energy of the fission-product nuclei, which is then converted to thermal energy (or heat) as the fission products slow down in the reactor fuel material. This thermal energy is removed from the reactor core and used to generate electricity or as process heat.

Energy is released during the nuclear-fission process because the total mass of the fission products and neutrons after the reaction is less than the total mass of the original neutron and the heavy nucleus that absorbed it. From Einstein's famous mass-energy equivalence relationship, $E = m c^2$, the energy released is equal to the tiny amount of mass that has disappeared multiplied by the square of the speed of light.

Nuclear fission can occur spontaneously in heavy elements but is usually caused when these nuclei absorb neutrons. In some circumstances, nuclear fission may also be induced by very energetic gamma rays (in a process called photofission) and by extremely energetic (GeV-class—that is, billion-electron-volt-class)—charged particles.

The most important fissionable (or fissile) materials are uranium–235, uranium–233 and plutonium–239.

See also: **space nuclear power; space nuclear propulsion**

fusion   In nuclear fusion, lighter atomic nuclei are joined together, or fused, to form a heavier nucleus. For example, the fusion of deuterium with tritium results in the formation of a helium nucleus and a neutron. Since the total mass of the fusion products is less than the total mass of the reactants (that is, the original deuterium and tritium nuclei), a tiny amount of mass has disappeared, and the equivalent amount of energy is released in accordance with Einstein's mass-energy equivalence formula:

$$E = m c^2$$

This fusion energy then appears as the kinetic (motion) energy of the reaction products. When isotopes of elements lighter than iron fuse together, some energy is liberated. Energy must be added to any fusion reaction involving elements heavier than iron.

The Sun is our oldest source of energy, the very mainstay of all terrestrial life. The energy of the Sun and other stars comes from thermonuclear-fusion reactions. Fusion reactions brought about by means of very high temperatures are called thermonuclear reactions. The actual temperature required to join, or fuse, two atomic nuclei depends on the nuclei and the particular fusion reaction involved. (Remember, the two nuclei being joined must have enough energy to overcome Coulombic, or "like-electric-charge," repulsion.) In stellar interiors, fusion occurs at temperatures of tens of millions of degrees Kelvin. When we try to develop useful controlled nuclear reactions (CTR) here on Earth, reaction temperatures of 50 million to 100 million degrees Kelvin are considered necessary.

Table 1 describes the major single-step thermonuclear reactions that are potentially useful in controlled fusion reactions for power and propulsion applications. Large space settlements and human-made "miniplanets" could eventually be powered by such CTR processes, while robot interstellar probes and giant space arks would use fusion for both power and propulsion. Helium-3 is a rare isotope of helium. Some space visionaries have already proposed mining the Jovian or Saturnian atmospheres for helium-3 to fuel our first interstellar probes.

At present, there are immense technical difficulties preventing our effective use of controlled fusion as a terrestrial or space energy source. The key problem is that

Table 1 Single-Step Fusion Reactions Useful in Power and Propulsion Systems

Nomenclature	Thermonuclear Reaction	Energy Released Per Reaction (MeV) [Q Value]	Threshold Plasma Temperature (keV)[a]
(D-T)	$^2_1D + ^3_1T \rightarrow ^4_2He + ^1_0n$	17.6	10
(D-D)	$^2_1D + ^2_1D \rightarrow ^3_2He + ^1_0n$	3.2	50
	$^2_1D + ^2_1D \rightarrow ^3_1T + ^1_1p$	4.0	50
(D-^{3}He)	$^2_1D + ^3_2He \rightarrow ^4_2He + ^1_1p$	18.3	100
(^{11}B-p)	$^{11}_5B + ^1_1p \rightarrow 3(^4_2He)$	8.7	300

where
    D is deuterium
    T is tritium
    He is helium
    n is neutron
    p is proton
    B is boron

[a]10 keV = 100 million degrees Kelvin.

the fusion gas mixture must be heated to tens of millions of degrees Kelvin and held together for a long enough period of time for the fusion reaction to occur. For example, a deuterium-tritium gas mixture must be heated to at least 50 million degrees Kelvin—and this is considered the easiest controlled fusion reaction to achieve! At 50 million degrees, any physical material used to confine these fusion gases would disintegrate, and the vaporized wall materials would then "cool" the fusion gas mixture, quenching the reaction.

There are three general approaches to confining these hot fusion gases, or plasmas: gravitational confinement, magnetic confinement and inertial confinement.

Because of their large masses, the Sun and other stars are able to hold the reacting fusion gases together by gravitational confinement. Interior temperatures in stars reach tens of millions of degrees Kelvin and use complete thermonuclear-fusion cycles to generate their vast quantities of energy. For main-sequence stars like or cooler than our Sun (about 10 million degrees Kelvin), the proton-proton cycle, shown in table 2, is believed to be the principal energy-liberating mechanism. The overall effect of the proton-proton stellar fusion cycle is the conversion of hydrogen into helium. Stars hotter than our Sun (those with interior temperatures of 16 million degrees Kelvin and higher) release energy through the carbon cycle, shown in table 3. The overall effect of this cycle is again the conversion of hydrogen into helium, but this time with carbon (carbon–12 isotope) serving as a catalyst.

Terrestrial scientists attempt to achieve controlled fusion through two techniques: magnetic-confinement fusion (MCF) and inertial-confinement fusion (ICF). In magnetic confinement strong magnetic fields are employed to "bottle up," or hold, the intensely hot plasmas

Table 2 Main Thermonuclear Reactions in the Proton-Proton Cycle

$^1_1H + ^1_1H \rightarrow ^2_1D + e^+ + \nu$

$^2_1D + ^1_1H \rightarrow ^3_2He + \gamma$

$^3_2He + ^3_2He \rightarrow 2(^4_2He) + 2(^1_1H)$

where
    ^1_1H is a hydrogen nucleus (that is, a proton, ^1_1p)

    ^2_1D is deuterium (an isotope of hydrogen)

    ^3_2He is helium–3 (a rare isotope of helium)

    ^4_2He is the main (stable) isotope of helium

    ν is a neutrino
    e^+ is a positron
    γ is a gamma ray

Table 3 Major Thermonuclear Reactions in the Carbon Cycle

$^{12}_6C + ^1_1H \rightarrow ^{13}_7N + \gamma$

$^{13}_7N \rightarrow ^{13}_6C + e^+ + \nu$ (radioactive decay)

$^1_1H + ^{13}_6C \rightarrow ^{14}_7N + \gamma$

$^1_1H + ^{14}_7N \rightarrow ^{15}_8O + \gamma$

$^{15}_8O \rightarrow ^{15}_7N + e^+ + \nu$ (radioactive decay)

$^1_1H + ^{15}_7N \rightarrow ^{12}_6C + ^4_2He$

where
    γ is a gamma ray
    ν is a neutrino
    e^+ is a positron

needed to make the various single-step fusion reactions occur (again, review table 1). In the inertial-confinement approach, pulses of laser light, energetic electrons or heavy ions are used to very rapidly compress and heat small spherical targets of fusion material. This rapid compression and heating of an ICF target allows the conditions supporting fusion to be reached in the interior of the pellet—before it blows itself apart.

Figure 1 shows the basic concept behind laser-fusion power generation. A very small sphere (less than one millimeter in diameter) containing a deuterium-tritium (D-T) mixture is uniformly illuminated with a short pulse of laser light. This rapid illumination compresses the sphere, heating it very quickly. The laser pulse is only about one nanosecond (10^{-9} second long.) Photons strike the D-T fuel pellet and are absorbed by its outer layer. The absorption of this laser energy causes the outer layer of the pellet to vaporize and blow away as a hot plasma (ionized gas), creating an inward force (from Newton's action-reaction principle) in the rest of the fuel pellet. This inward force compresses the fuel in the center of the pellet, heats it up to the fusion reaction temperature and holds it together by inertia just long enough for some useful fusion energy output to occur. The D-T fuel then undergoes fusion burn, ejecting energetic neutrons and helium nuclei over a 10- to 20-picosecond time interval. (A picosecond is 10^{-12} second.)

Figure 2 is a schematic diagram of a laser-fusion power plant that might be used on Earth or in a large space settlement. In order to generate useful power, the energy released by the fusion reactions must be converted to thermal energy (heat) to run conventional thermodynamic power cycles (for example, a Rankine cycle steam plant). The laser system must efficiently convert electricity into laser light, which can then be focused on inexpensive, yet

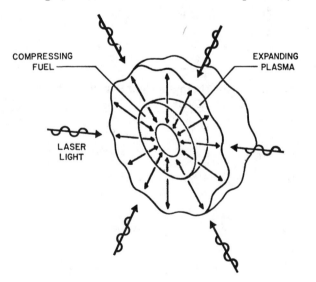

Fig. 1   Inertially confined fusion, using laser pulses to compress and heat a deuterium-tritium (D-T) pellet. (Drawing courtesy of Los Alamos National Laboratory.)

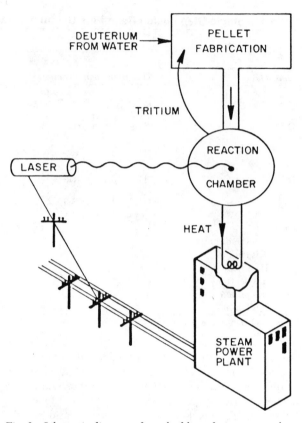

Fig. 2   Schematic diagram of a pulsed laser-fusion power plant. (Drawing courtesy of Los Alamos National Laboratory.)

precisely manufactured, D-T fuel pellets. (The manufacture of such fusion target pellets might even become a major space-based industry in the next century.) The interaction between the laser beam pulses and the fusion pellets would take place in a special reaction chamber, designed to capture all the fusion-reaction energy, producing more tritium for use in fuel pellets and gathering heat for conversion into electricity.

Additional tritium for these fusion power plants could be made available as part of the heat-generation process, through the interaction of fusion-released neutrons with lithium, as shown in table 4. Each laser pulse could contain perhaps one megajoule (one million joules) of energy, or about as much energy as is necessary to operate your color television set for one hour. When developed, laser-fusion power systems of this type should be able to burn somewhere between 10 and 50 fusion pellets per second, and each pellet burn should release about 100 times the input laser energy.

Although there are still many difficult technical issues

Table 4   Tritium Breeding Cycles

$$^{7}_{3}\text{Li} + ^{1}_{0}\text{n (fast)} \rightarrow ^{3}_{1}\text{T} + ^{4}_{2}\text{He} + ^{1}_{0}\text{n}$$

$$^{6}_{3}\text{Li} + ^{1}_{0}\text{n (slow)} \rightarrow ^{3}_{1}\text{T} + ^{4}_{2}\text{He} + 4.8 \text{ MeV}$$

to be resolved before we can achieve controlled fusion, it promises to provide a limitless terrestrial energy supply. Of course, fusion also represents the energy key to the full use of the resources of our Solar System and, possibly, to travel across the interstellar void.

See also: **interstellar travel; stars; Sun**

g   The symbol used for the acceleration due to gravity. For example, at the Earth's surface, g equals 9.8 meters per second squared.

See also: **g-force**

Gaia hypothesis   The hypothesis proposed by James Lovelock and Lynn Margulis that the Earth's biosphere has an important modulating effect on the terrestrial atmosphere. Because of the chemical complexity observed in the lower atmosphere, they suggested that life-forms within the terrestrial biosphere actually help control the chemical composition of the Earth's atmosphere—thereby ensuring the continuation of conditions suitable for life. Gas-exchanging microorganisms, for example, are thought to play a key role in this continuous process of environmental regulation. Without these "cooperative" interactions, in which some organisms generate certain gases and carbon compounds that are subsequently removed and used by other organisms, the planet Earth might also possess an excessively hot or cold planetary surface, devoid of liquid water and surrounded by an inanimate, carbon dioxide-rich atmosphere.

Gaia was the goddess of Earth in ancient Greek mythology. Lovelock and Margulis used her name to represent the terrestrial biosphere—namely, the system of life on Earth, including living organisms and their required liquids, gases and solids. Thus, their hypothesis simply states that "Gaia" (the Earth's biosphere) will struggle to maintain the atmospheric conditions suitable for the survival of terrestrial life.

If we use the Gaia hypothesis in our search for extraterrestrial life, we should look for alien worlds that exhibit variability in atmospheric composition. Extending this hypothesis beyond the terrestrial biosphere, a planet will either be living or else it will not! The absence of chemical interactions in the lower atmosphere of an alien world could be taken as an indication of the absence of living organisms.

While this interesting hypothesis is perhaps more metaphor than hard, scientifically verifiable fact, it is still quite useful in developing a sense of appreciation for the complex chemical interactions that have helped to sustain life in the Earth's biosphere. These interactions among microorganisms, higher-level animals and their mutually shared atmosphere might also have to be carefully considered in the successful development of effective closed life-support systems for use on permanent space stations, lunar bases and planetary settlements.

galactic   Of or pertaining to a galaxy, such as the Milky Way Galaxy.

galactic cluster   Collections or groups of galaxies. Galactic clusters may contain anywhere from a few galaxies up to several thousand member galaxies. Our Milky Way Galaxy is a member of a small, irregular galactic cluster called the Local Group.

See also: **galaxy; Local Group**

galactic cosmic rays   Extremely energetic atomic particles that originate outside our Solar System. These particles consist of approximately 85 percent protons (bare hydrogen nuclei), 14 percent alpha particles (bare helium nuclei) and 1 percent heavier nuclei. In addition, energetic electrons and positrons make up about 1 percent of the proton flux. Galactic cosmic rays can reach energies up to 10^{20} electron volts (eV) and appear to uniformly fill the interstellar medium.

See also: **cosmic rays**

galaxy   A very large accumulation of from 10^6 to 10^{12} stars. By convention, when the word is capitalized (Galaxy), it refers to a particular collection of stars, such as the Milky Way Galaxy. The existence of galaxies beyond our own Milky Way Galaxy was not firmly established by astronomers until 1924.

Galaxies—or "island universes," as they are sometimes called—come in a variety of shapes and sizes. They range from dwarf galaxies, like the Magellanic Clouds, to giant spiral galaxies, like the Andromeda Galaxy (see fig. C1 on color page A). Astronomers usually classify galaxies as either elliptical, spiral (or barred spiral) or irregular.

When we talk about galaxies, the scale of distances is truly immense. Galaxies themselves are tens to hundreds of thousands of light-years across; the distance between galaxies is generally a few million light-years! For example, the beautiful Andromeda Galaxy is approximately 20,000 light-years across and about 2 million light-years away.

See also: **Andromeda Galaxy; Magellanic Clouds; Milky Way Galaxy; stars**

Galilean satellites   Galileo Galilei, originator of the modern scientific method, shocked the world in 1610 when he announced the discovery of four satellites—Io, Europa, Ganymede and Callisto—orbiting the planet Jupiter (see fig. 1). His discovery provided proof of the heliocentric theory of the Solar System proposed by Nicolaus Copernicus—namely, that the Earth and other planets actually orbit the Sun and that the Earth is *not* the center of the Universe, as was thought up to that time.

Galileo told the story of his extraterrestrial discovery as follows:

> On the seventh day of January in the present year 1610, at the first hour of the night, when I was viewing the heavenly bodies with a telescope, Jupiter presented itself to me; and because I had prepared a very excellent instrument for myself, I perceived (as I had not before, on account of the weakness of my previous instrument) that beside the planet there were three starlets, small indeed, but very bright.

Galileo thought those "starlets" were just more of the fixed stars that his telescopes were allowing him to discover with astounding regularity. But the next night he saw they had changed position. The night after that was cloudy. Then on January 10 he saw only two "starlets," the third having disappeared behind Jupiter. On January 11 he reported:

> I had now decided beyond all question that there existed in the heavens three stars wandering around Jupiter as do Mercury and Venus around the Sun.

On January 13, 1610, Galileo spotted the fourth major Jovian moon. Though Galileo nearly paid for his observations and later writings with his life, he remained the most respected scientist of his time.

The "starlets" that Galileo saw are the four largest satellites of Jupiter, now known in his honor as the Galilean satellites. These moons are named Io, Europa, Ganymede and Callisto, after four of Jupiter's many lovers in Greek and Roman mythology.

Fig. 1   The Galilean satellites and their relative distances from the planet Jupiter. (Drawing courtesy of NASA.)

Figure C7 on color page C, shows Jupiter and its four planet-sized Galilean moons, as photographed in March 1979 by the *Voyager 1* spacecraft. The mosaic image is *not to scale*, but the satellites do appear in their relative positions. Innermost is Io; next is Europa; followed by the largest moon, Ganymede; and finally, outermost of the four, Callisto. There are 12 other, much smaller satellites now known to orbit Jupiter. Four lie within the orbit of Io, while the remaining 8 have orbits far outside that of Callisto.

The Galilean moons do not closely resemble any other celestial bodies in the Solar System and do not even resemble each other. Volcanic Io and icy Europa appear to have the youngest surfaces; Callisto, the oldest. Ganymede and Callisto may be as much as 50 percent water by weight; Io is waterless; while Europa may contain a giant

liquid water ocean beneath its icy crust. Io is the most volcanic body yet discovered in the Solar System; and Europa, the one with the smoothest surface. Scientists now speculate that if Europa does indeed possess this liquid water ocean, the Europan ocean itself may contain extraterrestrial life-forms!

See also: **Jupiter;** *Pioneer 10, 11;* **Voyager**

Galileo Project   An approved NASA planetary-exploration mission to orbit Jupiter and to send an instrumented probe into the giant planet's dense atmosphere. The project is named after the famous Italian astronomer Galileo Galilei (1564–1642), who in 1610 discovered the four major satellites of Jupiter: Io, Europa, Ganymede and Callisto. These four Jovian moons are also called the "Galilean satellites" in his honor. Galileo also helped confirm the theory that the Sun indeed was the center of our Solar System—a scientific view that almost cost him his life in 1633. Because of his pioneering astronomical work, Galileo became engaged in a bitter conflict with religious officials of the day, who supported the Ptolemaic (Earth-centered) concept of the Universe. Forced to renounce his belief in the Copernican (heliocentric, or Sun-centered) model of the Solar System, legend has it that the 70-year-old Galileo defiantly mumbled, while rising from his knees before his persecutors, "And yet it [the Earth] moves!"

The Galileo Project will enable space scientists to study, at close range and for almost two years, Jupiter, its satellites and its magnetospheric domain. Today, astronomers believe that Jupiter is composed of the original material from which stars form. This material, mainly hydrogen and helium, is largely unmodified by thermonuclear processes, since Jupiter—an "almost-star"—isn't supporting the nuclear-fusion burning found in stellar interiors. Close-up studies of the giant planet should therefore provide interesting information about the beginning and development of the Solar System as well as new insights into phenomena that directly relate to our understanding of all the planets.

This mission has as its main scientific objectives the study of the chemical composition and physical state of Jupiter's atmosphere, an examination of the chemical composition and physical state of selected Jovian satellites and an investigation of the structure and dynamics of Jupiter's magnetosphere.

The Galileo Project is scheduled for launch in May 1986, as the scientific successor to the spectacular Voyager program. The *Galileo* spacecraft will be placed in Earth orbit by the Space Shuttle and then sent toward Jupiter by a modified Centaur upper stage. About 150 days before the *Galileo* spacecraft's arrival at Jupiter (sometime in August 1988), the Jovian atmospheric probe will separate from the planetary orbiter. They will then follow independent flight paths to the giant planet. A few hours before probe entry into the Jovian atmosphere, the orbiter spacecraft will fly within 1,000 kilometers of the volcanic

Galilean satellite Io, making close-up observations of this interesting celestial body. Io's gravity will also help slow the orbiter, reducing the expenditure of propulsive energy required for capture by Jupiter itself.

Then, about the time the orbiting *Galileo* spacecraft is at its point of closest approach to Jupiter, the probe will arrive at the upper edges of the Jovian atmosphere and begin to penetrate the planet's frigid, swirling clouds.

Data from the probe instruments will be relayed back to Earth through the orbiter spacecraft, which will be passing by above the frozen clouds. As the probe strikes the upper layers of the Jovian atmosphere, traveling at approximately 180,000 kilometers an hour, it will slow down so rapidly that it will experience the effects of 400 times the Earth's surface gravity—that is, it will undergo a 400-g deceleration force. Once the strongest of these deceleration forces has passed, the probe will deploy a parachute, and the descent module will begin to take atmospheric measurements. Scientists want the instrument-laden probe to enter Jupiter's light-colored equatorial zone. They think the uppermost clouds of this portion of the Jovian atmosphere consist primarily of ammonia (NH_3). By entering at that location, the probe should be able to measure all the important cloud layers during its one-hour operation. Forty minutes after entry, it is expected that the probe will have descended below what are believed to be Jupiter's lowest water clouds. Then, after some 60 minutes of penetration, data relay from the probe to the orbiting *Galileo* spacecraft will most probably cease. At that point, the probe will have descended to depths where the surrounding atmospheric pressures correspond to some 15 to 20 Earth atmospheres, and it will be crushed, heated and eventually vaporized.

Unlike its short-lived probe, the orbiting *Galileo* spacecraft will enjoy a 20-month mission life around the giant planet—performing numerous close encounters with the Galilean moons. The orbiter spacecraft incorporates a new dual-spin design. Part of the spacecraft is three-axis stabilized so that the camera and other remote sensing instruments can be accurately pointed. The main portion of the spacecraft spins, allowing its instruments to continuously sweep across space to make their scientific measurements. The *Galileo* spacecraft will use a 4.8 meter diameter furlable antenna. Since Jupiter is too far from the Sun for solar cells to provide enough electric power, the *Galileo* spacecraft will use nuclear energy sources called radioisotope thermoelectric generators (RTGs), similar to the nuclear power supplies flown on the two highly successful *Voyager* spacecraft.

See also: **Galilean satellites; Jupiter; space nuclear power; Voyager**

Gamma Ray Observatory (GRO) A major NASA astronomy mission to be flown in the late 1980s. This spacecraft will carry instruments designed to detect gamma rays over an extensive range of energies. The GRO will be an extremely powerful tool for investigating some of the most puzzling mysteries in the extraterrestrial environment, including pulsars, quasars and active galaxies.

The GRO has adopted the following scientific objectives: (1) the study of discrete celestial objects, (2) a search for evidence of nucleosynthesis, (3) a gamma-ray survey of our Galaxy, (4) the study of other galaxies, (5) the observation of gamma-ray bursts and (6) a search for possible primordial black-hole emissions.

As presently planned, the Gamma Ray Observatory will carry four major instruments: a gamma-ray scintillation spectrometer, an imaging Compton telescope, an energetic gamma-ray telescope and a gamma-ray-burst and transient-source experiment. To achieve its scientific objectives, the GRO will be designed to conduct a survey of the "gamma-ray sky" over energy ranges extending from the upper end of existing X-ray observations (approximately tens of kiloelectron volts [keV] to the highest practical energies in the tens of GeV range. (A GeV represents a billion [10^9] electron volts.)

The GRO will be launched by the Space Shuttle into a 400 kilometer altitude circular orbit at 28.5 degrees inclination and have an observation period of two years.

See also: **astrophysics**

gamma rays [symbol: γ] High-energy, very-short-wavelength electromagnetic radiation. Gamma-ray photons are similar to X rays, except that they are usually more energetic and originate from processes and transitions within the atomic nucleus. The processes associated with gamma-ray emissions in astrophysical phenomena include: (1) the decay of radioactive nuclei, (2) cosmic-ray interactions, (3) curvature radiation in extremely strong magnetic fields and (4) matter-antimatter annihilation. Gamma rays are very penetrating and are best stopped or shielded against by dense materials, such as lead, tungsten or depleted uranium.

See also: **astrophysics**

g-force The force of gravity. On the surface of the Earth, for example, the g-force is one g.

See also: **g**

giant planets In our Solar System the large, gaseous outer planets: Jupiter, Saturn, Uranus and Neptune.

globular clusters Compact clusters of up to one million stars (see fig. C7a on color page C). These clusters are subsets, or subunits, of a galaxy and generally orbit around the galactic center.

graviton A quantum, or tiny packet, of gravitational radiation. It is analogous to the photon, which is a quantum of electromagnetic (EM) radiation.

See also: **astrophysics**

gravity The downward attractive force on a mass near a massive celestial body, like the Earth.

gravity anomaly   A region on a celestial body where gravity is lower or higher than expected. If the celestial object is assumed to have a uniform density throughout, then we would expect the gravity on its surface to have the same value everywhere.

See also: **mascon**

greenhouse effect   The general warming of the lower layers of a planet's atmosphere caused by the presence of carbon dioxide (CO_2) and water (H_2O) molecules. As happens on Earth, the greenhouse effect occurs whenever a planetary atmosphere is relatively transparent to visible light from the parent star (typically 0.3 to 0.7 micrometer wavelength) but is opaque to the longer-wavelength (typically 10.6-micrometer) thermal infrared radiation emitted by the planet's surface. Because of carbon dioxide in the atmosphere, this outgoing thermal radiation is blocked from escaping into space, and the absorbed energy then causes a rise in the atmospheric temperature. In the case of Earth, this atmospheric warming effect tends to increase with increasing atmospheric carbon dioxide content. (See fig. 1.)

In fact, some environmental scientists now warn that as we continue to burn vast amounts of fossil fuels, we are also dangerously increasing carbon dioxide levels in the Earth's atmosphere. We may even be creating the conditions for a runaway greenhouse, like that on Venus. Such an effect is a planetary climatic extreme in which all the water has been evaporated away from the surface of a life-bearing or potentially life-bearing planet. Scientists now postulate that Venus's present atmosphere allows some sunlight to reach the planet's surface, but thick clouds and a rich carbon dioxide content prevent surface heat from being radiated back into space. This condition has led to the evaporation of all surface water and produced current surface temperatures of approximately 485 degrees Celsius (900 degrees Fahrenheit)—temperatures hot enough to melt lead.

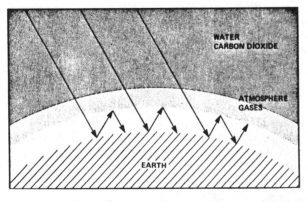

Fig. 1   Greenhouse effect traps solar energy. (Courtesy of NASA.)

"greening of the Galaxy"   A term describing the spreading of human life, technology and culture through interstellar space and eventually across the entire Galaxy.

See also: **extraterrestrial civilizations; humanization of space**

H

half-life   The time it takes for half the atoms of a particular radioactive substance to disintegrate to another nuclear species. Measured half-lives vary from millionths of a second to billions of years. The half-life ($T_{1/2}$) can be expressed mathematically as

$$T_{1/2} = (\ln 2)/\lambda$$

where λ is the decay constant for a particular radionuclide.

hazards to space workers   During a recent study performed by NASA and the U.S. Department of Energy, the issue of space-worker health and safety was addressed in the context of large space-construction operations involving Satellite Power Systems (SPSs). Current experience with human performance in space is mostly for individuals operating in low Earth orbit (LEO). The construction of large space settlements, lunar bases, orbiting factories and Satellite Power Systems would require human activities throughout cislunar space (the area between the Earth and the Moon). The maximum continuous time spent by humans in space is just a few hundred days (with Soviet cosmonauts holding the duration record), and people who have experienced spaceflight are generally a small number of highly trained and highly motivated individuals.

Medical and occupational experiments performed in space and operational life-support and monitoring systems used in manned spaceflight have been evaluated in great detail. These evaluations and analyses have been augmented by data obtained from experiments performed under simulated space conditions on Earth.

The available technical data base, although limited to essentially low-Earth-orbit spaceflight, suggests that with suitable protection, people can live and work in space safely and enjoy good health after returning to Earth. Data from the 84-day *Skylab 4* mission and several long-duration Soviet Salyut missions are especially pertinent to the question of the ability of relatively large numbers of people to live and work in space for 90 to 180 days at a time on permanent space stations.

Some of the major cause-effect factors related to space-worker health and safety are shown in figure 1. Many of these factors require "scaling up" from current medical, safety and occupational analyses to achieve the space technologies necessary to accommodate large groups of space travelers and permanent habitats. Some of these health and safety issues include: (1) preventing launch-abort, spaceflight and space-construction accidents; (2) preventing failures of life-support systems; (3) protecting space vehicles and habitats from collisions with space debris and meteoroids; and (4) providing habitats and good-quality living conditions that minimize psychological stress.

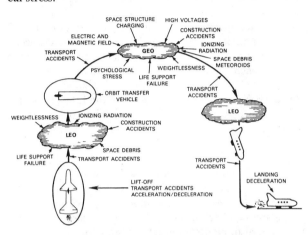

Fig. 1   Factors related to the health and safety of space workers. (Drawing courtesy of NASA and the U.S. Department of Energy.)

The biomedical effects of substantial acceleration and deceleration forces when leaving and returning to Earth, living and working in a weightless environment and the exposure to space radiation are the three main factors that must be dealt with if people are to live in cislunar space and eventually populate heliocentric (Sun-centered) space.

Astronauts and cosmonauts have adapted to weightlessness for extended periods of time in space and have experienced maximum acceleration forces equivalent to six times the Earth's gravity (6 g). No acute operational problems, significant physiological deficits or adverse health effects on the cardiovascular or musculoskeletal systems have yet been observed from these experiences.

The U.S. Space Transportation System, or Space Shuttle, can be regarded as a forerunner of more advanced "space-settler" launch vehicles. It has been designed to limit acceleration/deceleration loads to a maximum of 3 g, thereby opening space travel to a larger number of individuals.

Some physiological deviations have been observed in American astronauts and Soviet cosmonauts during and following extended space missions. Most of these observed effects appear to be related to adaption to microgravity conditions, with the affected physiological parameters returning to normal ranges either during the missions or shortly thereafter. No apparent persistent adverse consequences have been observed or reported to date. Neverthe-

less, some of these deviations could become chronic and might have important health consequences if they were experienced during long durations in space or in repeated long-term tours on a space station or at an orbiting construction facility.

The physiological deviations due to zero gravity have, as noted, usually returned to normal within a few days or weeks after return to Earth. Only bone calcium loss appears to require an extended period of recovery after a long-duration space mission.

Strategies are now being developed to overcome these physiological effects of weightlessness. An exercise regimen can be applied, and body fluid shifts can be limited by applying lower-body negative pressure. Antimotion medication is also useful for preventing temporary motion sickness or space sickness. Proper nutrition, with mineral supplements and regular exercise, appear to limit other observed effects. One way around this problem in the long term, of course, is to provide acceptable levels of "artificial" gravity in larger space bases and space settlements. In fact, very large space settlements will most likely offer the inhabitants a wide variety of gravity levels, ranging from microgravity up to normal terrestrial gravity levels. This multiple-gravity-level option will not only make space-settlement life-styles more diverse than on Earth but will also prepare planetary settlers for life on their new worlds or help other space settlers adjust to the "gravitational rigors" of returning to Earth.

The ionizing-radiation environment encountered by workers in cislunar space is characterized by fluxes of electrons, protons, neutrons and atomic nuclei. (See table 1.) In low Earth orbit (LEO), electrons and protons are trapped by the Earth's magnetic fields (forming the Van Allen belts). The amount of ionizing radiation in LEO varies with solar activity. The trapped radiation belts are of concern when space-worker crews transfer from low Earth orbit to geosynchronous orbit (GEO) or to lunar surface bases. In GEO locations solar-particle events represent a major radiation threat to space workers.

Table 1   Components of the Natural Space Radiation Environment

Galactic Cosmic Rays

Typically 85% protons, 13% alpha particles, 2% heavier nuclei
Integrated yearly fluence
  1×10^8 protons/cm² (approximately)
Integrated yearly radiation dose:
  4 to 10 rads (approximately)

Geomagnetically Trapped Radiation

Primarily electrons and protons
Radiation dose depends on orbital altitude
Manned flights below 300 km altitude avoid Van Allen belts

Solar-Particle Events

Occur sporadically; not predictable
Energetic protons and alpha particles
Solar-flare events may last for hours to days
Dose very dependent on orbital altitude and amount of shielding

Throughout cislunar space, space workers are also bombarded by galactic cosmic rays. These are very energetic atomic particles, consisting of protons, helium nuclei and heavy nuclei with an atomic number (Z) greater than two (HZE particles). Shielding, solar-flare warning systems and excellent radiation dosimetry equipment should help prevent any space worker from experiencing ionizing radiation doses in excess of occupational standards established for various extraterrestrial careers.

Space workers and settlers might also experience a variety of psychological disorders, including the solipsism syndrome and the shimanagashi syndrome. The solipsism syndrome is a state of mind in which a person feels that everything is a dream and is not real. It might easily be caused in an environment (such as a small space base) where everything is "artificial," or human-made. The shimanagashi syndrome is a feeling of isolation in which individuals begin to feel left out, even though life may be physically comfortable. Careful design of living quarters and good communication with Earth should relieve or prevent such psychological disorders.

Living and working in space in the next century should present some interesting challenges and may even pose some dangers and hazards. However, the rewards of an extraterrestrial life-style, for certain pioneering individuals, will more than outweigh any such personal risks.

See also: **Earth's trapped radiation belts; people in space; "shimanagashi" syndrome; solipsism syndrome; space settlement; space station**

heavy-lift launch vehicle (HLLV) A proposed launch vehicle (for use 1995–2005) that would be used to transport large masses (for example, 100 to 500 metric tons) of Satellite Power System (SPS) construction materials from the surface of the Earth to low Earth orbit. Figure 1 shows one design concept for this HLLV. It is an advanced-design, horizontal-takeoff (HTO), single-stage-to-orbit (SSTO) vehicle. The winged aerospace vehicle shown is a delta flying wing, consisting of a multicell pressure vessel. The wing contour is a supercritical airfoil with modified leading edge to improve supersonic and hypersonic performance.

See also: **Satellite Power System**

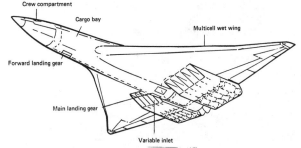

Fig. 1 One version of the heavy-lift launch vehicle, an advanced-design, horizontal-takeoff (HTO), single-stage-to-orbit (SSTO) "space freighter." (Drawing courtesy of the U.S. Department of Energy and NASA.)

heliocentric Relative to the Sun as a center, as in "heliocentric orbit" or "heliocentric space."

heliostat A mirrorlike device arranged to follow the Sun as it moves through the sky and to reflect the Sun's rays on a stationary collector.

Hellas (plural: Hellades) A quantity of information first proposed by the physicist Dr. Philip Morrison. It corresponds to 10^{10} bits of information—more or less the amount of information we know about ancient Greece. In considering interstellar communication with other intelligent civilizations, we would hope to send and receive something on the order of 100 Hellades of information or more at each contact.

hertz (Hz) The unit of frequency in the International or SI System of Units. One hertz is equal to one cycle per second.

$$1 \text{ Hz} = 1 \text{ cps}$$

This unit is often used in multiples, such as:
kilohertz (kHz) for one thousand (10^3) cycles per second
megahertz (MHz) for one million (10^6) cycles per second
gigahertz (GHz) for one billion (10^9) cycles per second.
See also: **electromagnetic spectrum**

HI region A diffuse region of neutral, predominantly atomic, hydrogen in interstellar space. Neutral hydrogen emits radio radiation at 1,420.4 megahertz, corresponding to a wavelength of approximately 21 centimeters. However, the temperature of the region (approximately 100 degrees Kelvin) is too low for optical emission.
See also: **HII region; nebula**

HII region A region in interstellar space consisting mainly of ionized hydrogen and existing mostly in discrete clouds. The ionized hydrogen of HII regions emits radio waves by thermal emissions and recombination-line emission, in comparison to the 21-centimeter radio-wave emission of neutral hydrogen in HI regions.
See also: **HI region; nebula**

hominid A humanlike primate belonging to the family Hominidae. Modern humans (*homo sapiens*) are the only nonextinct species of hominid presently on Earth. Most contemporary biology textbooks state that humans and apes belong to the same order (Primates) and to the same superfamily (Hominoidea). However, humans and apes actually belong to separate families within this superfamily—namely, Hominidae (humans) and Pongidae (apes).

Hubble's law The statement that redshifts of distant, extragalactic objects are directly proportional to their distances (D) from us. This also implies a linear proportionality to the relative velocity of recession (V) from us

Fig. C1   The Andromeda Galaxy (M31). (Courtesy of NASA.)
Fig. C2   The Earth as viewed from the *Apollo 11* spacecraft orbiting the Moon (July 1969). (Courtesy of NASA.)

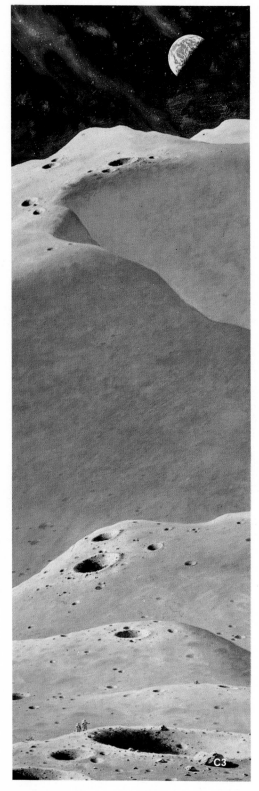

Fig. C3 *Astronauts Explore the Moon*, by artist Pierre Mion. (Courtesy of the NASA Art Program/NASA.)

Fig. C4 *Apollo 11* Astronaut Buzz Aldrin stepping down to the lunar surface on that historic July 20, 1969—the birth of our extraterrestrial civilization! (Courtesy of NASA.)

Fig. C5 Interior view of a large settlement in cislunar space. Many resupply materials for life-support systems and space-based industrial activities will eventually be exported into cislunar space from the Moon. (Courtesy of NASA.)

Fig. C6   Gemini IV Astronaut Edward H. White II walks in space, June 1965. (Courtesy of NASA.)

Fig. C7   A composite picture of Jupiter and the four Galilean satellites (Io, Europa, Ganymede and Callisto) taken by the *Voyager 1* spacecraft, March 1979. Io is in the upper left, Europa in the center, Ganymede front left and Callisto front lower right. Jupiter appears in the upper right background. (Courtesy of NASA/Jet Propulsion Laboratory.)

Fig. C7a   The Great Cluster in Hercules (M13). (Courtesy of NASA and the U.S. Naval Observatory.)

C9

C8

Fig. C8   A close-up view of mighty Jupiter and three of its satellites: Io (against planet), Europa (right) and Callisto (bottom). This image was taken by the *Voyager I* spacecraft when it was approximately 28 million km away from the planet. (Courtesy of NASA/Jet Propulsion Laboratory.)

Fig. C9   A close-up look at the Great Red Spot of Jupiter. (Courtesy of NASA/Jet Propulsion Laboratory.

C10

Fig. C10   A composite image of Ganymede, largest moon in the Solar System, taken by the *Voyager 2* spacecraft, July 1979, from a range of 1.2 million kilometers. (Courtesy of NASA/Jet Propulsion Laboratory.

Fig. C11   A four-picture mosaic of Jupiter's volcanic moon Io, as taken by the *Voyager I* spacecraft from a distance of 377,000 km in March 1979. (Courtesy NASA.)

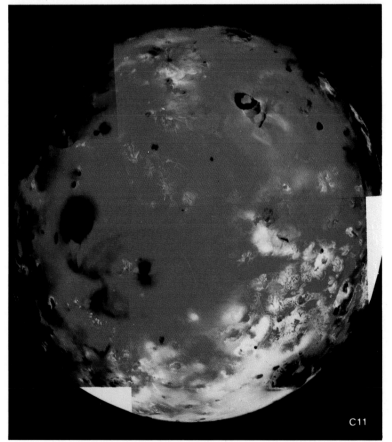

C11

E

Fig. C12   The Red Planet Mars, as imaged by the *Viking Orbiter 1* spacecraft in 1976. (Courtesy of NASA/Jet Propulsion Laboratory.)

Fig. C14   Painting by artist Kevin Davidson of a proposed unmanned, hydrazine-powered prop (propeller-driven) aircraft imaging the bottom of a Martian canyon. (Courtesy of the U.S. Air Force Art Collection.)

F

Fig. C13   The rock-laden red plains of Mars at the *Viking 2* Lander site. (Courtesy of NASA/Jet Propulsion Laboratory.)

Fig. C15   Artist's concept of the Mars Ascent Vehicle carrying its cargo of Martian soil samples for eventual analysis by scientists on Earth. (Courtesy of NASA.)

The *Echo* satellite leaves a "photographic trial" in the Milky Way. (Photograph courtesy of the U.S. Naval Observatory.

C17

Fig. C17 A beautiful full Moon as photographed from the *Apollo 17* spacecraft, 1972.
(Courtesy of NASA.)

(for velocities that are small when compared to the speed of light). The American astronomer Edwin Hubble (1889–1953) first proposed this relationship in 1929. Mathematically, Hubble's law can be expressed as:

$$V = H_o D$$

where H_o is the "Hubble constant," the constant of proportionality in the relation between the relative recession velocity of extragalactic objects and their distance from us. A currently favored value of H_o is approximately 50 kilometers per megaparsec per second, but higher values have also been used.

The inverse of H_o—namely, $1/H_o$—has the unit of time and is called the "Hubble time." It is a measure of the age of the Universe. It should also be noted that in an evolving Universe, the value of this "constant" will actually change with time.

humanization of space A two-part process in which we first learn to use the physical conditions (for example, high vacuum and microgravity), resources (for example, lunar and asteroid materials) and unique properties of outer space (for example, comprehensive view of the Earth and biological isolation from the planet) to better the quality of life for all in the upcoming decades and we (a selected few persons at first) then learn how to live in space.

The humanization of space identifies the start of humanity's extraterrestrial civilization—an age when the human-resource base is expanded into the Solar System. Table 1 lists some of the potential technical steps that might be taken during the creation of this most interesting phase of human civilization. In the projected sequence of technical achievements, we first learn how to permanently occupy near-Earth space and then expand our activities throughout cislunar space (that is, the space between the Earth and the Moon). As space-based industry grows, a subtle but very significant transition point is reached. A segment of the human race eventually becomes self-sufficient in cislunar space—this simply means that those individuals living in space habitats and in settlements on the surface of the Moon would no longer depend on the Earth for the "makeup" materials and energy supplies necessary for their survival. From that very historic moment forward, humankind would possess two distinct cultural subsets: terran and nonterran, or extraterrestrial! We would have met the extraterrestrials, and they would be us!

There are three basic building blocks on which the extraterrestrial component of human civilization will depend: (1) compact energy systems, especially power and propulsion modules; (2) the ability to process materials anywhere in the Solar System; and (3) space habitats to support a permanent human presence in outer space.

A selected few members of the Earth's population,

Table 1   Possible Steps in the Humanization of Space over the Next Century

STEP 1: Permanent Occupancy of Near-Earth Space

- Space Station/Space Operations Center [6–12 persons]
- Space Base [50–200 persons]
- Orbiting Propellant and Service Depot
- Near-Earth Orbital Launch Facility

STEP 2: Permanent Occupancy of Cislunar Space

- Large Powerplants (nuclear) at GEO [megawatt range]
- Manned Space Platform at GEO [6–20 persons]
- Orbiting Lunar Station [6–12 persons]
- Initial Lunar Base [6–12 persons]
- Cislunar Orbital Transfer Vehicles
- Permanent Lunar Settlements [200–300 persons]

STEP 3: Full Self-Sufficiency In Cislunar Space

- Space Communities in Earth Orbit
- Space Cities
- Extensive Lunar Settlements
- Settlements Throughout Cislunar Space
- Utilization of the Apollo/Amor Asteroids

STEP 4: Permanent Occupancy of Heliocentric Space

- Mars Orbiting Station
- Initial Martian Base
- Permanent Martian Settlements
- Asteroid Belt Exploration
- Manned Bases In Asteroid Belt
- Bases On Selected Outer Planet Moons [e.g. Titan, Ganymede]
- Planetary Engineering Programs [e.g. climate modification]
- Manmade "Planetoids" In Heliocentric Space
- First Interstellar Missions

SOURCE: Developed by the author.

responding to some deep evolutionary drive to expand into and explore the cosmos, will eventually establish permanent space settlements. Pursuing economic gain and following exploratory impulses, these 21st-century pioneers will spread first into cislunar space and then throughout the Solar System—all the while returning many quality-of-life-improving benefits to the terrestrial portion of human civilization.

Once self-sufficiency is achieved in cislunar space, the permanent occupancy of heliocentric, or interplanetary, space will occur. Human settlements will appear on Mars, in the asteroid belt and on selected moons of the giant outer planets. Finally, as humanity, or at least its extraterrestrial subset, starts to fill the ecosphere of our native star with human-made planetoids, a cosmic wanderlust will begin attracting certain Solar-System inhabitants toward journeys to the stars. With the launching of the first manned interstellar mission, the human race will indeed have come of age in the Universe. Perhaps we will be the first intelligent species to sweep through the Galaxy, or perhaps we are destined to meet other starfaring civilizations.

See also: **extraterrestrial civilizations; space industrialization**

humanoid   Literally, a creature that resembles a human being. As found in the science-fiction literature, *humanoid* is frequently used to describe an intelligent extraterrestrial being, while in anthropology the term refers to an early ancestor of *homo sapiens*.

hyperspace   A concept of convenience developed in science fiction to make "faster-than-light" travel appear credible. Hyperspace is frequently described as a special dimension or property of the Universe in which physical things are much closer together than they are in the normal space-time continuum.

In a typical science-fiction story, the crew of a spaceship simply switches into "hyperspace," and distances to objects in the "normal" Universe are considerably shortened. When the spaceship emerges out of hyperspace, the crew is where they wanted to be essentially instantly. Although this concept violates the speed-of-light barrier predicted by Einstein's special relativity theory, it is nevertheless quite popular in modern science fiction.

See also: **"faster-than-light" travel; interstellar travel; relativity; science fiction**

HZE particles   The most potentially damaging cosmic rays, with high atomic number (Z) and high kinetic energy (E). Typically, HZE particles are atomic nuclei with Z greater than 6 and E greater than 100 million electron volts. When these extremely energetic particles pass through a substance, they leave a large amount of energy deposited along their tracks. This deposited energy ionizes the atoms of the material and disrupts molecular bonds.

See also: **cosmic rays; hazards to space workers**

ice catastrophe   A planetary climatic extreme in which all the liquid water on the surface of a life-bearing or potentially life-bearing planet has become frozen or completely glaciated.

See also: **ecosphere**

IFO   An abbreviation for "identified flying object"—an unidentified flying object (UFO) that has subsequently been identified or explained. For example, a mysterious light in the night sky, originally reported to authorities as a UFO, might turn out to be the planet Venus or an airplane (an IFO).

See also: **unidentified flying object**

illumination from space   Just how can we presently take good advantage of the radiant energy of our parent star, the Sun? Well, in the early 1970s Dr. Krafft A. Ehricke and other far-thinking space engineers proposed that large mirrors be placed in orbit around the Earth to illuminate cities, agricultural regions, ice fields that were blocking navigation and terrestrial solar power installations. Ehricke called such mirrors "lunettas" if they reflected sunlight to localities on the dark (nighttime) side of the Earth and "solettas" if they reflected substantial amounts of solar energy (typically providing one solar constant) over limited regions of the Earth both day and night. (One solar constant is 1,371 watts per square meter of radiant solar energy.)

NASA investigations of solar-sail applications have also provided the technical characteristics for a large space mirror constructed out of aluminized mylar, possibly just 0.0025 millimeter (0.01 mil) thick. We might anticipate that such giant mirrors will eventually have areal densities as low as 6 grams per square meter, which corresponds to 6 metric tons per square kilometer.

Figure 1 shows the Sunblazer concept—a proposed illumination-from-space scheme in which a series of giant, mirrorlike reflector spacecraft are placed in an equatorial, geostationary orbit over North America. These large orbiting mirrors would be so positioned that they effectively extended daylight year-round to several continental regions, as well as to Hawaii. Sixteen such reflectors, for example, could add approximately two hours of illumination during peak evening and morning rush-hour periods. One or more of these colossal reflectors could also be diverted to illuminate Alaska, especially in the Fairbanks and Anchorage areas, extending their short winter days by approximately three hours.

Entire cities or other important regions could be illuminated by a single reflector spacecraft during a time of

Fig. 1   The Sunblazer concept. (Courtesy of NASA.)

power blackout or other emergency. Because of the costs of building and deploying such giant solar reflector space-craft, these large mirrors would probably not be dedicated to just emergency illumination operations. However, if a series of these platforms were placed in operation (as, for example, in the Sunblazer concept), then up to four of these giant reflectors could be diverted in a nighttime emergency to illuminate any location between the Virgin Islands and the Hawaiian Islands to a level equal to streetlight intensity, or 15 full moons.

There are many exciting terrestrial applications for such colossal mirrors in space. These include: nighttime illumi-nation for urban areas; nighttime illumination for agricul-tural and industrial operations; nighttime illumination for disaster relief and emergency operations; frost-damage protection; local climate manipulation; increased solar flux to enhance solar energy conversion processes on the ground; ocean cell warming for climate control; enhanced agriculture through the stimulation of photosynthesis processes; and controlled snowpack melting. Can you think of any others?

Of course, such giant mirrors placed in appropriate orbits around the Moon could help illuminate mining and exploration operations during the long lunar night (ap-proximately 14 Earth days' duration) and could be used to prevent excessive facilities and equipment "cold soak" during these extended periods of lunar darkness. In orbit around Mars, such large reflector mirrors represent one of the major tools of planetary engineering. They would help bring more sunlight to the polar regions, promoting con-trolled melting of the Martian polar caps. Such mirrors could also provide more benign growing environments for genetically engineered plants—some of which would be used to help "terraform" the thin, carbon dioxide-rich Martian atmosphere.

images from space   How can we learn more about our extraterrestrial environment? One very effective way is to send robot spacecraft with imaging systems to target celestial objects throughout our Solar System. In the last 17 years, spacecraft imaging systems have made most previous visual planetary data obsolete. Taking advantage of close flybys, orbits and landings, robot spacecraft have provided scientists with exceptionally clear and close views of alien worlds as far away as Saturn and its complement of moons. In fact, our spacecraft have visited all the planets known to ancient astronomers. Pictures telemetered back from space have allowed planetary sci-entists to discover a Moonlike surface on Mercury and circulation patterns in the thick Venusian atmosphere. Images of Mars have shown craters, giant canyons and volcanoes on the planet's surface. Interesting details about the circulation patterns in the Jovian atmosphere have been revealed; active volcanoes have been discovered on Jupiter's moon Io; and a ring has even been found circling the King of the Planets. New moons were discovered in orbit around Saturn; known moons have been viewed close up; and rings once thought to be just four in number have now been resolved into a complex configuration consisting of more than a thousand concentric ring features. (See fig. 1.)

These new discoveries about the planets and their moons, as well as many other exciting observations about our extraterrestrial environment, were made possible by the development of a spacecraft technology for imagery collection, processing and transmission. The spectacular views we've enjoyed of the Martian surface or of Saturn's complex rings through this spacecraft imagery technology are facsimile images and not true photographs. A scanning optical system on board the spacecraft converts sunlight reflected from a planet or moon into numerical data. The numerical data are then telemetered to Earth via electro-magnetic (radio) waves. On Earth, giant tracking stations collect these electromagnetic signals, and processing com-puters assemble the received information into useful im-ages of alien worlds.

Most planetary spacecraft produce images of the planets and their moons by using slow-scan television cameras. These cameras take a much longer time to form and transmit images than do the commercial television system cameras used here on Earth. Although they take longer to generate, the extraterrestrial images are of a much higher quality and contain more than twice the amount of infor-mation contained in the picture appearing on your home television screen.

Since the successful flyby of Mars by the *Mariner 4* spacecraft in 1964, major improvements have occurred in spacecraft imaging systems. An entire week was required for the *Mariner 4* spacecraft to transmit enough informa-tion to create just 21 images of Mars, since the early planetary-exploration craft could only transmit its data at a rate of under 10 bits per second. (A bit is a unit of information that can be represented by a 1 or a 0.) By comparison, in 1979 the same amount of data was con-tained in just one *Voyager* spacecraft image of Jupiter and

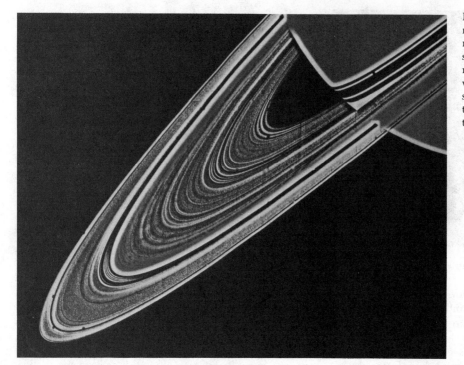

Fig. 1 This computer-assembled mosaic image of Saturn's magnificent rings was taken by the *Voyager 1* spacecraft on November 6, 1980, at a range of 8 million kilometers. It revealed an extraordinarily complex structure of concentric rings. (Courtesy of NASA/Jet Propulsion Laboratory.)

was transmitted in only 48 seconds.

Using the *Voyager* spacecraft imaging system as our example, the process of generating finished extraterrestrial images of alien worlds can be accomplished in five general steps:

> *On the spacecraft*
> Step one: image scanning
> Step two: data storage and transmission
>
> *On earth*
> Step three: data reception
> Step four: data storage
> Step five: image reconstruction

The *Voyager* spacecraft carries a dual television camera system on a scientific instrument platform that can be tilted in any direction for precise aiming. On command, a celestial object can be viewed with either wide-angle or narrow-angle telephoto lenses. Reflected light from the extraterrestrial target enters the lenses and falls on the surface of a selenium-sulfur vidicon television tube, 11 millimeters square. Unlike most standard television cameras, a shutter controls the amount of light reaching the tube. Exposure periods can vary from 0.005 second for very bright targets to 15 seconds or more for very faint objects or when searches are being conducted for previously unknown, and very dim, satellites. The television tube temporarily retains the image until it can be scanned or measured for brightness levels. During the scanning process, the vidicon-tube surface is divided into 800 lines, each line consisting in turn of 800 points. These individual points are called pixels. (The word *pixel* is a contraction of the term *picture element*.) The total number of pixels into

which a *Voyager* image is divided is then 800 × 800, or 640,000. (See fig. 2.)

As each pixel is scanned for brightness, it is assigned a number from 0 to 255. The measured range from black to white is 256, or the number 2 to the eighth power (2^8). To express the assigned pixel brightness in terms a computer can understand, each number is converted into binary

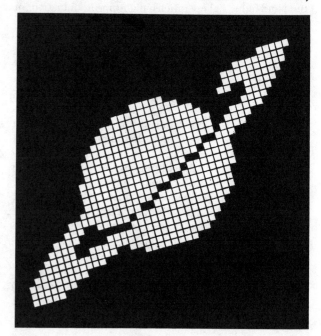

Fig. 2   A representative planetary image broken up into pixels. (Drawing courtesy of NASA/and modified by the author.)

Table 1   Binary Bit Sequence

Binary values	1	2	4	8	16	32	64	128
Binary bits	0	0	0	0	0	0	0	0

Table 2   Examples of Pixel Brightness

Binary values	1	2	4	8	16	32	64	128
Brightness no.								
0	0	0	0	0	0	0	0	0
9	1	0	0	1	0	0	0	0
58	0	1	0	1	1	1	0	0
183	1	1	1	0	1	1	0	1
255	1	1	1	1	1	1	1	1

language. Eight binary digits, or bits, are needed to represent each number. Each of these eight bits, in sequence, represents a doubling of numerical value. The eight bits are arranged according to their numerical values in table 1. During information transmission, each bit is given as either a 0 or a 1. A 0 means the numerical value of that bit is zero, while a 1 means the numerical value of the bit corresponds to the bit sequence value shown in table 1. Then, to convert from binary bits back to the brightness number, we simply have to sum, or total, the values of each bit in the eight-bit sequence, or "word." Table 2 provides several examples of pixel brightness expressed in both numerical and binary form. After carefully reviewing tables 1 and 2, can you tell what brightness number would be assigned to a pixel that had the following eight-bit binary sequence: 0 0 1 0 1 0 0 0?

If you had estimated a pixel brightness value of 20, you are correct and are well on your way to understanding how these magnificent extraterrestrial images are gathered by robot spacecraft, transmitted to Earth and computer-processed for use by scientists.

After the *Voyager* spacecraft imaging system scans the pixels and converts the light levels into binary form, the bit information is either stored on tape for later transmission or sent directly back to Earth in real time. These data are transmitted at a rate of more than 100,000 bits per second. For each *Voyager* image, 5,120,000 bits (640,000 × 8 binary bits) must be sent! Spacecraft data-storage capability is used when the vehicle passes out of sight behind a planet or moon and radio communications are temporarily eclipsed. With the *Voyager* spacecraft, data equivalent to 100 images can be stored for later transmission.

On Earth, radio signals from the spacecraft are received by one of three large radio antennas of NASA's Deep Space Network (DSN). Each dish-shaped antenna has a diameter of 64 meters. A motor-driven system precisely aims the antenna toward the distant spacecraft and compensates for the Earth's rotation. To maintain contact with robot spacecraft in deep space, these antennas are located approximately 120 degrees apart around the Earth at Goldstone, California; Madrid, Spain; and Canberra, Australia. As one antenna loses contact due to the Earth's rotation, another DSN antenna comes into view and takes over the job of receiving spacecraft data. While not in contact with interplanetary spacecraft and probes, these antennas are directed to other tasks, such as radio astronomy experiments and the search for extraterrestrial intelligence (SETI).

When the spacecraft data are received on Earth, computers simultaneously store the data for future use and reassemble them into images. In generating these images, the computer converts the binary bit sequences for each pixel into small squares of light, as was shown in figure 2. The brightness of the squares is determined by the numerical value assigned to the pixel. The squares are then displayed on a television screen and assembled into a grid 800 × 800 pixels in dimension. The resulting image formed by all the lighted squares on the high-resolution television screen is a black-and-white facsimile picture of the target extraterrestrial object.

If we want color images of alien worlds, considerably more information is needed from the robot spacecraft. In the *Voyager* spacecraft, a wheel with a variety of colored filters is rotated in front of the television tube when the images are being acquired. In rapid succession, three separate images are taken of the same celestial object, through blue, green and orange filters. The television tube is scanned for each image, and the resulting binary bits are then transmitted to Earth. By the time the scanning of the third "color-filtered" image is complete, over 15 million binary bits (3 × 640,000 × 8) are traveling toward Earth at the speed of light in the data-transmission signal.

Each color filter affects the amount of light reaching the television tube. The orange filter, for example, is transparent to orange light, but blue light appears much darker than normal. On Earth, computer processing is used to give color to the three filtered images and blend them together to form a "true" color image.

One final step may be added in the creation of these spectacular extraterrestrial images. To the unaided eye, some images might appear not to be particularly descriptive. The shading differences in planetary surfaces or cloud tops may be a little too subtle to be detected by a cursory visual examination. Scientists use a selective computer enhancement of portions of the image to bring out such subtle details. With computer processing, pixels of a particular numerical value can be assigned an unusual color to make them clearly stand out. If we have two almost identical shades of yellow, for instance, they can be colored red and blue, thereby greatly exaggerating their differences.

Images of the planets and moons of our Solar System have proved to be the most valuable source of spacecraft-acquired information available to planetary scientists. Data gathered at close range (and above the filtering effects of the terrestrial atmosphere) have produced views that are far better in quality and detail than pictures taken through telescopes on Earth. The unprecedented quality of these images from space has greatly assisted scientists in developing better theories of the nature and origin of our Solar System. Through spacecraft imaging systems in the short space of just two decades, we have "discovered" over a dozen new worlds and expanded our cosmic horizon at a rate unmatched in all previous human history!

See also: **Deep Space Network; Planetary Image Facility; remote sensing; Voyager**

"infective" theory of life   The belief that some primitive form of life—perhaps selected, hardy bacteria or "engineered" microorganisms—was placed on an ancient Earth by members of a technically advanced extraterrestrial civilization. This planting or "infecting" of simple microscopic life on a then-lifeless planet could have been intentional (that is, "directed panspermia") or accidental (for example, through the arrival of a "contaminated" space probe or from "space garbage" left behind by extraterrestrial visitors).

See also: **life in the Universe; panspermia**

inferior planets   Planets that have orbits that lie inside the Earth's orbit around the Sun—namely, Mercury and Venus.

infrared (IR) astronomy   The branch of astronomy dealing with infrared (IR) radiation from celestial objects. Most celestial objects emit some quantity of infrared radiation. However, when a star is not quite hot enough to shine in the visible portion of the electromagnetic spectrum, it emits the bulk of its energy in the infrared. Infrared astronomy, consequently, involves the study of relatively cool celestial objects, such as interstellar clouds of dust and gas (typically about 100 degrees Kelvin) and stars with surface temperatures below about 6,000 degrees Kelvin.

Many interstellar dust and gas molecules emit characteristic infrared signatures that astronomers use to study chemical processes occurring in interstellar space. This same interstellar dust also prevents astronomers from viewing visible light coming from the center of our Milky Way Galaxy. However, infrared radiation from the galactic nucleus is not as severely absorbed as radiation in the visible portion of the electromagnetic spectrum and infrared astronomy therefore enables scientists to study the dense core of the Milky Way.

Unfortunately, water and carbon dioxide in the Earth's atmosphere absorb most of the interesting infrared radiation arriving from celestial objects. There are only a few narrow IR spectral bands or windows that can be used by Earth-based astronomers in observing the Universe; and even these IR windows are distorted by "sky noise" (undesirable infrared radiation from atmospheric molecules).

However, the Space Age has provided astronomers a solution to this problem and has opened up an entirely new region of the electromagnetic spectrum to detailed observation. We can now place sophisticated infrared telescopes in space, above the limiting and disturbing effects of the Earth's atmosphere. The Infrared Astronomical Satellite (IRAS) launched in January 1983 represented the first extensive scientific effort to explore the Universe in the infrared portion of the electromagnetic spectrum. IRAS was an international effort involving the United States, the United Kingdom and the Netherlands. This spacebased IR telescope completed the first all-sky survey in a wide range of IR wavelengths with a sensitivity 100 to 1,000 times greater than any previous telescope. Scientists are now using IRAS data to produce a comprehensive catalog and maps of significant infrared sources in the Universe. These sources include stars that may possess planetary systems or at least planetary systems under formation.

Infrared astronomy also allows astrophysicists to observe stars (called protostars) as they are being formed in giant clouds of dust and gas (nebula), long before their thermonuclear furnaces have ignited and they have "turned on" in visible light.

Figure 1 is an infrared map of the Swan Nebula, created from data acquired by NASA's Kuiper Airborne Observatory. The Swan Nebula, also called M17, is located in the southern sky in the constellation Sagittarius. This region contains a cloud of ionized, interstellar gas excited by recently formed stars. Astrophysicists call this type of cloud an HII region.

Maybe one of the most exciting discoveries awaiting our use of future infrared telescopes in space is the detection and identification of an advanced extraterrestrial civilization through its telltale astroengineering activities. In 1960 the physicist Freeman J. Dyson suggested that we search for evidence of extraterrestrial beings by looking for artificial cosmic sources of IR radiation. He postulated that intelligent beings of an advanced civilization might eventually desire to capture all the radiant energy output of their parent star. They would subsequently construct a huge cluster of habitats and space platforms, called a Dyson sphere, around their star. This Dyson sphere, perhaps at a size comparable to the Earth's orbit around the Sun, would lie within the ecosphere of the parent star and would intercept all its radiant energy output. The intercepted starlight, after useful energy extraction, would then be reradiated to outer space at approximately ten micrometers wavelength. This infrared wavelength corresponds to a heat rejection surface temperature for the

Fig. 1   An infrared map of the Swan Nebula (M17). (Courtesy of NASA.)

Dyson sphere of approximately 200 to 300 degrees Kelvin. Thus, according to Dyson, if we detect an infrared radiation source at about 250 degrees Kelvin which is approximately one or two astronomical units in diameter, it just might represent the astroengineering handiwork of an advanced extraterrestrial civilization!

See also: **astrophysics; extraterrestrial civilizations; Dyson sphere; infrared radiation; nebula**

infrared (IR) radiation   That portion of the electromagnetic (EM) spectrum lying between the optical (visible) and radio wavelengths. It is frequently taken as spanning three decades of the EM spectrum, from 1 micrometer to 1,000 micrometers wavelength. The English-German astronomer Sir William Herschel (1738–1822) is credited with the discovery of infrared radiation. In 1800 this famous astronomer was experimenting with sunlight broken into its separate colors (that is, the visible spectrum) by a prism. Herschel used a black-bulb thermometer to measure the temperatures of the individual colors in sunlight. To his amazement, when he placed this thermometer just beyond the red end of the spectrum, where no visible light was falling, the thermometer registered higher-than-anticipated temperatures. He immediately realized that an invisible form of radiation existed and called it infrared—meaning "below, or inferior to, red."

See also: **astrophysics; infrared astronomy; Shuttle Infrared Telescope Facility**

inner planets   The terrestrial planets: Mercury, Venus, Earth and Mars. These planets all have orbits around the Sun that lie inside the main asteroid belt.

intergalactic   Between or among the galaxies. Although no place in the Universe is truly "empty," the space between clusters of galaxies comes very close. These intergalactic regions contain less than one atom in every 10 cubic meters. Even though the galaxies continually supply new matter to intergalactic space, the continued expansion of the Universe makes the overall effect negligible. In fact, intergalactic space is very empty and is getting more empty every moment as the Universe expands.

interplanetary   Between the planets; within the Solar System.

interplanetary dust (IPD)   Particles of matter found within the Solar System between individual planets. By convention, the term *interplanetary dust*, or *IPD*, applies to all solid bodies ranging in size from submicrometer diameter to tens of centimeters diameter, with corresponding masses ranging from 10^{-17} gram to approximately 10 kilograms. Near the Earth the IPD flux is taken as approximately 10^{-13} to 10^{-12} gram per square meter per second (g/m²/s). It is roughly estimated that the Earth collects over 10,000 metric tons of IPD per year. Astronomers now believe that the entire IPD cloud, with its estimated total mass of between 10^{+16} and 10^{+17} kilograms, is of cometary origin.

See also: **comet; meteoroids; zodiacal light**

interstellar   Between or among the stars.

interstellar communication   One of the fundamental aspects of being human is our desire to communicate. In recent years we have begun to respond to a deep cosmic yearning to reach beyond our own Solar System to other star systems—hoping not only that someone or something is out there but that they will eventually "hear us" and perhaps even return our message.

Because of the vast distances between even nearby stars, when we say "interstellar communication" we are not talking about communication in "real time." (Communication in real time does not involve a perceptible time lag—that is, messages and responses are received immediately after they are sent.) Rather, our initial attempts at interstellar communication have actually been more like putting a message in a bottle and tossing it in the "cosmic sea," or perhaps even placing a message in a time capsule or "cosmic safety deposit box" for some future generation of human or alien beings to find and learn about life on Earth in the 20th century.

Attempts to "communicate" with alien civilizations that might exist among the stars is often called *CETI*, an acronym that means "communication with extraterrestrial intelligence." If, on the other hand, we quietly watch the skies for "signs" of some super-extraterrestrial civilization (for example, looking for the infrared signatures from Dyson spheres) or patiently listen for intelligent radio messages transmitted by advanced alien races (mainly in the microwave region of the spectrum), then we call the process SETI, or simply the "search for extraterrestrial intelligence."

Since 1960 there have been over 35 serious SETI observation efforts, the vast majority involving listening to selected portions of the microwave spectrum, in hopes of detecting "radio signals" indicative of the existence of intelligent extraterrestrial civilizations among the stars. To date, none of these efforts has provided any positive evidence that such "intelligent" radio signals exist, carrying messages to other advanced or even developing galactic races. However, SETI observers have only examined a few of the billions of stars in our Galaxy and have only listened to a few rather narrow portions of the spectrum within which such intelligent signals might be transmitted. Furthermore, it is only within the last few decades that we have developed the technology, largely radio astronomy–related, to enable us to be at even a minimum "interstellar communications horizon." A century ago, for example, the Earth could have been "bombarded" with many alien signals—but no one would have had the technology to receive and interpret them.

We have also deliberately attempted to communicate with alien civilizations by sending messages out beyond the Solar System on several of our spacecraft and by sending a very powerful radio message to the stars using the world's largest radio-telescope facility, the Arecibo Observatory in Puerto Rico (see fig. 1 on page 7). Since the age of radio and television, we have also unintentionally been leaking radio-frequency signals (now about 30 light-years out) into the Galaxy. Imagine the impact some of our early television shows would have on an alien civilization capable of intercepting and reconstructing these signals!

Our three most important attempts at interstellar communication (from Earth to the stars) to date are: (1) the special message plaque placed on both the *Pioneer 10* and *11* spacecraft departing the Solar System on interstellar trajectories; (2) the "Sounds of Earth" record included on the *Voyager 1* and *2* spacecraft, which will also depart the Solar System on interstellar trajectories; and (3) the famous Arecibo Interstellar Message, transmitted on November 16, 1974, by the world's most powerful radio telescope.

> See also: **Arecibo Interstellar Message; consequences of extraterrestrial contact; interstellar contact; Pioneer plaque; search for extraterrestrial intelligence; Voyager record; What do you say to a little green man?**

interstellar contact Several methods of achieving contact with intelligent extraterrestrial life-forms have been suggested. These methods include: (1) interstellar travel by means of starships, leading to physical contact between different civilizations; (2) indirect contact through the use of robot interstellar probes; (3) serendipitous contact; (4) interstellar communication involving the transmission and reception of electromagnetic signals; and (5) exotic techniques involving perhaps information transfer through the modulation of gravitons, neutrinos or streams of tachyons; the use of some form of telepathy; and matter transfer

through the use of hyperspace or distortions in the space-time continuum that help "beat" the speed-of-light barrier.

INTERSTELLAR TRAVEL/PHYSICAL CONTACT

The classic method of interstellar contact in science fiction is the starship. With this class of spaceship, an intelligent civilization would be capable of eventually sweeping through the Galaxy, finding and contacting other life-forms wherever they existed and planting life wherever it didn't, but could, exist. Probably nothing would be more exciting, and even a little frightening, to a technically emerging planetary society than to have its sky suddenly fill with an armada of giant starships. The inhabitants of the planet would be advanced enough to appreciate the great technology levels required to bring the alien visitors across the interstellar void. This physical contact could also prove a very humbling experience for a planetary civilization, like our own, that had just struggled to achieve interplanetary spaceflight capabilities. A variety of contact scenarios can be found in science fiction. These scenarios range from a friendly welcome into a galactic community to a hostile attempt to "capture the planet." In the belligerent scenarios, those beings on the starship play the role of "invaders," while the planet's inhabitants become the "defenders." Depending on the level of technology mismatch and any literary gimmicks the S/F writers include (such as a "biological Achilles' heel" for the invading species), the battle for the planet or star system goes either way in the story.

However, even though we have successfully begun to master interplanetary flight with chemical propulsion systems and can complement these propulsion systems with more advantageous nuclear-fission- and eventually nuclear-fusion-powered propulsion systems—the energetic demands of interstellar flight simply overwhelm any propulsion technology we can extrapolate as 21st-century engineering and beyond.

One example might explain these "hard" circumstances a little better. Let's ignore *all* current engineering and materials-science limitations and construct (at least on paper) the *very best propulsion system "physics can buy"*; that is, we're going to build the most advanced propulsion system our current understanding of physics will allow—despite the fact that the actual engineering technology to accomplish this construction task may be centuries away, if ever! We would construct a *photon rocket*, whose propellant is a mixture of equal parts of matter and antimatter. This photon rocket uses the annihilation reaction that occurs when we blend matter and antimatter. This extremely energetic reaction turns every kilogram of propellant into pure energy, mainly a gamma radiation. These gamma rays would then be directed out of the rocket's special "thrust chamber" in a perfectly collimated radiation stream (that is, a radiation stream in which the rays are parallel) that provides a reaction (retrodirected) thrust to the starship and its payload. The complete conversion of just one gram of matter-antimatter in an annihilation reaction would release some 9×10^{13} joules of energy! (In comparison, a one-kiloton [kT] nuclear-explosion yield amounts to a release of some 4×10^{12} joules.) For the moment, we have neglected all the engineering problems of obtaining and containing antimatter and of preventing nuclear-radiation leakage into the crew compartment.

Let's now use this "best-we-can-possibly-build" photon-powered starship on a 10-year round-trip journey to a nearby star system (for example, Alpha Centauri, which is 4.23 light-years away). To further optimize this exercise, let us also assume that the entire starship, except for the matter-antimatter propellant, has a mass of only 1,000 tons and that the starship can achieve a cruising speed of 99 percent of the speed of light (0.99 c) after a reasonably short period of acceleration (about one year at a constant acceleration rate of one g). According to one set of calculations for this hypothetical round-trip interstellar mission, we would need 33,000 tons of matter-antimatter propellant (16,500 tons of each type) to annihilate en route. The total energy release associated with 33,000 tons of mass converted into pure energy is approximately 3×10^{24} joules. As a point of reference, our Sun's energy output is approximately 4×10^{27} joules per second.

A few other "engineering" details are also worth mentioning here. During the initial acceleration period, our matter-antimatter-powered starship must achieve power levels of about 10^{18} watts. If only one part in a million of this energy release leaks into the ship, it would experience a one-million-megawatt (10^{12}-watt) heat flux. A very elaborate and heavy cooling and heat-rejection system would be needed to keep the starfarers and their equipment from melting. The same, if not worse, constraints apply to radiation leakage into the crew compartment. Extensive shielding will be needed to protect the crew and their equipment both from "engine-room leakage" and also from the radiation spall (erosion of solid surfaces) that will occur when a starship moving at near-light speed hits interstellar dust and molecules.

The sobering conclusion of this paper exercise, although contrary to the bulk of popular science fiction, is that, based on our current understanding of the physical laws of the Universe—interstellar starships carrying human crews on round-trip journeys within a crew's life span appear out of the question, not only for the present but for an indefinitely long time into the future. Interstellar travel is *not* a physical impossibility, but for today and for many tomorrows, it appears technically out of our reach. Many breakthroughs, most unimaginable at present, would have to occur before human beings can realistically think about boarding a starship and making a round-trip (or perhaps one-way) journey to another star system.

On the basis of this conclusion, we cannot initiate interstellar contact using the physical travel of human beings across the interstellar void. If contact is made by starship in the next few centuries, it will most likely be "them" visiting "us." Perhaps, some technically powerful alien society has had better luck unraveling nature's secrets and has discovered forms of energy and physical laws that are currently far beyond our intellectual horizon. If this is the case, and if these alien beings also have a societal commitment to interstellar exploration, and if they decide that our rather common G–spectral-class star is worth visiting—then perhaps . . . just perhaps, physical contact by starship will be made. But we cannot seriously include this possibility in our own attempts at interstellar contact because we have no control over the circumstances. This "starfaring" civilization either exists, or it doesn't; it is committed to interstellar exploration, or it isn't; and it either decides to explore our particular Solar System, or it bypasses us in favor of a more interesting galactic region.

In the absence of credible evidence that alien starships exist or are en route to visit us, we must be content within the next few decades to attempt interstellar contact through one of the alternative techniques suggested here.

INDIRECT CONTACT BY MEANS OF INTERSTELLAR ROBOTIC PROBES

Instead of sponsoring round-trip interstellar missions with "manned" starships, an advanced civilization might elect to send one-way robotic probes to explore neighboring star systems. The use of these interstellar probes has several potential advantages over a fully outfitted starship. First, the interstellar probe can be much smaller, since it does not require elaborate life-support equipment, crew quarters or a propellant supply for the journey back. This reduction in size greatly reduces the overall expense of the mission and eases the demands placed upon the propulsion system. Second, the probes can take a much longer time to reach their destination. A 50- or 100-year flight time for a robotic spaceship does not involve the same design complications that a mission of similar length would have on a "manned" starship. Again, this eases the technology demands on the propulsion system. We might now consider a propulsion system that reaches only one-tenth the speed of light (0.1 c) as a maximum cruising speed. While this is still a very challenging technology development, it is several orders of magnitude less of a challenge than designing a propulsion system to drive a starship at 99 percent of the speed of light or better.

A robotic interstellar probe would have to possess an advanced form of machine intelligence to execute repair functions en route, scan ahead for possible dangers in interstellar space and then execute a meaningful exploration program in a totally alien star system.

In reviewing the technical literature, three general types of interstellar-probe missions appear: (1) the flyby, (2) the sentinel and (3) the self-replicating machine (SRM). (Each of these major classes of robot-probe missions also has several possible variations, which will not be discussed here.)

The flyby interstellar robotic probe represents a one-way, one-shot attempt at interstellar contact and exploration. The probe and its complement of instruments would be launched toward the target star system and then be accelerated by the propulsion system up to a minimum velocity of about 10 percent of the speed of light (0.1 c). Since the probe does not need to decelerate when it gets to the new star system, only propellant for the initial acceleration is required. This is then the simplest, and perhaps "easiest," interstellar mission to develop. The probe would take several decades to cross the interstellar void. As it neared the target star system (perhaps at a distance of one light-year away), it would initiate an extensive long-range scanning operation to discover whether the star system contained planets that demanded special examination. If planets were detected, smaller robot scout ships would then be sent ahead. They would be launched from the mothership and powered by advanced nuclear-fission engines. Traveling at perhaps 10 percent of the speed of light, the larger interstellar-probe mothership would only briefly encounter the new star system. It would gather as much data as possible with its on-board sensor arrays and would also collect the data transmitted from any scout ships that were sent ahead. These smaller ships might have placed themselves in orbit around the new star or landed on a planet of interest in search of life. All these data would then be transmitted back to the home civilization. The probe, its mission complete, would disappear into the interstellar void.

If any of the scout ships had detected intelligent life on a planet in the new star system, the mothership might transmit a message or even deploy a special scout ship that contained appropriate greetings and detailed information about the sponsoring civilization.

This type of one-way robot-probe mission was studied in Project Daedalus, the first detailed engineering study involving the feasibility of interstellar travel. A pulsed nuclear-fusion system, using deuterium/helium–3 as the thermonuclear fuel, was proposed as the Daedalus propulsion unit.

In a sentinel interstellar-probe mission, the propulsion system must be capable of accelerating the probe to at least 10 percent of the speed of light (0.1 c). It must also be capable of decelerating the probe spacecraft from "light-speeds" to speeds that permit gravitational capture by the new star. This sentinel probe, now orbiting the target star, might spend years, decades or perhaps centuries searching the planets for signs of life. If life is detected, this probe may monitor its development, sending back information to the home civilization at selected time increments.

If the robot probe discovered an emerging technical civilization on one of the planets, it might be designed to execute a special contact protocol (procedure). For example, it could announce its presence when it was "triggered" by the detection of a certain level of technology. This robot sentinel might silently watch the planet's civilization emerge from an agrarian society to an industrial one. The development of radio-wave communication, atomic energy or spaceflight might be technology levels in the emerging planetary civilization that would trigger the

sentinel to action. Suddenly, the somewhat startled younger civilization would receive a "message from the stars"! Unlike the physical arrival of "little green men" in their starship, this form of interstellar contact would be indirect, with the smart robot probe serving as a surrogate alien visitor. Some individuals have already suggested that such sentinel or monitoring probes (somewhere out there in the Solar System) are monitoring our development right now! However, our explorations of the Moon, Mars, Venus, Mercury, Jupiter and Saturn have not yet discovered their suggested presence.

The third general class of robot interstellar-probe mission involves the use of a self-replicating machine (SRM). In this case, the propulsion requirements start approaching the demands of a full starship. The SRM probe must accelerate to some fraction of the speed of light, decelerate when it arrives at the target star system and then accelerate again to light-speed as it searches for another star system to explore. In the general scenario for the use of an SRM probe, the robot spaceship encounters a star system and begins the search for "suitable planets." In this case, a suitable planet is one that has the resources the probe needs to build an exact replica of itself—including computer systems, propulsion system, sensors for exploration and so on. This class of probe must also be capable of detecting and identifying life-forms, especially intelligent life-forms. If intelligent life is detected, the probe may execute a special contact protocol. This could involve the presentaton of simple messages or the construction from native materials of replicas of special devices and objects from the advanced civilization. If the inhabitants of the planet have matured to the level of interplanetary spaceflight, the SRM probe might even be programmed to replicate a copy of itself to leave behind as a "technological gift." The initial self-replicating machine probe would refurbish itself with planetary resources and then make an appropriate number of robot-probe replicas. All of these SRM probes would then depart to explore other star systems. This would create an exponentially growing population of smart machine explorers passing like a wave through the Galaxy. Imagine the surprised response of the inhabitants of an emerging civilization as one of these robot probes entered their system, provided messages and replicated token objects of "friendship," and then devoured a few choice asteroids and small moons to replicate itself several times. As the alien robot armada whizzed off into the "sunset," residents of the "contacted" star system might ask "Who was that little green man and his robot companion?"

To get into the interstellar-probe business, an advanced civilization would have to make a serious financial and technical commitment to interstellar exploration. For example, just to visit and possibly "bug" all the star systems within 1,000 light-years would require about 1 million flyby or sentinel probes—or one very sophisticated and reliable self-replicating machine probe. In the case of flyby or sentinel probes, if the advanced civilization then launched one of these probes a day, just the launching process alone would take three millennia!

SERENDIPITOUS CONTACT

We cannot rule out the possibility that we might suddenly stumble on some evidence of extraterrestrial intelligence. Perhaps an archaeologist exploring ancient Mayan temples in the Yucatan will discover the wreckage of a small alien scout ship; or maybe an astronomer pondering over Space Telescope data will come across an image of an alien interstellar probe or the unmistakable signature of some great feat of astroengineering, like the construction of a Dyson sphere; or possibly, sometime in the next century, an asteroid prospector, chasing down an interesting one-kilometer-size object on her radar screen, will come face-to-face with a derelict alien starship or, even more exciting perhaps, a robot sentinel probe—which might then begin to "speak."

As the word *serendipity* implies, these would be totally unexpected but nonetheless very exciting forms of interstellar contact. However, the chances of such accidental discoveries appear astronomically small. We cannot ignore this possibility in considering interstellar contact; but we do not exercise any real control over the situation either.

TRANSMISSION AND RECEPTION OF ELECTROMAGNETIC SIGNALS

An advanced civilization might like to minimize the expense of searching for other intelligent civilizations. One very dominant factor is the amount of energy an intelligent alien society must expend in announcing its existence to the Galaxy. As we discussed previously, sending a starship or even a robot probe across interstellar distances is a very energy-intensive activity. Today, many scientists involved in the search for extraterrestrial intelligence (SETI) believe that we cannot expect to find other intelligent life by "tossing tons of metal across the interstellar void." Even if we could build a starship or a sophisticated interstellar robot probe—the undertaking would still be extremely costly in terms of time, energy and financial resources. This is because we might have to search many star systems, perhaps out hundreds of light-years from our Sun, before we achieved contact with the nearest intelligent civilization. These SETI scientists often recommend an alternative to matter transfer: They suggest that we send some form of "radiation" instead.

Regardless of what form of radiation a civilization decides to use in trying to achieve interstellar contact by signaling, the signal itself should have the following desirable properties: (1) the energy expended to deliver each bit or piece of information should be minimized; (2) the velocity of the signal should be as high as possible; (3) the particles or waves making up the signal should be easy to generate, transmit and receive; (4) the particles or waves should not be appreciably absorbed or deflected by the interstellar medium or planetary atmospheres; (5) the number of particles or waves transmitted and received should be much greater than the natural background; (6)

the signal should exhibit some property not found in naturally occurring radiations; and (7) the radiation of such signals should be indicative of the activities of a technically advanced civilization.

Carefully reviewing these suggested requirements, SETI scientists eliminated charged particles, neutrinos, gravitons and so on in favor of spatially and temporally coherent electromagnetic waves. For example, the kinetic energy of an electron traveling at 50 percent of the speed of light (0.5 c) is about 10^8 times the total energy of a 150-gigahertz (microwave) photon. All other factors taken as equal, an interstellar communications system using electrons would need 100 million times as much power as one using microwaves (photons). More exotic particles, such as neutrinos, are not easily generated, modulated (to put a message on them) or collected.

SETI scientists have searched over the entire electromagnetic spectrum for suitable regions in which to conduct an interstellar conversation. Their generally unanimous conclusion is that the microwave region appears to be the most obvious choice for advanced civilizations to communicate with each other and even to attempt to communicate with emerging planetary civilizations. These SETI observers have also identified a special band within the microwave region called the "water hole." This narrow region lies between the spectral lines of hydrogen (1,420 megahertz [MHz]) and the hydroxyl radical (1,662 megahertz). At present, the water hole is one of the highly preferred bands used by terrestrial scientists in their search for radio signals generated by intelligent alien civilizations.

EXOTIC SIGNALING TECHNIQUES

While often suggested in science fiction as ways of rapidly transferring matter or information throughout the Universe, telepathy, the manipulation of gravitons or neutrinos, and the use of wormholes or hyperspace are "techniques" well beyond our own communications horizon. If such exotic alternatives exist and are now being used to signal the Earth, we would have no real way of knowing about it. We simply couldn't receive and interpret such messages. Remember, the terrestrial atmosphere is saturated with human-made radio and television signals—but without an antenna and the proper receiver, a person would be totally unaware of their presence. Our understanding of the Universe and "how things work in nature" is not sophisticated enough for us to think about using such exotic techniques in attempting interstellar contact in the next few decades or even centuries.

See also: **ancient astronauts; consequences of extraterrestrial contact; Dyson sphere; extraterrestrial civilizations; interstellar communication; Project Daedalus; robotics in space; search for extraterrestrial intelligence**

interstellar gas and dust   Material found in space between the stars. Low-density hydrogen and other gases have been detected from their absorption and emission of specific wavelengths of light and radio waves. Fine dust particles in interstellar space can scatter starlight in much the same way as smog scatters light here on Earth.

See also: **interstellar medium**

interstellar medium   The gas and dust particles that are found between the stars in our Galaxy. The gas mostly consists of relatively cool clouds of hydrogen (at 50 to 100 degrees Kelvin), such as the HI region, as well as regions of hot gas (typically 1,000 to 10,000 degrees Kelvin). But don't worry. Because of their very low densities, these interstellar gases do not represent any high-temperature threat to a starship or interstellar probe. In fact, the biggest problem facing a spacecraft in interstellar space will be to avoid getting too cold.

Recent astrophysical discoveries have given scientists new insights about the interstellar medium. For example, if we got on board a hypothetical starship and departed from our Solar System, we would first encounter a region of "warm" interstellar gas for about 10 light-years in all directions. This gas has an estimated density of 100,000 atoms per cubic meter—a very hard vacuum by terrestrial standards. We would then enter a much "hotter" region, where the interstellar gas has an equivalent temperature in excess of 100,000 degrees Kelvin. However, overheating of our starship is not a problem, because there are less than 10 atoms per cubic meter in this new region. As our starship continued to travel away from the Solar System, we would learn that this new "bubble of fine vacuum" extended in all directions for about 100 light-years.

Astronomers have only recently discovered such large, hot bubbles of gas in interstellar space. Using space-based observatories (especially ultraviolet astronomy devices), they have now identified two possible sources of these very hot interstellar gases: (1) intense stellar winds that boil off hot stars, and (2) the rarer, but more violent, blasts of matter from supernova explosions. These hot gas bubbles pile up colder matter throughout interstellar space, sweeping the colder matter into very large clouds that eventually become stellar nurseries. Some of these large, "piled-up" cold clouds contain enough matter to form a million Suns.

The interstellar medium also contains cosmic dust, mostly in the form of tiny, sooty sand grains about 0.1 micrometer in diameter. This cosmic dust reddens and dims the light coming from stars behind it.

Finally, the interstellar medium is saturated by galactic cosmic rays, which spiral along weak magnetic field lines at nearly the speed of light.

See also: **cosmic rays; galactic cosmic rays; stars**

interstellar probes   Automated interstellar spacecraft launched by advanced civilizations to explore other star systems. Such probes would most likely make use of very smart machine systems capable of operating independently for decades or centuries. Once the robot probe arrives at a new star system, it begins an exploration procedure. The target star system is scanned for possible life-bearing planets, and if any are detected, they become the object of more intense investigation. Data collected by the "mother" interstellar probe and any miniprobes (deployed to explore individual objects of interest within the new star system) are transmitted back to the home star system. There, after light-years of travel, the signals are intercepted by scientists, and interesting discoveries and information are used to enrich the civilization's understanding of the Universe.

Robot interstellar probes might also be designed to protectively carry specially engineered microorganisms, spores and bacteria. If a probe encounters ecologically suitable planets on which life has not yet evolved, then it could "seed" such barren but potentially fertile worlds with primitive life-forms or at least life precursors. In that way, the sponsoring civilization would not only be exploring neighboring star systems; it would also be spreading life itself through the Galaxy.

See also: **extraterrestrial civilizations; interstellar travel; life in the Universe; panspermia; Project Daedalus; robotics in space** (extraterrestrial impact of self-replicating systems)

interstellar spaceship   A space vehicle capable of spanning the great distances between stars in a galaxy. This vehicle could be a fully automated robot probe or a starship occupied by intelligent creatures from an alien civilization. (The word *starship* is frequently used in the future science literature to describe an interstellar spaceship that carries intelligent beings.)

Based on our present understanding of the laws of physics, including the speed-of-light barrier that emerges out of Einstein's theory of relativity, the construction of an interstellar spaceship appears to be an extremely challenging technical and economic undertaking—well beyond our current space technologies or any we can reasonably extrapolate as being available in the 21st century. To update the lament of ancient terrestrial sailors: "Oh Lord, the interstellar distances are so vast, and our ships (especially their propulsion systems) so small!"

See also: **Project Daedalus; robotics in space** (extraterrestrial impact of self-replicating systems); **starship**

interstellar travel   Matter transport between star systems in a galaxy. The "matter" transported may be: (1) a robot interstellar probe on an exploration or "prelife-seeding" mission; (2) an automated spacecraft that carries a summary of the cultural and technical heritage of an alien civilization; (3) a starship "manned" by intelligent extraterrestrial creatures who are on a voyage of exploration and "contact" with other life-forms; or perhaps even (4) a giant interstellar ark that is carrying an entire alien civilization away from its dying star system in search of suitable planets around other stars.

See also: **interstellar contact** (interstellar travel/

physical contact); **Project Daedalus; starship**

ion   An electrically charged atom or molecule. A positive ion (or cation) contains fewer electrons than the normal atom or molecule, while a negative ion (or anion) contains more electrons than found in the electrically neutral atom or molecule.

ion engine   A rocket engine that produces thrust by expelling high-velocity (accelerated) atomic particles.
See also: **nuclear electric propulsion system**

ionization   The process by which one or more electrons are either added to or removed from an atom or molecule, thereby creating an ion. High temperatures, electrical discharges or nuclear radiations can cause ionization.
See also: **ion**

ionizing radiation   Radiation capable of producing ions by adding electrons to, or removing electrons from, an electrically neutral atom, group of atoms or molecule.

jansky [symbol: Jy]   A unit used to describe the strength of an incoming electromagnetic wave signal. The jansky is frequently used in radio and infrared astronomy. It is named after the American radio engineer, Karl G. Jansky (1905–1950), who discovered extraterrestrial radio-wave sources in the 1930s—a discovery generally regarded as the birth of radio astronomy.

1 jansky (Jy) $= 10^{-26}$ watts per meter squared per hertz
1 Jy $= 10^{-26}$ W/m^2-Hz
See also: **radio astronomy**

Jovian   Of or relating to the planet Jupiter; (in science fiction) a native of the planet Jupiter.

Jovian planets   The giant, gaseous outer planets of our Solar System. These include: Jupiter, Saturn, Uranus and Neptune.

Jupiter   In Roman mythology Jupiter (called Zeus by the ancient Greeks) was lord of the heavens, the mightiest of all the gods. So, too, Jupiter is first among the planets in our Solar System. Jupiter has even been called a "near-star." If it had been only about 100 times larger, nuclear burning could have started in its core, and Jupiter would have become a star and a rival to the Sun itself. Not only is

Jupiter the largest of the planets, with 318 times the mass of the Earth, but its interesting complement of 16 known satellites resemble a miniature solar system.

The largest of these moons are called the Galilean satellites. They are: Io, Europa, Ganymede and Callisto. These moons were discovered in 1610 by Galileo—an event that helped spark the birth of modern observational astronomy and the overall use of the scientific method. The discovery of these four moons provided strong support for the then-revolutionary Copernican theory (that the Sun, *not* the Earth, is at the center of our Solar System). Galileo came under bitter personal attack by ecclesiastical authorities for his "earthshaking" discoveries. Through Galileo's pioneering efforts, centered in part on his early observations of Jupiter and its four major moons, the scientific method rapidly became understood and accepted. This, in turn, resulted in the exponential growth of scientific knowledge and technology—the foundation of our modern world and the stepping-stone to our extraterrestrial civilization.

Interest in Jupiter and its moons has been greatly stimulated in recent years by the Jovian encounters of four American spacecraft: *Pioneer 10* (December 1973), *Pioneer 11* (December 1974), *Voyager 1* (March 1979) and *Voyager 2* (July 1979). These spacecraft provided a wealth of new information and spectacular photographs of Jupiter and its complement of natural satellites (see figs. C7 and C8 on color pages C and D).

In March 1972 NASA sent the first of these four historic space probes to survey the outer planets. For each spacecraft, Jupiter was the first port of call.

Pioneer 10, which was launched from Cape Canaveral on March 2, 1972, was the first spacecraft to penetrate the main asteroid belt and travel to the outer regions of our Solar System. In December 1973 it returned the first close-up pictures of Jupiter as it flew within 130,000 kilometers of the giant planet's banded cloud tops. *Pioneer 11* followed a year later. *Voyagers 1* and *2* were launched in the summer of 1977 and returned spectacular photographs of Jupiter and its 16 satellites during flybys in 1979.

During their visits, these robot explorers found Jupiter to be a whirling ball of liquid hydrogen, topped with a uniquely colorful atmosphere, which is mostly hydrogen and helium (see fig. 1). The Jovian atmosphere also contains small amounts of methane, ammonia, ethane, acetylene, phosphene, germanium tetrahydride and possibly hydrogen cyanide. Jupiter's clouds also contain water crystals. Exobiologists think it is possible that between the planet's frigid cloud tops and the warmer hydrogen ocean that lies below, there might be regions in the Jovian atmosphere where methane, ammonia, water and other gases could react to form organic molecules. Because of Jupiter's atmospheric dynamics, however, these organic compounds—if they exist—are probably quite short-lived.

The Great Red Spot of Jupiter has been observed for centuries by terrestrial astronomers. It is now believed to be a tremendous atmospheric storm, similar to a hurricane

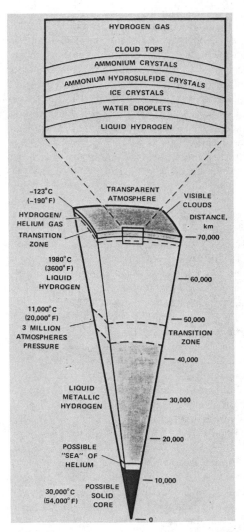

Fig. 1   A current model of Jupiter based on *Pioneer* and *Voyager* data. (Drawing courtesy of NASA.)

on Earth, which rotates counterclockwise (see fig. C9 on color page D).

The *Pioneer* and *Voyager* spacecraft detected lightning in Jupiter's upper atmosphere and observed auroral emissions in the Jovian polar regions that appeared similar to the northern and southern lights occurring here on Earth.

Voyager 1 produced the first clear evidence of a ring encircling Jupiter. Photographs returned by this spacecraft and its sistership (*Voyager 2*) showed a narrow ring too faint to be observed with Earth-based telescopes (see fig. 2).

Jupiter is the fifth planet from the Sun and is separated from the terrestrial planets by the main asteroid belt. The giant planet rotates at a dizzying pace—once every 9 hours, 50 minutes and 30 seconds. It takes Jupiter almost 12 Earth-years to complete a journey around the Sun. Its mean distance from the Sun is 5.2 astronomical units (AU), or 7.78×10^8 kilometers. When in opposition to the Sun, Jupiter is about 6.0×10^8 kilometers from Earth,

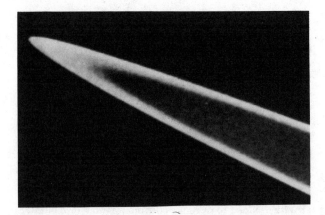

Fig. 2   An exciting view of Jupiter's ring as captured by the *Voyager 2* spacecraft, July 1979. (Courtesy of NASA/Jet Propulsion Laboratory.)

while at conjunction this distance is about 9.6×10^8 kilometers. Table 1 provides a summary of the physical and dynamic characteristics of this giant planet.

A new NASA mission to Jupiter, called the Galileo Project, is being readied for the later 1980s. It consists of an atmospheric probe and an orbiting spacecraft to help unlock even more of the secrets of this interesting planet and its many moons.

THE MOONS OF JUPITER

Jupiter has 16 known moons, four of which (Io, Europa, Ganymede and Callisto) are called the Galilean satellites. Tables 2 and 3 provide physical and dynamic data for these satellites. Very little was actually known about the Jovian moons until the *Pioneer* and *Voyager* spacecraft encountered Jupiter between 1973 and 1979. These flybys provided a great deal of valuable information and imagery, including the discovery of three new moons (1979J1, 1979J2 and 1979J3), active volcanism on Io and a possible liquid water ocean underneath the icy surface of Europa.

With the exception of the four Galilean satellites, all the other moons of Jupiter are relatively small. The four outermost satellites are called Sinope, Pasiphae, Carme and Ananke. They have retrograde orbits of high inclination and circle Jupiter about once every two Earth-years. These outermost moons lie at distances of 20 million to 24 million kilometers and measure less than 50 kilometers in diameter. Space scientists believe that these outermost Jovian moons are probably captured Trojan asteroids.

As we move inward, the next four Jovian satellites encountered are Elara, Lysithea, Himalia and Leda. These moons are clustered between 11 million and 12 million kilometers away from the planet. These moons also have high-inclination orbits but move in a normal prograde (counterclockwise) direction. The largest two, Elara and Himalia, are known to have a rocky appearance with low albedo. Some scientists speculate that this cluster of Jovian satellites may also be captured Trojan asteroids.

Moving in further toward Jupiter, we next encounter Callisto, one of the Galilean satellites. Callisto is outermost

Table 1   Physical and Dynamic Properties of Jupiter

Mass (excluding the 16 moons)	1.9×10^{27} kg
Diameter (equatorial)	142,796 km
Density (mean)	1.32 g/cm³
Period of revolution around Sun (sidereal period)	11.862 years
Period of rotation	9 hr, 55 min, 30 sec (approximate)
Mean orbital velocity	13.06 km/sec
Eccentricity	0.04845
Inclination to the ecliptic	1.3 degrees
Escape velocity	61 km/sec
Semimajor axis of orbit (average distance to Sun)	7.78×10^8 km (5.20 AU) [43.25 light-min]
Aphelion distance	8.15×10^8 km (5.45 AU)
Perihelion distance	7.41×10^8 km (4.95 AU)
Earth–Jupiter distances Closest (Jupiter in opposition to Sun)	6.0×10^8 km (4.0 AU) [33.3 light-min]
Farthest (Jupiter at conjunction with Sun)	9.6×10^8 km (6.4 AU) [53.4 light-min]
Albedo (average)	0.70
Number of known natural satellites	16

SOURCE: NASA.

Table 2   Properties of Known Jovian Moons

Moon	Diameter (km)	Semimajor Axis (km)	Sidereal Period (days)	Eccentricity	Inclination (degrees)	Year Discovered/ Discoverer
Sinope	6–36	23,700,000	758	0.28	156 (retrograde)	1914/Nicholson
Pasiphae	8–46	23,300,000	735	0.38	147 (retrograde)	1908/Melotte
Carme	8–40	22,350,000	692	0.21	163 (retrograde)	1938/Nicholson
Ananke	6–28	20,700,000	617	0.17	147 (retrograde)	1951/Nicholson
Elara	80 ± 20	11,740,000	260.1	0.207	24.8	1905/Perrine
Lysithea	6–32	11,710,000	260	0.130	29.0	1938/Nicholson
Himalia	170 ± 20	11,470,000	250.6	0.158	27.6	1904/Perrine
Leda	2–14	11,110,000	240	0.146	26.7	1974/Kowal
Callisto	4,820	1,880,000	16.69	0.01	0.2	1610/Galileo
Ganymede	5,270	1,070,000	7.16	0.001	0.2	1610/Galileo
Europa	3,130	670,900	3.55	0	0.5	1610/Galileo
Io	3,640	421,600	1.77	0	0	1610/Galileo
1979J2	80 (est)	222,400	0.675	?	?	1980/Synnott
Amalthea	270 × 155	181,300	0.489	0.003	0.4	1892/Barnard
1979J1	40 (est)	129,000	0.298	?	?	1979/Jewett
1979J3	40 (est)	127,600	0.295	?	?	1980/Synnott

SOURCE: NASA.

Table 3   Physical Data for the Galilean Satellites

Galilean Moon	Diameter (km)	Mass (kg)	Density (g/cm³)	Albedo
Callisto	4,820	1.074×10^{23}	1.8	0.17
Ganymede	5,270	1.490×10^{23}	1.9	0.43
Europa	3,130	0.487×10^{23}	3.0	0.64
Io	3,640	0.891×10^{23}	3.5	0.63

SOURCE: NASA.

and least dense of the Galilean moons. It has a diameter of 4,820 kilometers and an average distance of 1.88 million kilometers from Jupiter's center of mass. (See fig. 3.) Callisto is a dark (0.17 albedo), heavily cratered celestial body. It has a density of 1.8 grams per cubic centimeter (g/cm³), suggesting that water (ice) makes up perhaps 50 percent of its bulk composition. Scientists believe that Callisto is the oldest, most heavily cratered object yet observed in the Solar System. It appears to be a tectonically dead world, whose surface displays no history of geologic activity. There appear to be no cracks, no volcanoes and no mountains—only thousands and thousands of impact craters. *Voyager* spacecraft measured surface temperatures at low latitudes that ranged from 155 degrees Kelvin in the daytime to 80 degrees Kelvin at night.

Ganymede is Jupiter's largest moon. It orbits Jupiter at a distance of about 1 million kilometers. Ganymede's diameter of 5,270 kilometers makes it the largest moon in our Solar System. (The Saturnian moon Titan is slightly smaller, with a diameter of about 5,150 kilometers.) Ganymede's estimated density of 1.9 grams per cubic centimeter suggests that it, like Callisto, has a bulk composition of about 50 percent water by weight. *Voyager* photographs indicate that approximately half of Ganymede's surface is exposed (water) ice and half is darker rock. (See fig. C10 on color page E.) Current structural models for this Jovian moon imply a silicate-rich core, an ice crust and either a liquid water mantle or a solid, warm, convecting ice mantle.

Spectroscopic studies indicate that Ganymede's crust may be mostly water in the form of ice mixed with silicates. Such a crust is unable to support high vertical relief because creep or flow of the crustal ice eventually erases such structures.

Voyager photographs have identified at least four types of surface areas on Ganymede. Each type of surface may represent a distinct stage in the history of this fascinating satellite.

These four surface features of Ganymede are believed to have formed within the first billion years of the moon's existence and then remained frozen in their current position ever since. *Voyager* spacecraft sensors recorded surface temperatures (at low latitudes) that ranged from 85 to 145 degrees Kelvin.

Europa (see fig. 4) is the smallest of the Galilean moons. It has a 3,130-kilometer diameter (about 10 percent smaller than our Moon) and an average density of 3 grams per cubic centimeter, suggesting a primary composition of 80 percent silicate rock and 20 percent water. Europa is about 671,000 kilometers from Jupiter's center of mass. Its albedo of 0.64 is the highest observed value for any Jovian satellite, suggesting a surface covered by (water) ice and frost.

Europa has the smoothest surface found to date on any celestial body in the Solar System. No large-scale physical relief (such as mountains, valleys or even large impact craters) can be seen on its surface. However, there are long linear structures that resemble cracks in the shell of a hard-boiled egg. These linear features crisscross the incredibly smooth surface and are thought to be shallow cracks in an icy crust.

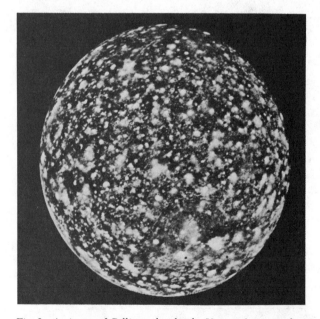

Fig. 3   An image of Callisto taken by the *Voyager 2* spacecraft at a range of 1.1 million kilometers. Scientists think that Callisto's surface is a mixture of ice and rock that dates back to the final stages of planetary formation (over 4 billion years ago)—when its surface was bombarded by a torrent of meteorites. Younger craters show up as bright spots, probably because they expose fresh ice. (Courtesy of NASA.)

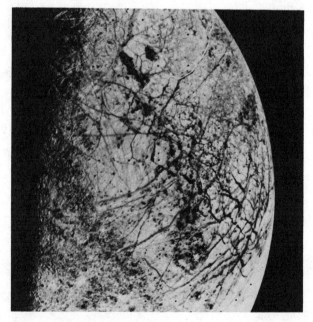

Fig. 4   Europa, smallest of the Galilean moons, as imaged by the *Voyager 2* spacecraft at a distance of 241,000 kilometers. (Courtesy of NASA.)

One of the most exciting possibilities, strongly hinted at by the *Voyager* data, is that Europa may have an ocean of liquid water beneath its surface layer of ice and frost. Figure 5 shows a possible structural model for Europa, including this liquid water ocean. The *Voyager* imagery depicted Europa as a world covered with cracked ice sheets. This moon is composed of rock and ice, with the denser silicate rock thought to form a core. Remnant internal heat from radioactive decay or tidal pumping (or both) may allow this water ocean to exist as a liquid at some depth beneath the protective ice surface. As shown in figure 5, when a crack in the ice occurs, the liquid water will flash-boil into the vacuum of space, and then it will fall back on Europa as frost.

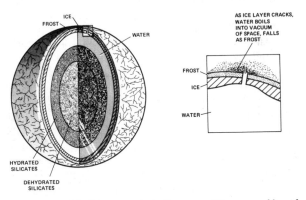

EUROPAN OCEAN

Fig. 5   A model of Europa—including a possible ocean of liquid water beneath its outer layer of ice. (Courtesy of NASA.)

Some exobiologists now speculate that if the Europan ocean exists, it may also contain extraterrestrial life-forms! They assume that Europa, like its parent planet Jupiter, had methane, ammonia and water all present as primordial volatiles. An earlier, much warmer Jupiter may have stimulated the chemical evolution of life in Europa's ancient ocean. Then, as Jupiter cooled and the environmental conditions on Europa changed, these alien life-forms, once started, may have tenaciously evolved into hardier creatures that could now be lurking in the depths of this extraterrestrial ocean. These Europan life-forms—should they exist—might cluster around the cracks or thin spots in the surface ice, desperately trying to gather the feeble, but life-supporting, rays of the Sun and even "planetshine" from nearby Jupiter. (At Jupiter's distance from the Sun, the solar flux is only about 1/27th what it is above the Earth's atmosphere.)

Perhaps submarine volcanic activity on Europa provides the energy necessary to heat sulfur compounds that are then used by extraterrestrial microorganisms in a process called chemosynthesis. (Chemosynthesis is a chemical parallel to photosynthesis.) Exobiologists are quick to point out that life, once started on Earth, has now spread over the planet to even the most hostile locations, such as under the polar ice pack and in the darkest depths of the oceans.

They therefore reason that it is possible that alien life, once initiated billions of years ago in a more benign ancient Europan ocean, may still cling to existence in the current dark, watery world beneath the moon's protective layer of surface ice. Of course, only additional exploration will resolve this most intriguing line of speculation.

Data collected by the *Voyager* spacecraft indicate that Europa has a maximum daytime surface temperature of 125 degrees Kelvin in its equatorial regions.

Io is the innermost and densest of the Galilean satellites. It has a distinctly orange color and possesses an albedo of 0.63. Io lies at a distance of 422,000 kilometers from Jupiter's center of mass and has a diameter of 3,640 kilometers. Over geologic time, Io has been dominated by a powerful heating mechanism caused by the modulation of Jupiter's intense tidal force due to the orbital motions of other Galilean satellites.

Most space scientists now believe that Io's internal energy is released by this tidal pumping phenomenon. Tidal pumping may be explained as follows: If Io were Jupiter's only moon, it would keep the same face toward Jupiter all the time (just as our Moon does when it travels around the Earth). Jupiter's immense gravitational field would pull Io's crust upward about a hundred meters—approximately the length of a football field. This crustal bulge, or "tide," would be stationary. However, Io is not alone in orbit around Jupiter. As Europa and Ganymede pass Io, their gravitational fields create perturbations, or changes, in the shape of Io's orbit. These perturbations then cause the amplitude, or height, of Io's tidal bulge to rise and fall, generating tremendous frictional heat in Io's interior.

Without a doubt, Io surpasses Earth as the most volcanically active body in the Solar System. Eruptions on Io appear to last longer than those on Earth. Io's volcanic eruptions also appear to be more violent than Earth's. *Voyager* spacecraft measurements indicated that materials are fired from Io's volcanoes at speeds as high as 3,600 kilometers per hour. In contrast, the Earth's most violent volcano, Mt. Aetna, ejects material at about 180 kilometers per hour. (See fig. C11 on color page E.)

Io is a waterless, rocky body about the size of the Moon. On Earth, volcanoes are thought to be generated by heat from decaying radioactive materials in the terrestrial lithosphere, and the major driving force of the Earth's explosive volcanoes is steam.

Since Io is waterless, what replaces steam as the major driving force of Io's volcanoes? The *Voyager* spacecraft detected sulfur dioxide gas coming from Io. Planetary scientists currently believe that gases from molten sulfur deep beneath Io's crust (melted by the effects of tidal heating) provide the driving force for the satellite's volcanoes. Gaseous sulfur erupting from Io's volcanoes would cool quickly and condense on the moon's surface. This material could then explain the satellite's vivid colors.

Io's surface is estimated to be about 1 million years old, making it the youngest and most dynamically active in the entire Solar System. Its appearance and high density (3.5

grams per cubic centimeter) suggest that most of the lighter volatiles, such as water and carbon dioxide, have been driven off, while the heavier, less volatile compounds, such as sulfur and sulfur dioxide, have been volcanically recycled into its crust—creating its bright red, orange, yellow and black appearance.

Io's eruptions may send particles throughout the Jovian magnetosphere. (The magnetosphere is the volume of space around a planet that traps or deflects charged atomic particles.) Jupiter's immense magnetosphere was detected by the *Pioneer* and *Voyager* spacecraft as far out as approximately 15 million kilometers on Jupiter's day side and beyond Saturn's orbit (more than 600 million kilometers away) on Jupiter's night side (see fig. 6).

The *Voyager* spacecraft detected particles of sulfur and oxygen, thought to be from Io, in a doughnut-shaped cloud, or torus, of ions that wobbles around Jupiter at the same distance as Io's orbit. These ions may follow magnetic field lines from the orbiting torus not only to the outer edges of the Jovian magnetosphere but also inward to Jupiter's north and south poles, causing the brilliant auroras also seen by the *Voyager* spacecraft. Some space scientists speculate that the volcanic eruptions on Io may be the source of the faint ring discovered around Jupiter. Some estimates indicate that Io may be ejecting up to three metric tons of material each day into space. At this rate, Io would lose less than 0.1 percent of its mass in the next billion years.

Moving inward from the Galilean satellites, we find the four innermost known satellites of Jupiter: 1979J2, Amalthea, 1979J1 and 1979J3.

The Jovian moon now tentatively called 1979J2 was the second satellite to be discovered by the *Voyager* mission. It is about 80 kilometers in diameter and lies some 222,400 kilometers from Jupiter's center of mass (just 151,000 kilometers above the top of the swirling Jovian atmosphere). It circles Jupiter in about 16 hours and 11 minutes and appears to consist of carbonaceous chondrite material mixed with sulfur (possibly supplied by Io).

The diamond-shaped Amalthea was discovered in 1892. From *Voyager* imagery, it appears to be about 270 kilometers along its major axis, 170 kilometers on its intermediate axis and 155 kilometers on its polar axis. It has a mean distance of 181,300 kilometers from Jupiter's center of mass. Amalthea is dark and red, possibly due to sulfur compounds deposited from Io. It has an albedo of approximately 0.05.

The satellite 1979J1 is located at the outer edge of the Jovian ring system. It circles Jupiter in about 0.3 day and is 129,000 kilometers away from Jupiter's center of mass. This satellite has an approximate diameter of 40 kilometers and is one of the three moons discovered by the *Voyager* spacecraft.

Finally, the innermost known Jovian satellite is tentatively called 1979J3. It, too, was discovered by *Voyager* and is approximately 40 kilometers in diameter. It has an orbital velocity of 113,600 kilometers per hour (31.6 kilometers per second), making it the fastest-moving moon yet discovered in our Solar System. It takes about seven hours to orbit Jupiter at a mean altitude of 56,200 kilometers above the top of the Jovian clouds (or some 127,000 kilometers from Jupiter's center of mass).

See also: **Galilean satellites; Galileo Project;** *Pioneer 10, 11;* **Voyager**

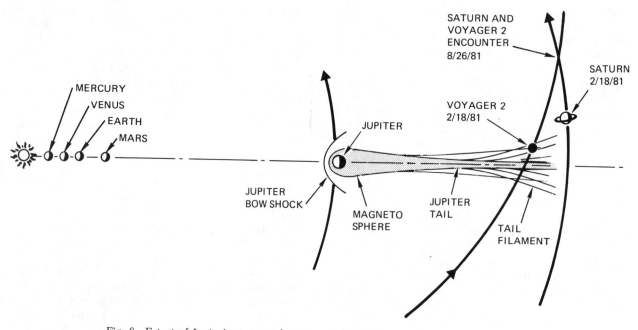

Fig. 6   Extent of Jupiter's magnetosphere/magnetotail. (*Voyager 2* trajectory to Saturn is also shown.) (Courtesy of NASA.)

H

Kirkwood gaps Gaps or "holes" in the main asteroid belt between Mars and Jupiter where essentially no asteroids are located. These gaps, initially explained in 1857 by Daniel Kirkwood, an American astronomer (1814–95), are caused by the gravitational attraction of Jupiter. They occur at distances from the Sun that correspond to orbital periods that are harmonics (that is, $1/2$, $2/5$, $1/3$, $1/4$, and so on) of Jupiter's orbital period.

See also: **asteroid**

Klystron An electron tube used to generate and amplify microwave current.

See also: **Satellite Power System**

L

Laboratory hypothesis A variation of the Zoo hypothesis response to the Fermi paradox. This particular hypothesis postulates that the reason we cannot detect or interact with technically advanced extraterrestrial civilizations in the Galaxy is because they have set the Solar System up as a "perfect" laboratory. These hypothesized extraterrestrial experimenters want to observe and study us but do not want us to be aware of or influenced by their observations.

See also: **Fermi paradox; Zoo hypothesis**

Lagrangian libration points The five points in outer space where a small object can have a stable orbit in spite of the gravitational attractions exerted by two much more massive celestial objects when they orbit about a common center of mass. The existence of such points was first postulated by the French mathematician Joseph Louis Lagrange (1736–1813). The Trojan group of asteroids, which occupy such Lagrangian points 60 degrees ahead of and 60 degrees behind the planet Jupiter in its orbit around the Sun, are one example.

In cislunar space—that is, the region of space associated with the Earth-Moon system—the five Lagrangian points arise from a balancing of the gravitational attractions of the Earth and Moon with the centrifugal force that an observer would feel in the rotating Earth-Moon coordinate system. The main feature of these points or regions in cislunar space (see fig. 1) is that an object placed there will

keep a fixed relation with respect to the Earth and Moon as the entire system revolves around the Sun.

The Lagrangian points called L_1, L_2 and L_3 in figure 1 are saddle-shaped "gravity valleys," with the interesting property that if you move an object at right angles (that is, perpendicularly) to a line connecting the Earth and the Moon (called the Earth-Moon axis), this object will slide back toward the axis; however, if you displace it along this axis, it will move away from the Lagrangian point indefinitely. Because of this, these three Lagrangian points are called points of unstable equilibrium.

In contrast, the Lagrangian points L_4 and L_5 represent bowl-shaped "gravity valleys." If an object at L_4 or L_5 is moved slightly in any direction, it returns to the Lagrangian point. These two points are therefore called points of stable equilibrium. Lagrangian points L_4 and L_5 are located on the Moon's orbit about the Earth, at equal distance from both the Moon and the Earth. They have been proposed as the sites for large human settlements in cislunar space.

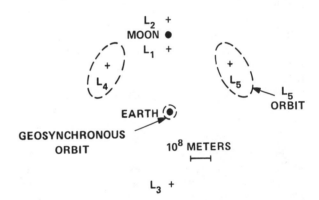

Fig. 1 Five Lagrangian libration points in the Earth-Moon system. (Drawn by the author.)

large space structures The building of very large and complicated structures in space will be one of the hallmarks of our extraterrestrial civilization. Although enormous, these structures will actually help to shrink the total cost of using space, and their construction and operation will support the "humanization of space." The Space Shuttle Orbiter's cargo bay can carry up to 29,500 kilograms of payload from the surface of the Earth to low Earth orbit. Transportation costs are assessed by both the mass of the payload and its volume. So it is very wise to design space hardware that is both light and modest in dimensions. Structures that would have been too fragile to stand up under their own weight on Earth can now be compactly stored in the Orbiter's cargo bay and then safely deployed in their final, extensive configuration in the microgravity environment of space.

The new capability to supervise deployment and construction operations in orbit is a crucial factor in the effective use of space. The Shuttle can carry a work force of up to seven astronauts per flight and will remain close at hand while the early construction jobs are performed.

All of this creates exciting new possibilities for the engineering of space hardware and poses an entirely new set of technical challenges to space technologists. What are the strongest, lightest and most stable materials to use in space construction? How do you load the Space Shuttle so as to build these colossal objects with the fewest trips into space? What are the best ways to assemble these structures once the materials are delivered to the orbiting "construction sites"? And perhaps the most obvious question: What kinds of structures will we want to build?

Television viewers in the remote regions of Alaska, the Rocky Mountains and Appalachia rapidly joined the world of modern space-based electronic communications and entertainment in 1974. The reason was simple—ATS-6 (standing for Advanced Technology Satellite-6), a nine-meter-wide dish antenna that relayed TV signals down to small receivers in previously isolated areas of the United States, was successfully launched. ATS-6 had the largest civilian communications antenna launched in the pre-Shuttle era.

Some of the new superantennas now being considered will be 10 times that size, or bigger than a football field! This tremendous size means a boost in transmitting power as well as increased sensitivity to weak signals from the ground. As a result, instead of needing massive antenna dishes on the Earth straining to hear weak messages from space as we do today, the roles will be reversed. (This reversal of roles is also called a technology inversion.) A few giant antennas placed in geostationary orbits covering the globe will replace numerous smaller communications satellites now in space. And millions of inexpensive satellite-receiving dishes on the rooftops of homes and businesses will receive signals now picked up by very large and powerful ground stations. The true "electronic cottage" will have been born.

The implications of this coming boom in antenna performance for change in our daily lives are profound. Many exciting ideas and technologies consistent with this "large-antenna revolution" already exist. These include: working models of personal "Dick Tracy" wrist radios, designs for electronic mail systems with direct home delivery via satellite and 300-channel television sets tuned to stations all over the globe.

And there will be other exciting and vital space-based information services economically emerging as new space industries as a result of large antenna farms and platforms in orbit. For example, figure 1 shows the large antenna system of the proposed Earth Observation Spacecraft (EOS) being assembled in orbit. When folded, this system would be only 4.1 meters wide by 17.8 meters long and could easily be carried in the Shuttle Orbiter's cargo bay. However, when erected in orbit, it is actually some 120 meters by 60 meters and has a 116-meter-long mast. This advanced antenna system is designed to observe Earth resources, including measuring soil moisture, sea-surface conditons, and ice and its boundaries.

Fig. 1 In-orbit assembly of an intricate antenna system for a proposed Earth Observation Spacecraft (EOS). (Courtesy of NASA.)

The first large space antennas will most likely be deployables. They will fold into compact containers on Earth, go up whole in one Shuttle flight, then deploy automatically in space in a single operation. The key, obviously, is to have the largest possible antenna dish unfolding from the smallest and lightest possible package.

One type of deployable—the hoop-column, or "maypole," antenna—would open up once in orbit much like an umbrella does. A cylinder no bigger than a school bus can be transformed within an hour into a huge antenna dish 100 meters across. Depending on the length of the various strings that stretch the fabric taut inside its stiff outer hoop, this type of antenna can be designed into many shapes—that is, the bowl of the dish could be made flat, more hollowed out or even made of four different surfaces, each focusing a beam in its own different direction. Multibeam feeds could also allow one antenna to do the work of several by pointing signals toward different areas of the Earth's surface.

In another type of deployable antenna, called the offset wrap-rib type, the dish fabric is attached to flexible ribs that wrap around a central hub. The whole package is quite compact initially, but once in space, another marvelous transformation in size takes place. A long (some 150 meters for a 100-meter-diameter antenna dish) mast telescopes out from the core and turns a corner so that the dish is offset and not blocked by the mast. This is an advantage in sensitive radar and radiometry missions. Then, like a pinwheel coming to life, the ribs unfurl and straighten until they fully extend to stretch and support a round dish.

Whatever their ultimate shape, these large space structures will place great demands on the materials from which they are made. Even though they will be free from the weight stresses imposed by gravity on the surface of the Earth, there will be other strains from their tight packaging and from the space environment itself. Space technologists and engineers will need to build these structures with new materials—materials that are at the same time light, very strong, thermally stable and either flexible or rigid (depending on their application). Telescoping masts must be light, yet remain very stiff. Antenna ribs, on the other hand, need to be strong but should be flexible enough to wrap around their hub. Furthermore, the configuration needs to remain fixed in position equally well in the hot Sun as in the frigid shadows encountered in orbit, because if a structure were to expand with heat or shrink with cold, it would upset the extremely precise shape of an antenna (some of which can be out of tolerance no more than a few millimeters in a total diameter of 100 meters).

One substance that appears to meet these rigorous demands quite well is the graphite-epoxy composite now used in lightweight tennis rackets, golf clubs, airplane parts and the Space Shuttle. A three-meter-long hollow tube of this material can be lifted with one finger yet, in its particular applications, is ten times stronger than steel.

Other materials are tailored for specific jobs. The hundreds of threads that pull and stretch a hoop-column antenna into shape might be made of quartz filament, because quartz is very stable. The antenna dishes themselves can be made of fabrics that fold like cloth before they are deployed. These would be metal meshes woven like nylon stockings or soft patio screening and coated with a very thin layer of gold for reflectivity. A finer mesh will be used for dishes that handle smaller wavelengths. For very small wavelengths there are ultrathin membranes made of transparent films coated with metals that look and feel like sheets of Christmas tinsel.

Let us now imagine that six different groups want to fly different remote sensing instruments in Earth orbit, all at about the same altitude and inclination. Instead of crowding and cluttering low Earth orbit with six different spacecraft, why not just build one large platform to which all six instruments are attached? They could then share the cost of the power and communications systems, stability control and cooling devices. Shuttle astronauts or space-station workers would need to visit just one place in space (instead of six) to repair and maintain the systems. This space platform appears to be a technically sound and economically good idea.

Some of these large space platforms, especially those dedicated to communications and information services, will need to hover in geostationary Earth orbits (GEOs) about 35,900 kilometers high in order to look down on large sections of the globe or to stay fixed in one spot (as seen by an observer on Earth). Since the Space Shuttle Orbiter ascends no higher than a few hundred kilometers above the surface of the Earth, orbital transfer vehicles must be attached either to an undeployed package (as taken right out of the Orbiter's cargo bay) or to an already assembled structure to boost it to a higher orbit.

Eventually, no matter how cleverly the platforms and antennas are packed, they will be too large to unfold in a single deployable unit. At that point, we will have to send up "erectable" space structures in pieces. These pieces can be loaded in the Shuttle's cargo bay on Earth, lifted into low Earth orbit, unfolded and finally assembled by space workers into a single, giant structure.

What kinds of building blocks will we use on these floating construction devices? Ideally, they should be basic, simple and adaptable to many different types of structure. These "erectables" have their roots in common household objects—in collapsible cardboard boxes, folding chairs, telescoping car radio antennas, accordion baby gates—anything we have tried to make smaller and more portable. Masts for dish antennas will telescope into their full lengths from small cylinders. Latticed trusses will store as flat packages, unfold first into diamond shapes and then finally into tetrahedrons.

But in each case, no matter how flexible their hinges when stored, the modules must hold stiff when deployed in space, as would the hexagonal pieces for large antennas. Looking a bit like minitrampolines when unfolded, these hexagons will be attached precisely and rigidly to form great reflecting surfaces many city blocks in area.

Not all of these building blocks will need to unfold. Some of them will store quite easily just as they are, such as the light graphite-epoxy tubes that will stack inside one another like ice cream cones and sit on racks in the cargo bay like arrows in a quiver. These tubes would then be attached to form struts that can themselves be joined to build larger beams or trusses. Or they might be used to form a thin hoop for a space antenna.

Unfolding with the push of a button, deployable antennas will, in a sense, build themselves. "Erectables," on the other hand, will not. Space workers or very smart automated machines controlled by astronauts will have to snap the separate pieces together. Ongoing assembly projects will therefore mean having the first construction sites in space, giving rise to a new type of work for human beings.

Many important factors must be taken into account in planning and conducting a space construction project. These include: safety and fatigue of the space workers; speed in moving from one space location to another; the requirement for simple tools and the need to restrain them so they don't float away; and how much time the Space Shuttle or manned orbital transfer vehicle would lose lingering at the construction site.

In one method of assembly, space workers tethered to the Shuttle would simply move from beam to column to module, snapping, locking or latching everything together. Their travel time could be shortened by wearing the jet-packs called manned maneuvering units (MMUs). It is not yet, however, entirely certain how we will combine space worker and sophisticated machine in space deployment and assembly operations. For some projects it might be more efficient to move the astronauts around on a scaffold in some type of mobile work station instead of having them free-flying all over the construction site. The scaffold would rest on a frame in the Shuttle Orbiter's cargo bay and move either up-down or right-left. As sections of the structure were completed, they would be moved away from the work station so that the part to be built is always in reach. Astronauts could also stand in open cherry pickers attached to the Shuttle's 15-meter remote manipulator arm and be moved from beam joint to beam joint like telephone line workers working on high wires. Even more sophisticated operations would involve the use of closed cherry pickers, where space workers inside a comfortable chamber would operate sophisticated manipulator arms in a highly mobile space system.

Or, for repetitious and perhaps very dangerous construction tasks, unmanned free-flying teleoperators—essentially smart robots—could do the work with their own dexterous manipulator arms. There could also be assembler devices to form three-dimensional structures from struts by following simple, repeatable steps and maneuvering television units that would transmit pictures to technicians in the Shuttle or space-station control room so that they could direct the assembly work by remote control. These advanced construction devices would most likely be used later in the Shuttle era, perhaps in conjunction with permanently inhabited space platforms. In the meantime, astronauts will have to learn to erect structures the size of large stadiums in the unusual and challenging world of microgravity. Seemingly easy tasks will become complicated. For example, workers trying to turn ordinary bolts will be as likely to turn themselves as the bolts, thanks to the lack of leverage that accompanies the free-fall condition of orbiting objects.

After deployable and erectable systems, the next logical step will be to build large structures completely from scratch by fabricating the construction elements in space. A machine for that very purpose has already been designed. Called the automated beam builder, it can be deployed at one end of the Shuttle's cargo bay. Spools of ultralight material, most likely graphite-epoxy or metal matrix composites, would be loaded into the machine on Earth and carried into orbit. Once at the space construction site, this beam builder would heat, shape and weld the material into meter-wide triangular beams that might be cut to any length, then latched together to build large structures. By loading the Orbiter's payload bay with extra spools, enough material could be carried up in one trip to build thousands of meters of beams. With the beam builder, the dreams for humanity's extraterrestrial civilization as found in the science-fiction literature would be converted to practical blueprints for colossal structures that would dwarf the Space Shuttle docking with them (see fig. 2).

As such platforms and structures grow in size, they will become even more complicated. The need to control and maintain a perfectly fixed attitude is crucial to antennas and remote sensing instruments, which would be useless unless pointed precisely. This means that these mammoth structures will not be able to wobble or bend out of alignment. Several things will combine to distort their orbital positions, because as large as these structures will be, they will also be relatively light and delicate. For example, a 50-meter-diameter space antenna would have a mass of approximately 4.5 metric tons. In comparison, NASA's 64-meter antenna dish in Goldstone, California, with its heavy supports and concrete base, has a mass of 7,260 metric tons. An entire large space structure could easily be pushed out of alignment by the steady, streaming pressure of the solar wind; in addition, every time an Orbiter docked or made physical contact with one of these large platforms, the delicate balance of forces controlling an orbiting object would again be upset. Some future structures will be so very large and extensive that they will

Fig. 2   Colossal space structure being assembled in low Earth orbit. (Courtesy of NASA.)

even experience "tidal effects," as if they were minimoons with gravity (Earth's in particular) tugging harder on one edge than on the other.

Obviously, to successfully build and use such large structures in space, we will need precise and sophisticated controls for stability, starting with sensors to indicate just when the structure is moving out of alignment. On-board computers could then determine how to compensate; and finally, small gas jets located around the structure would fire to make the necessary corrections. All of this would be a constant, self-regulating process in an age when mammoth space structures served humanity throughout cislunar space.

See also: **microgravity; space construction; space industrialization; space platform; space station**

laser A term that stands for: *l*ight *a*mplification by *s*timulated *e*mission of *r*adiation. The laser is a device for generating coherent beams of (monochromatic) optical radiation.

See also: **Satellite Power System**

L₅ pathway to space settlements One proposed pathway to developing space settlements involves the use of large habitats positioned at libration point 5 (L₅). These human communities in space would then orbit the Sun in fixed relation to the Earth and the Moon. This concept uses the space transportation pathways shown in figure 1. The Moon would be mined for oxygen, aluminum, silica and the undifferentiated matter needed to shield the habitat from the ionizing radiations found in outer space. Lunar workers would ship millions of metric tons of lunar

material each year to libration point 2 (L₂) by means of an electromagnetic mass driver. At L₂, also called Lagrangian point 2, the lunar material is gathered with a mass catcher and transshipped to L₅ to be refined and processed. In this scheme small amounts of special materials, plastics and organic substances would be shipped to the space settlement at L₅ from the Earth. At L₅ the space settlers would build and assemble giant Satellite Power Systems and deliver these structures to their operational locations at geosynchronous Earth orbit (GEO). In addition to fabricating Satellite Power Systems that would supply energy to Earth, the L₅ inhabitants would establish a stable economy in space by raising their own food (in space farming facilities) and would also work on the construction of other large space habitats. In this way, once firmly established above the Earth's atmosphere, human settlements would start to replicate themselves and begin to inhabit the majority of appropriate regions within cislunar space.

See also: **Lagrangian libration points; Satellite Power System; space settlement**

life in the Universe The history of life in the Universe can be explored in the context of a grand, synthesizing scenario called cosmic evolution (see fig. 1). This sweeping scenario links the development of galaxies, stars, planets, life, intelligence and technology and then speculates on where the ever-increasing complexity of matter is leading.

And where *does* all this lead to in the cosmic-evolution scenario? Well, we should first recognize that all living things are extremely interesting pieces of matter. Life-forms that have achieved intelligence and have developed technology are especially interesting and valuable in the cosmic evolution of the Universe! Intelligent creatures with technology, including human beings here on Earth, can exercise conscious control over matter in progressively more effective ways as the level of their technology grows. Ancient cave dwellers used fire to provide light and warmth. Modern humans harness solar energy, control falling water and split atomic nuclei to provide energy for light, warmth, industry and entertainment. People in the

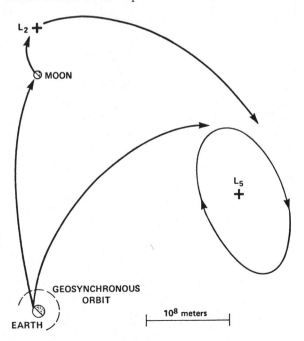

Fig. 1 Paths through space for space settlements at L₅. (Courtesy of NASA.)

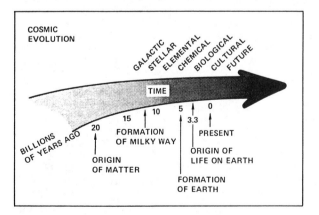

Fig. 1 The scenario of cosmic evolution. (Courtesy of NASA [from the work of Eric J. Chaisson].)

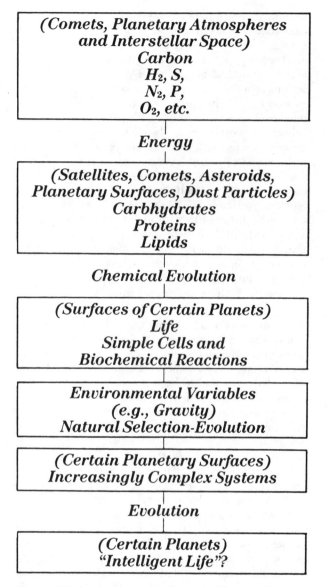

Fig. 2   The chemical evolution of life on Earth placed within the context of the cosmic evolution of our Solar System. (Drawing courtesy of NASA.)

next century will "join atomic nuclei" (controlled fusion) to provide light, warmth, industry, entertainment and interplanetary communications for their settlements on the Moon and Mars! The trend should be obvious. Some scientists even speculate that if technologically advanced civilizations throughout the Galaxy can learn to live with the awesome powers unleashed in such advanced technologies—then it may be the overall destiny of advanced intelligent life-forms (including perhaps humans) to ultimately exercise control over all the matter in the Universe!

Let's now examine what scientists currently think is the process of the chemical evolution of terrestrial life. Figure 2 describes the theory of the chemical origin of life on

Earth, within the overall scenario of cosmic evolution (as previously described in fig. 1). Many scientists now believe that the origin of life on Earth was closely tied to the origin of the Solar System. Figure 3 suggests a relationship between various celestial objects found in our Solar System and its chemical evolution. Just as the term *biological evolution* implies that all terrestrial organisms have a common ancestry, so does the term *chemical evolution* (in a general sense) imply that all matter in the Solar System has a common heritage—the primordial cloud of interstellar dust and gas.

The chain of events involving the overall chemical evolution of our Solar System may have been triggered by an ancient supernova. The violent detonation from this nearby supernova would have sent a burst of matter and radiation forward, causing our ancestral nebula to start condensing. The solid arrows in figure 3 describe contributions of matter from one celestial object to another. The dashed line suggests that comets may preserve intact the original interstellar dust and gas material from which the entire Solar System evolved.

A general model for the rise of life on Earth has also been developed. It starts (almost immediately after the original condensation, or accretion, of the Earth) with a period of chemical evolution on the planet that resulted in the formation of the small organic molecules essential for life. These prebiotic (prelife) molecules were synthesized from the gases that made up the Earth's primitive atmo-

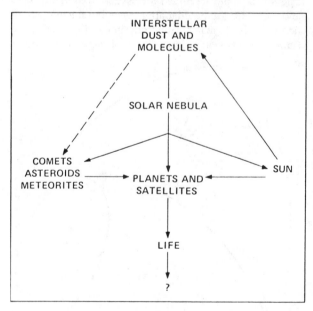

Fig. 3   The relationships between various celestial objects in the chemical evolution of our Solar System. (Solid lines indicate the transfer of matter from one celestial source to another; dashed lines indicate that comets may be composed of primordial matter from the original solar nebula; the arrow from the term *Life* indicates the possible flow of life from Earth as part of our extraterrestrial civilization.) (Drawing courtesy of NASA [from the work of Sherwood Chang].)

sphere. This initial atmosphere is believed to have been reducing in nature. That is, the first terrestrial atmosphere contained no free oxygen and was composed mostly of hydrogen (H_2), methane (CH_4), ammonia (NH_3) and water vapor (H_2O).

As millions of years passed, these gases were activated by natural energy sources, such as sunlight (especially ultraviolet radiation) and lightning. Under the influence of these energy sources, the original molecules broke apart, and their atoms recombined to form new, more complex molecules. These processes marked the start of the prebiotic synthesis of organic matter.

Then, under the influence of primitive planetary environmental conditions, these prebiotic organic molecules combined to make even larger, still more complex molecules. Ultimately, hundreds of thousands to millions of these molecules formed into structures that displayed the fundamental characteristics of living things: metabolism, respiration, reproduction and the transfer of genetic information. This stage was reached about 3½ billion to 4 billion years ago on Earth, or some ½ billion to 1 billion years after the Earth and the Solar System were formed from the initial cloud of interstellar dust and gas.

Over the next 3.5 billion years, these primitive living organisms slowly but continuously evolved into the vast array of living systems we find here on "Spaceship Earth" today. The fossilized record locked up in the terrestrial lithosphere provides the raw material for many laboratory studies. In fact, the Earth's rocks are a repository that describes chemical evolution from the first living cells up through the arrival of today's complex life-forms. As scientists now unravel the details of this process for the chemical evolution of terrestrial life, we must also ask ourselves another very intriguing question: If it happened here, did it or can it happen elsewhere? In other words, what are the prospects for finding extraterrestrial life—in this Solar System or perhaps on Earthlike planets around distant stars?

According to the principle of mediocrity (frequently used by exobiologists), there is nothing "special" about the Solar System or the planet Earth. Therefore, as these scientists speculate, if similar conditions have existed or are now present on "suitable" worlds around alien suns, the chemical evolution of life will also occur.

Perhaps a more demanding question is: Does alien life develop to a level of intelligence? And even more interesting: Do intelligent alien life-forms acquire advanced technologies and learn to live with these vast powers over nature? It is just possible that we here on Earth are the only life-forms anywhere in the Galaxy to acquire intelligence and high technology! Just think, for a moment, about the powerful implications of this simple conjecture. Are we the best the Universe has been able to produce in over 15 billion years of cosmic evolution? If so, a human being—any human being—is something very special.

Our preliminary search on the Moon and on Mars for extraterrestrial life in the Solar System has to date been unsuccessful. However, the giant, gaseous outer planets and their constellations of intriguing moons present some tantalizing possibilities. Who cannot get excited about a possible ocean of liquid water beneath the Jovian moon Europa and the (remote) chance that this extraterrestrial ocean might contain alien life-forms! Of course, even the final verdict concerning life on Mars (past or present) will not be properly resolved until more detailed investigations of the Red Planet have occurred. Perhaps a terrestrial explorer will stumble upon a remote exobiological niche in some deep Martian canyon; or possibly a team of miners, searching for certain ores on Mars, will uncover the fossilized remains of an ancient creature that roamed the surface of the Red Planet in more hospitable environmental eras! Speculation, yes—but not without reason.

All we can say now with any degree of certainty is that our overall understanding of the cosmic prevalence of life will be significantly influenced by the exobiological discoveries (pro and con) that will occur in the next few decades. In addition to looking for extraterrestrial life on other worlds in our Solar System, exobiologists can also search for life-related molecules in space in order to determine the cosmic nature of prebiotic chemical synthesis.

Recent discoveries, for example, show that comets may represent a unique repository of information about chemical evolution and organic synthesis at the very outset of the Solar System. (Scientists think that comets have remained unchanged since the formation of the Solar System.) Exobiologists now have evidence that the organic molecules considered to be the molecular precursors to those essential for life are prevalent in comets. These discoveries have provided further support for the hypothesis that the chemical evolution of life has occurred and is now occurring widely throughout the Galaxy. Some scientists even suggest that comets have played a significant role in the chemical evolution of life on Earth. They speculate that significant quantities of important life-precursor molecules could have been deposited in an ancient terrestrial atmosphere by cometary collisions.

Meteoroids are solid chunks of extraterrestrial matter. As such, they represent another source of interesting information about the occurrence of prebiotic chemistry beyond the Earth. In 1969, for example, meteorite analyses provided the first convincing proof of the existence of extraterrestrial amino acids. (Amino acids are a group of molecules necessary for life.) Since that time, a large amount of information has been gathered showing that many more of the molecules considered necessary for life are also present in meteorites. As a result of this line of investigation, it now seems clear to exobiologists that the chemistry of life is not unique to the Earth. Future work in this area should greatly help our understanding of the conditions and processes that existed during the formation of the Solar System. These studies should also provide clues concerning the relations between the origin of the Solar System and the origin of life.

The basic question, Is life—especially intelligent life—

unique to the Earth? lies at the very core of our concept of self and where we fit in the cosmic scheme of things. If life is precious and rare, then we have a truly serious obligation to the entire (as yet "unborn") Universe to carefully preserve the organic heritage that has taken over 4 billion years to evolve on this tiny planet. If, on the other hand, life (including intelligent life) is abundant throughout the Galaxy, then we should eagerly seek to learn of its existence. Fermi's famous question, "Where are they?" then takes on even more significance in the Age of Space.

See also: **amino acid; biological evolution; Drake equation; exobiology; extraterrestrial civilizations; extraterrestrial contamination; Fermi paradox; Viking Project**

light flash   A momentary flash of light seen by astronauts in space, even with their eyes closed. Scientists believe that there are probably at least three causes for these light flashes. First, energetic cosmic rays passing through the eye's "detector" (the retina) ionize a few atoms or molecules, resulting in a signal in the optic nerve. Second, HZE particles can produce Cerenkov radiation in the eyeball. (Cerenkov radiation is the bluish light emitted by a particle traveling very near the speed of light when it enters a medium in which the velocity of light is less than the particle's speed.) Finally, alpha particles from nuclear collisions caused by very energetic Van Allen belt protons can produce ionization in the retina, again triggering a signal in the optic nerve. Astronauts have reported seeing these light flashes in a variety of sizes and shapes.

See also: **cosmic rays; Earth's trapped radiation belts; hazards to space workers; HZE particles**

light-minute   A unit of length equal to the distance traveled by a beam of light (or any electromagnetic wave) in the vacuum of outer space in one minute. Since the speed of light (c) is 299,792.5 kilometers per second (km/sec) in free space, a light-minute corresponds to a distance of approximately 18 million kilometers.

light-second (ls)   A unit of length equal to the distance traveled by a beam of light (or any electromagnetic wave) in the vacuum of outer space in one second. Since the speed of light (c) in free space is 299,792.5 kilometers per second (km/sec), a light-second corresponds to a distance of approximately 300,000 kilometers.

light-year   The distance that light travels at 3×10^8 meters per second in one year (3.15×10^7 seconds). One light-year is equal to a distance of 9.46×10^{12} kilometers, or about 63,000 times the distance from the Earth to the Sun.

Local Group   A small grouping of about 20 galaxies, of which the Milky Way (our Galaxy) and the Andromeda Galaxy are dominant members.

Luna   A family of Soviet spacecraft intended for lunar exploration.

See also: **Moon; planetary exploration**

lunar   Of or pertaining to the Moon.

lunar bases and settlements   When human beings return to the Moon, it will not be for a brief moment of scientific inquiry as occurred in the Apollo program, but rather as permanent inhabitants of a new world. They will build bases from which to completely explore the lunar surface, establish science and technology laboratories that take advantage of the special properties of the lunar environment and exploit the Moon's mineral resources in support of humanity's extraterrestrial expansion.

In the first stage of a possible lunar-development scenario, men and women, along with their smart machines, would return to the Moon to conduct more extensive site explorations and resources evaluations. These efforts would pave the way for the first permanent lunar base.

The next critical stage in humanity's use of the Moon will be the establishment of the first permanent lunar base camp. The base shown in figure C29 on color page L has a not unintentional similarity to remote scientific bases and outposts found here on Earth. In this base a team of from 10 up to perhaps 100 "permanent" lunar workers will set about the task of fully investigating the Moon. They will take particular advantage of the Moon as a "science in space platform" and perform the fundamental scientific and engineering studies needed to confirm the specific roles that the Moon will play in the full development of cislunar space. For example, the discovery of frozen volatiles (including water) in the perpetually frozen recesses of the Moon's polar regions could change strategies for orbital-transfer-vehicle and space-station resupply as well as open up new, more rapid pathways for the development of a full-scale lunar civilization.

Many lunar base and settlement applications, both scientific and industrial, have been proposed since the Apollo program. Some of these concepts include: (1) a lunar scientific laboratory complex; (2) a lunar industrial complex to support space-based manufacturing; (3) an astrophysical observatory for Solar-System and deep-space surveillance; (4) a "fueling" station for orbital transfer vehicles that travel throughout cislunar space; (5) a training site and assembly point for manned Mars expeditions; (6) a nuclear waste repository for the very long-lived radioisotopes (such as the transuranic nuclides) originating in terrestrial nuclear fuel cycles and for spent space nuclear power plants used throughout cislunar space; and (7) the site of innovative political, social and cultural developments—essentially rejuvenating humanity's concept of itself and its ability to change its destiny.

All these lunar base and settlement applications are very exciting and definitely deserve expanded study—especially in the context of an operating Space Shuttle and a planned permanent space station in low Earth orbit. The question space planners are already beginning to ask is: Where do we go from a space station in low Earth orbit? The two most popular responses are: the Moon or Mars. A human expedition to Mars can be well served by the capabilities and technologies associated with a flourishing lunar base complex.

As lunar activities mature, the initial lunar base will grow into an early settlement of about 1,000 more or less "permanent" residents. Then, as the lunar industrial complex expands and lunar raw materials, food and manufactured products start to support space industrialization throughout cislunar space, the lunar settlement itself will expand to a population of around 10,000, with electric energy demands on the order of several megawatts.

Mature settlements on the Moon will continue to grow, reaching a population of about 500,000 and attaining a social and economic "critical mass" that supports true self-sufficiency from Earth. This moment of self-sufficiency for the lunar civilization will also be a very historic moment in human history. For from that time on, the human race will exist in two distinct and separate "biological niches"—we will be terran and nonterran, or extraterrestrial.

With the rise of a self-sufficient, autonomous lunar civilization, future generations will have a "choice of worlds" on which to live and prosper. Of course, such a major social development will also produce its share of "cultural backlash." Citizens of the 21st century may start seeing personal ground vehicles with such bumper-sticker slogans as: "This is my world—love it or leave it!"; "Terran go home"; or perhaps "Protect terrestrial jobs—ban lunar imports."

All major lunar-development strategies include the use of the Moon as a platform from which to conduct "science in space." Scientific facilities on the Moon will take advantage of its unique environment to support platforms for astronomical, solar and space science (plasma) observations. The unique environmental characteristics of the lunar surface include low gravity (one-sixth that of the Earth), high vacuum (about 10^{-12} torr [a torr is a unit of pressure equal to 1/760 of an atmosphere]), seismic stability, low temperatures (50 to 80 degrees Kelvin at the poles) and a low radio noise environment on the far side. More advanced astronomical observations will be made from space in the future, mainly to escape from the distortional effects of the Earth's atmosphere and ionosphere.

Astronomy from the lunar surface offers the distinct advantages of a low radio noise environment and a stable platform in a low-gravity environment. The far side of the Moon is permanently shielded from direct terrestrial radio emissions. As future radio-telescope designs approach their ultimate (theoretical) performance limits, this uniquely quiet lunar environment may be the only location in all cislunar space where sensitive radio-wave detection instruments can be used to full advantage, both in radio astronomy and in our search for extraterrestrial intelligence (SETI). In fact, radio astronomy, including

extensive SETI efforts, may represent one of the main "lunar industries" of the next century. In a figurative sense, SETI performed by lunar-based scientists will be "extraterrestrials" searching for other extraterrestrials.

The Moon also provides a solid, seismically stable, low-gravity, high-vacuum platform for conducting precise interferometric and astrometric observations. For example, the availability of ultrahigh-resolution (micro-arc-second) optical, infrared and radio observatories will allow us to search for extrasolar planets encircling nearby stars.

A lunar scientific base also provides life scientists with a unique opportunity to extensively study biological processes in reduced gravity (1/6 g) and in low magnetic fields. Genetic engineers can conduct their experiments in comfortable facilities that are nevertheless physically isolated from the Earth's biosphere. Exobiologists can experiment with new types of plants and microorganisms under a variety of simulated alien-world conditions. Genetically engineered "lunar plants," grown in special facilities, will become a major food source and also supplement the regeneration of the atmosphere in the lunar habitats.

The true impetus for large, permanent lunar settlements will most likely arise from the desire for economic gain—a time-honored stimulus that has driven much technical, social and economic development on Earth. The ability to provide useful products from native lunar materials will have a controlling influence on the overall rate of growth of the lunar civilization. Some "early lunar products" can now easily be identified. These products would support overall space-industrialization efforts. They include: (1) the production of oxygen for use as a propellant for orbital transfer vehicles used throughout cislunar space (for example, low Earth orbit to lunar orbit or lunar orbit to geostationary Earth orbit); (2) the use of "raw" lunar soil and rock materials for radiation shielding—a critical, mass-intensive component of space stations, space settlements and personnel transport vehicles; and (3) the production of ceramic and metal products to support the construction of large structures and habitats in space.

The initial lunar bases can be used to demonstrate industrial applications of native Moon resources and to operate small pilot factories that provide selected raw and finished products for use both on the Moon and in Earth orbit.

The Moon has large supplies of silicon, iron, aluminum, calcium, magnesium, titanium and oxygen. Lunar soil and rock can be melted to make glass—in the form of fibers, slabs, tubes and rods. Sintering (a process whereby a substance is formed into a coherent mass by heating [but without melting]) can produce lunar bricks and ceramic products. Iron metal can be melted and cast or converted to specially shaped forms using powder metallurgy. These lunar products would find a ready "market" as shielding materials, in habitat construction, in the development of large space facilities and in electric power generation and transmission systems.

Lunar mining operations and factories can then be

expanded to meet growing demands for lunar products throughout cislunar space. With the rise of lunar agriculture, the Moon may even become our "extraterrestrial breadbasket"—providing the majority of all food products consumed by humanity's extraterrestrial citizens.

One interesting space-industrialization scenario involves an extensive lunar surface mining operation that provides the required quantities of materials in a preprocessed condition to a giant space manufacturing complex located at Lagrangian libration point 4 or 5 (L₄ or L₅). These exported lunar materials would consist primarily of oxygen, silicon, aluminum, iron, magnesium and calcium locked into a great variety of complex chemical compounds. It is anticipated by space-technology visionaries that the Moon will become the chief source of materials for space-based industries in the latter part of the 21st century.

Numerous other tangible and intangible advantages of lunar settlements will accrue as a natural part of their creation and evolutionary development. For example, the high-technology discoveries originating in a complex of unique lunar laboratories could be channeled directly into appropriate economic and technical sectors on Earth, as new ideas, techniques, products and so on. The existence of "another human world" will create a permanent open-world philosophy for all human civilization. Application of space technology, especially lunar base–generated technology, will trigger a terrestrial renaissance, leading to an overall increase in the creation of wealth, the search for knowledge and even the creation of beauty by all humanity.

Our present civilization—as the first to venture into cislunar space and to create permanent lunar settlements—will long be admired, not only for its great technical and intellectual achievements but also for its innovative cultural accomplishments. Finally, it is not too remote to speculate that the descendants of the first lunar settlers will become first the interplanetary, then the interstellar, portion of the human race! The Moon and its emerging civilization will become our gateway to the Universe.

See also: **heliostat; Lagrangian libration points; Moon; Satellite Power System; space industrialization; space settlement**

lunar crater   A depression, usually circular, on the surface of the Moon. It frequently occurs with a raised rim called a ringwall. Craters range in size up to 250 kilometers in diameter. The largest lunar craters are sometimes called "walled plains." The smaller craters—say, 15 to 30 kilometers across—are often called "craterlets"; while the very smallest, just a few hundred meters across, are called "beads." Many lunar craters have been named after famous people, usually astronomers.

See also: **Moon**

lunar day   The period of time associated with the rotation of the Moon about the Earth. It is equal to 27.322 "Earth"-days. The lunar day is also equal in length to the sidereal month.

Lunar Geoscience Orbiter   Are there volatile materials such as water ice trapped in permanently darkened, frigid polar regions of the Moon? The exploitation of the lunar resource base and the development of our extraterrestrial civilization could be greatly enhanced by such a marvelous discovery! NASA's Lunar Geoscience Orbiter would assess the Moon's resources, including a search for frozen volatiles at the poles. It would also measure the Moon's elemental and mineralogical surface composition, surface topography and gravity field on a global basis. This mission can be launched at almost any time by the Space Shuttle/Centaur and inserted into an initial elliptical orbit around the Moon for certain very close remote sensing measurements, such as gamma-ray spectroscopy. This would be followed by a one-year operating period in which the spacecraft would travel in a 50- to 100- kilometer circular polar orbit, allowing all the regions of the Moon to be observed and assessed.

See also: **lunar bases and settlements**

lunar rovers   Manned and automated (robot) vehicles used to help explore the Moon's surface. The Lunar Rover Vehicle (LRV), shown in figure 1, was also called a space buggy and the "Moon car." It was used by American astronauts during the *Apollo 15, 16* and *17* expeditions to the Moon. This vehicle was designed to climb over steep slopes, go over rocks and move easily over sandlike lunar surfaces. It was able to carry more than twice its own mass (about 210 kilograms) in passengers, scientific instruments and lunar material samples. This electric-powered vehicle could travel about 16 kilometers per hour (10 miles per hour) on level ground. The vehicle's power came from two 36-volt silver-zinc batteries that drove independent ¼-horsepower electric motors in each wheel. Apollo astronauts used their space buggies to explore well beyond their initial lunar-landing sites. With these vehicles, they were able to gather Moon rocks and travel much farther and quicker across the lunar surface than if they had had to explore on foot. For example, during the *Apollo 17* expedition, the rover traveled 19 kilometers (12 miles) on just one of its three excursions. The informal four-wheeled-vehicle lunar speed record is now approximately 17.6 kilometers per hour (11 miles per hour) and was set by the *Apollo 17* astronauts.

Figure 2 shows a design concept for an automated, or robot, lunar rover vehicle. Automated surface rovers such as these would gather soil and rock samples from remote areas of the Moon—for example, the polar regions—and return these materials for analysis at an automated lunar base (shown in the background of fig. 2) or at a manned settlement. These surface rovers would help identify the

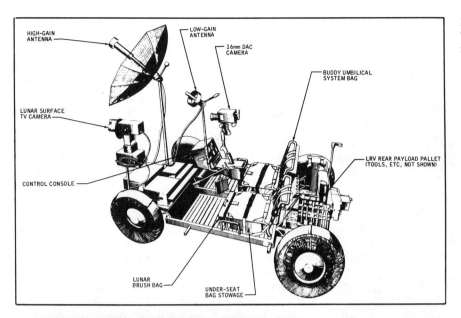

Fig. 1   The Lunar Rover Vehicle, or "Moon car," used by the Apollo astronauts. (Courtesy of NASA.)

Fig. 2   Artist's concept of an automated lunar rover exploring the remote regions of the Moon's south pole. Soil and rock samples gathered by the robot vehicle would be returned to an automated lunar base (*seen in the background*) for analysis. (Courtesy of NASA.)

location of valuable mineral deposits and pave the way for lunar mining operations.

See also: **extraterrestrial expeditions; lunar bases and settlements; Mars surface rovers; Moon; robotics in space** (contemporary space robots)

lunar soil and rock samples   Have you ever looked at a lunar rock very closely? Have you ever had the urge to touch something extraterrestrial? Maybe NASA's eductional program involving lunar materials is just right for you.

The Apollo Moon landings were a magnificent triumph of technology and human courage. These manned expeditions to another world also helped establish a new era in our study of the Solar System and the Universe. For the first time in human history, we could study another world at close range. Astronauts walked on the Moon, photographed the fine details of its once mysterious surface and set up instruments to probe its interior. From orbit around the Moon, other sensitive instruments in the Apollo spacecraft were used to map the Moon's chemical composition, gravity and magnetism.

The physical return of lunar soil and rock samples to the Earth was another major technical achievement of the Apollo program. Terrestrial scientists were able to investigate these precious pieces of extraterrestrial matter and gently pry from these long-silent rocks the complicated history of the Moon and the Solar System within which it resides. (See fig. 1.)

The six successful Apollo Moon-landing missions returned more than 2,000 different rock and soil samples from our celestial companion. The samples had a total mass of 342 kilograms (842 pounds). By carefully studying these lunar materials, space scientists learned that the Moon is not a uniform and monotonous world, but rather a complex and individual "planet" with its own unique history.

But these Moon rocks are more than just important scientific specimens. They are the tangible symbol of a great achievement in human history. They are extremely interesting and exciting to look at. And in a very special way, they bring us close to alien worlds.

Recognizing these significant benefits, the Office of Public Affairs at NASA's Johnson Space Center (JSC) maintains a public display and education program involving the loan of lunar samples. This Lunar Sample Display Program is divided into three subprograms: the Regular Display Program, the Educational Disk Program and the Thin Section Program.

The JSC Public Services Branch manages a traveling lunar materials display program consisting of display sam-

Fig. 1 *Apollo 12* lunar sample number 12039 (approximately 0.25 kilogram). It is a porphyritic rock, containing pyroxene (40 percent), plagioclase (40 percent) and olivine (20 percent) and is estimated to be 3.2 billion years old. (Courtesy of NASA.)

ples that range from 70 to 160 grams in size and are encapsulated in clear lucite pyramids. These displays are distributed to the NASA Centers throughout the United States. (Each NASA Center then services its regional area.) The display samples are available on loan to museums and planetariums or to any nonprofit organization sponsoring a community or civic event. The lunar-display loan period ranges from two weeks to two months. NASA imposes the following general requirements to qualify for this type of Moon-rock display: the sample must be hand-carried to and from locations; it must be secured in a safe or vault-type facility when not on display; and it must be kept under constant surveillance while on display. All requests for the Regular Display Program must be coordinated through the Public Services Branch, NASA Johnson Space Center, Houston, Texas 77058. Regional NASA Centers can provide additional information about this program.

The Educational Disk Program consists of six samples of lunar material (three different soils and three different rocks) encapsulated in a 15-centimeter (6-inch) diameter clear lucite disk. This disk is accompanied by written and pictorial descriptions of each sample in the disk, a film, a

sound and slide presentation, a teacher workbook and supporting printed materials. The program was especially designed for use as a science teaching aid in a classroom environment. Science teachers may qualify for the use of an educational disk in their own classroom by attending one of the many workshops sponsored by NASA's Space Science Education Specialists. These workshops are scheduled throughout the year at various locations around the United States. Museums and planetariums that have educational programs can also request the disk for use. The basic requirements imposed by NASA for using an educational disk are: the disk must be secured, while not in use, in a safe or vault-type facility; the package may be sent via registered mail to and from locations; and the disk must be kept under constant surveillance while on display. For additional information, school officials should contact: Educational Coordinator, NASA Johnson Space Center, Houston, Texas 77058; while museum and planetarium personnel should contact: Exhibits Coordinator, NASA Johnson Space Center, Houston, Texas 77058.

As a complement to the lunar sample disk, NASA has also developed a meteorite disk that is available for use and study by museums and planetariums. The Meteorite Disk Program is essentially the same as the Lunar Disk Program. For further information contact: Public Services Branch, NASA Johnson Space Center, Houston, Texas 77058.

Finally, NASA has developed a set of lunar material thin sections for instructive and study purposes at colleges and universities. This program consists of 12 samples of soils and rocks. College and university instructors can obtain additional information about this particular program by contacting: Curator's Office, NASA Johnson Space Center, Houston, Texas 77058.

See also: **extraterrestrial expeditions; lunar bases and settlements; Moon; NASA facilities**

Magellanic Clouds   The two dwarf galaxies that are closest to our Milky Way Galaxy. The Large Magellanic Cloud (LMC) is about 150,000 light-years away and the Small Magellanic Cloud (SMC) approximately 170,000 light-years distant. Both are visible to observers in the Southern Hemisphere and resemble luminous clouds several times the size of the Full Moon. Their presence was first recorded in 1519 by the Portuguese explorer Ferdinand Magellan, after whom they are named.

In March 1979 a huge gamma-ray burst was detected by

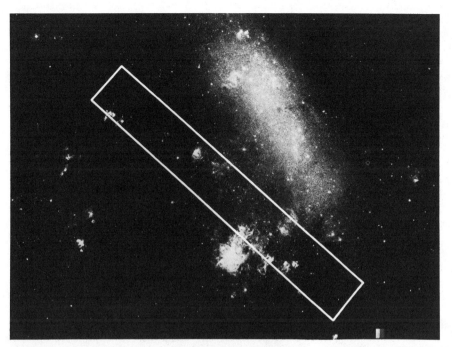

Fig. 1 A photograph of the Large Magellanic Cloud. (Courtesy of NASA and the United Kingdom's Schmidt Observatory in Australia.)

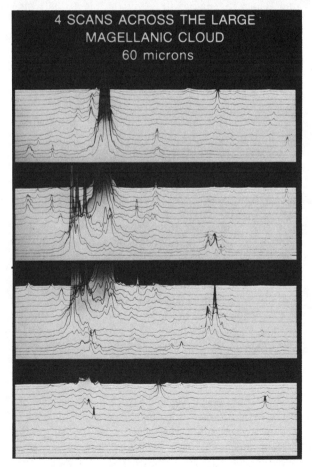

Fig. 2   Infrared Astronomical Satellite (IRAS) scans across the Large Magellanic Cloud, January 1983. (Courtesy of NASA.)

Earth-orbiting spacecraft. Scientists now think this burst originated from the Large Magellanic Cloud. If this giant gamma-ray burst actually did come from a source over 150,000 light-years away, it would represent an astrophysical event involving an enormous release of energy.

Figure 1 is a photograph of the Large Magellanic Cloud taken by the United Kingdom's Schmidt Telescope in Australia. The white rectangle superimposed on this picture identifies the region also scanned by the Infrared Astronomical Satellite (IRAS) in 1983. Figure 2 presents the results of four overlapping scans across this region of the Large Magellanic Cloud by IRAS. Each scan shows the strength of the signals recorded by IRAS's fifteen 60-micrometer-wavelength (that is, far-infrared) detectors as a function of time as the orbiting telescope tracked across the sky. The four pictures shown in figure 2 each depict a 75-second stretch of observations that were made on different orbits during January 1983. As can be seen, many infrared sources were detected, including possibly stars in their "birthing process," emerging out of clouds of interstellar dust and gas. Infrared maps of the Large Magellanic Cloud, such as these, help astronomers solve many puzzling questions about the birth and death of stars in regions of the Universe much less complex and confused than our own Galaxy.

See also: **dwarf galaxy; Milky Way Galaxy; stars**

magnetopause   The boundary layer that separates the relatively strong geomagnetic field from the magnetosheath. It is a part of the Earth's magnetosphere.

See also: **Earth's trapped radiation belts; magnetosheath; magnetosphere**

magnetosheath   The transition zone in which the interplanetary field and the solar wind are deflected after passing through the bow shock. It is part of the Earth's magnetosphere.

See also: **bow shock; Earth's trapped radiation belts; magnetosphere; solar wind**

magnetosphere   The region around a planet in which charged atomic particles are influenced by the planet's own magnetic field rather than by the magnetic field of the parent star, as projected by the stellar (solar) wind. The magnetosphere generally includes any trapped radiation belts that might encircle the planet. In our Solar System, the Earth, Jupiter, Saturn and Mercury are currently known to possess magnetospheres. As shown in figure 1, studies by spacecraft and space probes have now mapped much of the region of magnetic field structures and streams of trapped particles around the Earth. The solar wind, streaming out from the Sun, shapes the Earth's magnetosphere into a teardrop, with a long magnetotail stretching out opposite the Sun.

See also: **Earth's trapped radiation belts**

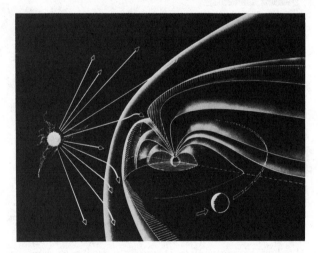

Fig. 1   Spacecraft and space probes have helped us map much of the Earth's magnetosphere—the region of magnetic field structures and streams of trapped radiation belts around our planet. (Courtesy of NASA.)

magnetron   A magnetically controlled tube used to generate and amplify microwave radiation.

See also: **Satellite Power System**

Mainbelt Asteroid Multiple Orbiter/Flyby Mission
What do asteroids really look like? What are they made of? Are some asteroids the broken remains of larger Solar-System bodies? Space scientists think that asteroids have remained relatively unchanged since their formation early in the evolution of the Solar System. To help answer these and many other intriguing questions about the minor planets, NASA is considering a Mainbelt Asteroid Multiple Orbiter/Flyby Mission. The specific objectives of this mis-

sion are to characterize the various types of asteroids and to conduct a more detailed study of one or two selected main-belt asteroids, with particular emphasis on mineral content and surface features.

By convention, space scientists usually say *rendezvous missions* when they are referring to comets and Earth-approaching asteroids that have gravitational spheres of influence no more than a few tens of kilometers. When they describe encounters with larger targets among the main-belt asteroids (for example, objects greater than 50 kilometers diameter), the term *orbiter spacecraft* or *orbiter mission* is considered more appropriate.

The asteroid orbiter/flyby spacecraft would be launched by a Space Shuttle/Centaur configuration between 1995 and 2000 and reach its first main-belt asteroid target after a flight through interplanetary space of about four years. During the outbound flight, the spacecraft would take advantage of one or more gravity assists as might be provided by the planets Mars and Jupiter. Then, after a propulsive maneuver, the spacecraft would match orbit with the target asteroid—which, if large enough, would permit an orbital capture. Remote sensing observations would then be made over a several-week period, after which the spacecraft would be placed on a new trajectory. This new trajectory would take the spacecraft past several other representative asteroids on its way to a second orbital encounter with a principal minor planet. During the flyby portions of this mission, remote sensing observations would be performed. At the second major asteroid target, the spacecraft would match orbits and remain with this celestial object until the end of the mission—several months later. The total mission time from launch would be approximately six years.

See also: **asteroid; asteroid mining;**

Mariner Mark II spacecraft   A new class of modular spacecraft proposed by NASA for Solar-System and planetary exploration. It will come in a variety of configurations. The basic spacecraft will weigh about 600 kilograms. This would include about 100 kilograms for scientific instruments but would exclude the propulsion tanks, fuel and any probes. The central module, or "bus," supports the external modules and houses spacecraft electronics. The radio-frequency (RF) module includes a fixed high-gain antenna along with its feed and receiver. The antenna size can be changed from mission to mission. The power module can be either a radioisotope thermoelectric generator (RTG) or a solar panel. An RTG is the power source of choice for outer-planet missions, which must operate for many years at great distances from the Sun. However, because the nuclear-powered RTG also represents a small external radiation source, it often cannot be used on a spacecraft that also carries very sensitive radiation detectors, such as gamma-ray spectrometers. To avoid excessive interference with such scientific radiation-detection instruments, either the RTGs can be placed on extended booms, or else a solar panel can be used in the power module. The propulsion module provides both the impulse for trajectory changes (called "delta v" [or Δv, for short]) and the reaction control required for maintaining the proper orientation of the spacecraft. In the case of this flyby/probe carrier, the propellant tanks required for interplanetary maneuvers would be jettisoned before the probe is released.

See also: **planetary exploration**

Mars Throughout human history Mars, the Red Planet, has been at the center of astronomical thought. (See fig. C12 on color page F.) The ancient Babylonians, for example, followed the motions of this wandering red light across the night sky and named it after Nergal, their god of war. In time, the Romans, also honoring their own god of war, gave the planet its present name. The presence of an atmosphere, polar caps and changing patterns of light and dark on the surface caused many pre-Space Age astronomers and scientists to consider Mars an "Earthlike planet"—the possible abode of extraterrestrial life.

Of all the other planets in the Solar System, Mars has been the leading candidate for possessing extraterrestrial life-forms. Pre-Space Age scientists and astronomers observed the Red Planet through Earth-based telescopes and saw what appeared to be straight lines crisscrossing its surface. These observations (later determined to be optical illusions) led to the very popular notion that an intelligent race of Martians had constructed a large system of irrigation canals to support life on a "dying planet." In fact, when Orson Welles broadcast a radio drama in 1938 based on H. G. Wells' science-fiction classic *War of the Worlds*, enough people believed the report of invading Martians to create a near-panic in some areas.

Another reason for exobiologists to anticipate the existence of life on Mars was the apparent seasonal color changes of the planet's surface. These observed changes led to speculation that environmental conditions on Mars might cause certain Martian vegetation to bloom in the warmer months and become dormant during the colder periods.

Within the last two decades, however, sophisticated robot spacecraft—flybys, orbiters and landers (see fig. C13 on color page G)—have shattered these romantic myths of a race of ancient Martians struggling to bring water to the more productive regions of a dying world. Spacecraft-derived data have shown instead that the Red Planet is actually a "halfway" world. (See table 1.) Part of the Martian surface is ancient, like the surfaces of the Moon and Mercury, while part is more evolved and Earthlike. Contemporary information about Mars is presented in tables 2 and 3.

In August and September 1975, two *Viking* spacecraft were launched on a mission to help answer the question, Is there life on Mars? Each *Viking* spacecraft consisted of an orbiter and a lander. While scientists did not expect these spacecraft to discover Martian cities bustling with intelligent life, the exobiology experiments on the lander were designed to find evidence of primitive life-forms, past or present. Unfortunately, the results sent back by the two robot landers were teasingly inconclusive.

Scientists still don't know whether there is life on Mars. The small samples of Martian soil analyzed by the Viking Landers were specially treated in three different experimental protocols (procedures) designed to detect biological processes. While some of the tests indicated that biological activities were occurring, the same responses could be explained by Martian soil chemistry. There was also a very notable absence of evidence that organic molecules (life precursors) exist on Mars.

Despite the inconclusive results of the Viking Project exobiology experiments, we now know more about Mars than any other planet (except Earth) in our Solar System. A large number of missions to Mars have been conducted by the United States and the Soviet Union (review table 1). Before the Viking Project missions, for example, four *Mariner* spacecraft missions were sent to Mars by NASA. Three of these *Mariner* missions were flybys, while one involved an orbiter that conducted an extensive initial survey of the planet.

Mariner 4 was launched by NASA late in 1964. It flew past the Red Planet on July 14, 1965, and approached to within 9,656 kilometers of its surface. Looking at the 22 close-up photographs of Mars taken by *Mariner 4*, scientists found no evidence of Martian cities, Martian canals or even flowing water!

The *Mariner 6* and 7 missions to Mars followed during the summer of 1969. These flyby spacecraft returned approximately 200 photographs that showed a diversity of surface conditions. *Mariner 9* was launched on May 30, 1971, and arrived at Mars 5½ months after lift-off. *Mariner 9* orbited the planet, returning over 7,000 images that revealed previously unknown Martian surface features, including evidence of ancient rivers and possible seas that could have existed at one time on the Red Planet. (See fig. 1.)

The Viking Project was the first mission to successfully soft-land a robot spacecraft on another planet (excluding here, of course, the Earth's Moon). All four *Viking* spacecraft (two orbiters and two landers) exceeded by considerable margins their design goal lifetime of 90 days. The four spacecraft were launched in 1975 and began to operate around or on the Red Planet in 1976. When the *Viking 1* Lander touched down on the Plain of Chryse on July 20, 1976, it found a bleak landscape. Several weeks later, its twin, the *Viking 2* Lander, set down on the Plain of Utopia and discovered a more gentle, rolling landscape. One by one these robot explorers finished their highly successful visits to Mars. The *Viking Orbiter 2* spacecraft ceased operation in July 1978; the *Lander 2* fell silent in April 1980; *Orbiter 1* managed at least partial operation until August 1980; and the *Viking Lander 1* made its final transmission on November 11, 1982. NASA officially ended the Viking mission May 21, 1983.

As a result of these interplanetary missions, we now

Table 1   Missions to Mars (1962–84)

Mission (Payload)	Country of Origin	Launch Date	Remarks
Mars 1	USSR	Nov. 1, 1962	Passed Mars at 193,000 km on June 19, 1963, but communications failed on March 21, 1963.
Mariner 3	USA	Nov. 5, 1964	Mission failed (shroud malfunction).
Mariner 4	USA	Nov. 28, 1964	Successful planetary and interplanetary exploration; encounter (flyby) occurred on July 14, 1965; 22 images taken of Mars.
Zond 2	USSR	Nov. 30, 1964	Passed Mars at 1,500 km on Aug. 6, 1965; however, communications/ batteries failed in May 1965.
Mariner 6	USA	Feb. 25, 1969	Successful planetary exploration; Mars flyby within 3,300 km on July 31, 1969; studied atmosphere of Mars; returned TV images of surface.
Mariner 7	USA	March 27, 1969	Successful planetary exploration; Mars flyby at 3,518 km on Aug. 5, 1969; spacecraft identical to *Mariner 6.*
Mars 2 (Orbiter and Lander)	USSR	March 28, 1971	Orbiter circled Mars on Nov. 27, 1971; capsule (lander) ejected at 47°E and landed.
Mars 3 (Orbiter and Lander)	USSR	May 28, 1971	Orbited Mars on Dec. 2, 1971; capsule ejected and soft-landed at 45°S 158°W.
Mariner 9	USA	May 30, 1971	Orbited Mars on Nov. 13, 1971; transmitted over 6,900 images of Martian surface; all scientific instruments operated successfully; mission terminated on Oct. 27, 1972.
Mars 4	USSR	July 21, 1973	Passed Mars at 2,200 km on Feb. 10, 1974, but failed to enter Mars orbit as planned.
Mars 5	USSR	July 25, 1973	Orbited Mars on Feb. 2, 1974; gathered Martian data and served as relay station.
Mars 6 (Orbiter and Lander)	USSR	Aug. 5, 1973	Lander soft-landed on Mars at 24°S 25°W on March 12, 1974; returned atmospheric data during descent.
Mars 7 (Orbiter and Lander)	USSR	Aug. 9, 1973	Missed Mars by 1,300 km (aimed at 50°S 28°W); contact lost on March 9, 1974.
Viking 1 (Orbiter and Lander)	USA	Aug. 20, 1975	Nuclear-powered lander successfully soft-landed on July 20, 1976, on the Plain of Chryse at 47.97°W 22.27°N. First on-site analysis of surface material of another planet.
Viking 2 (Orbiter and Lander)	USA	Sept. 9, 1975	Nuclear-powered lander successfully soft-landed on Sept. 3, 1976, on the Plain of Utopia at 47.67°N 225.75°W; returned scientific data.
			Viking 1 and 2 Orbiter spacecraft returned over 40,000 high-resolution images of the surface of Mars; also collected gravity-field data, monitored atmospheric water levels and thermally mapped selected sites on the surface.

SOURCE: Author, NASA.

Table 2   Physical Data for Mars

Mean distance from Sun	228×10^6 km
Period of revolution (sidereal year)	1.88 years
Rotation period (sidereal day)	24 hr, 37 min, 23 sec
Inclination of axis	25°
Inclination of orbit to ecliptic	1.9°
Eccentricity of orbit	0.093
Diameter	
Equatorial	6,794 km
Polar	6,751 km
Mass	6.42×10^{23} kg
Mean density	3.933 g/cm^3
Gravity at surface	372.52 cm/sec^2
Escape velocity	5.024 km/sec
Normal albedo	0.1–0.4
Surface temperature extremes	130 K–300 K
Surface atmospheric pressure (varies seasonally)	5.9–15.0 millibars
Atmosphere (main components)	carbon dioxide (CO_2) 95.32% volume
	nitrogen (N_2) 2.7%
	argon (Ar) 1.6%
	oxygen (O_2) 0.13%
	carbon monoxide (CO) 0.07%
	water vapor (H_2O) 0.03% [a]
Satellites	2 (Phobos and Deimos)

[a]Variable.

SOURCE: Author, NASA.

Table 3   *Viking 1* Lander Site Chemistry

Location: Chryse Planitia 47.97°W 22.27°N

Local topography: low, rolling hills, mostly covered by fine-grained debris.

Remarks: Surface samples analyzed at both *Viking 1* and *2* landing sites are similar; it appears that these are representative of fine debris everywhere on the planet.

Compound[a]	Percentage Mass
SiO_2	44.7
Al_2O_3	5.7
Fe_2O_3	18.2
MgO	8.3
CaO	5.6
K_2O	<0.3
TiO_2	0.9
SO_3	7.7
Cl	0.7

[a]Except for chlorine (Cl), the elements silicon (Si), aluminum (Al), iron (Fe), magnesium (Mg), calcium (Ca), potassium (K), titanium (Ti), and sulfur (S) are all present in the Martian soil as their common oxides, such as silicon dioxide (SiO2).

SOURCE: NASA.

condense out and form clouds that ride high in the Martian atmosphere or form local patches of morning fog in valleys. There is also evidence that Mars had a much denser atmosphere in the past—one capable of permitting liquid water to flow on the planet's surface. Physical features resembling riverbeds, canyons and gorges, shorelines and even islands hint that large rivers and maybe even small seas once existed on the Red Planet.

THE SEARCH FOR EXTRATERRESTRIAL LIFE ON MARS

Each Viking Lander had cameras capable of "capturing" any large Martian life-forms that might inhabit the regions near either landing site. In thousands of images taken on the surface of Mars, no large life-forms were detected. The Viking Landers also carried three biology experiments to search for extraterrestrial life (especially at the microorganism level). These experiments were: (1) the gas-exchange experiment, which was designed to detect metabolization and related gaseous-exchange products in the presence of a nutrient solution; (2) the labeled-release experiment, designed to detect the release of radioactivity from labeled carbon (using radioisotope-tracer techniques) bound in a nutrient solution; and (3) the pyrolitic-release experiment, based on the exobiological hypothesis that any Martian life-form would be able to incorporate atmospheric carbon (tagged with a radioisotope tracer) in the presence of sunlight—that is, by photosynthesis.

All three exobiology experiments exhibited signs of activity, but many scientists now feel that this activity was probably due to chemical, and not biological, reactions in the Martian soil. In addition, the Viking's gas chromatograph/mass spectrometer (GCMS) failed to detect any organic material, even at very low concentrations.

know that Martian weather changes very little. For example, the highest atmospheric temperature recorded by either Viking Lander was minus 21 degrees Celsius (midsummer at the *Viking 1* site); while the lowest recorded temperature was minus 124 degrees Celsius (at the more northerly *Viking 2* site during winter).

The atmosphere of Mars was found to be primarily carbon dioxide. Nitrogen, argon and oxygen are present in small percentages, along with trace amounts of xenon and krypton. The Martian atmosphere contains only a wisp of water (about 1/1000th as much as found in the Earth's atmosphere). But even this tiny amount of water can

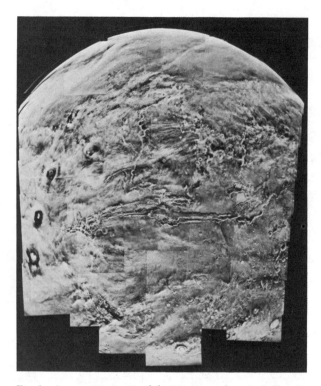

Fig. 1   A composite image of the Martian surface, including the Valles Marineris, which is as long as the North American continent from coast to coast. (Taken by *Viking Orbiter 1*, February 1980.) (Courtesy of NASA/Jet Propulsion Laboratory).

Therefore, despite intensive initial investigations, the greatest Martian mystery of all—Is there life?—still remains a mystery. More sophisticated expeditions to Mars are needed in the next few decades before exobiologists can state that life does (or perhaps did) exist on Mars—or else positively conclude that Mars is a barren, lifeless world. (See table 4.) In either circumstance, the implications for our models depicting the cosmic prevalence of life will be profound.

THE MOONS OF MARS

Mars has two small, irregularly shaped moons called Phobos ("fear") and Deimos ("terror"). These natural satellites were discovered in 1877 by Asaph Hall. They both have ancient, cratered surfaces with some indication of regoliths to depths of possibly five meters or more. (See figs. 2 and 3.) The physical properties of these two Martian moons are presented in table 5. Since both satellites appear similar to water-rich, carbonaceous chondrite asteroids, some scientists now speculate that they may have originally been minor planets that were gravitationally captured by the Red Planet.

WILL THERE BE A MARTIAN IN YOUR FUTURE?

As we enter the next millennium, human development will be highlighted by the establishment of our extraterrestrial civilization. We will first learn how to permanently occupy near-Earth space and then expand human activities throughout cislunar space (the area between Earth

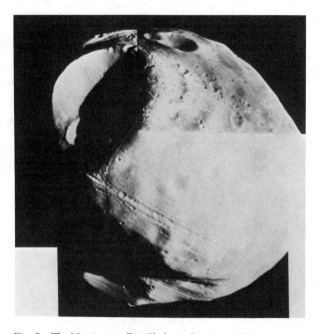

Fig. 2   The Martian satellite Phobos, photographed from a range of 612 kilometers by the *Viking Orbiter 1* spacecraft. Stickney is the largest crater on Phobos; it measures about 10 kilometers across and can be seen at the left near the morning terminator (the border between the illuminated and unilluminated areas of Phobos's surface). (Courtesy of NASA/Jet Propulsion Laboratory.)

Fig. 3   The tiny Martian satellite Deimos, as imaged by the *Viking 1* Orbiter. (Courtesy of NASA.)

and the Moon). With the rise of self-sufficient space settlements on the Moon and in Earth orbit, humankind will then expand into heliocentric (Sun-centered) space. Mars will become the focus of extensive exploration and resource-evaluation missions, using robotic spacecraft and human expeditions. Table 6 describes one possible grand scenario for the exploration and development of the Red Planet in the next century and beyond.

According to the Solar System Exploration Committee (SSEC) of the NASA Advisory Council:

Mars is a key member of the triad Earth-Venus-Mars and is closely linked to the Earth by virtue of

Table 4   Some of the Properties of the Martian Environment that Appear Unfavorable to Life as We Know It

• Lack of liquid water at the surface.

• The cold temperatures and extreme temperature variations that occur both diurnally (day/night) and seasonally.

• The presence of strong oxides in the soil.

• Very low atmospheric pressure at the surface.

• The apparent absence of nitrogen in the upper Martian regolith.

• The penetration of solar ultraviolet radiation all the way to the surface of Mars.

SOURCE: NASA.

Table 5   Physical Properties of the Martian Moons, Phobos and Deimos

Property	Phobos	Deimos
Orbital Elements		
Semimajor axis	9,378 km	23,459 km
Eccentricity	0.015	0.00052
Inclination	1.02°	1.82°
Sidereal period (approximately)	7 hr, 39 min	30 hr, 18 min
Physical Parameters		
Longest axis	13.5 km	7.5 km
Intermediate axis	10.7 km	6.0 km
Shortest axis	9.6 km	5.5 km
Rotation	Synchronous	Synchronous
Mass	9.8×10^{15} kg	2.0×10^{15} kg
Density	2.0 g/cm^3	1.9 g/cm^3
Albedo	0.05	0.06
Surface gravity	1 cm/sec^2	0.5 cm/sec^2

SOURCE: NASA.

Table 6   Scenario for the Exploration and Settlement of Mars in the 21st Century

STAGE 1: Advanced exploration with sophisticated spacecraft
  • Mars Geoscience/Climatology Orbiter
  • Mars Aeronomy Orbiter

STAGE 2: Robotic surface exploration
  • Mars Network Mission (penetrators)
  • Mars airplane
  • Surface rover(s)
  • Mars Sample Return Mission (MSRM)

STAGE 3: Human exploration of Mars
  • Nuclear-electric propulsion (NEP) expedition

STAGE 4: Development of Martian resources
  • Site preparation (automated)
  • Initial base (10–30 persons)
  • Early Martian settlement (10^3–10^4 persons)
    –Initiation of planetary engineering projects
  • Autonomous Martian civilization (more than 10^5 persons)
    –Full-scale planetary engineering projects
    –Permanent human presence in heliocentric space
    –Independence of Earth-Moon system

SOURCE: Developed by the author.

Table 7   Prioritized Scientific Objectives for the Continued Exploration of Mars

1. Intensive study of local areas to:
    • Establish the chemical, mineralogical, and petrological character of the Martian surface.
    • Establish the nature and chronology of the major surface-forming processes.
    • Determine the distribution, abundance, sources, and sinks of volatile material.[a]
    • Establish the interaction of the surface material with the atmosphere and radiation environment.

2. Explore the structure and the general circulation of the Martian atmosphere.

3. Explore the nature and dynamics of Mars' interior.

4. Establish the nature of the Martian magnetic field and the character of the upper atmosphere and its interaction with the solar wind.

5. Establish the global chemical and physical characteristics of the Martian surface.

[a]Includes an assessment of current and past biological potential of Martian environment.

SOURCE: NASA.

the volcanic, erosional, and climatic phenomena that it is known to exhibit. The study of Mars is essential for our understanding of the evolution of the Earth and the inner Solar System.

This committee's recently prioritized scientific objectives for the continued exploration of Mars are presented in table 7. The Mars Geoscience/Climatology Orbiter, the Mars Aeronomy Orbiter and the Mars Network Mission (using planetary penetrators) are currently viewed as supporting these scientific objectives over the next few decades. Beyond this period, or perhaps complementing its later phases, we could witness the use of very smart machines of exploration such as autonomous Mars surface rovers, a Mars airplane and a Mars Sample Return Mission (MSRM).

The Mars Geoscience/Climatology Orbiter would be inserted into a polar orbit around the Red Planet. From a Sun-synchronous, nearly circular 300-kilometer orbit, it would observe the planet for almost a full Martian year (approximately two Earth-years), making global maps of elemental and mineralogical surface compositions. It would also measure the seasonal cycles of carbon dioxide, water and dust in the Martian atmosphere and study the interactions between volatile reservoirs (such as the polar caps) and the Martian atmosphere. This spacecraft would pave the way for more extensive exploration programs, robotic and manned, that would take place in the 21st century.

Instruments on board the Mars Aeronomy Orbiter would help determine the daily and seasonal variation of the upper Martian atmosphere and ionosphere. This proposed mission would also measure the escape rates of the constituents of the Martian atmosphere, such as hydrogen, oxygen and nitrogen.

Planetary scientists believe that experiments performed from a network of surface penetrators can provide the important facts needed to start understanding the evolution, geologic history and nature of a planetary body such as Mars. "Network science" can, in fact, be defined as a set of systematic measurements made over a relatively long period of time (such as one Martian year) at widely distributed locations on a planet's surface. The measurements typically would include seismic, meteorologic, heat flow, water content, geochemical and imagery—as needed to fully characterize each site. A Mars Probe Network would help establish the chemical composition of near-surface materials, evaluate general atmospheric-circulation patterns and investigate the internal structure and seismicity of the planet.

Some planetary scientists think that a lightweight, robotic Mars airplane represents a very versatile tool for exploring the Red Planet. This type of airplane could perform high-resolution surface mapping missions, carry and distribute special sensor packages or play an important scouting role for a manned expedition or robotic Mars Sample Return Mission. These hydrazine-engine-powered aircraft would be capable of flying at altitudes of 500 meters up to 15,000 meters, with respective ranges of 26 kilometers and 6,700 kilometers.

Smart surface rovers also have a very useful role to play in the overall exploration of Mars in the next century. These smart autonomous rovers can gather samples of surface materials, deploy instruments on or beneath the Martian surface and perform detailed site investigations. A rover can work independently, or perhaps be operated in teams of two or four, transmitting data back to an orbiting mothership, which, in turn, relays the findings back to Earth.

An automated Mars Sample Return Mission would, as the name implies, return one or more samples of Martian surface material to Earth or to an Earth-orbiting quarantine facility, where extensive scientific investigations can be made by terrestrial scientists.

A human expedition to Mars early in the next century would represent our physical penetration into heliocentric space. One proposed mission scenario, involving a nuclear-electric-propelled spacecraft, involves a single spacecraft with a Mars lander that would carry a crew of five on a 2.6-year duration mission to the Red Planet. The nuclear-electric propulsion (NEP) system would be powered by a megawatt-class (2.6 megawatts) advanced-design space nuclear reactor. Closed air and water life-support systems and artificial gravity would sustain the crew throughout the flight through interplanetary space.

The Mars expedition vehicle would be assembled at a space station in low Earth orbit. It would then be transferred to geosynchronous Earth orbit, where its crew would board it for the long, spiraling outward journey to Mars (about 510 days). It would take another 39 days to perform a capture spiral maneuver around Mars, ending up in a circular, 3,000-kilometer-altitude orbit above the Red Planet. The crew would then engage in a 100-day reconnaissance mission, including a 30-day surface exploration mission by three of the five crew members. A 23-day Mars departure spiral would start the electrically propelled vehicle back to Earth. Mars-to-Earth transfer would take about 229 days under optimum coast conditions. The vehicle would then execute a 16-day capture spiral to geosynchronous orbit around the Earth, and the crew would transfer to a special quarantine (if necessary) and debriefing facility on the space station before eventually returning to Earth.

Exactly what happens after the first human expedition to Mars is, of course, open to wide speculation at present. People here on Earth could simply marvel at "another outstanding space-technology first" and then settle back to their "more pressing" terrestrial pursuits. (This pattern unfortunately followed the spectacular Apollo Moon landing missions of 1969–72.) On the other hand, if this first human expedition to Mars were widely recognized and accepted as the technical precursor to our permanent occupancy of heliocentric space, then Mars would truly become the central object of greatly expanded space activities (perhaps complementing the rise of a self-sufficient lunar civilization).

Very sophisticated surface rovers would be used to prepare suitable sites for the first permanent bases, each housing perhaps 10 to 100 people. These early bases on the Red Planet would focus their activities on detailed exploration and resource identification and would most likely be supported by a Mars-orbiting space station. Another important objective for the first "Martians" will be to conduct basic science and engineering projects that take advantage of the Martian environment and native Martian materials. Then, as prospects for Martian-based industries grow, these early bases will also expand, reaching the size of modest settlements, with upwards of 1,000 Martians each—all committed to discovery, adventure and profit on the Red Planet.

As the early settlements mature and are economically nourished by the planet's resource base, Martian fuels, food, water, metals and manufactured products will support the wave of human expansion to the mineral-rich asteroid belt and to the giant outer planets and their intriguing complement of moons.

During this growth and expansion process, a point will be reached when the Martian population, for all practical purposes, becomes fully self-sufficient. The very thought of a neighboring planet inhabited by intelligent beings has always been stimulating to the human race.

An autonomous Martian civilization will most likely be characterized by the initiation of planetary engineering projects. Planetary engineering involves the large-scale modification or reconfiguration of a planet to provide a more habitable ecosphere for the terrestrial settlers. On Mars the human settlers would most likely first seek to make the atmosphere more dense (and eventually even breathable) and to alter its temperature extremes to more "Earthlike" ranges. These planetary engineering efforts could include melting the polar caps and transporting large quantities of liquid water to the equatorial regions of the planet—perhaps by means of a large series of open canals. Those who speculated in the early 20th century that Mars was inhabited by intelligent creatures who constructed canals to transport water from the polar regions may only have been a century or two early in their bold assumptions!

Mars, properly explored and used, not only opens up the remainder of heliocentric space for human development; it also establishes the technological pathways for the first interstellar missions. The development of very smart machines, the ability to modify the ecosphere of a planet and the technology to control and manipulate large quantities of energy are all necessary if human explorers are ever to venture across the interstellar void in search of new worlds around distant suns. The conquest of Mars in the 21st century will provide a large portion of the technology base needed to seriously consider flights beyond our own Solar System.

See also: **exobiology; Mars Aeronomy Orbiter; Mars Geoscience/Climatology Orbiter; Mars penetrator; Mars Sample Return Mission; Mars surface rovers; planetary engineering; planetary exploration; Viking Project**

Mars Aeronomy Orbiter   A proposed NASA mission to Mars to be launched in 1988 or at two-year intervals thereafter during appropriate planetary "launch windows." Aeronomy is the science that involves the study of the upper atmosphere of a planet, especially regions of ionized gas. The Mars Aeronomy Orbiter would satisfy many scientific objectives. First, its data will determine the diurnal (day/night) and seasonal variation of the upper Martian atmosphere and ionosphere. Its instruments will observe the interaction of the solar wind with the Martian atmosphere and verify whether Mars possesses an intrinsic magnetic field. This orbiter would also measure the escape rates of the constituents of the Martian atmosphere, such as hydrogen, oxygen and nitrogen.

The spacecraft will be placed into a highly elliptical orbit (three Mars radii by 150 kilometers) with a 77.5-degree inclination. During a full Martian year (which is approximately 687 "Earth-days" long), this particular orbit would provide "periapsis sampling" at all Martian latitudes and local times. (Periapsis refers to the point in an orbit at which the spacecraft is closest to the celestial body being orbited.) The Mars Aeronomy Orbiter will have a mission lifetime of at least one Martian year to provide a complete seasonal survey of the upper Martian atmosphere.

This proposed spacecraft will supply very useful data about Martian atmospheric processes—data that would support more sophisticated missions of exploration, robotic and eventually human. Any future plans for "terraforming" the Red Planet in the next century and beyond will rely extensively on the data base created by this advanced explorer spacecraft and others like it.

See also: **Mars; orbits of objects in space; planetary engineering; robotics in space**

Mars airplane   How would we use an unmanned airplane like the one shown in figure C14 on color page F in the further exploration of Mars? Some scientists think that such lightweight robot vehicles represent a very versatile means of transportation around the Red Planet. They could either carry experiment packages or play an important support role for special missions like a Mars Sample Return (MSR) project. This Mars airplane could be used, for example, to deploy a network of science stations, such as seismometers or meteorology stations, at selected Martian sites within an accuracy of a few kilometers. With a 40-kilogram payload capacity, the flying platform could perform high-resolution imagery (say, less than 0.5-meter ground-spot size) or conduct magnetic, gravity and geochemical surveys. It would be capable of flying at altitudes of 500 meters to 15 kilometers, with ranges of 25.5 kilometers to 6,700 kilometers.

The Mars airplane has many characteristics of a terrestrial competition glider. It would have a very lightweight airframe made of carbon fiber composites and weigh less than 40 kilograms. The wings, fuselage and tail sections would fold, allowing it to fit into a protective aeroshell for its initial descent into the Martian atmosphere after deployment from its "aircraft-carrier" spaceship. This Mars airplane would be powered by a 15-horsepower hydrazine airless engine that is used to drive a three-meter propeller.

This type of extraterrestrial airplane offers two basic options. First, it can be employed as a powered flyer that performs aerial surveys, atmospheric soundings and so on and then crashes when its fuel supply is exhausted. Or else it can be built as a plane equipped with a variable-thrust rocket so that it may soft-land and then take off again.

One basic mission concept would be to send several of these "aircraft-carrier" spaceships to Mars on an extensive exploration program. Each spaceship would carry up to four Mars airplanes (folded and tucked in their respective aeroshells). While the carrier spacecraft orbits Mars, individual airplanes would be deployed on command from Earth into the Martian atmosphere. Each Mars airplane would enter the atmosphere of the Red Planet in its protective aeroshell. While descending through the upper Martian atmosphere, this aeroshell would be discarded, parachutes deployed and the plane unfolded. It would continue to descend to its operational altitude and then fly off on its mission of exploration, instrument deployment or sample collection. Once its hydrazine fuel was exhausted, each Mars airplane would acquire an eternal resting place among the shifting red dunes of the planet. Some would land softly and be found nearly intact by later teams of terrestrial explorers, while others would crash and become crumpled piles of "extramartian" debris. Communications with scientists on Earth would be maintained through a network of strategically located relay or communications satellites orbiting around Mars. This Mars airplane, or bigger versions of it, could also be used in direct support of a human expedition to Mars. The terrestrial explorers would use such robot scouts to find suitable sites for more detailed scientific investigation. In time, the thin Martian atmosphere could become host to a squadron of such flyers—each gliding across the surface of the Red Planet in response to scientific targeting instructions from a human explorer of the 21st century.

See also: **Mars; Mars Sample Return Mission; Mars surface rovers**

Mars Geoscience/Climatology Orbiter A planned NASA planetary-exploration mission to Mars. It would be launched in 1990 or during other appropriate "launch windows" in 1992, 1994 and so on. After a one-year flight, the orbiter spacecraft would be inserted into a polar orbit about the Red Planet. Following a drift in this orbit to the desired Sun angle, a small orbital plane change (approximately three degrees) will be performed, creating a nearly circular, 300-kilometer-altitude, Sun-synchronous mapping orbit. From this orbital location observations of the planet will be made for at least one Martian year (approximately two Earth-years). The final spacecraft orbit will be determined by planetary-quarantine (protection) requirements.

There are several major scientific objectives for this mission: (1) to obtain global maps of elemental and mineralogical surface compositions; (2) to explore and define the nature of the Martian magnetic field (if any exists); (3) to measure the seasonal cycles of carbon dioxide (CO_2), water (H_2O) and dust; (4) to study the interactions between volatile reservoirs, such as the polar caps, and the Martian atmosphere; and (5) to create global maps of the Martian

gravitational field and surface. The data acquired by this spacecraft will support more extensive Martian exploration programs, robotic and manned, that will occur in the 21st century.

See also: **Mars; Mars Sample Return Mission; orbits of objects in space; planetary exploration**

Mars penetrator The overall goal of planetary-exploration missions is to gather data that will help in understanding the formation, evolution and present state of planetary bodies and prepare the way for human expeditions, resource exploitation and permanent habitation, if appropriate. Planetary bodies are complex and are characterized by both surface and interior irregularities. Precise measurements of key physical properties of these celestial bodies, when taken from a network of sites distributed over the planet's surface, can provide a data base for constructing a composite picture of that planet. In addition, such a network provides simultaneous measurements of transient phenomena of the planet's interior, surface and atmosphere. In fact, a network of complementary stations is one very effective way of obtaining meaningful data as a function of time and location. These data will provide the information necessary to develop sound working hypotheses and models about planetary processes and eventually pave the way for more extensive exploration and possible exploitation by human explorers.

Space scientists have concluded that experiments performed from a network of penetrators can provide essential facts needed to begin understanding the evolution, history and nature of a planetary body, such as Mars. "Network science," in fact, can be defined as a set of systematic measurements over a relatively long time duration (one Martian year at a minimum) at locations distributed widely over the planet's surface. The scientific measurements should necessarily include seismic, meteorologic and, if possible, other experiments to characterize the local site—such as heat flow, water content, geochemistry and imaging.

There are several main reasons for using penetrators to deploy science experiments on Mars and other celestial bodies. First, the penetrators establish a global network of stations to measure transient phenomena over widely scattered regions on the planet's surface. Second, the penetrators actually place some of the experiments beneath the planet's surface. Finally, the penetrators provide an effective way of characterizing local sites in regions that could not be safely reached by larger, more sophisticated surface rovers. In fact, penetrators can be useful in any future Mars mission, since their deployment immediately establishes a planetary monitoring network.

When such penetrators are used in a planetary or global monitoring network, they possess inherent advantages for certain science experiments. For example, penetrators provide a much better coupling to the ground, thereby supporting a broad spectrum of seismic-event measurements free from noise introduced by meteorologic activity. The acquisition of time-varying measurements of wind direction, velocity, atmospheric pressure and humidity can be made at many more sites than would be economically possible using sophisticated surface vehicles. When such meteorological data are combined with image observations of transient events over widely separated regions, a more general understanding of the planet's atmospheric processes can be gained.

A typical penetrator system consists of four major subassemblies: (1) the launch tube, (2) the deployment motor, (3) the decelerator (usually a two-stage device) and (4) the penetrator itself. The launch tube attaches to the host spacecraft and houses the penetrator, deployment motor and two-stage decelerator. The deployment motor is based on well-proven solid-rocket motor technology and provides the required deorbit velocity. The two-stage decelerator includes a furlable umbrella heat shield for the first stage of hypersonic deceleration. The second stage consists of a small drogue (parachute), which ensures proper impact conditions. The penetrator itself is a steel device, shaped like a rocket, with a blunt ogive (curved) nose and conical-flared body. The afterbody of the penetrator remains at the planet's surface, with the forebody penetrating the subsurface material. Figure 1 shows the penetrator with typical network instruments—here, seismic and meteorologic sensors. Penetrator subsystems include structure, data processing and control, communications, power, thermal control and umbilical cable in addition to scientific instruments and sensors. The penetrator structure is designed to penetrate a variety of soils and rocks and to withstand the effects of the way it enters the ground (called inclination and angle of attack) at impact. The afterbody includes a deployable boom for the meteorologic instruments and an antenna. Penetrator power is provided

by a nuclear energy source, called a radioisotope thermoelectric generator (or RTG, for short).

Before being attached to the "carrier spacecraft," each penetrator system will be assembled and checked out. If sterilization is necessary to support a planetary quarantine protocol (procedure), each penetrator system will be subjected to terminal sterilization procedures after assembly. It will then be attached to the carrier, or "mother spacecraft," in a biologically shielded launch tube. During the flight to Mars, housekeeping information and instrument-status data will be sent to the spacecraft through an umbilical connection and then transmitted back to Earth.

The penetrators will be monitored during this time by scientists on Earth, who will review the housekeeping data. When the carrier spacecraft arrives at Mars, the penetrators will be individually targeted, with the spacecraft positioned so as to properly "aim" the launch tube for propulsive separation. Separation from the mothership involves a sequence of actions that include venting pressure, opening the launch-tube covers and firing the deployment motor.

After separation from the mother spacecraft, one by one each penetrator will independently enter the Martian atmosphere behind a deployable heat shield and then float down on its parachute. Upon impact, the probe will bury itself in the Martian soil, leaving some instrumentation and an antenna at the surface. Communications with scientists on Earth from these surface/subsurface extraterrestrial monitoring sites will be accomplished by means of the orbiting mothership, which now interrogates each penetrator at least once a Martian day. A very large network of penetrators is considered necessary to obtain a general atmospheric-circulation model of Mars; but many other planetary science objectives can be satisfied, at least

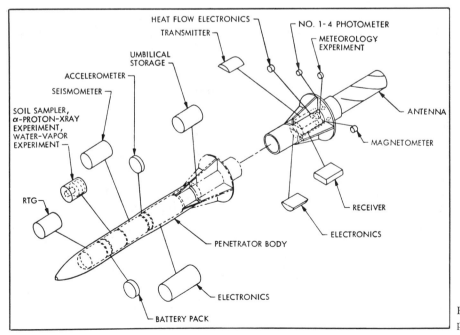

Fig. 1   Components of a typical Mars penetrator. (Courtesy of NASA.)

partially, with a minimum of three to six probe stations.

The overall scientific objective in creating a Mars Probe Network would be to establish a planetary network of seismic stations, meteorological stations, and geochemical and geophysical observation sites that can remotely and automatically operate on Mars for an extended period of time. The penetrator and its afterbody would contain a wide variety of instruments emplaced in the Martian soil and on its surface. Specific scientific objectives for this penetrator network on Mars include: (1) a determination of the chemical composition of Martian near-surface materials, (2) a study of the internal structure and seismicity of Mars, (3) an evaluation of the general circulation patterns of the Martian atmosphere and (4) a characterization of local atmospheric conditions in a variety of Martian locales—many inaccessible to larger, surface rover vehicles. Individual penetrators as well as an extensive network can be used as part of any future Mars mission involving widespread exploration, robotic or human.

See also: **Mars; planetary engineering; robotics; in space; space nuclear power**

Mars Sample Return Mission (MSRM)   The purpose of an automated Mars Sample Return Mission (as the name implies) is to return one or more samples of Martian surface material to Earth, where extensive scientific investigation of these materials can be undertaken. Figure C15 on color page G shows a Mars Ascent Vehicle (MAV) lifting off from the surface of the Red Planet with its precious extraterrestrial cargo of Martian soil and rock samples. In this particular sample-return mission scenario, a "carrier" spacecraft would ferry a sample-gathering lander to Mars and place it in a prelanding orbit around the Red Planet. The lander would then touch down on the surface and collect soil samples. The lander craft might be assisted in its exogeological hunt by a small minirover system. The Mars Ascent Vehicle (a part of the original sample-gathering lander) would then lift off from the Martian surface and rendezvous and dock with a separate Earth-return spacecraft. While the coupled spacecraft are orbiting the Red Planet, the canisters of Martian soil samples would be automatically transferred from the ascent vehicle to the Earth-return spacecraft. This spacecraft would then depart for Earth.

To avoid potential problems of extraterrestrial contamination of the Earth by alien microorganisms that might be contained in the Martian soil, these samples would most likely be analyzed first in an Earth-orbiting quarantine facility. If no hazards were discovered, the Martian soil samples would then be allowed to enter the Earth's biosphere so that extensive investigations could be conducted on them in terrestrial laboratories.

See also: **extraterrestrial contamination; Mars**

Mars surface rovers   Automated rovers and mobility systems can be used to satisfy a number of exploration needs on the surface of Mars. For example, they can acquire specific samples of surface materials for automated evaluation in on-board laboratories; they can deploy instruments on or beneath the Martian surface; or they can perform extensive site investigations. The mobility of such surface rovers can also be used to extend the range of landers, Mars airplanes or a Mars Sample Return Mission. In addition, surface rovers could also serve as an independent mission, operating perhaps in teams of two or four and transmitting data back to Earth via a Mars-orbiting mothership. Two pairs of surface rovers might travel up to five kilometers each Martian day and also assist one another as needed.

Planetary surface rovers can be designed to meet a full range of mission requirements. They can vary in mass from 20 kilograms to approximately 2,000 kilograms. Large, autonomous full-capability rovers shown in figure C24 on color page J would typically have a total mass of between 400 and 500 kilograms, including a scientific-payload capacity of 80 to 100 kilograms. For operation in hostile planetary environments (including operations in darkness), these robot rovers would be powered by radioisotope thermoelectric generator (RTG) systems. They would be capable of autonomously traveling approximately 400 meters per Martian day and would have a total range of several hundred kilometers. Full-capability, automated rovers would have a mission design lifetime of at least one Martian year.

See also: **lunar rovers; Mars; Mars airplane; Mars Sample Return Mission; robotics in space** (contemporary space robots); **space nuclear power**

Martian   Of or relating to the planet Mars; (currently in science fiction and once a permanent settlement is established) a native of the planet Mars.

mascon   A term meaning "mass concentration." An area of mass concentration or high density within a celestial body, usually near the surface. In 1968, data from five U.S. lunar orbiter spacecraft indicated that regions of high density or mass concentration existed under circular maria (extensive dark areas) on the Moon. The Moon's gravitational attraction is somewhat higher over such mascons, and their presence perturbs (causes variations in) the orbits of spacecraft around the Moon.

See also: **Moon**

mass driver   An electromagnetic device that can accelerate payloads to a very high terminal velocity. Small, magnetically levitated vehicles, called "buckets," are used to carry the payloads. These "buckets" would contain superconducting coils and be accelerated by pulsed magnetic fields along a linear track or guideway. When these buckets reach an appropriate terminal velocity, they release their payloads and are then decelerated for reuse.

Because lunar materials will play a prime role in the development of space settlements and extraterrestrial industries, the mass-driver concept has frequently been considered a vital component of a lunar-materials transportation system. It can be used to accelerate buckets of lunar materials into space—most likely to Lagrangian libration point 2. When the bundles of lunar materials reach this point in cislunar space, the tug of the Moon's gravity will have slowed them down sufficiently to permit collection and shipment to space-based manufacturing facilities.

See also: **Lagrangian libration points; lunar bases and settlements; Moon; space settlement**

materials processing in space (MPS)   Materials processing in space, first using the Space Shuttle/Spacelab configuration (see fig. 1) and later through modules on permanent space stations and dedicated space industrial platforms, offers a great potential for significantly advancing materials science and for creating new products for use on Earth.

Materials-science experiments conducted on Skylab, the Apollo-Soyuz Test Project (ASTP) and Space Shuttle missions have demonstrated that new knowledge can be gained through the effective use of the microgravity environment found in orbit. Metals, for example, can be solidified without the disturbing effects of gravity-driven convection. Crystallographers can watch crystals form in microgravity and learn why certain dopant materials (small amounts of impurities) that are needed to improve electronic characteristics distribute themselves far more uniformly in orbit than when crystals are formed on Earth. These valuable materials-science insights can be used to improve terrestrial manufacturing techniques or to establish the framework for space-based manufacturing.

For the field of materials science and technology, one of the main advantages of the Space Age is the ability to

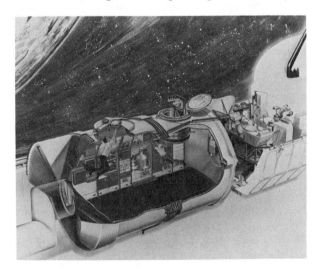

Fig. 1 A cutaway view of the European Space Agency's Spacelab, the manned, reusable multimission platform for research and testing in space. Spacelab consists of a pressurized module and U-shaped platforms to take advantage of the unique properties of outer space. It is carried into orbit and operates inside the cargo bay of the Space Shuttle. (Courtesy of MBB/ERNO.)

escape from terrestrial gravitational effects that adversely impact materials-processing operations on Earth. These gravitational effects include convection, sedimentation and buoyancy. In addition, the microgravity environment of space provides the opportunity to conduct "containerless processing." By using the long-duration microgravity environment of an orbiting space platform to overcome these adverse gravitational effects, both new materials and new manufacturing processes can be developed.

Convection is the spontaneous mixing or stirring in a liquid or gas as (fluid) currents flow between temperature gradients. Convective phenomena are unpredictable and chaotic and often lead to undesirable structural and compositional differences in a material after it has solidified. Both crystal growth and solidification processes are enhanced if convective disturbances are suppressed. A microgravity environment gives materials scientists and engineers the opportunity to reduce or completely eliminate such undesirable convective phenomena.

On Earth, gravity causes heavier components of a mixture to settle to the bottom (sedimentation) while less dense materials rise to the top (buoyancy). As a result, sedimentation or buoyancy effects complicate terrestrial manufacturing processes involving different-density alloys or composite materials. There are hundreds of potentially interesting metallic combinations that, like oil and water, just won't properly mix on Earth. As long as these metallic mixtures remain essentially separated through sedimentation or buoyancy effects, they are not particularly useful. But when combined in microgravity, the lighter-density components of such mixtures or alloys will remain suspended for indefinitely long periods of time and therefore permit the formation of essentially uniform solid composites or alloys whose constituents have large density differences. When uniformly mixed and properly solidified, many of these new composite materials take on unusual properties, such as very high strength, excellent semiconductor behavior or perhaps outstanding performance as a superconductor.

Hydrostatic pressure places a strain on materials during solidification processes on Earth. Certain crystals are sufficiently dense and delicate that they are subject to strain under the influence of their own weight during growth. Such strain-induced deformations in crystals degrade their overall performance. In microgravity, heat-treated, melted and resolidified alloys can be developed free of such deformations.

Containerless processing in microgravity eliminates the problems of container contamination and wall effects. These are often the greatest source of impurities and imperfections when a molten material is formed here on Earth. But in space a material can be melted, manipulated and shaped, free of contact with a container or crucible, by acoustic, electromagnetic or electrostatic fields. In microgravity the surface tension of the molten material helps hold it together, while on Earth this cohesive force is overpowered by gravity.

One promising new space-based materials-manufacturing technology is the manufacture of certain high-purity pharmaceuticals. One example is the production of urokinase, a lifesaving drug in blood-clot treatments. Urokinase is an enzyme produced from a specific kidney cell. However, until recently, it has been virtually impossible to isolate this kidney cell from others. Cells may be separated from one another by a process called electrophoresis. The basis on which electrophoresis works is that living cells suspended in a liquid have a small negative electric charge on their surfaces. Cells vary not only in electric charge but also in size and weight. When an electric charge is passed through the liquid, various cells start drifting at different rates because of their varied charges, sizes and shapes. Individual cells of the same type flock together in a band, or layer.

One difficulty with electrophoresis on Earth is that the electric current heats the water, causing convection currents in the solution. This prevents clear lines of demarcation between bands of nearly identical cells, such as kidney cells.

Electrophoresis experiments performed on board the Apollo spacecraft (in the ASTP) and on board the Space Shuttle have demonstrated that in microgravity, where gravity-induced convection currents do not occur, the urokinase-producing cell can be isolated from others. By improving techniques for separation of the urokinase-producing cell, manufacturers hope to extract many times as much urokinase as they could when the cell was not isolated.

Alloys are mixtures of two or more metallic or metallic and nonmetallic materials that provide a new material with useful properties of both or all constituents. For example, steel is an alloy of iron and carbon, while brass is an alloy of copper and zinc. Many alloys cannot be made on Earth or at least are extremely difficult to make on Earth, because the heavier element tends to settle too fast when the mixture is solidifying. An example of this situation is aluminum antimonide, which is an alloy of aluminum and antimony. One attribute of aluminum antimonide is that it may be 30 percent to 50 percent more efficient than silicon when used in solar cells or computer circuit chips. Antimony is more than three times denser and tends to settle faster than aluminum when the two are mixed in a molten state. This causes a lack of uniformity in the alloy, severely limiting its usage in solar cells or computer circuit chips. In microgravity experiments, however, a largely uniform mixture has been obtained.

Crystals are solids in which most of the atoms are arranged in a regular pattern. Crystals such as silicon and germanium have extensive applications in the electronics and computer industries. These crystals are chemically grown under carefully controlled conditions. In experiments on board Skylab, the Apollo spacecraft used in the ASTP mission and the Space Shuttle/Spacelab configuration, higher-quality crystals were obtained than those made under similar conditions here on Earth.

Monodisperse (identically sized) latex spheres, whose diameters are smaller than the width of a human hair, are used in both medicine and industry to conduct research and to calibrate instruments. A variety of sizes are required to meet different needs. However, terrestrial gravity has limited the sizes that could be produced to about three micrometers. Beyond this size, gravitational influences give the particles a tendency to settle or clump. Experiments in microgravity have resulted in the development of larger latex spheres than is possible on Earth.

Materials processing in space offers the potential of not only stimulating space industrialization but also positively influencing the quality of life for all on Earth in the next century.

See also: **microgravity; space industrialization**

Megasphere   A truly enormous, Dyson sphere-like structure, tens of parsecs across, that captures the energy output of millions of stars clustered tightly together at the center of a galaxy. (A parsec is 3.26 light-years.) While a Dyson sphere represents the astroengineering feat of a mature Kardashev Type II civilization (solar-system level), the Megasphere represents the spectacular astroengineering accomplishment of an extremely powerful, galactic-level, Kardashev Type III extraterrestrial civilization.

See also: **Dyson sphere; extraterrestrial civilizations**

Mercurian   Of or relating to the planet Mercury; (in science fiction) a native of the planet Mercury.

Mercury   The innermost planet in our Solar System, orbiting the Sun at approximately 0.4 astronomical unit (AU). This planet, named for the messenger god of Roman mythology, is a scorched, primordial world that is only 40 percent larger in diameter than the Earth's Moon.

Before the Space Age, astronomers had attempted to identify its surface features, but because Mercury is so small and so close to the Sun, and is usually lost in its glare, this innermost planet remained only a featureless white blur in their telescopes. The first detailed observations were made in the late 1960s, when scientists bounced radar signals off its surface. Analysis of the returning radar signals revealed a rough surface and permitted the first accurate determination of Mercury's rotation rate. Prior to these pioneering radar observations, astronomers believed that the planet always kept the same face toward the Sun. These scattered signals indicated, however, that Mercury actually turns on its axis within a period of approximately 59 days. It takes the planet approximately 88 days to orbit around the Sun.

NASA's *Mariner 10* spacecraft provided the first close-up views of Mercury (see fig. 1). This spacecraft was launched from the Kennedy Space Center in November 1973. After traveling almost five months (including a flyby of Venus), this spacecraft passed within 805 kilome-

Fig. 1 This mosaic of several hundred images from the *Mariner 10* spacecraft provides an enhanced view of Mercury's southern hemisphere. (Courtesy of NASA/Jet Propulsion Laboratory.)

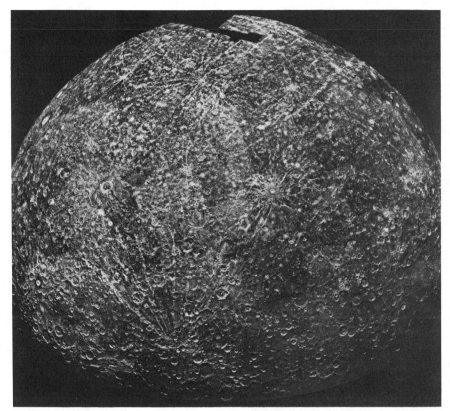

ters (500 miles) of Mercury on March 29, 1974. The *Mariner 10* spacecraft then looped around the Sun and made another rendezvous with Mercury on September 21, 1974. This encounter process was repeated a third time on March 16, 1975, before the control gas used to orient the spacecraft was exhausted. This triple flyby of the planet Mercury by the *Mariner 10* spacecraft is referred to as Mercury I, II and III in the technical literature.

The images of Mercury transmitted back to Earth by *Mariner 10* revealed an ancient, heavily cratered world that closely resembled the Earth's Moon. Unlike the Moon, however, Mercury has huge cliffs (called lobate scarps) that crisscross the planet. These great cliffs were apparently formed when Mercury's interior cooled and shrank, compressing the planet's crust. The cliffs are as high as 2 kilometers (1.2 miles) and as long as 1,500 kilometers (932 miles).

To the surprise of scientists, instruments on board *Mariner 10* discovered that Mercury has a weak magnetic field. It also has a wisp of an atmosphere—a trillionth of the density of the Earth's atmosphere and made up mainly of helium, hydrogen, neon and argon.

Temperatures on the sunlit side of Mercury reach approximately 700 degrees Kelvin (427 degrees Celsius)—a temperature that exceeds the melting point of lead; while on the dark side, temperatures plunge to a frigid 100 degrees Kelvin (minus 173 degrees Celsius). Quite literally, Mercury is a world seared with intolerable heat in the daytime and frozen at night.

The "days" and "nights" on this planet are quite long by terrestrial standards, since it takes about 59 Earth-days for Mercury to make a single rotation about its axis. The planet spins at a rate of approximately 10 kilometers per hour (6 miles per hour), measured at its equator. For comparison, the Earth spins at about 1,600 kilometers per hour (1,000 miles per hour) at its equator.

As shown in figure 1, Mercury's surface features include impact craters like those found on the Moon, as well as many secondary craters. A large number of these craters are surrounded by blankets of ejecta (material thrown out at the time of a meteorite impact) and secondary craters that were created when chunks of ejected material fell back down to the planet's surface. Because Mercury has a higher gravitational attraction than the Moon, these secondary craters are not spread as widely from each primary crater as occurs on the Moon. Most of Mercury looks a lot like the lunar highlands. However, Mercury appears not to have the dark deposits of lava that engulfed much of the Moon about 3 billion to 4 billion years ago. Mercury is also denser than the Moon. Scientists now speculate that Mercury has a large, iron-rich core—the source of its weak, but detectable, magnetic field.

Table 1 presents some contemporary physical- and dynamic-property data about the Sun's closest planetary companion.

Because Mercury lies deep in the Sun's gravity field, its detailed exploration with sophisticated orbiters and landers will require the development of powerful, advanced planetary spacecraft. For now, it remains just out of physical reach—except for flyby encounters like the highly successful *Mariner 10* mission.

See also: **Moon**

Table 1 Physical and Dynamic Properties of the Planet Mercury

Radius (mean equatorial)	2,439 km
Mass	3.30×10^{23} kg
Mean density	5.44 g/cm³
Acceleration of gravity (at the surface)	3.70 m/sec²
Escape velocity	4.25 km/sec
Normal albedo (averaged over visible spectrum)	0.125
Surface temperature extremes	$-173°$ C to $+427°$ C
Atmosphere	negligible
Number of natural satellites	none
Flux of solar radiation	
Aphelion	6,290 W/m²
Perihelion	14,490 W/m²
Semimajor axis	5.79×10^{7} km (0.387 AU)
Perihelion distance	4.60×10^{7} km (0.308 AU)
Aphelion distance	6.98×10^{7} km (0.467 AU)
Eccentricity	0.20563
Orbital inclination	7.004 degrees
Mean orbital velocity	47.87 km/sec
Sidereal day (a Mercurean "day")	58.646 Earth-days
Sidereal year (a Mercurean "year")	87.969 Earth-days
Earth-to-Mercury distance	
Maximum	20.25×10^{7} km (11.26 light-min)
Minimum	8.94×10^{7} km (4.97 light-min)

SOURCE: NASA.

metagalaxy The entire system of galaxies, including our Milky Way Galaxy; the entire contents of the Universe together with the region of space it occupies.

meteoroids An all-encompassing term that refers to solid objects found in space, ranging in diameter size from micrometers to kilometers and in mass from less than 10^{-12} grams to more than 10^{+16} grams. If these pieces of extraterrestrial material are less than 1 gram, they are often called micrometeoroids.

When objects of more than approximately 10^{-6} grams reach the Earth's atmosphere, they are heated to incandescence (that is, they glow with heat) and produce the visible effect popularly called a meteor.

If some of the original meteoroid survives its glowing plunge into the Earth's atmosphere, the remaining unvaporized chunk of space matter is then called a meteorite.

Scientists currently think that meteoroids originate primarily from asteroids and comets that have perihelia (portions of their orbits nearest the Sun) near or inside the Earth's orbit. The parent celestial objects are assumed to have been broken down into a collection of smaller bodies by numerous collisions. Recently formed meteoroids tend to remain concentrated along the orbital path of their parent body. These "stream meteoroids" produce the well-known meteor showers that can be seen at certain dates from Earth.

Meteoroids are generally classified by composition as stony, iron and (sometimes) "icy." Astronomers use the composition of a meteoroid to make inferences about the parent celestial body. Meteoroids are attracted by the Earth's gravitational field, so that the meteoroid flux from allowed directions in near-Earth space is actually increased up to approximately 1.7 over the interplanetary meteoroid flux value. The Earth also shields certain meteoroid arrival directions. Both of these factors—the defocusing factor and the shielding factor—must be considered.

How much extraterrestrial material falls on the Earth each year? Space scientists estimate that about 10^{+7} kilograms (or 10,000 metric tons) of "cosmic rocks" now fall on our planet annually.

What is the meteoroid hazard to an astronaut or lunar miner on the surface of the Moon? Since the Moon does not have a protective atmosphere, its surface is constantly bombarded by meteoroids and micrometeoroids. These particles and rocks have impact velocities between 2.4 and 72 kilometers per second. However, the actual meteoroid hazard to an individual astronaut on the lunar surface is quite small. The impact rate per square centimeter appears to lie between 1.1 and 50 craters per million years for holes or pits greater than 500 micrometers in diameter. So unless you're very unlucky, the odds are quite low that you will "catch a falling star" as you wander across the lunar surface searching for mineral resources or for a place to locate a new radio telescope.

See also: **asteroid; comet; hazards to space workers**

microbes Microscopic organisms, some of which cause disease. Many microbes are found in and on the human body.

microgravity Because the inertial trajectory of a spacecraft (for example, the Space Shuttle) compensates for the force of the Earth's gravity, an orbiting spacecraft and all its contents approach a state of free-fall. In this state of free-fall, all objects inside the spacecraft appear "weightless."

It is important to understand how this condition of weightlessness, or the apparent lack of gravity, develops.

Newton's universal law of gravitation tells us that any two objects have a gravitational attraction for each other that is proportional to their masses and inversely proportional to the square of the distance between their centers of mass. It is also interesting to recognize that a spacecraft orbiting the Earth at an altitude of 400 kilometers is only 6 percent farther from the center of the Earth than it would be if it were on the Earth's surface. Using Newton's law, we find that the gravitational attraction at this particular altitude is only 12 percent less than the attraction of gravity at the surface of the Earth. In other words, an Earth-orbiting spacecraft and all its contents are very much under the influence of the Earth's gravity! The phenomenon of weightlessness occurs because the spacecraft and its contents are in a continual state of free-fall.

Figure 1 describes the different orbital paths a falling object may take when dropped from a point above the Earth's sensible atmosphere. With no tangential-velocity component, an object would fall straight down (trajectory 1) in our simplified demonstration. As we give the object an increasing tangential-velocity component, it still "falls" toward the Earth under the influence of terrestrial gravitational attraction, but the velocity component now gives the object a trajectory that is a segment of an ellipse. As shown in trajectories 2 and 3 in figure 1, as we give the object a larger tangential velocity, the point where it finally hits the Earth moves farther and farther away from the release point. If we keep increasing this velocity component, the object eventually "misses the Earth" completely (trajectory 4). As the tangential velocity is further increased, the object's trajectory takes the form of a circle (trajectory 5) and then a larger ellipse, with the release point representing the point of closest approach to the Earth (or "perigee"). Finally, when the initial tangential-velocity component is about 41 percent greater than that needed to achieve a circular orbit, the object follows a parabolic, or escape, trajectory and will never return (trajectory 6).

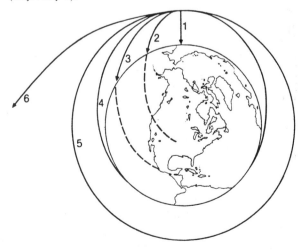

Fig. 1   Various orbital paths of a falling body around the Earth. (Drawing courtesy of NASA.)

Einstein's principle of equivalence tells us that the physical behavior inside a system in free-fall is identical to that inside a system far removed from other matter that could exert a gravitational influence. Therefore, the term *zero gravity* (also called "zero g") or *weightlessness* is frequently used to describe a free-falling system in orbit.

Sometimes people ask what is the difference between mass and weight. Why do we say, for example, "weightlessness" and not "masslessness"? *Mass* is the physical substance of an object—it has the same value everywhere. *Weight*, on the other hand, is the product of an object's mass and the local acceleration of gravity (in accordance with Newton's second law of motion, $F = ma$). For example, you would weigh about one-sixth as much on the Moon as here on Earth, but your mass remains the same in both places.

A "zero-gravity" environment is really an ideal situation that can never be totally achieved in an orbiting spacecraft. The venting of gases from the space vehicle, the minute drag exerted by a very thin terrestrial atmosphere at orbital altitudes and even crew motions create nearly imperceptible forces on people and objects alike. These tiny forces are collectively called "microgravity." In a microgravity environment, astronauts and their equipment are almost, but not entirely, weightless.

Microgravity represents an intriguing experience for space travelers. You can perform slow-motion somersaults and handsprings. You can float with ease through a space cabin (see fig. 2). You can push off one wall of a space station and drift effortlessly to the other side. You can lift or move "heavy" objects, which are essentially weightless. And if you're just a little bit clumsy, you don't need to worry about dropping things—whatever slips from your hand will simply float away.

However, life in microgravity is not necessarily easier than life on Earth. For example, the caloric (food-intake) requirements for people living in microgravity are the same as those on Earth. Living in microgravity also calls for special design technology. A beverage in an open container, for instance, will cling to the inner or outer walls and, if shaken, will leave the container as free-floating droplets or fluid globs. Such free-floating droplets are not merely an inconvenience. They can annoy crew members (no one wants to get "slimed" in orbit), and they represent a definite hazard to equipment, especially sensitive electronic devices and computers.

Therefore, water is usually served in microgravity through a specially designed dispenser unit that can be turned on or off by squeezing and releasing a trigger. Other beverages, such as orange juice, are typically served in sealed containers through which a plastic straw can be inserted. When the beverage is not being sipped, the straw is simply clamped shut.

Microgravity living also calls for special considerations in handling solid foods. Crumbly foods are provided only in bite-sized pieces to avoid having crumbs floating around the space cabin. Gravies, sauces and dressings

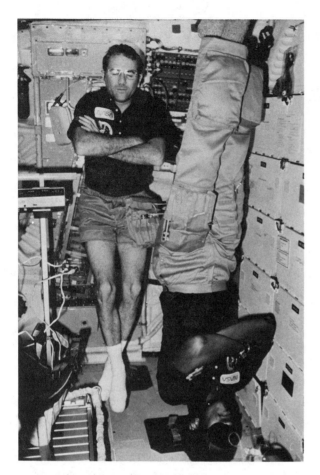

Fig. 2   STS-8 astronauts (*left*) Truly and (*right*) Bluford enjoy a "floating" rest session in the Shuttle *Challenger's* cabin. (Courtesy of NASA.)

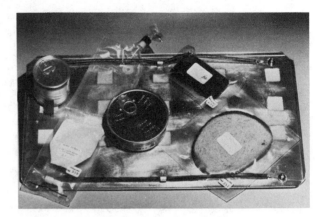

Fig. 3   Specially devised Shuttle astronaut food tray. (Courtesy of NASA.)

have a viscosity (stickiness) that prevents them from simply lifting off the food trays and floating away. Typical space food trays are equipped with magnets, clamps and double-adhesive tape to hold metal, plastic and other utensils (see fig. 3). Astronauts are provided with forks and spoons. However, extraterrestrial diners must learn to eat without sudden starts or stops if they expect the solid food to stay on their utensils.

Personal hygiene is also a bit challenging in microgravity. Waste water in the Shuttle's galley from utensil cleanup or an astronaut's washing (sponge bath) is directed away by a flow of air (which provides a force substituting for gravity) to a drain that then leads to a sealed tank. Shuttle astronauts must take sponge baths rather than showers or regular baths. However, space-station crews will most likely have more "Earthlike" shower facilities, perhaps something like the apparatus used on Skylab.

Because water adheres to the skin in microgravity, perspiration can also be annoying, especially during strenuous activities. In the absence of proper air circulation, perspiration can accumulate layer by layer on an astronaut's skin.

Waste elimination in microgravity represents another challenging design problem. To help Shuttle astronauts go to the bathroom, a special toilet device has been engineered to closely resemble the normal sanitary procedures performed here on Earth. The main differences are that the astronaut must use a seat belt and foot restraints to keep from drifting. The wastes themselves are flushed away by a flow of air and a mechanical device.

The Space Shuttle's waste-collection system has a set of controls that are used to configure the system for various operational modes, including: urine collection only, combined urine and feces collection and emesis collection (vomit collection). The overall microgravity toilet system consists of a commode (or waste collector) to handle solid wastes and a urinal assembly to handle fluids.

The urinal is used by both male and female astronauts—with the individual either holding the urinal while standing or sitting on the commode with the urinal mounted to the waste-collection system. Since the urinal has a contoured cup with a spring assembly, it provides a good seal with the female crew member's body. During urination, a flow of air creates a pressure differential that draws the urine off into a fan separator/storage tank assembly.

The microgravity commode is used for collecting both feces and emesis. When properly functioning, it has a capacity for storing the equivalent of 210 person-days of vacuum-dried feces and toilet tissue. This device may be used up to four times per hour, and it may be used simultaneously with the urinal. To operate the waste collector during defecation, the astronaut positions himself or herself on the commode seat. Handholds, foot restraints and waist restraints help the individual maintain a good seal with the seat. The crew member uses this equipment like a normal terrestrial toilet, including tissue wipes. Used tissues are disposed of in the commode. Everything stored in the waste collector—feces, tissues and fecal and emesis bags—are then subjected to vacuum-drying in the collector.

Shaving can also cause problems in microgravity, if whiskers end up floating around the cabin. These free-floating whiskers could damage delicate equipment (especially electronic circuits and optical instruments) or else irritate the eyes and lungs of space travelers. One solution is to use a safety razor and shaving cream or gel. The whiskers will adhere to the cream until wiped off with a disposable towel. Another approach is to use an electronic razor with a built-in vacuum device that sucks away and stores the whiskers as they are cut.

For long-duration space missions, other personal hygiene tasks that might require some special procedure or device in microgravity include nail trimming and hair cutting. Special devices have also been developed for female astronauts to support personal hygiene requirements associated with the menstrual cycle.

Microgravity living is definitely different from the lifestyles permitted at the bottom of a one-g gravity well on the surface of the Earth. Furniture, for example, must be bolted in place—or else it will simply float around the cabin. Tether lines, belts, Velcro™ anchors and handholds enable astronauts to move around and to keep themselves and other objects in place.

Sleeping in space is also an interesting experience. Shuttle astronauts can sleep either horizontally or vertically while in orbit. Their fireproof sleeping bags attach to rigid padded boards for support. But the astronauts themselves literally sleep "floating in air."

Special tools (such as torqueless wrenches), handholds and foot restraints help an astronaut turn a nut or tighten a screw while in orbit. These devices are needed to balance or neutralize reaction forces. If these special devices were not available, a space worker might find himself or herself helplessly rotating around a "work piece" or work station.

Exposure to microgravity also causes a variety of physiological (bodily) changes. For example, space travelers appear to have "smaller eyes," because their faces have become puffy. They also get rosy cheeks and distended veins in their foreheads and necks. They may even be a little bit taller than they are on Earth, because their body masses no longer "weigh down" their spines.

Leg muscles shrink, and anthropometric (measurable postural) changes also occur. Astronauts tend to move in a slight crouch, with head and arms forward.

Some space travelers suffer from a temporary condition resembling motion sickness. This condition is called space sickness or space-adaptation syndrome. In addition, sinuses become congested, leading to a condition similar to a cold.

Many of these microgravity-induced physiological effects appear to be caused by fluid shifts from the lower to the upper portions of the body. So much fluid goes to the head that the brain may be fooled into thinking that the body has too much water. This can then result in an increased production of urine.

Extended stays in microgravity tend to shrink the heart,

decrease production of red blood cells and increase production of white blood cells. A process called resorption occurs. This is the leaching of vital minerals and other chemicals (such as calcium, phosphorus, potassium and nitrogen) from the bones and muscles into body fluids that are then expelled as urine. Such mineral and chemical losses can have adverse physiological and psychological effects. In addition, prolonged exposure to a microgravity environment might cause bone loss and a reduced rate of bone-tissue formation.

While a brief stay (say, up to 30 days) in microgravity may prove to be an exhilarating, nondetrimental experience for most space travelers, permanent space bases and large space settlements will resort to "artificial gravity" (created through the slow rotation of these habitats and facilities) to provide a more "Earthlike" home in cislunar space and to avoid any serious health effects that might arise from permanent exposure to a microgravity environment. These space habitats might even offer the inhabitants the very exciting possibility of life in a multiple-gravity-level world, with different modules or zones simulating gravity conditions from microgravity up to normal terrestrial gravity.

Besides providing an interesting new dimension for human experience, the microgravity environment of an orbiting space system offers the ability to manufacture new and improved materials and products that simply cannot be made on Earth. Although microgravity can be simulated from here on Earth using drop towers, special airplane trajectories and sounding rocket flights, these simulations are only short-duration activities (seconds to minutes) that are frequently "contaminated" by vibrations and other undesirable effects. However, the long-term microgravity environment found in orbit provides an entirely new dimension for materials-science research, life-science research and manufacturing processes. Today, we can only partially speculate on the overall impact that access to microgravity will have on our 21st-century lifestyles. Through the use of permanent space stations and platforms, we will be the first human generation that can regularly examine material behavior, physical processes, manufacturing techniques and life processes in the absence of Earth's loving but firm one-g grasp! The potential for revolutionary breakthroughs, unanticipated discoveries and unusual developments in a great number of technical areas is simply astounding!

See also: **materials processing in space; Newton's laws of motion; orbits of objects in space; space industrialization; space settlement; space station; Space Transportation System**

microorganism A very tiny plant or animal, especially a protozoan or a bacterium.

See also: **extraterrestrial contamination**

microwave A comparatively short-wavelength electromagnetic (EM) wave in the radio-frequency portion of the

EM spectrum. The term *microwave* is usually applied to those EM wavelengths that are measured in centimeters, approximately 30 centimeters to 1 millimeter (with corresponding frequencies of 1 GHz [gigahertz] to 300 GHz). Present-day microwave amplifiers and oscillator tubes operate in a frequency range up to about 40 GHz, while solid-state microwave devices can reach frequencies of up to 100 GHz.

See also: **electromagnetic spectrum; Satellite Power System**

Milky Way Galaxy   Our home Galaxy. The immense band of stars stretching around the night sky (see fig. C16 on color page G) represents our "inside view" of the Milky Way. The Milky Way Galaxy is a giant, rotating disk of stars, gas and dust. Current estimates indicate that our Galaxy may contain over 600 billion solar masses. (A solar mass is a unit used for comparing stellar masses, with 1 solar mass being equal to the mass of the Earth's Sun.) Astronomers now divide the Milky Way Galaxy up into four major components: the spherical *central bulge*, or galactic nucleus (possibly containing a black hole); the thin, spiraling *disk*, which contains stars, dust and gas (including the Sun and the Earth); the *spheroidal component*, which contains old stars and extends for approximately 81,540 light-years (25 kiloparsecs) from the galactic center (this region was formerly called the "halo"); and finally, the newly discovered *corona*, a tenuous but very large region. The corona is believed to have a minimum radius of 195,700 light-years (60 kiloparsecs). It apparently does not contain stars but because of its huge extent may be where the "missing mass" of our Galaxy can be found.

See also: **black holes; galaxy; stars**

moon   A natural satellite of any planet.

Moon   The Moon (the term is capitalized when used in this sense) is the Earth's only natural satellite and closest celestial neighbor (see fig. C17 on color page H). While life on Earth is made possible by the Sun, it is also regulated by the periodic motions of the Moon. For example, the months of our year are measured by the regular motions of the Moon around the Earth, and the tides rise and fall because of the gravitational tug-of-war between the Earth and the Moon. Throughout history, the Moon has had a significant influence on human culture, art and literature. Even in the Age of Space, it has proved to be a major sociotechnical stimulus. It was just far enough away to represent a real technological challenge to reach it; yet it was still close enough to allow us to be successful on the first concentrated effort. The Apollo expeditions to the Moon from 1969 to 1972 made us extraterrestrial travelers.

Scientific interest in the Moon dates back to the earliest periods of recorded history, but it wasn't until very recently, with the birth of the Space Age, that we finally developed the technical tools needed to examine our celes-

tial companion firsthand. As evidenced by the "sensational" news clippings describing the pioneering rocket work of Dr. Robert Goddard (the father of American rocketry), the Moon was considered totally inaccessible by the vast majority of people. Today, however, it has become a "planet" to visit, to explore and eventually to inhabit.

The Moon was the first extraterrestrial object surveyed by spacecraft. Table 1 provides a detailed summary of all the major unmanned missions to the Moon, while table 2 provides information about the manned Apollo program expeditions into cislunar space and on to the lunar surface itself.

From evidence gathered by the early unmanned missions (such as Ranger, Surveyor and the Lunar Orbiter spacecraft) and by the Apollo missions, lunar scientists have learned a great deal more about the Moon and have been able to construct a geologic history dating back to its infancy. Table 3 provides selected physical and dynamic properties of the Moon.

Because the Moon does not have any oceans or other free-flowing water and lacks a sensible atmosphere, appreciable erosion, or "weathering," has not occurred there. In fact, the Moon is actually a "museum world." The primitive materials that lay on its surface billions of years ago are in an excellent state of preservation. (See fig. 1.) Scientists believe that the Moon was formed some 4.6 billion years ago and then differentiated quite early, perhaps only 100 million years later. Tectonic activity ceased eons ago on the Moon. The lunar crust and mantle are quite thick, extending inward to more than 800 kilometers. However, the deep interior of the Moon is still unknown. It may contain a small iron core at its center, and there is some evidence that the lunar interior may be hot and even partially molten. Moonquakes have been measured within the lithosphere and interior, most being the result of tidal stresses. Chemically, the Earth and the Moon are quite similar, though compared to the Earth, the Moon is depleted in the more easily vaporized materials. The lunar surface consists of highlands composed of alumina-rich rocks that formed from a globe-encircling molten sea and maria made up of volcanic melts that surfaced about 3.5 billion years ago. However, despite all we have learned in the past two decades about our nearest celestial neighbor, lunar exploration has really only just started. Several puzzling mysteries still remain, including an explanation for the remnant magnetism measured in the rocks despite the absence of a lunar dynamo (a hot, molten core that conducts electric currents) and the origin of the Moon itself.

There are three general theories concerning the origin of the Moon that are currently popular. The least "exotic" hypothesis suggests that the Moon and the Earth accreted (accumulated) from a nebula of gas and dust that surrounded the primordial Sun in much the same relative positions that they occupy today. Another theory proposes that the Moon formed in a different part of the solar nebula but was later captured by the Earth. In this lunar-

Fig. 1 The crater Tsiolkovsky as photographed from the *Apollo 15* Command Module, 1971. (Courtesy of NASA.)

origin scenario, the Moon may have either survived intact or else been fragmented by numerous collisions and then reaccumulated in Earth orbit. The third general lunar-origin theory suggests that the Moon separated from a partially differentiated Earth during a rapid rotational instability that occurred shortly after accretion.

As previously mentioned, the surface of the Moon has two major regions with distinctive geologic features and evolutionary histories. First is the relatively smooth, dark areas that Galileo originally called "maria" (because he thought they were seas or oceans). Second is the densely cratered, rugged highlands (uplands), which Galileo called "terrae." The highlands occupy about 83 percent of the Moon's surface and generally have a higher elevation (as much as five kilometers above the Moon's mean radius). In other places the maria lie about five kilometers below the mean radius and are concentrated on the near side of the Moon.

The main external geologic process modifying the surface of the Moon is meteoroid impact. Craters range in size from very tiny pits only micrometers in diameter to gigantic basins hundreds of kilometers across (see fig. 2).

The lunar highlands, or terrae, are the bright (high-albedo) regions on the Moon's surface. They are primarily ancient surfaces that have become extensively cratered during the early history of the Moon, when the meteoroid flux was high. The anorthositic gabbros (consisting mainly of plagioclase, olivine and pyroxene) are the most abundant rocks in the lunar highlands. Almost all the samples returned by the Apollo expeditions have been extensively brecciated (fragmented), metamorphosed and chemically contaminated by repeated meteoroid impact.

The lunar maria are concentrated in the topographic basins on the near side of the Moon. The average normal albedo of the maria is a rather low 7 percent. These maria are vast plains of basaltic lava flows that were erupted

Fig. 2   The heavily cratered far side of the Moon as seen from the *Apollo 16* Command Module, 1972. (Courtesy of NASA.)

after the highlands and the impact basins formed. The basalt in the middle of the basin may be several kilometers thick but still represents only a small part of the crust, which is approximately 60 kilometers thick. Lunar basalts are chemically and mineralogically similar to their terrestrial counterparts. They differ only in their high abundance of calcium-plagioclase and titanium and in their total absence of hydrous minerals.

The surface of the Moon is strongly brecciated, or fragmented. This mantle of weakly coherent debris is called regolith. It consists of shocked fragments of rocks, minerals and spherical pieces of glass formed by meteoroid impact. The thickness of the regolith is quite variable and depends on the age of the bedrock beneath and on the proximity of craters and their ejecta blankets. Generally, the maria are covered by 3 to 16 meters of regolith, while the older highlands have developed a "lunar soil" at least 10 meters thick. Table 4 provides a summary of the major chemical elements found on the lunar surface.

The Moon has a very tenuous atmosphere, which is more properly termed an "exosphere" (see table 5). It is a nearly collisionless gas. There are no "aeolean," or wind, effects on the lunar surface from this wisp of an atmosphere. As measured by Apollo mission instruments, the nighttime lunar atmosphere at the surface contains approximately 200,000 molecules per cubic centimeter. Therefore, without a protective atmosphere, astronaut explorers and lunar workers will be exposed to raw, unfiltered sunlight, atomic-particle radiation from the Sun and the hazard of meteoroid bombardment (the extent of this hazard is discussed below).

On the lunar surface, incoming sunlight has an intensity of approximately 1,371 watts per square meter and a spectral distribution resembling a blackbody radiator at 5,760 degrees Kelvin.

The Moon does not possess a magnetic field strong enough to prevent the direct impingement of atomic particles from space onto the lunar surface. There are three sources for these charged particles at the surface of the Moon: (1) the solar wind, (2) solar cosmic rays and (3) high-energy galactic cosmic rays. Table 6 summarizes the charged-particle environment at the lunar surface.

The Moon is also constantly bombarded by meteoroids and micrometeoroids. Since it does not possess a protective atmospheric blanket, these meteoroids have impact velocities ranging from 2.4 to 72 kilometers per second. However, the actual meteoroid-bombardment hazard for individual astronauts and lunar workers is quite small. The impact rate per square centimeter appears to fall between 1.1 and 50 craters per million years for pits greater than 500 micrometers in diameter. A direct measurement of the incident meteoroid flux has been accomplished for particles larger than about 100 grams by using the lunar seismic network deployed during the Apollo missions. Between 70 and 150 events were recorded each year until

the experiment was discontinued. The equation given below describes the present-day meteoroid flux on the lunar surface:

$$\log N = -1.62 - 1.16 \log m$$

where N is the number of impacting bodies per square kilometer per year with a mass greater than m
m is the mass (in grams)

This proposed distribution appears to result from two populations of impacting bodies—one that varies little throughout the year and the other made up of larger objects that intersect the Moon's orbit during April, May and June, producing a higher-than-average meteoroid flux.

Because of its relatively close proximity to Earth and its mineral-resource potential, the Moon will play a very critical role in the development of our extraterrestrial civilization. To initiate the further exploration and eventual use of the Moon, we can first send sophisticated machines. For example, an unmanned Lunar Geoscience Orbiter Mission could continue the scientific investigations started by the Apollo astronauts and would produce extensive maps of the entire lunar surface. Other robot spacecraft, like the Soviet *Luna 16* and *Luna 20* landers, could be used to return additional soil samples from previously unvisited regions such as the far side and the poles.

Then, when human beings return to the Moon, it will not be for another brief moment of scientific inquiry but rather to establish a permanent presence on another world. Within the next few decades, we will establish permanent bases and settlements on the surface of our celestial companion and perform the basic scientific and engineering experiments necessary to develop a complete industrial complex. The use of lunar materials has frequently been suggested by space technologists as an essential pathway in the industrialization of space. With its mineral wealth, strategic location and reduced surface gravity, the Moon could easily become our gateway to the entire Solar System. We can even speculate that the Moon settlements will serve as the technical and social "training ground" for the spacefaring portion of the human race.

See also: **extraterrestrial expeditions; lunar bases and settlements; lunar soil and rock samples**

Table 1   A Summary of Moon Missions

SPACECRAFT	LAUNCH DATE; NATIONALITY; WEIGHT; MISSION/RESULTS
Pioneer 0	Aug. 17, 1958; USA; 38 kg; attempt to orbit Moon/launch vehicle exploded at an altitude of 16 km.
Pioneer 1	Oct. 11, 1958; USA; 38 kg; attempt to orbit Moon/launch vehicle reached an altitude of 113,830 km, then fell back into the South Pacific.
Pioneer 2	Nov. 8, 1958; USA; 39 kg; attempt to orbit Moon/launch vehicle reached an altitude of 1,550 km, then fell back to Earth near Africa.
Pioneer 3	Dec. 6, 1958; USA; 6 kg; attempt to fly-by Moon/launch vehicle reached an altitude of 102,320 km, then fell back to Earth over Africa.

Luna 1[a] Jan. 2, 1959; USSR; 361 kg; attempt to impact Moon/partial success—missed Moon by some 5,000 km, then entered solar orbit.

Pioneer 4 March 3, 1959; USA; 6 kg; lunar flyby/successful—passed Moon at 60,500 km, then entered solar orbit.

Luna 2 Sept. 12, 1959; USSR; 390 kg; impact Moon/successful—first lunar impact, struck 335 km from visible center.

Pioneer P-1 Sept. 24, 1959; USA; 170 kg; attempt to orbit Moon/launch vehicle and spacecraft destroyed in explosion during static test before launch.

Luna 3 Oct. 4, 1959; USSR; 435 kg; photography of far side of Moon/successful—returned pictures of 70 percent of lunar far side.

Pioneer P-3 Nov. 26, 1959; USA; 169 kg; attempt to orbit Moon/failure—launch vehicle shroud tore away during ascent, payload impacted near Africa.

Pioneer P-30 Sept. 25, 1960; USA; 176 kg; attempt to orbit Moon/failure—launch vehicle malfunction, payload impacted in Africa.

Pioneer P-31 Dec. 15, 1960; USA; 176 kg; attempt to orbit Moon/failure—launch vehicle climbed to an altitude of 13 km, then exploded.

Ranger 1 Aug. 23, 1961; USA: 306 kg; high-Earth-orbit test of spacecraft/failure—intended to climb to 1,102,850 km but remained in low Earth orbit.

Ranger 2 Nov. 18, 1961; USA; 306 kg; high-Earth-orbit test of spacecraft/failure—intended to climb to 1,102,850 km but remained in low Earth orbit.

Ranger 3 Jan. 26, 1962; USA; 330 kg; attempted TV reconnaissance of Moon and hard landing/partial success—spacecraft missed Moon by 36,808 km, no TV pictures or landed instruments.

Ranger 4 April 23, 1962; USA; 331 kg; attempted TV reconnaissance of Moon and hard landing/partial success—timer failure, spacecraft fell on far side of Moon, no TV pictures.

Ranger 5 Oct. 18, 1962; USA; 342 kg; attempted TV reconnaissance of Moon and hard landing/partial success—power failure caused spacecraft to miss Moon by 725 km, entered solar orbit.

Luna
(unannounced) Jan. 4, 1963; USSR; 1,400 kg (?); attempted soft landing on Moon/failure—spacecraft achieved only Earth orbit.

Luna 4 April 2, 1963; USSR; 1,422 kg; attempted soft landing on Moon/partial success—lunar flyby at 8,500 km, entered solar orbit.

Ranger 6 Jan. 30, 1964; USA; 365 kg; attempted TV reconnaissance of Moon and hard landing/partial success—impacted on target but no TV pictures returned.

Ranger 7 July 28, 1964; USA: 366 kg; TV reconnaissance of Moon and hard landing/success—returned over 4,300 high-resolution images of Moon before impacting on target at Mare Nubium.

Ranger 8 Feb. 17, 1965; USA; 367 kg; TV reconnaissance of Moon and hard landing/success—returned 7,137 high-resolution images of Moon before impacting on target at Mare Tranquillitatis.

Kosmos 60 March 12, 1965; USSR; 1,470 kg; attempted soft landing on Moon/failure—spacecraft achieved only Earth orbit.

Ranger 9 March 21, 1965; USA; 366 kg; TV reconnaissance of Moon and hard landing/successful—returned 5,814 high-resolution images of Moon before impacting inside Crater Alphonsus.

Luna 5 May 9, 1965; USSR; 1,476 kg; attempted soft landing on Moon/partial success—retrofire failure caused spacecraft to crash-land on Mare Nubium.

Luna 6 June 8, 1965; USSR; 1,442 kg; attempted soft landing on Moon/partial success—missed Moon by 160,000 km, entered solar orbit.

[a]Soviet data are from TASS bulletins.

Zond 3	July 18, 1965; USSR; 890 kg (?); photography of Moon's far side/success—returned 25 pictures of lunar far side, then entered solar orbit.
Luna 7	Oct. 4, 1965; USSR; 1,506 kg; attempted soft landing on Moon/partial success—retrofired early, spacecraft crashed in Oceanus Procellarum.
Luna 8	Dec. 3, 1965; USSR; 1,552 kg; attempted soft landing on Moon/partial success—retrofired late, crashed in Oceanus Procellarum.
Luna 9	Jan. 31, 1966; USSR; 1,583 kg; soft landing on Moon/success—first lunar soft landing, 100-kg capsule returned photographs from lunar surface at Oceanus Procellarum.
Kosmos 111	March 1, 1966; USSR; 1,600 kg (?); attempt to orbit Moon/failure—spacecraft only achieved Earth orbit.
Luna 10	March 31, 1966; USSR; 1,600 kg; to orbit Moon/successful—first spacecraft to achieve lunar orbit; returned physical data from lunar surface.
Surveyor 1	May 30, 1966; USA; 995 kg; soft landing on Moon/success—touchdown north of Flamsteed, returned 11,237 pictures from lunar surface.
Explorer 33	July 1, 1966; USA; 93 kg; attempt to orbit Moon/partial success—spacecraft failed to approach Moon at proper velocity, achieved Earth orbit instead.
Lunar Orbiter 1	Aug. 10, 1966; USA; 387 kg; to orbit Moon/successful—photographic mapping of lunar surface (Apollo landing sites).
Luna 11	Aug. 24, 1966; USSR; 1,604 kg (?); to orbit Moon/success—spacecraft achieved lunar orbit but did not return lunar surface pictures.
Surveyor 2	Sept. 20, 1966; USA; 1,000 kg; attempt to soft-land on the Moon/partial success—stabilization failure, spacecraft crashed southeast of Crater Copernicus.
Luna 12	Oct. 22, 1966; USSR; 1,625 kg (?); to orbit Moon/success—spacecraft returned images of lunar surface.
Lunar Orbiter 2	Nov. 6, 1966; USA; 390 kg; to orbit Moon/successful—photographic mapping of lunar surface (Apollo landing sites and far side).
Luna 13	Dec. 21, 1966; USSR; 1,595 kg (?); soft landing on Moon/successful—soft landing achieved on Oceanus Procellarum, returned pictures of lunar surface, studied lunar soil density.
Lunar Orbiter 3	Feb. 5, 1967; USA; 385 kg; to orbit Moon/successful—photographic mapping of lunar surface (Apollo landing sites).
Surveyor 3	April 17, 1967; USA; 1,035 kg; soft landing on Moon/success—touch down in Oceanus Procellarum, returned images of surface, dug lunar soil with shovel.
Lunar Orbiter 4	May 4, 1967; USA; 390 kg; to orbit Moon/successful—photographic mapping of large areas of Moon.
Surveyor 4	July 14, 1967; USA; 1,039 kg; attempted soft landing on Moon/partial success—signals ceased at lunar impact on Sinus Medii.
Explorer 35	July 19, 1967; USA; 104 kg; to orbit Moon/successful—returned physical data from lunar orbit.
Lunar Orbiter 5	Aug. 1, 1967; USA; 390 kg; to orbit Moon/successful—photographic mapping of Moon, including much of far side.
Surveyor 5	Sept. 8, 1967; USA; 1,005 kg; soft landing on Moon/success—soft-landed on Mare Tranquillitatis and returned 18,006 pictures of lunar surface, performed first chemical analysis of lunar soil.
Surveyor 6	Nov. 7, 1967; USA; 1,008 kg; soft landing on Moon/success—soft-landed on Sinus Medii, returned 30,065 pictures of lunar surface and performed chemical and mechanical studies of lunar soil.
Surveyor 7	Jan. 7, 1968; USA; 1,040 kg; soft landing on Moon/success—landed near north rim of crater Tycho, returned 21,274 images of lunar surface and performed chemical analysis of lunar soil from trench it dug.

Zond 4	March 2, 1968; USSR; 5,800 kg (?); spacecraft test mission/partial success—flew to lunar distance but recovery in doubt.
Luna 14	April 7, 1968; USSR; 1,615 kg (?); to orbit Moon/success—achieved lunar orbit, returned data on lunar mass distribution.
Zond 5	Sept. 14, 1968; USSR; 5,800 kg (?); circumlunar flight/successful—ballistic reentry with biological specimens and pictures.
Zond 6	Nov. 10, 1968; USSR; 5,800 kg (?); circumlunar flight/successful—lifting reentry with biological specimens and pictures.
Luna 15	July 13, 1969; USSR; 5,800 kg (?); soft landing on Moon/partial success—lunar orbit achieved, but crash landing occurred on Mare Crisium.
Zond 7	Aug. 7, 1969; USSR; 5,800 kg (?); circumlunar flight/successful—photographs of lunar far side taken and lifting reentry accomplished.
Kosmos 300	Sept. 23, 1969; USSR; 5,800 kg (?); soft landing on Moon/failure—spacecraft only achieved Earth orbit.
Kosmos 305	Oct. 22, 1969; USSR; 5,800 kg (?); soft landing on Moon/failure—spacecraft only achieved Earth orbit.
Luna 16	Sept. 12, 1970; USSR; 5,800 (?); soft landing on Moon/successful—landed on Mare Faecunditatis, performed automated lunar soil sample collection and returned it to Earth.
Zond 8	Oct. 20, 1970; USSR; 5,800 kg (?); circumlunar flight/successful—took photographs and accomplished ballistic reentry.
Luna 17	Nov. 10, 1970; USSR; 5,800 (?); soft landing on Moon/successful—landed on Mare Imbrium, included *Lunokhod 1* automated rover vehicle, which conducted long-term exploration program.
Luna 18	Sept. 2, 1971; USSR; 5,800 kg (?); soft landing on Moon/partial success—spacecraft achieved lunar orbit but crashed on landing at Mare Faecunditatis.
Luna 19	Sept. 28, 1971; USSR; 5,800 kg (?); to orbit Moon/success—returned photographs and data.
Luna 20	Feb. 14, 1972; USSR; 5,800 kg (?), soft landing on Moon/successful—landed on Mare Crisium, made automated soil sample collection and returned it to Earth.
Luna 21	Jan. 8, 1973; USSR; 5,800 kg (?); soft landing on Moon/successful—landed near Le Monnier, included *Lunokhod 2* automated rover vehicle.
Explorer 49	June 10, 1973; USA; 328 kg; to orbit Moon/successful—performed radioastronomy experiments from far side of Moon.
Luna 22	May 29, 1974; USSR; 5,800 kg (?); to orbit Moon/success—achieved lunar orbit, took photographs and collected data.
Luna 23	Oct. 28, 1974; USSR; 5,800 kg (?); soft landing on Moon/partial success—soft landing in southern part of Mare Crisium but drill damaged so no soil sample returned to Earth.
Luna 24	Aug. 9, 1976; USSR; 5,800 kg (?); soft landing on Moon/successful—performed automated soil sample collection in Mare Crisium and returned to Earth.

SOURCE: NASA.

Table 2   Apollo Program Summary

Spacecraft Name	Crew	Date	Flight Time (Hrs., Min., Sec.)	Revolutions	Remarks
Apollo 7	Walter H. Schirra Donn Eisele Walter Cunningham	10/11–22/68	260:8:45	163	First manned Apollo flight demonstrated the spacecraft, crew and support elements. All performed as required.

Apollo 8	Frank Borman James A. Lovell, Jr. William Anders	12/21–27/68	147:00:41	10 rev. of Moon	History's first manned flight to the vicinity of another celestial body.
Apollo 9	James A. McDivitt David R. Scott Russell L. Schweickart	3/3–13/69	241:00:53	151	First all-up manned Apollo flight (with Saturn V and command, service and lunar modules). First Apollo EVA. First docking of CSM with LM.
Apollo 10	Thomas P. Stafford John W. Young Eugene A. Cernan	5/18–26/69	192:03:23	31 rev. of Moon	Apollo LM descended to within 14.5 km of Moon and later rejoined CSM. First rehearsal in lunar environment.
Apollo 11	Neil A. Armstrong Michael Collins Edwin E. Aldrin, Jr.	7/16–24/69	195:18:35	30 rev. of Moon	First landing of men on the Moon. Total stay time: 21 hrs., 36 min.
Apollo 12	Charles Conrad, Jr. Richard F. Gordon, Jr. Alan L. Bean	11/14–24/69	244:36:25	45 rev. of Moon	Second manned exploration of the Moon. Total stay time: 31 hrs., 31 min.
Apollo 13	James A. Lovell, Jr. John L. Swigert, Jr. Fred W. Haise, Jr.	4/11–17/70	142:54:41	—	Mission aborted because of service module oxygen tank failure.
Apollo 14	Alan B. Shepard, Jr. Stuart A. Roosa Edgar D. Mitchell	1/31–2/9/71	216:01:59	34 rev. of Moon	First manned landing in and exploration of lunar highlands. Total stay time: 33 hrs., 31 min.
Apollo 15	David R. Scott Alfred M. Worden James B. Irwin	6/26–7/7/71	295:11:53	74 rev. of Moon	First use of lunar roving vehicle. Total stay time: 66 hrs., 55 min.
Apollo 16	John W. Young Thomas K. Mattingly II Charles M. Duke, Jr.	3/16–27/72	265:51:05	64 rev. of Moon	First use of remote controlled television camera to record lift-off of the LM ascent stage from the lunar surface. Total stay time: 71 hrs., 2 min.
Apollo 17	Eugene A. Cernan Ronald E. Evans Harrison H. Schmitt	12/7–19/72	301:51:59	75 rev. of Moon	Last manned lunar landing and exploration of the Moon in the Apollo program returned 110 kg of lunar samples to Earth. Total stay time: 75 hrs.

SOURCE: NASA.

Table 3   Physical and Astrophysical Properties of the Moon

Diameter (equatorial)	3,476 km	Sidereal month (rotation period)	27.322 days
Mass	$7.350 \ 10^{22}$ kg		
Mass (Earth's mass = 1.0)	0.0123	Albedo (mean)	0.07
Average density	3.34 g/cm^3	Mean visual magnitude (at full)	−12.7
Mean distance from Earth (center-to-center)	384,400 km		
		Surface area	37.9×10^6 km^2
Surface gravity (equatorial)	1.62 m/sec^2	Volume	2.20×10^{10} km^3
Escape velocity	2.38 km/sec	Atmospheric density (at night on surface)	2×10^5 molecules/cm^3
Orbital eccentricity (mean)	0.0549		
Inclination of orbital plane (to ecliptic)	5° 09′	Surface temperature	102 K–384 K

SOURCE: NASA.

Table 4   Major Chemical Elements Found on the Lunar Surface

| | Highland Rocks | | Mare Basalts | |
	Anorthositic Gabbro	Gabbroic Anorthosite	Olivine Basalt (A12)	Green Glass (A15)
SiO_2	44.5	44.5	45.0	45.6
TiO_2	0.39	0.35	2.90	0.29
Al_2O_3	26.0	31.0	8.59	7.64
FeO	5.77	3.46	21.0	19.7
MnO	—	—	0.28	0.21
MgO	8.05	3.38	11.6	16.6
CaO	14.9	17.3	9.42	8.72
Na_2O	0.25	0.12	0.23	0.12
K_2O	—	—	0.064	0.02
P_2O_5	—	—	0.07	—
Cr_2O_3	0.06	0.04	0.55	0.41
Total	99.9	100.2	99.77	99.4

SOURCE: NASA.

Table 5   The Lunar Atmosphere

| | Concentration (molecules/cm³) | |
Gas	Day	Night
H_2	$<6 \times 10^3$	$<4 \times 10^4$
^4He	2×10^3	4×10^4
^{20}Ne	—	$<10^5$
^{36}Ar	—	$<3 \times 10^3$
^{40}Ar	—	4×10^4

SOURCE: NASA.

Table 6   Charged-Particle Environment at the Lunar Surface

Source	MeV/ Nucleon	Proton Flux $(Ncm^{-2} \sec^{-1})$	Penetration Depth (cm)
Solar wind	10^{-3}	10^8	10^{-6}
Solar cosmic rays	1 to 10^2	10^2	1 to 10^{-3}
Galactic cosmic rays	10^2 to 10^4	1	1 to 10^3

SOURCE: NASA.

multibiotic   An environmental system consisting of many types of plants and animals.

mutation   A change in the genetic (DNA) code, usually caused by ionizing radiation, such as cosmic rays, passing through the nucleus of a cell that controls development. The developed organism, called a mutant, differs from others in its species.

N

NASA facilities   The National Aeronautics and Space Administration (NASA) carries on space and aeronautical activities for peaceful purposes for the benefit of all humanity. Since NASA's creation in October 1958, its network of centers and facilities has grown across the United States. It is at these field installations that NASA conducts many scientific programs, ranging from aerodynamic research to make civilian aviation safer here on Earth to sending very sophisticated spacecraft throughout the Solar System exploring new worlds and searching for extraterrestrial life.

The major NASA installations include: NASA headquarters, the Ames Research Center (ARC), the Dryden Flight Research Facility (DFRF), the Goddard Space Flight Center (GSFC), the Jet Propulsion Laboratory (JPL), the Johnson Space Center (JSC), the Kennedy Space Center (KSC), the Langley Research Center (LaRC), the Lewis Research Center (LeRC), the Marshall Space Flight Center (MSFC), the Michoud Assembly Facility (MAF), the Slidell Computer Complex (SCC), the National Space Technology Laboratories (NSTL) and the Wallops Flight Facility (WFF). Each of these facilities will be briefly described.

NASA HEADQUARTERS

NASA headquarters manages the spaceflight centers, research centers and other NASA installations. Planning, direction and management of NASA's research and development (R&D) programs are the responsibility of individual program offices that report to and are directed by headquarters officials. Responsibilities at NASA headquarters include: the determination of projects and programs; the establishment of management policies, procedures and performance criteria; and the review and evaluation of all phases of the U.S. civilian space program.

NASA headquarters is located in Washington, D.C., in very close proximity to the National Air and Space Museum.

AMES RESEARCH CENTER (ARC)

The NASA Ames Research Center is located at the

southern end of San Francisco Bay near Mountain View, California. The U.S. Naval Air Station, Moffett Field, is adjacent to the south and east of Ames.

The programs at this center are directed toward research and development programs in the fields of aeronautics, space science, life science and spacecraft technology. The Ames Research Center operates specially equipped aircraft that serve as airborne laboratories for use by scientists from all over the world in conducting both terrestrial and space studies.

ARC also has management responsibility for the Pioneer series of spacecraft. Six of these spacecraft are in orbit around the Sun; two others are now becoming the first human-made objects to leave the Solar System after providing close-up images of Jupiter and Saturn; while another Pioneer spacecraft orbits the planet Venus.

In life-science laboratories Ames scientists study the origin of life as part of an investigation of the cosmic prevalence of life and a search for extraterrestrial life. Other Ames scientists are involved in human-factors engineering and assist in establishing medical criteria for manned spaceflight.

The Ames Research Center also exercises management control for the Dryden Flight Research Facility at Edwards, California.

HUGH L. DRYDEN FLIGHT RESEARCH FACILITY (DFRF)

The Dryden Flight Research Facility, a directorate of the Ames Research Center, is located at Edwards (Air Force Base), California, on the edge of the Mojave Desert, approximately 130 kilometers (80 miles) north of Los Angeles.

This facility was established as the major NASA installation for high-speed flight tests. The primary research tools for conducting the programs and missions of the Dryden Facility are its aircraft. They range from Century series fighters to advanced supersonic and hypersonic aircraft and aerospace flight research vehicles, such as wingless lifting bodies. There are also special ground-based facilities, including a Flight Test Range with a fully instrumented tracking station; a high-temperature-loads calibration laboratory and a remotely piloted research vehicle (RPRV) facility.

The mission of the Dryden Facility involves research and test activities on all phases of flight, including low-speed, supersonic and hypersonic; takeoff and landing; and space-vehicle reentry behavior. For example, this facility conducted the original approach and landing tests for the Space Shuttle Orbiter.

ROBERT H. GODDARD SPACE FLIGHT CENTER (GSFC)

The Goddard Space Flight Center is named for Dr. Robert H. Goddard, the father of American rocketry. The main GSFC facility is located at Greenbelt, Maryland, about 16 kilometers (10 miles) northeast of Washington, D.C. Other facilities include Wallops Island, Virginia, the Goddard Institute for Space Studies in New York City (which conducts much of the theoretical research for the

Goddard Space Flight Center) and 16 tracking stations around the world.

GSFC is responsible for automated spacecraft and sounding-rocket experiments in support of basic and applied research. (A sounding rocket is used to study atmospheric conditions.) Satellite and sounding-rocket projects at Goddard provide data about the Earth's environment, Sun-Earth relationships and the Universe. GSFC projects advance technology in such areas as communications, meteorology, navigation and Earth resources monitoring. Goddard is also the home of the National Space Science Data Center—the central repository of data collected from NASA's spaceflight experiments.

Goddard personnel are stationed around the globe as part of the Space Tracking Data Network (STDN) team. GSFC workers are also located at facilities of the NASA Communications Network, which links the STDN together.

The Goddard Visitor Center is a major tourist attraction and represents one of the largest and most attractive of all NASA visitor facilities. Exhibits include: Delta rockets; Mercury, Gemini and Apollo capsule models; a satellite-linked telephone; Moon rocks; a weather station; and a collection of communications satellites.

JET PROPULSION LABORATORY (JPL)

The Jet Propulsion Laboratory is located in Pasadena, California, approximately 32 kilometers (20 miles) northeast of Los Angeles. JPL is a government-owned facility that is staffed and managed by the California Institute of Technology. In addition to the Pasadena site, JPL also operates the Deep Space Communications Complex—a station of the worldwide Deep Space Network (DSN). This station is located at Goldstone, California.

JPL is engaged in activities associated with deep-space automated scientific missions. The laboratory is also involved with spacecraft tracking, data acquisition and the data reduction and evaluation required by deep-space flight. JPL personnel study advanced propulsion systems for interplanetary spacecraft, as well as new methods for guidance, navigation and control. Complete spacecraft, such as the famous Voyager and Viking systems, are designed and tested at the laboratory. In addition, JPL personnel manage and direct the worldwide Deep Space Network (DSN).

The JPL Visitor Center features displays of full-size spacecraft from Explorer to Voyager. One of the most popular attractions is the Space Flight Operations Facility, where spacecraft tracking and scientific data are received and processed from NASA's Deep Space Network.

LYNDON B. JOHNSON SPACE CENTER (JSC)

The Johnson Space Center is located on NASA Highway 1, adjacent to Clear Lake, Texas, and is approximately 32 kilometers (20 miles) southeast of downtown Houston.

JSC was founded in September 1961 as NASA's primary center for the design, development and manufacture of manned spacecraft; the selection and training of astronauts; and the ground control of manned space missions. In addition, Johnson Space Center personnel manage many of the medical, engineering and scientific experiments carried on board manned space flights. JSC is currently the lead NASA center for the management of the Space Shuttle.

One of JSC's most famous facilities is the Mission Control Center (MCC), from which U.S. manned flights from the *Gemini IV* mission through the Apollo program and Skylab, and continuing now into the Shuttle era, have been monitored. (See fig. 1.)

The Johnson Space Center is also responsible for directing operations at the White Sands Test Facility (WSTF). This facility is located on the western edge of the U.S. Army White Sands Missile Range in southeastern New Mexico near the city of Las Cruces. WSTF supports the Space Shuttle propulsion system, power-system testing and materials testing. The White Sands facility also serves as an alternate Shuttle landing site, as occurred on the second test flight of the Orbiter *Columbia*.

The JSC Visitor Center is one of the major tourist attractions in Texas. Visitors can see actual flight vehicles from almost every U.S. manned space program, including Mercury, Gemini and Apollo. Visitors can see the control room from which all manned U.S. missions are monitored, a display of lunar rocks, a full-size mock-up of the Space Shuttle and the Mission Simulation and Training Facility.

JOHN F. KENNEDY SPACE CENTER (KSC)

The Kennedy Space Center, America's spaceport, is located on the east central coast of Florida about 80 kilometers (50 miles) from Orlando. The NASA facility is adjacent to Cape Canaveral Air Force Station. KSC is some 55 kilometers (34 miles) long and varies in width from 8 kilometers (5 miles) to 16 kilometers (10 miles). This land

Fig. 1   The Mission Control Center at the Johnson Space Center helped guide and monitor the activities of the Apollo astronauts on the Moon. (Courtesy of NASA.)

area with its adjoining bodies of water provides sufficient separation from the surrounding civilian communities to launch space vehicles without endangering the public.

KSC serves as the primary NASA center for the test, checkout and launch of rockets and space vehicles. This responsibility currently includes the launch of manned and unmanned space vehicles from launch complexes at both the Kennedy Space Center and Cape Canaveral Air Force Station (Air Force Eastern Space and Missile Center) in Florida and from the Air Force Western Space and Missile Center at Vandenberg Air Force Base in California. Currently, personnel at KSC are concentrating on the servicing, checkout and launch of Space Shuttle vehicles. (See fig. 2.)

Fig. 2   The Space Shuttle *Discovery* lifts off from launchpad 39A at the Kennedy Space Center on its maiden flight, August 30, 1984. (Courtesy of NASA.)

The KSC Visitor Center is one of the most popular tourist attractions in the state of Florida. Visitors can see indoor and outdoor exhibits of space vehicles and rockets, plus displays of space equipment, facilities and products. Among the outdoor exhibits are Mercury-Redstone, Mercury-Atlas, Saturn 1B and Gemini-Titan II space vehicles, as well as rocket engines and a full-scale Apollo Lunar Module. Indoors, tourists may view exciting exhibits concerning the Space Shuttle, lunar rocks, space suits, the space station and an Apollo program Lunar Rover. For a modest fee, visitors can also take a guided bus tour past many of the historic and current launch facilities at both KSC and Cape Canaveral Air Force Station.

All but the actual operational areas of this NASA spaceport are designated as the Merritt Island National Wildlife Refuge. Much of this wildlife refuge also forms part of the Cape Canaveral National Seashore.

LANGLEY RESEARCH CENTER (LaRC)

The Langley Research Center, NASA's oldest center, is located in Hampton, Virginia, approximately 160 kilometers (100 miles) south of Washington, D.C. LaRC is situated in the Tidewater area of Hampton Roads, between Norfolk and Williamsburg, Virginia. Langley's primary mission is the development of advanced concepts and technology for future aircraft and spacecraft systems, with particular emphasis placed on environmental effects, performance, range, safety and economy.

Langley's aeronautical research program is aimed at identifying and pursuing basic and applied research opportunities that offer the greatest potential for increasing aircraft performance, efficiency and capability. LaRC facilities in this area include the National Transonic Facility, a new, cryogenic, high-pressure wind tunnel that provides a unique capability for conducting special aerodynamic research at subsonic and transonic speeds.

The Langley Visitor Center features more than 40 exhibits that chronicle human achievements in both aeronautics and spaceflight. In the Aeronautics Gallery visitors can view models and photographs that trace the history of aerial flight. The Space Gallery contains a Moon rock, Mercury and Apollo space capsules, space suits and exhibits on our Solar System and the mysteries of the Universe.

LEWIS RESEARCH CENTER (LeRC)

The Lewis Research Center is located on the west side of Cleveland's Hopkins Airport in Cuyahoga County, Ohio. LeRC is the primary NASA center for research and technology development in aircraft propulsion, space propulsion, space power systems and satellite communication. This center also conducts research on new systems for converting chemical and solar energy into electric power. Major research facilities at LeRC range from atmospheric wind tunnels to large space-environment test chambers.

Personnel at Lewis also manage the Centaur upper stage vehicle, which can be used with the expendable Atlas launch vehicle and the reusable Space Shuttle.

This center has management responsibility for the Plum Brook Station, located about 80 kilometers (50 miles) away from Lewis, near Sandusky, Ohio. The Plum Brook Station is an adjunct facility to LeRC and provides very-large-sized special test installations. At present these Plum Brook installations are being maintained in a standby mode.

The Lewis Visitor Center serves as the nucleus for visitor programs aimed at the aerospace interests and needs of the educational community and the general public. For example, the Teacher Resource Room can provide educators with slides, videotapes, publications and lesson plans that describe NASA programs and activities.

GEORGE C. MARSHALL SPACE FLIGHT CENTER (MSFC)

The Marshall Space Flight Center is located on the U.S. Army's Redstone Arsenal, just outside Huntsville, Alabama. MSFC is one of the nation's pioneering space centers. It was established in 1960 by a team of former Army rocket experts, headed by Dr. Wernher von Braun. The center provides project management, as well as scientific and engineering support, for many of NASA's major space programs and endeavors. Marshall has a wide spectrum of technical facilities and laboratories. Although it was originally founded as NASA's primary center for propulsion-system development, it has now diversified and become a center for the development of payloads and space-science activities (such as the Space Telescope) as well.

Marshall is responsible for managing the development of the Space Shuttle main engines, its solid rocket boosters and the massive external tank. (See fig. 3.) MSFC is also responsible for the Space Telescope, NASA's use of Spacelab and a variety of other key R&D programs.

In addition to its main facilities at Huntsville, MSFC operates the Michoud Assembly Facility (MAF) in New Orleans, Louisiana, and the Slidell Computer Complex (SCC) at Slidell, Louisiana.

The Marshall Visitor Center is located at the site of a very large space museum, the state-operated Alabama Space and Rocket Center. This museum features displays of major historic space systems, including a giant Saturn V Moon rocket.

NATIONAL SPACE TECHNOLOGY LABORATORIES (NSTL)

The National Space Technology Laboratories facility is located in Hancock County, near Bay St. Louis, Mississippi. Formerly designated the Mississippi Test Facility (MTF), NSTL was given full field installation status by NASA in 1974 because of its capabilities in space applications and Earth resources activities. Its Saturn rocket test stands have been modified to accommodate Space Shuttle main engine testing, which is now the primary mission of NSTL.

Fig. 3   The massive Space Shuttle external tank (ET) is prepared for launch at the Kennedy Space Center. The Marshall Space Flight Center is responsible for the Shuttle's main engines, solid rocket boosters and this external propellant tank. (Courtesy of NASA/Kennedy Space Center.)

WALLOPS FLIGHT FACILITY (WFF)

The Wallops Flight Facility, a part of the Goddard Space Flight Center, is located on Virginia's Delmarva Peninsula on the Atlantic coast. It is approximately 64 kilometers (40 miles) southeast of Salisbury, Maryland, and some 116 kilometers (72 miles) north of the Chesapeake Bay Tunnel. This facility includes three separate areas on the Atlantic Coast: the main installation, the Wallops Island launch site and about 1,140 acres of marshland.

The Wallops Flight Facility is responsible for managing NASA's Suborbital Sounding Rocket Projects. This responsibility includes: mission and flight planning, flight operations, and payload landing and recovery. WFF also performs payload and payload-carrier design, development, fabrication and testing; sounding-rocket experiment support; launch operations; and tracking and data acquisition. In addition, this facility manages, monitors and schedules balloon activities conducted for the NASA balloon program. WFF is responsible for managing the National Scientific Balloon Facility at Palestine, Texas.

The GSFC/Wallops Visitor Center provides self-guided walking tours. Visitors can see a collection of spacecraft, spaceflight articles and aerospace exhibits.

See also: **spaceport**

National Air and Space Museum   Just where do you go on Earth to see, hear and touch "things extraterrestrial"? Well, one very popular place is the Smithsonian Institution's National Air and Space Museum in Washington, D.C. Opened in 1976, the National Air and Space Museum provides over nine million visitors each year with a dazzling array of spacecraft and flying machines. Twenty-three well-prepared exhibit areas house artifacts that range from the Wright brothers' original 1903 Flyer to a touchable Moon rock brought back by the *Apollo 17* astronauts. There is even a full-sized Skylab Orbital Workshop that you may enter (see fig. 1). The museum has dozens of space-technology artifacts, including rockets built by Dr. Robert Goddard; historic unmanned spacecraft such as *Explorer 1*, *Pioneer 10* and the Viking Mars Lander; as well as numerous spacecraft from the Mercury, Gemini and Apollo manned space programs.

All aircraft and spacecraft displayed in the museum were actually flown or were at least used as backup flight vehicles and hardware (unless a particular exhibit label specifically identifies an exception).

Constantly changing, high-quality presentations in the museum's Albert Einstein Sky Theater and Langley Theater complement the outstanding exhibit areas and provide an added dimension of enjoyment to your visit.

See also: **extraterrestrial museum**

nebula   A cloud of interstellar gas or dust. It can be seen either as a dark hole or band against a brighter background (this is called a dark nebula) or as a luminous patch of light (this is called a bright nebula). Figure C18 on color page I shows the Orion Nebula, also called M42.

Nemesis   A postulated dark stellar companion to our Sun, whose close passage every 26 million years is thought to be responsible for the cycle of mass extinctions that seem to have also occurred on Earth at 26-million-year intervals. This "death star" companion has been named for the Greek goddess of retributive justice or vengeance. If it really does exist, it might be a white dwarf, a rogue star that was captured by the Sun or possibly a tiny but gravitationally influential neutron star.

The passage of such a death star through the Oort Cloud (a postulated swarm of comets surrounding our Solar System) could provoke a massive shower of comets into the Solar System. One or several of these comets impacting on Earth would then trigger massive extinctions and catastrophic environmental changes within a very short period of time.

A detailed astronomical search for Nemesis over the next few years, using such advanced equipment as the Space Telescope, should let us know whether our Solar System is really being stalked by a "dark," potentially deadly celestial companion to our Sun.

See also: **extraterrestrial catastrophe theory; rogue star; stars**

Fig. 1   Visitors wait to enter the backup Skylab Orbital Workshop in the Space Hall of the National Air and Space Museum, Washington, D.C. (Courtesy of the National Air and Space Museum.)

Neptune   The outermost of the Jovian planets and the first planet to be discovered using theoretical predictions. Neptune's discovery was made by J.G. Galle at the Berlin Observatory in 1846. This discovery was based on independent orbital perturbation (disturbance) analyses by the French astronomer U. Le Verrier and the British scientist J. Adams. It is considered to be one of the triumphs of 19th-century theoretical astronomy.

Although Neptune is called the eighth planet from the Sun, at the present time (and until 1999), it is actually the most distant planet from the Sun in our Solar System! Pluto's eccentric orbit has now brought that planet within the orbit of Neptune for a period of 20 years. But don't worry—there is no chance for a planetary collision between these two frigid worlds.

Neptune is frequently treated as a "twin" of Uranus—but it is really as dissimilar to Uranus as Saturn is to Jupiter. For example, Neptune is more massive than Uranus and has a higher mean, or average, density. Neptune also appears to have an internal heat source, while Uranus does not. In fact, Neptune radiates about twice the energy that it receives from the Sun!

However, these two Jovian worlds also have some areas of similarity. Both Neptune and Uranus are believed to possess deep atmospheres that are mainly composed of hydrogen and helium. Their atmospheres are also known to contain methane (CH_4), a gas that gives them a distinctive blue-green color when viewed through terrestrial telescopes. Scientists currently theorize that both planets have iron-silicate cores that are surrounded (from the core outward) by layers of ice, liquid hydrogen and gaseous hydrogen.

Neptune has two known natural satellites, Triton and Nereid. If we compare Neptune's system of moons to those of the other Jovian planets, Neptune's satellite system can best be described by the word *chaotic*. The larger moon, Triton (which is also one of the largest moons in the Solar System), has an unusual, decaying retrograde orbit with a low eccentricity. Scientists believe that Triton is locked in a synchronous rotation about Neptune—a rapidly decaying orbit that will cause its destruction when it falls into Neptune in 10 million to 100 million years. Current data hint at the possibility that Triton may be "rocky" (that is, containing various terrestrial, lunar and meteoritic materials) rather than "icy." Neptune's other moon, tiny Nereid, has the most eccentric orbit of any known planetary moon in the Solar System. Scientists think Nereid is asteroidal in nature, although very little is presently known about its physical characteristics.

The discovery of rings around Jupiter (in 1979) and around Uranus (in 1977), leaves Neptune as the only Jovian planet without a known ring system.

At a distance of some 4.5 billion kilometers (2.8 billion miles) from the Sun, Neptune will remain a featureless, blue-green blur in terrestrial telescopes until it is visited by the *Voyager 2* spacecraft in 1989. (See fig. C19 on color page I.) This planetary encounter should revolutionize our knowledge about Neptune. Tables 1 and 2 present the currently known information about Neptune and its two moons. In many areas, however, because of Neptune's great distance from the Earth, these data are indefinite and subject to a wide degree of variation.

Table 1   Physical and Dynamic Properties of Neptune

Radius (mean)	24,700 km
Mass	1.03×10^{26} kg
"Surface" gravity	13.8 m/sec²
Mean density	1.66 g/cm³
Effective temperature	60 K
Magnetic field	none observed
Number of natural satellites	2
Semimajor axis	4.5×10^9 km (30.06 AU)
Eccentricity	0.0086
Orbit inclination	1.776°
Orbital velocity (mean)	5.4 km/sec
Sidereal year ("Neptunian" year)	164.8 Earth-years
Rotation period	18 hours (?)
Inclination of equator to the orbit	28.8°
Atmosphere Mainly Other components (blue-green color)	hydrogen and helium methane (CH_4)
Rings	none yet observed
Flux of solar radiation	1.55 W/m²
Earth-to-Neptune distances Maximum Minimum	 4.64×10^9 km (31 AU) [258 light-min] 4.34×10^9 km (29 AU) [242 light-min]

Table 2   Physical Data for Neptune's Moons

	Triton	*Nereid*
Orbital period	5.88 days	359.9 days
Semimajor axis	355.3×10^3 km	$5,567 \times 10^3$ km
Eccentricity	<0.01	0.748
Orbit inclination (to Neptune's equator)	160°	27.6°
Mass (estimated)	1.34×10^{23} kg	2.1×10^{19} kg
Radius	~1,900 km	~150 km (?)
Density (mean)	~4.9 g/cm³	(?)
Average temperature	44 K	44 K
Composition (suspected)	"rocky"	"asteroidal"
Year discovered	1846	1949

Neptunian   Of or relating to the planet Neptune; (in science fiction) a native of the planet Neptune.

neutrino [symbol: ν]   An electrically neutral elementary particle with a negligible mass. It interacts very weakly with matter and is therefore very difficult to detect. The neutrino is produced in many nuclear reactions, such as beta decay, and has extremely high penetrating power. For example, neutrinos from the Sun, called solar neutrinos, usually pass right through the Earth without interacting.

neutron [symbol: n]   An uncharged elementary particle with a mass slightly greater than that of the proton. It is found in the nucleus of every atom heavier than hydrogen. A free neutron is unstable, with a half-life of about 12 minutes, and decays into an electron, a proton and a neutrino. Neutrons sustain the fission chain reaction in a nuclear reactor.

Newton's laws of motion   A set of three fundamental postulates that form the basis of the mechanics of rigid bodies. These were formulated in 1687 by the English scientist and mathematician Sir Isaac Newton (1642–1727).

Newton's first law is concerned with the principle of inertia and states that if a body in motion is not acted upon by an external force, its momentum remains constant. This can also be called the law of conservation of momentum.

The second law states that the rate of change of momentum of a body is proportional to the force acting upon the body and is in the direction of the applied force. A familiar statement of this law is the equation $\vec{F} = m\vec{a}$, where $\vec{F}$ is the vector sum of the applied forces, m is the mass and $\vec{a}$ the vector acceleration of the body.

Newton's third law is the principle of action and reaction. It states that for every force acting upon a body, there is a corresponding force of the same magnitude exerted by the body in the opposite direction.

nonionizing radiation   Electromagnetic radiation too low in energy to expel an electron from an atom or molecule. Microwave radiation is an example of nonionizing radiation.

nuclear-electric propulsion system (NEPS)   A propulsion system in which a nuclear reactor is used to produce the electricity needed to operate the electric propulsion engine(s). A nuclear-electric propulsion (NEP) system, like the one shown in figure 1, provides shorter trip times and greater payload capacity than any of the competitive chemical propulsion technologies that might be used in the next two decades for detailed exploration of the outer planets, especially Saturn, Uranus, Neptune and their respective moons. An orbital NEP vehicle can also be effectively used in cislunar space to gently transport large cargoes and structures from low Earth orbit to geosynchronous orbit or to lunar orbit in support of expanding space-industrialization activities.

See also: **space nuclear power**

nuclear energy   The energy liberated by a nuclear reaction. These reactions include fission, fusion and radioactive decay.

See also: **space nuclear power**

nucleic acids   The long, chainlike molecules DNA and RNA, which are the building blocks of the genetic code in living matter.

See also: **life in the Universe**

nucleosynthesis   The production of heavier chemical elements from hydrogen nuclei or protons in thermonuclear reactions in stellar interiors.

See also: **astrophysics; fusion; stars**

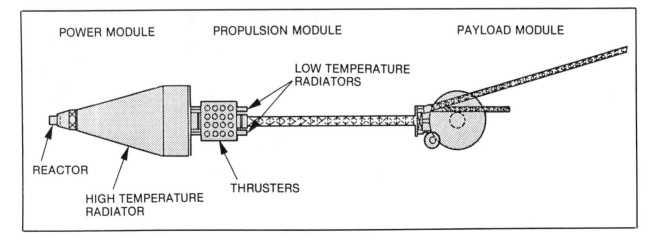

Fig. 1 Nuclear-electric propulsion spacecraft configuration for planetary-exploration missions. (Courtesy of NASA/Jet Propulsion Laboratory.)

occult, occulting   The disappearance of one celestial object behind another. For example, a solar eclipse is an occulting of the Sun by the Moon; that is, the Moon comes between the Earth and the Sun, temporarily blocking the Sun's light and darkening regions of the Earth.

one g   The downward acceleration of gravity at the Earth's surface.
　　See also: **gravity**

open universe   A model in cosmology in which it is assumed that the Universe is infinite and will expand forever. Space is taken as unbounded, containing no edges and possessing negative curvature. We say that space has negative curvature if parallel lines eventually diverge. In cosmology, negatively curved spaces are open and infinite.

orbital station   A term found in the Soviet astronautical literature to describe a space station.
　　See also: **space station**

orbital-transfer vehicle (OTV)   A propulsion system used to transfer a payload from one orbital location to another—as, for example, from low Earth orbit (LEO) to geostationary Earth orbit (GEO). Orbital-transfer vehicles can be expendable or reusable and may involve chemical, nuclear or electric propulsion systems. An expendable orbital-transfer vehicle is frequently referred to as an upper-stage unit, while a reusable OTV is often called a "space tug." OTVs can be designed to move people and cargo between different destinations in cislunar space.

Two current orbital-transfer vehicles are the Payload Assist Module (PAM) and the Inertial Upper Stage (IUS).

The Payload Assist Module-D is one of a new generation of OTVs used to boost Space Shuttle-carried payloads from low Earth orbit to higher altitudes (see fig. 1). The PAM-D is used to transport medium-mass payloads (up to about 1,250 kilograms) from a shallow, circularized orbit to a highly elliptical transfer orbit with an apogee of about 35,888 kilometers (22,300 miles). At a selected apogee the spacecraft's on-board apogee kick motor is fired to position the spacecraft or payload in geostationary orbit at its operational location. The PAM-D is the first of three solid-fueled upper stages commercially developed by McDonnell Douglas Astronautics for use with the Space Shuttle. The PAM-D is also designed for flight on board the Delta expendable launch vehicle.

The PAM-D measures approximately 1.2 meters (48 inches) in diameter by about 1.8 meters (72 inches) in height. This upper stage and its "airborne" support equipment have an overall mass of 3,319 kilograms (7,318

pounds). PAM-D propulsion is provided by a Thiokol Star-48 solid rocket motor, nominally loaded with about 1,738 kilograms (3,833 pounds) of HTPB (hydroxyl terminated polybutadiene) propellant. This propellant charge provides approximately 66,700 newtons (15,000 pounds-force) of thrust during a nominal burn time of 84.7 seconds.

The Inertial Upper Stage (IUS) is another OTV designed to boost Space Shuttle-carried spacecraft to higher-altitude orbits. The IUS is used to transfer heavier spacecraft (up to 2,268 kilograms [5,000 pounds]) from a circularized low Earth orbit to an operational location at GEO. The IUS was developed by the U.S. Air Force under contract with Boeing Aerospace Company for use on board the Space Shuttle and the Air Force's expendable Titan launch vehicles. A Shuttle-deployed IUS is capable of boosting spacecraft that have a mass of up to 2,268 kilograms into geosynchronous orbit. A Titan 34D-deployed IUS is capable of boosting a 1,814-kilogram (4,000-pound) payload to GEO.

The standard two-stage IUS measures approximately 2.9 meters (9.5 feet) in diameter by about 4.9 meters (16 feet) in height. It has a mass of approximately 14,515 kilograms (32,000 pounds). Propulsion for the standard IUS is provided by two solid rocket motors. The first-stage solid rocket motor contains 9,707 kilograms (21,400 pounds) of solid propellant, composed of a mixture of

Fig. 1   The first deployment of a commercial satellite from an orbiting space vehicle. The Satellite Business Systems (SBS)-3 satellite is launched from the Space Shuttle *Columbia*'s cargo bay during the STS-5 mission, November 1982. The PAM-D OTV will place the spacecraft in its operational (higher-altitude) orbit. (Courtesy of NASA/Kennedy Space Center.)

ammonium perchlorate, aluminum and HTPB. The first stage generates an average thrust of 185,000 newtons (41,600 pounds-force). The second-stage solid rocket motor contains 2,722 kilograms (6,000 pounds) of propellant and generates about 78,285 newtons (17,600 pounds-force) of thrust. The second stage has an extendable exit cone for increased performance on Shuttle missions.

Many other OTV concepts are also being considered for use with both the Space Shuttle and a permanent space station in low Earth orbit. For example, the Aeroassisted Orbital Transfer Vehicle (AOTV) would be a totally reusable liquid-fueled space vehicle that has performance capabilities beyond those of the IUS. The AOTV would take Shuttle-carried payloads or spacecraft serviced at the space station to higher orbits and deploy them in their operational locations. The AOTV could also retrieve other payloads in high-altitude orbits and return them to either the Shuttle or the space station in LEO. Upon return to lower orbital altitudes, the AOTV would use the Earth's upper atmosphere as a "natural brake" rather than using propellants to achieve deceleration.

Another concept is the Future Orbital Transfer Vehicle (FOTV). The FOTV would be a totally reusable, liquid-fueled space vehicle capable of delivering heavy payloads to GEO and then returning to its hangar at the space station for servicing and reuse. Between orbital-transfer missions, the space hangar serves as both a maintenance bay and a shelter protecting the FOTV from space debris.

See also: **space launch vehicles; space station; Space Transportation System**

Orbiting Very Long Baseline Interferometry Observatory

Very large arrays of radio telescopes can provide high angular resolution if the signals they receive can be added coherently, preserving the relative phases of the electromagnetic (EM) waves that make up the incoming radio-wave signal. This coherent combination of incident signals is the basis of interferometry, a technique now being used in ground-based radio astronomy networks. Because of the large size of these arrays, the technique is called "very-long-baseline interferometry," or VLBI.

The angular resolution of radio maps made by ground-based VLBI networks can be greatly increased by adding one or more space-based radio telescopes to the network. The addition of a single orbiting antenna, some 10 to 30 meters in diameter, as a complement to an existing ground-based network can provide unprecedented angular resolutions of 10^{-3} to 10^{-5} arc-seconds.

With an Orbiting Very Long Baseline Interferometry Observatory, the resulting high-angular-resolution radio maps will help scientists better explain the physics of quasars, galactic centers and the dynamics of stellar evolution. The observatory is planned for 1995-2000.

See also: **astrophysics; radio astronomy**

orbits of objects in space

We must know about the science and mechanics of orbits to launch, control and track spacecraft and to predict the motion of objects in space. An *orbit* is the path in space along which an object moves around a primary body. Common examples of orbits include the Earth's path around its celestial primary (the Sun) and the Moon's path around the Earth (its primary body). A single orbit is a complete path around a primary as viewed from space. It differs from a revolution. A single *revolution* is accomplished whenever an orbiting object passes over the primary's longitude or latitude from which it started. For example, the Space Shuttle *Challenger* completed a revolution whenever it passed over approximately 80 degrees west longitude on Earth. However, while the *Challenger* was orbiting from west to east around the globe, the Earth itself was also rotating from west to east. Consequently, the *Challenger's* period of time for one revolution was actually longer than its orbital period (see fig. 1). If, on the other hand, the *Challenger* were orbiting from east to west (not a practical flight path from a propulsion-economy standpoint), then because of the Earth's west-to-east spin, its period of revolution would be shorter than its orbital period. An east-to-west orbit is called a *retrograde orbit* around the Earth, while a west-to-east orbit is called a *posigrade orbit*. If *Challenger* were traveling in a north-south orbit, or *polar orbit*, it would complete a period of revolution whenever it passed over the latitude from which it started. Its orbital period would be about the same as the revolution period, but not identical, because the Earth actually wobbles slightly north and south.

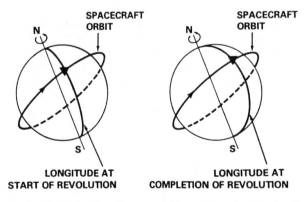

Fig. 1 An illustration of a spacecraft's west-to-east orbit around the Earth and how the Earth's west-to-east rotation moves longitude ahead. As shown here, the period of one revolution can be longer than the orbital period. (Courtesy of NASA.)

There are other terms used to describe orbital motion. The *apoapsis* is the farthest distance in an orbit from the primary; the *periapsis*, the shortest. For orbits around the planet Earth, the comparable terms are *apogee* and *perigee* (see fig. 2).

For objects orbiting the Sun, *aphelion* describes the point on an orbit farthest from the Sun; *perihelion*, the point nearest to the Sun.

Another term we frequently encounter is the *orbital*

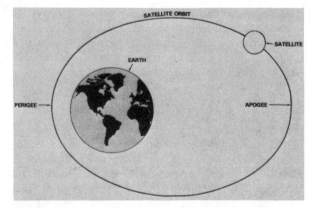

Fig. 2   The expressions *apogee* and *perigee* described in terms of a spacecraft's orbit around the Earth. (Drawing courtesy of NASA.)

plane. An Earth satellite's orbital plane can be visualized by thinking of its orbit as the outer edge of a giant, flat plate that cuts the Earth in half. This imaginary plate is called the orbital plane.

Inclination is another orbital parameter. This term refers to the number of degrees the orbit is inclined away from the equator. The inclination also indicates how far north and south a spacecraft will travel in its orbit around the Earth. If, for example, a spacecraft has an inclination of 56 degrees, it will travel around the Earth as far north as 56 degrees north latitude and as far south as 56 degrees south latitude. Because of the Earth's rotation, it will not, however, pass over the same areas of Earth on each orbit. A spacecraft in a polar orbit has an inclination of about 90 degrees. As such, this spacecraft orbits the Earth traveling alternately in north and south directions. A polar-orbiting satellite eventually passes over the entire Earth because the Earth is rotating from west to east beneath it. NASA's Landsat is an example of a spacecraft whose cameras and multispectral sensors observe the entire Earth from a nearly polar orbit, providing valuable information about the terrestrial environment and resource base.

A satellite in an equatorial orbit around the Earth has zero inclination. The Intelsat communications satellites are examples of satellites in equatorial orbits. By placing such spacecraft into near-circular equatorial orbits at just the right distance above the Earth, these spacecraft can be made to essentially "stand still" over a point on the Earth's equator. Such satellites are called *geostationary.* They are in *synchronous orbits,* meaning they take as long to complete an orbit around the Earth as it takes for the Earth to complete one rotation about its axis (that is, approximately 24 hours). A satellite at the same "synchronous" altitude but in an inclined orbit may also be called synchronous. While this particular spacecraft would not move much east and west, it would move north and south over the Earth to the latitudes indicated by its inclination. The terrestrial ground track of such a spacecraft resembles an elongated figure eight, with the crossover point on the equator.

All orbits are elliptical, in accordance with Kepler's first law of planetary motion (described shortly). However, a spacecraft is generally considered to be in a circular orbit if it is in an orbit that is nearly circular. A spacecraft is taken to be in an elliptical orbit when its apogee and perigee differ substantially.

Two sets of scientific laws govern the motions of both celestial objects and human-made spacecraft. One is Newton's law of gravitation; the other, Kepler's laws of planetary motion.

The brilliant English scientist and mathematician Sir Isaac Newton observed the following physical principles:

1. All bodies attract each other with what we call gravitational attraction. This applies to the largest celestial objects and to the smallest particles of matter.
2. The strength of one object's gravitational pull upon another is a function of its mass—that is, the amount of matter present.
3. The closer two bodies are to each other, the greater their mutual attraction.

These observations can be stated mathematically as:

$$F = \frac{G\, m_1\, m_2}{r^2}$$

where F is the gravitational force acting along the line joining the two bodies (N)

m_1, m_2 are the masses (in kilograms) of body one and body two, respectively

r is the distance between the two bodies (m)

G is the universal gravitational constant (6.6732 × 10^{-11} newton—meter^2/kilogram^2)

Specifically, Newton's law of gravitation states that two bodies attract each other in proportion to the product of their masses and inversely as the square of the distance between them. This physical principle is very important in launching spacecraft and guiding them to their operational locations in space and is frequently used by astronomers to estimate the masses of celestial objects. For example, Newton's law of gravitation tells us that for a spacecraft to stay in orbit, its velocity (and therefore its kinetic energy) must balance the gravitational attraction of the primary object being orbited. Consequently, a satellite needs more velocity in low than in high orbit. For example, a spacecraft with an orbital altitude of 250 kilometers (150 miles) will have an orbital speed of about 28,000 kilometers per hour (17,500 miles per hour). Our Moon, on the other hand, which is about 442,170 kilometers (238,857 miles) from Earth, has an orbital velocity of approximately 3,660 kilometers per hour (2,287 miles per hour). Of course, to boost a payload from the surface of the Earth to a high-altitude (versus low-altitude) orbit requires the expenditure of more energy, since we are in effect lifting the object further out of the Earth's "gravity well."

Any spacecraft launched into orbit moves in accordance with the same laws of motion that govern the motions of the planets around our Sun and the motion of the Moon around the Earth. The three laws that describe these planetary motions, first formulated by Johannes Kepler (1571–1630), may be stated as follows:

1. Each planet revolves around the Sun in an orbit that is an ellipse, with the Sun as its focus, or primary body.
2. The radius vector—such as the line from the center of the Sun to the center of a planet, from the center of the Earth to the center of the Moon or from the center of the Earth to the center (of gravity) of an orbiting spacecraft—sweeps out equal areas in equal periods of time.
3. The square of a planet's orbital period is equal to the cube of its mean distance from the Sun. We can generalize this last statement and extend it to spacecraft in orbit about the Earth by saying that a spacecraft's orbital period increases with its mean distance from the planet.

In formulating his first law of planetary motion, Kepler recognized that purely circular orbits did not really exist—rather, only elliptical ones were found in nature, being determined by gravitational perturbations (disturbances) and other factors. Gravitational attractions, according to Newton's law of gravitation, extend to infinity, although these forces weaken with distance and eventually become impossible to detect. However, spacecraft orbiting the Earth, while primarily influenced by the gravitational attraction of the Earth (and anomalies in the Earth's gravitational field), are also influenced by the Moon and the Sun and possibly other celestial objects, such as the planet Jupiter.

Kepler's third law of planetary motion states that the greater a body's mean orbital altitude, the longer it will take for it to go around its primary. Let's take this principle and apply it to a rendezvous maneuver between a Space Shuttle Orbiter and a satellite in low Earth orbit. To catch up with and retrieve an unmanned spacecraft in the same orbit, the Space Shuttle must first be decelerated. This action causes the Orbiter vehicle to descend to a lower orbit. In this lower orbit, the Shuttle's velocity would increase. When properly positioned in the vicinity of the target satellite, the Orbiter would then be accelerated, raising its orbit and matching orbital velocities for the rendezvous maneuver with the target spacecraft.

Another very interesting and useful orbital phenomenon is the Earth satellite that appears to "stand still" in space with respect to a point on the Earth's equator. Such satellites were first envisioned by the English scientist and writer Arthur C. Clarke, in a 1945 essay in *Wireless World*. Clarke described a system in which satellites carrying telephone and television would circle the Earth at an orbital altitude of approximately 35,580 kilometers (22,240 miles) above the equator. Such spacecraft move around the Earth at the same rate that the Earth rotates on its own axis. Therefore, they neither rise nor set in the sky like the planets and the Moon but rather always appear to be at the same longitude, synchronized with the Earth's motion. At the equator the Earth rotates at about 1,600

kilometers per hour (1,000 miles per hour). Satellites placed in this type of orbit are called "geostationary" or "geosynchronous" spacecraft.

It is interesting to note here that the spectacular Voyager missions to Jupiter, Saturn and beyond used a "gravity-assist" technique to help speed them up and shorten their travel time. How is it, you may wonder, that a spacecraft can be speeded up while traveling past a planet? It probably seems obvious that a spacecraft will increase in speed as it approaches a planet (due to gravitational attraction), but the gravity of the planet should also slow it down as it begins to move away again. So where does this increase in speed really come from?

Let us first consider the three basic possibilities for a spacecraft trajectory when it encounters a planet (see fig. 3). The first possible trajectory involves a direct hit or hard landing. This is an *impact trajectory*. The second type of trajectory is an *orbital-capture trajectory*. The spacecraft is simply "captured" by the gravitational field of the planet and enters orbit around it (see trajectories b and c). Depending upon its precise speed and altitude (and other parameters), the spacecraft can enter this captured orbit from either the leading or trailing edge of the planet. In the third type of trajectory, a *flyby trajectory*, the spacecraft remains far enough away from the planet to avoid capture but passes close enough to be strongly affected by its gravity. In this case, the speed of the spacecraft will be increased if it approaches from the trailing side of the planet (see trajectory d) and diminished if it approaches from the leading side (see trajectory e). In addition to changes in speed, the direction of the spacecraft's motion also changes.

It may be obvious to you by now that the increase in speed of the spacecraft actually comes from a decrease in speed of the planet itself! In effect, the spacecraft is being "pulled" by the planet. Of course, this has been a greatly

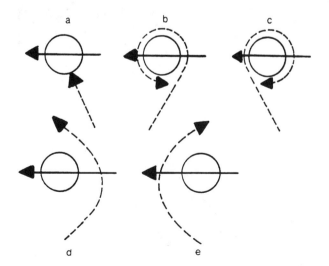

Fig. 3   Possible trajectories of a spacecraft encountering a planet. (NASA drawing, modified by the author.)

simplified discussion of complex encounter phenomena. A full account of spacecraft trajectories must consider the speed and actual trajectory of the spacecraft and planet, how close the spacecraft will come to the planet, and the size (mass) and speed of the planet in order to make even a simple calculation.

Perhaps an even better understanding of gravity assist can be obtained if we use vectors in a more mathematical explanation. The way in which speed is added to the flyby spacecraft during close encounters with the planet Jupiter is shown in figure 4. During the time that spacecraft, such as *Voyager 1* and *2*, are near Jupiter, the heliocentric (Sun-centered) path they follow in their motion with respect to Jupiter closely approximates a hyperbola.

The heliocentric velocity of the spacecraft is the vector sum of the orbital velocity of Jupiter (V_j) and the velocity of the spacecraft with respect to Jupiter (that is, tangent to its trajectory—the hyperbola). The spacecraft moves toward Jupiter along an asymptote, approaching from the approximate direction of the Sun and with asymptotic

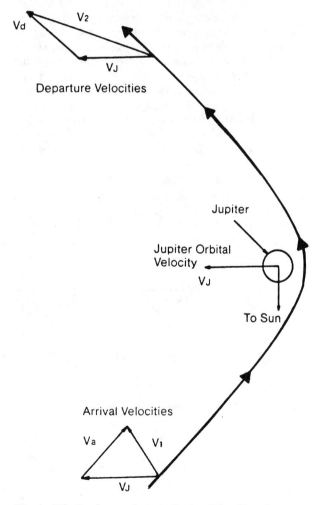

Departure Velocities

Jupiter

Jupiter Orbital Velocity

V_J

To Sun

Arrival Velocities

Fig. 4   Velocity changes during a Jupiter flyby. (Drawing courtesy of NASA.)

velocity (V_a). The heliocentric arrival velocity (V_1) is then computed by vector addition (see fig. 4):

$$V_1 = V_j + V_a$$

The spacecraft departs Jupiter in a new direction, determined by the amount of bending that is caused by the effects of the gravitational attraction of Jupiter's mass on the mass of the spacecraft. The asymptotic departure speed (V_d) on the hyperbola is equal to the arrival speed. Thus, the length of V_a equals the length of V_d. For the heliocentric departure velocity, $V_2 = V_j + V_d$. This vector sum is also depicted in figure 4.

During the relatively short period of time that the spacecraft is near Jupiter, the orbital velocity of Jupiter (V_j) changes very little, and we assume that V_j is equal to a constant.

The vector sums in figure 4 illustrate that the deflection, or bending, of the spacecraft's trajectory caused by Jupiter's gravity results in an increase in the speed of the spacecraft along its hyperbolic path, as measured relative to the Sun. This increase in velocity reduces the total flight time necessary to reach Saturn and points beyond. This "indirect" type of deep-space mission to the outer planets saves two or three years of flight time when compared to "direct-trajectory" missions, which do not take advantage of gravity assist.

Of course, while the spacecraft gains speed during its Jovian encounter, Jupiter loses some of its speed. However, because of the extreme difference in their masses, the change in Jupiter's velocity is negligible.

other world enclosures (OWEs)   Special facilities on large, orbiting space complexes used to simulate the conditions, especially gravity levels and atmospheres, encountered on other celestial bodies in our Solar System. These modular extraterrestrial laboratories would offer exobiologists, space scientists, interplanetary explorers and planetary settlers the unique opportunity of totally experiencing the alien-world conditions before actual expeditions or settlement activities are undertaken.

outer planets   The planets in our Solar System with orbits larger than that of Mars. These are: Jupiter, Saturn, Uranus, Neptune and Pluto. Of these outer planets, all except Pluto are also called the "giant planets."

oxygen chauvinism   A strongly prejudiced opinion that if a planet's atmosphere does not contain oxygen, the planet is essentially uninhabitable and cannot support life. It should be noted, however, that life actually originated on Earth in the absence of oxygen in the ancient terrestrial atmosphere. In fact, some living systems are even poisoned by this gas. We call living systems that occur in the absence of free oxygen "anaerobic" systems.

See also: **extraterrestrial life chauvinisms**

P

panspermia The general hypothesis that microorganisms, spores or bacteria attached to tiny particles of matter have diffused through space, eventually encountering a suitable planet and initiating the rise of life there. The word itself means "all-seeding."

In the 19th century the British scientist Lord Kelvin (William Thompson) suggested that life may have arrived here on Earth from outer space, perhaps carried inside meteorites. Then, in 1903, the Nobel Prize-winning Swedish chemist Svante Arrhenius put forward the idea that is now generally regarded as the panspermia hypothesis. Arrhenius said that life really didn't start here on Earth but rather was "seeded" by means of extraterrestrial spores (seed-like germs), bacteria or microorganisms. According to his hypothesis, these microorganisms, spores or bacteria originated elsewhere in the Galaxy (on a planet in another star system where conditions were more favorable for the chemical evolution of life) and then wandered through space attached to tiny bits of cosmic matter that moved under the influence of stellar radiation pressure.

The greatest difficulty most scientists today have with Arrhenius's original panspermia concept is simply the question of how these "life seeds" can wander through interstellar space for up to several billion years, receive extremely severe radiation doses from cosmic rays and still be "vital" when they eventually encounter a solar system that contains suitable planets. Even on a solar-system scale, the survival of such microorganisms, spores or bacteria would be difficult. For example, "life seeds" wandering from the vicinity of the Earth to Mars would be exposed to both ultraviolet radiation from our Sun and ionizing radiation in the form of solar-flare particles and cosmic rays. This interplanetary spore migration might take several hundred thousand years in the airless, hostile environmental conditions of outer space.

Dr. Francis Crick and Dr. Leslie Orgel attempted to resolve this difficulty by proposing the directed-panspermia hypothesis. Feeling that the overall concept of panspermia was too interesting to abandon entirely, in the early 1970s they suggested that an ancient, intelligent alien race could have constructed suitable interstellar robot spacecraft; loaded these vehicles with an appropriate cargo of microorganisms, spores or bacteria; and then proceeded to "seed the Galaxy" with life, or at least the precursors of life. This "life-seed" cargo would have been protected during the long interstellar journey and then released into suitable planetary atmospheres or oceans when favorable planets were encountered by the robot starships.

Why would an extraterrestrial civilization undertake this type of project? Well, it might first have tried to communicate with other races across the interstellar void; then, when this failed, it could have convinced itself that it was *alone!* At this point in its civilization, driven by some form of "missionary zeal" to "green" (or perhaps "blue") the Galaxy with life as it knew it, the alien race might have initiated a sophisticated directed-panspermia program. Smart robot spacecraft containing well-protected spores, microorganisms or bacteria were launched into the interstellar void to seek new "life sites" in neighboring star systems. This effort might have been part of an advanced-technology demonstration program, a form of planetary engineering on an interstellar scale. These life-seeding robot spacecraft may also have been the precursors of an ambitious colonization wave that never came—or is just now on its way!

In their directed-panspermia discussions, Crick and Orgel identified what they called the theorem of detailed cosmic reversibility. This "theorem" suggests that if we can now contaminate other worlds in this Solar System with microorganisms hitchhiking on terrestrial spacecraft, then it is also reasonable to assume that an advanced, intelligent extraterrestrial civilization could have used its robot spacecraft to contaminate or seed our world with spores, microorganisms or bacteria sometime in the very distant past.

Others have even suggested that life on Earth might have evolved as a result of microorganisms inadvertently left here by ancient astronauts themselves. It is most amusing to speculate that we may be here today because ancient space travelers were "litterbugs," scattering their garbage on a then-lifeless planet. (This line of speculation is sometimes called "the extraterrestrial garbage theory of the origin of life.")

Sir Fred Hoyle and N.C. Wickramasinghe have also recently explored the issue of directed panspermia and the origin of life on Earth. In several publications they argue convincingly that the biological composition of living things on Earth has been and will continue to be radically influenced by the arrival of "pristine genes" from space. They further suggest that the arrival of these cosmic microorganisms, and the resultant complexity of terrestrial life, is not a random process, but one carried out under the influence of a greater cosmic intelligence.

This brings up another interesting point. As we here on Earth develop the technology necessary to send smart machines and humans to other worlds in our Solar System (and eventually even to other star systems), should we initiate a program of directed panspermia? If we became convinced that we might really be alone in the Galaxy, then strong intellectual and biological imperatives might urge us to "green the Galaxy," or to seed life as we know it where there is now none! Perhaps, late in the next century, robot interstellar explorers will be sent out from our Solar System, not only to search for extraterrestrial life but also to plant life on potentially suitable extrasolar planets when none is found. This may be one of our higher cosmic callings, to be the first intelligent species to rise to a level of technology that permits the expansion of life itself within the Galaxy. Of course, our directed-panspermia effort might only be the next link in a cosmic chain of events that was started eons ago by a long-since-extinct alien civilization. Perhaps millions of years from now, on an Earthlike planet around a distant Sunlike star, other intelligent beings will start wondering whether life on their world started spontaneously or was seeded there by an ancient civilization (in this case, us) that has long since disappeared from view in the Galaxy!

While the panspermia or directed-panspermia hypotheses do not address how life originally started somewhere in the Galaxy, they certainly provide some intriguing concepts regarding how, once started, life might "get around."

See also: **ancient astronauts; extraterrestrial contamination; life in the Universe; robotics in space; search for extraterrestrial intelligence**

parsec (pa)   A unit of distance frequently encountered in astronomical studies. The parsec is defined as a parallax shift of one second of arc. The term itself is a shortened form of "*par*allax *sec*ond." The parsec is the extraterrestrial distance at which the mean radius of the Earth's orbit (one astronomical unit, by definition) subtends an angle of one arc-second. It is therefore also the distance at which a star would exhibit an annual parallax of one arc-second.

1 parsec = 3.26 light-years (or 206,265 AU)

The kiloparsec (kpa) represents a distance of 1,000 parsecs (or 3,260 light-years); and the megaparsec (Mpa), a distance of 1 million parsecs (or 3.26 million light-years).

See also: **Appendix A** (table A-2)

people in space   In just a single generation, human beings have extended their physical domain from the upper limits of the Earth's atmosphere to the mountains and valleys of the Moon. The first orbital flight by a human being occurred on April 12, 1961, when Soviet cosmonaut Yuri Gagarin flew around the Earth in the *Vostok 1* spacecraft. Precisely two decades later, American astronauts John Young and Robert Crippen piloted the Space Shuttle *Columbia*—the world's first spaceship—on its maiden orbital flight. Table 1 describes the manned flights that occurred in these two historic decades, during which we learned that people could live and work in space.

Throughout the U.S. space program, there has been continued discussion concerning the relative value of "manned" versus "unmanned" missions. In a real sense, the question is essentially one involving the ultimate location of the human being (that is, on the ground or in space), since automated payloads are ultimately controlled by people. In the Shuttle era, an era of reusable spaceships and routine access to low Earth orbit, this apparent dichotomy of "manned" versus "unmanned" space mission is rapidly vanishing.

The human role in American space efforts has gradually expanded through five highly successful initial manned programs: Mercury, Gemini, Apollo, Skylab and the Apollo-Soyuz Test Project (ASTP).

Project Mercury, the first U.S. manned spaceflight program, was established on October 5, 1958, only five days after NASA was created. During Project Mercury, America performed its first human journey into the extraterrestrial environment. We learned that with proper equipment, people could indeed survive and function normally there.

The Gemini program extended U.S. manned spaceflight activities through the development of a two-person spacecraft that was designed for longer-duration flights (versus the Mercury program). From March 1965 to November 1966, 10 manned Earth-orbital missions were flown. The Gemini program demonstrated that humans could gainfully live, move about and work effectively in space. Techniques such as trajectory shaping and precise maneuvering for spacecraft rendezvous and docking were also developed. These sophisticated space operations techniques were then directly used in the Apollo program.

In the late 1960s and early 1970s, the American manned spaceflight program was dominated by the Apollo program. With the flight of *Apollo 8* (December 1968), human beings first circled the Moon and returned safely to Earth. Starting with the *Apollo 11* mission (July 20, 1969), terrans walked on the lunar surface, and our extraterrestrial civilization was born! In all, 12 human beings had the opportunity to walk on an alien world in the Apollo program. In December 1972 this highly successful extraterrestrial-expedition program came to an end with the triumphant completion of the *Apollo 17* mission. Throughout the Apollo program, people played useful, often essential, roles in the operation of complex, multipurpose vehicles and in the conduct of sophisticated mission activities. In the Apollo program, humans continually exercised a high degree of judgment, selectivity and discrimination. Other situations (both planned and unplanned) took advantage of our analytical capabilities, manual dexterity and ability to respond to the unexpected. Finally, the Apollo program marked the first planned use of a human as a scientist in space. *Apollo 17* astronaut Dr. Harrison H. Schmitt became the world's first practicing exogeologist!

Skylab, the first American space station, was launched in 1973. This program expanded our knowledge of Earth-orbital operations and supported the performance of over 50 scientific, technological and medical experiments. The 100-ton *Skylab* space station was placed in orbit by a Saturn V rocket, while three-person crews were carried into space in Apollo spacecraft lifted by Saturn 1B rockets. The first *Skylab* crew remained in orbit for 28 days; the second, for 59 days; and the third, for 84 days. The *Skylab* missions clearly showed that human beings can adapt well and function properly in the microgravity environment of low Earth orbit for long periods of time, provided they have a proper diet and adequate exercise, sleep, work and recreation. (As shown in table 1, Soviet cosmonauts have stayed in orbit for even longer periods of time.) In *Skylab* it was clearly demonstrated that: (1) people can function effectively in space for long periods of time; (2) there are many worthwhile experiments, tasks and investigations that can *best* be accomplished through manned orbital operations (see fig. C20 on color page I) and (3) there are beneficial services (planned or emergency) that people can more advantageously perform in space.

The Apollo-Soyuz Test Project (ASTP) in July 1975 was a cooperative U.S.-USSR space mission to test compatible rendezvous, docking and crew-transfer systems. Other goals included the performance of in-orbit experiments and Earth observations. Results from the Earth-observation and photography experiments confirmed the ability of a trained observer in orbit to greatly improve knowledge about the terrestrial environment and the major physical processes that help shape and drive it. ASTP results have clearly indicated that a trained astronaut-observer can expertly describe terrestrial features and phenomena (especially those that take place or change quickly); assimilate and interpret what has been seen (here the person exercises both judgment and recall); and rapidly select observational targets and modify observation-experiment protocols (procedures) as required. A human's unique ability to perform in this way is a direct complement to Earth-observing automated spacecraft. The Space Shuttle/Spacelab configuration can serve (in part) as a highly versatile continuation of this manned Earth-observation activity.

The previous manned U.S. space programs have successfully demonstrated a variety of roles for people in space.

These roles include: (1) space-vehicle pilot, (2) long-term inhabitant, (3) payload/experiment manager, (4) extravehicular activity (EVA), (5) mission planner and subsystem maintainer, (6) in-orbit scientific investigator, (7) in-orbit equipment operator and experimenter, and (8) in-orbit engineer and technician. In all of these functions and roles, the performance of human beings is far superior to that of automated systems.

In the Shuttle era, the human role in space not only takes advantage of these demonstrated capabilities but routinely places people, both male and female, in orbit with very flexible, highly sophisticated equipment that guarantees a continually expanded role in extraterrestrial activities, including satellite retrieval and repair, scientific investigations (Spacelab) and in-orbit assembly and construction. The Space Shuttle will also be used to assemble the first permanent U.S. space station in low Earth orbit. We cannot talk about our extraterrestrial civilization unless there are people who become long-term inhabitants of outer space. The Space Shuttle and the space station will help us learn more fully how to live, work and play in the extraterrestrial environment. Effective utilization of these magnificent space-technology tools over the next few decades will greatly accelerate the rate at which space ultimately benefits the overall human condition.

It is almost certain that some of the children born on Earth since Yuri Gagarin's first orbital flight will walk on the surface of Mars! And they and their children will live in space settlements orbiting the Earth, on the surface of the Moon and sprinkled throughout heliocentric (Sun-centered) space.

See also: **extraterrestrial careers; extraterrestrial expeditions; hazards to space workers**

Table 1   A Summary of American and Soviet Manned Spaceflight Experience (April 1961–April 1981)

Launch Date	Astronaut(s)/ Cosmonaut(s)	Mission	Length of Flight	Points of Biomedical Interest
12 April 1961	Gagarin	Vostok 1	1 hr. 48 min.	First manned orbital flight
5 May 1961	Shepard	Mercury MR-3	15 min.	Suborbital flights. No significant physiological problems noted.
21 July 1961	Grissom	Mercury MR-4	16 min.	
6 August 1961	Titov	Vostok 2	25 hr. 18 min.	First reports of space motion sickness, relieved by restriction of head and body movement.
20 February 1962	Glenn	Mercury MA-6	4 hr. 55 min.	First U.S. manned orbital flights. No significant physiological effects noted.
24 May 1962	Carpenter	Mercury MA-7	4 hr. 56 min.	
11 August 1962	Nikolayev	Vostok 3	94 hr. 22 min.	First dual mission. Objective to study human capabilities to function as an operator in space.
12 August 1962	Popovich	Vostok 4	70 hr. 57 min.	
3 October 1962	Schirra	Mercury MA-8	9 hr. 13 min.	First episode of orthostatic intolerance noted postflight.
15 May 1963	Cooper	Mercury MA-9	34 hr. 20 min.	
14 June 1963	Bykovskiy	Vostok 5	119 hr. 6 min.	

Launch Date	Astronaut(s)/ Cosmonaut(s)	Mission	Length of Flight	Points of Biomedical Interest
16 June 1963	Tereshkova	Vostok 6	70 hr. 50 min.	First flight of a woman in space. Labile cardiovascular responses inflight.
12 October 1964	Komarov Feoktistov Yegorov	Voskhod 1	24 hr. 17 min.	Crew operated in "shirtsleeve" environment. First medical examination in space by a physician. Better characterization of vestibular dysfunction.
18 March 1965	Belyayev Leonov	Voskhod 2	26 hr. 2 min.	First USSR extravehicular activity (EVA). Neurovestibular function studies performed.
23 March 1965	Grissom Young	Gemini GT-3	4 hr. 53 min.	Cardiopulmonary monitoring inflight utilizing ECG, blood pressure, respiration rate measurements.
3 June 1965	McDivitt White	Gemini GT-4	97 hr. 48 min.	First U.S. extravehicular activity (EVA). Overheating inside the EVA suit.
21 August 1965	Cooper Conrad	Gemini GT-5	190 hr. 59 min.	
4 December 1965	Borman Lovell	Gemini GT-7	13 days, 18 hr. 35 min.	Comprehensive medical evaluations.
15 December 1965	Schirra Stafford	Gemini GT-6	25 hr. 51 min.	First U.S. rendezvous flight.
16 March 1966	Armstrong Scott	Gemini GT-8	10 hr. 42 min.	First docking in space with a target satellite.
3 June 1966	Stafford Cernan	Gemini GT-9	72 hr. 22 min.	Two hour, seven minute walk in space performed. EVA suit visor fogging.
18 July 1966	Young Collins	Gemini GT-10	70 hr. 46 min.	EVA performed without heat or work rate problems.
12 September 1966	Conrad Gordon	Gemini GT-11	71 hr. 17 min.	High and exhausting work loads during EVA.
11 November 1966	Lovell Aldrin	Gemini GT-12	94 hr. 34½ min.	Extravehicular activity includes astronauts working with tools. No biomedical problems encountered during EVA.
23 April 1967	Komarov	Soyuz 1	27 hr.	Reentry braking system failed. Vehicle destroyed, resulting in death of cosmonaut.
11 October 1968	Cunningham Schirra Eisele	Apollo 7	260 hr. 9 min.	Crew experienced symptoms of upper respiratory viral infection inflight.
26 October 1968	Beregovoy	Soyuz 3	94 hr. 51 min.	First Soviet attempt to manually dock in space with an unmanned target, Soyuz 2. Consistently high heart rates during flight.
21 December 1968	Borman Lovell Anders	Apollo 8	147 hr.	First circumnavigation of the moon. First U.S. report of symptoms of motion sickness. Two-week preflight crew health stabilization program (HSP) instituted.
14 January 1969 15 January 1969	Shatalov Volynov Khrunov Yeliseyev	Soyuz 4 Soyuz 5	71 hr. 14 min. 72 hr. 46 min.	Spacecraft docked for approximately four hours. Two cosmonauts transferred to Soyuz 4 utilizing EVA procedures.

Launch Date	Astronaut(s)/ Cosmonaut(s)	Mission	Length of Flight	Points of Biomedical Interest
3 March 1969	McDivitt Scott Schweickart	Apollo 9	241 hr. 1 min.	Launch postponed for three days because of viral infection. Flight plans for EVA revised because of symptoms of motion sickness.
18 May 1969	Stafford Cernan Young	Apollo 10	192 hr. 3 min.	Fiberglass insulation produced skin, eyes, and upper respiratory passages irritation. No impact on mission functions.
16 July 1969	Armstrong Aldrin Collins	Apollo 11	195 hr. 18 min.	The first walk on the moon. Occurrence of bends reported by one crewman. Lunar quarantine instituted postflight. Three weeks of HSP initiated for this and subsequent missions.
11 October 1969	Shonin Kubasov	Soyuz 6	118 hr. 42 min.	First mission with multiple crews. The intent was to study human performance capabilities inflight, welding of metals in space, feasibility of building a manned space station.
12 October 1969	Filipichenko Volkov Gorbatko	Soyuz 7	118 hr. 41 min.	
13 October 1969	Shatalov Yeliseyev	Soyuz 8	118 hr. 41 min.	
14 November 1969	Conrad Bean Gordon	Apollo 12	244 hr. 36 min.	Contact dermatitis from biosensor electrolyte paste. Lunar quarantine postflight.
11 April 1970	Lovell Haise Swigert	Apollo 13	142 hr. 52 min.	Forced to return to Earth because of explosion in service module. Circumnavigation of moon completed. Urinary tract infection occurred due to the combined effects of cold, dehydration, and prolonged wearing of the urine collecting device.
1 June 1970	Nikolayev Sevastyanov	Soyuz 9	424 hr. 59 min.	Extensive biomedical investigations to determine cardiovascular and musculoskeletal responses. Test of different exercise regimen. Protracted recovery period postflight.
2 February 1971	Shepard Roosa Mitchell	Apollo 14	216 hr. 1 min.	
22 April 1971	Shatalov Yeliseyev Rukavishnikov	Soyuz 10	48 hr.	Docked with Salyut 1 for 5½ hours.
6 June 1971	Dobrovolsky Volkov Patsayev	Soyuz 11	552 hr.	Docked with Salyut 1. During return to Earth, pressure leak in Soyuz vehicle hatch resulted in decompression and death of all three crewmen. Focal areas of atrophy in antigravity muscles noted.
26 July 1971	Scott Worden Irwin	Apollo 15	295 hr. 12 min.	Cardiac arrhythmias and extrasystoles observed during flight. Postflight lunar quarantine discontinued. First use of Lunar Rover vehicle.
16 April 1972	Mattingly Duke Young	Apollo 16	254 hr. 51 min.	Study of light flash phenomenon.
7 December 1972	Cernan Evans Schmitt	Apollo 17	301 hr. 51 min.	Record stay on the lunar surface (75 hours) and 34 km travel utilizing the Lunar Rover vehicle.

Launch Date	Astronaut(s)/ Cosmonaut(s)	Mission	Length of Flight	Points of Biomedical Interest
25 May 1973	Conrad Kerwin Weitz	Skylab 2	672 hr. 49 min.	First detailed metabolic studies. First U.S. physician as a crewmember.
28 July 1973	Bean Garriott Lousma	Skylab 3	59 days	Reversal of red cell mass loss noted.
27 September 1973	Lazarev Makarov	Soyuz 12	2 days	Flight of the second Soviet physician/cosmonaut. Spacecraft modified to hold two crewmen in space suits rather than three in coveralls.
16 November 1973	Carr Gibson Pogue	Skylab 4	84 days	Apparent beneficial effects of vigorous exercise in minimizing cardiovascular deconditioning post-flight. First inflight determination of lung vital capacity changes.
19 December 1973	Klimuk Lebedev	Soyuz 13	8 days	Test of countermeasures for cardiovascular deconditioning: anti-g suits, loading suits, bungee exercises. Studies of cerebral circulation utilizing rheography.
4 July 1974	Popovich Artyukhin	Soyuz 14	16 days	Docked with Salyut 3 space station.
26 August 1974	Sarafanov Demin	Soyuz 15	2 days	Failed to dock with Salyut 3.
2 December 1974	Filipichenko Rukavishnikov	Soyuz 16	7 days	Test flight to verify Soyuz systems prior to the joint US/USSR ASTP mission.
9 January 1975	Gubarev Grechko	Soyuz 17	30 days	Docked with Salyut 4 space station.
5 April 1975	Lazarev Makarov	Soyuz X	Aborted	Cosmonauts suffered injuries and exposure. No fatalities.
25 May 1975	Klimok Sevastyanov	Soyuz 18	63 days	Docked with Salyut 4. Continuation of studies to test human endurance to weightlessness.
15 July 1975	Kubasov Leonov	Soyuz 19 (ASTP)	6 days ⎫	First international space mission. US/USSR joint spaceflight experiment. Cosmonauts and astronauts transferred from respective spacecrafts. U.S. crewmen exposed to nitrogen tetroxide accidentally.
15 July 1975	Stafford Brand Slayton	Apollo-Soyuz Test Project	9 days ⎭	
6 July 1976	Volynov Zholobov	Soyuz 21	48 days	Docked with Salyut 5 space station.
15 September 1976	Bykovskiy Aksenov	Soyuz 22	8 days	USSR/GDR joint experiments.
14 October 1976	Zudov Rozhdestvenskiy	Soyuz 23	2 days	Failed to dock with Salyut 5.
7 February 1977	Gorbatko Glazkov	Soyuz 24	18 days	Docked with Salyut 5. Evaluation of medical countermeasures to prevent deconditioning in long duration missions. Space processing.
9 October 1977	Kovalenok Ryumin	Soyuz 25	2 days	Failed to dock with Salyut 6 space station.

Launch Date	Astronaut(s)/ Cosmonaut(s)	Mission	Length of Flight	Points of Biomedical Interest
10 December 1977	Romanenko Grechko	Soyuz 26	96 days	First prime crew of Salyut 6. Significant cardiovascular deconditioning postflight.
10 January 1978	Dzhanibekow Makarov	Soyuz 27	6 days	First visiting crew to Salyut 6.
2 March 1978	Gubarev Remek	Soyuz 28	8 days	Second visiting and first international crew to Salyut 6.
15 June 1978	Kovalenok Ivanchenko	Soyuz 29	140 days	Second prime crew of Salyut 6. New EVA suit with minimum prebreathing requirements introduced.
27 June 1978	Klimuk Hermaszewski	Soyuz 30	8 days	
26 August 1978	Bykovskiy Jaehn	Soyuz 31	8 days	
25 February 1979	Lyakhov Ryumin	Soyuz 32	175 days	Third prime crew of Salyut 6. First reports of recurrence of vestibular symptoms inflight and postflight.
9 April 1979	Rakavishnikov Ivanov	Soyuz 33	2 days	Fourth international flight. Failed to dock with Salyut 6.
9 April 1980	Popov Ryumin	Soyuz 35	185 days	Longest period in space to date.
26 May 1980	Kubasov Farkash	Soyuz 36	8 days	
5 June 1980	Malyshev Aksenov	Soyuz T-2	5 days	Docked with Salyut 6. Manned test of a new orbital transfer vehicle.
22 July 1980	Gorbatko Tuan	Soyuz 37	8 days	
17 September 1980	Romanenko Mendez	Soyuz 38	8 days	
27 November 1980	Kizim Makarov Strekalov	Soyuz T-3	13 days	
12 March 1981	Kovalenok Savinykh	Soyuz T-4	75 days	Last prime crew onboard Salyut 6 space station. Extensive use of mechanical countermeasures for motion sickness and cardiovascular deconditioning.
22 March 1981	Dzhanibekov Gurragcha	Soyuz 39	9 days	
12 April 1981	Crippen Young	STS-1	54 hr. 21 min.	First demonstration of successful man-piloted hypersonic flight. First experience with G_z forces on reentry. First runway landing.

Source: NASA,.

personnel launch vehicle (PLV)   A proposed vehicle that would be used to transport space workers and their supplies from the surface of the Earth to low Earth orbit (LEO) to support the construction of Satellite Power System units.

See also: **Satellite Power System**

personnel orbital-transfer vehicle (POTV)   An orbital-transfer vehicle used to transfer people and priority cargo between low Earth orbit and geosynchronous Earth orbit (GEO) in support of the construction of Satellite Power System units.

See also: **Satellite Power System**

photoionization   The ionization of an atom or molecule by a photon of electromagnetic radiation. Incoming radiation from the Sun, for example, causes photoionization processes to occur in the Earth's ionosphere (upper atmosphere).

photoklystron   A device for directly converting visible light to microwave radiation.

See also: **Satellite Power System**

photon   According to quantum theory, the elementary bundle or packet of electromagnetic radiation, such as a photon of light. Photons have no mass and travel at the speed of light. The energy of the photon is equal to the product of the frequency of the electromagnetic radiation (v) and Planck's constant (h).

$$E = hv$$

where h is equal to 6.626×10^{-34} joule-sec; v is the frequency (hertz)

photosynthesis   The process by which photons (light energy) and chlorophyll manufacture carbohydrates out of carbon dioxide (CO_2) and water (H_2O).

See also: **life in the Universe**

Pioneer plaque   On June 13, 1983, the *Pioneer 10* spacecraft became the first human-made object to leave the Solar System. In an initial attempt at interstellar communication, the *Pioneer 10* spacecraft and its sistercraft (*Pioneer 11*) were equipped with identical special plaques (see fig. 1). The plaque is intended to show any intelligent alien civilization that might detect and intercept either spacecraft millions of years from now when the spacecraft was launched, from where it was launched and by what type of intelligent beings it was built. The plaque's design is engraved into a gold-anodized aluminum plate, 152 millimeters by 229 millimeters (or 6 by 9 inches). The plate is approximately 1.27 millimeters (0.05 inch) thick. It is attached to the Pioneer spacecraft's antenna support struts in a position that helps shield it from erosion by interstellar dust.

Let's now review the message contained in the Pioneer

plaque. Numbers have been superimposed on the plaque illustrated in figure 1 to assist in this discussion. At the far right, the bracketing bars (1) show the height of the woman compared to the Pioneer spacecraft. The drawing at the top left of the plaque (2) is a schematic of the hyperfine transition of neutral atomic hydrogen—a universal "yardstick" that provides a basic unit of both time and space (length) throughout the Galaxy. This figure illustrates a reverse in the direction of the spin of the electron in a hydrogen atom. The transition depicted emits a characteristic radio wave with an approximately 21-centimeter wavelength. Therefore, by providing this drawing, we are telling any technically knowledgeable alien civilization finding it that we have chosen 21 centimeters as a basic length in the message. While extraterrestrial civilizations will certainly have different names and defining dimensions for their basic system of physical units, the wavelength size associated with the hydrogen radio-wave emission will still be the same throughout the Galaxy. Science and commonly observable physical phenomena represent a general galactic language—at least for starters.

The horizontal and vertical ticks (3) represent the number 8 in binary form. Hopefully, the alien beings pondering over this plaque will eventually realize that the hydrogen wavelength (21 centimeters) multiplied by the binary number representing 8 (indicated alongside the woman's silhouette) describes her overall height—namely, 8 × 21 centimeters = 168 centimeters, or approximately 5 feet 5 inches tall. Both human figures are intended to represent the intelligent beings that built the Pioneer spacecraft. The man's hand is raised as a gesture of goodwill. These human silhouettes were carefully selected and drawn to maintain ethnic neutrality. Furthermore, no attempt was made to explain terrestrial "sex" to an alien culture—that is, the plaque makes no specific effort to explain the potentially "mysterious" differences between the man and woman depicted.

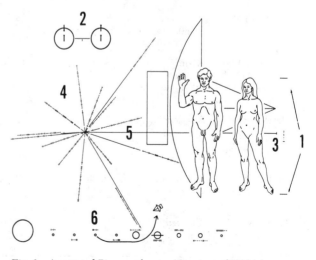

Fig. 1   Annotated Pioneer plaque. (Courtesy of NASA.)

The radial pattern (4) should help alien scientists locate our Solar System within the Milky Way Galaxy. The solid bars indicate distance, with the long horizontal bar (5) with no binary notation on it representing the distance from the Sun to the galactic center, while the shorter solid bars denote directions and distances to 14 pulsars from our Sun. The binary digits following these pulsar lines represent the periods of the pulsars. From the basic time unit established by the use of the hydrogen-atom transition, an intelligent alien civilization should be able to deduce that all times indicated are about 0.1 second—the typical period of pulsars. Since pulsar periods appear to be slowing down at well-defined rates, the pulsars serve as a form of galactic clock. Alien scientists should be able to search their astrophysical records and identify the star system from which the Pioneer spacecraft originated and approximately when it was launched, even if each spacecraft isn't found for hundreds of millions of years. Consequently, through the use of this pulsar map, we have attempted to locate ourselves, both in galactic space and in time.

As a further aid to identifying the Pioneer's origin, a diagram of our Solar System (6) is also included on the plaque. The binary digits accompanying each planet indicate the relative distance of that planet from the Sun. The Pioneer's trajectory is shown starting from the third planet (Earth), which has been offset slightly above the others. As a final clue to the terrestrial origin of the Pioneer spacecraft, its antenna is depicted pointing back to Earth.

This message was designed by Drs. Frank Drake and Carl Sagan, and the artwork was prepared by Linda Salzman Sagan.

When the *Pioneer 10* spacecraft sped past Jupiter in December 1973, it acquired sufficient kinetic energy (through the gravity-assist technique) to carry it completely out of the Solar System. Table 1 describes some of the "near" star encounters that *Pioneer 10* will undergo in the next 860,000 years. Sometime, perhaps a billion years from now, it may pass through the planetary system of a distant stellar neighbor, one whose planets may have evolved intelligent life. If that intelligent life has also developed the technology capable of detecting the (by then derelict) *Pioneer 10* spacecraft, it may also possess the curiosity and technical systems needed to intercept it and eventually decipher the message from Earth!

The *Pioneer 11* spacecraft carries an identical message. After that spacecraft's encounter with Saturn, it also acquired sufficient velocity to escape the Solar System, but in almost the opposite direction to *Pioneer 10*. In fact, *Pioneer 11* is departing in the same general direction in which our Solar System is moving through Space.

As some scientists have philosophically noted, the Pioneer plaque represents "at least one intellectual cave painting, a mark of humanity, which might survive not only all the caves on Earth, but also the Solar System itself!"

See also: **interstellar communication;** *Pioneer 10, 11*

Pioneer 10, 11   The *Pioneer 10* and *11* spacecrafts, as their names imply, have been true extraterrestrial explorers—the first human-made objects to navigate the main asteroid belt, the first to encounter Jupiter and its fierce radiation belts, the first to encounter Saturn and its magnificent ring system and the first spacecraft to leave the Solar System.

The Pioneer spacecraft investigated magnetic fields, cosmic rays, the solar wind and the interplanetary dust concentrations as they flew through interplanetary space. At Jupiter and Saturn, scientists used the spacecraft to

Table 1   Near Star Encounters Predicted for the *Pioneer 10* Spacecraft

Star No.	Name	Information
1	Proxima Centauri	Red dwarf "flare" star. Closest approach is 6.38 light-years after 26,135 years.
2	Ross 248	Red dwarf star. Closest approach is 3.27 light-years after 32,610 years.
3	Lamda Serpens	Sun-type star. Closest approach is 6.9 light-years after 173,227 years.
4	G 96	Red dwarf star. Closest approach is 6.3 light-years after 219,532 years.
5	Altair	Fast-rotating white star (1.5 times the size of the Sun and 9 times brighter). Closest approach is 6.38 light-years after 227,075 years.
6	G 181	Red dwarf star. Closest approach is 5.5 light-years after 292,472 years.
7	G 638	Red dwarf star. Closest approach is 9.13 light-years after 351,333 years.
8	D + 19 5036	Sun-type star. Closest approach is 4.9 light-years after 423,291 years.
9	G 172.1	Sun-type star. Closest approach is 7.8 light-years after 847,919 years.
10	D + 25 1496	Sun-type star. Closest approach is 4.1 light-years after 862,075 years.

[a]SOURCE: NASA/Kennedy Space Center.

investigate the giant planets and their interesting complement of moons in four main ways: (1) by measuring particles, fields and radiation; (2) by spin-scan imaging the planets and some of their moons; (3) by accurately observing the paths of the spacecraft and measuring the gravitational forces of the planets and their major satellites acting on them; and (4) by observing changes in the frequency of the S-band radio signal before and after occultation (the temporary "disappearance" of the spacecraft caused by their passage behind these celestial bodies) to study the structures of their ionospheres and atmospheres.

The *Pioneer 10* spacecraft was launched from Cape Canaveral by an Atlas-Centaur rocket on March 3, 1972. It became the first spacecraft to cross the main asteroid belt and the first to make close-range observations of the Jovian system. Sweeping nearby Jupiter in December 1973, it discovered no solid surface under the thick layer of clouds enveloping the giant planet—an indication that Jupiter was a liquid hydrogen planet. *Pioneer 10* also explored the giant Jovian magnetosphere, made close-up pictures of the intriguing Red Spot and observed at relatively close range the Galilean satellites: Io, Europa, Ganymede and Callisto. When *Pioneer 10* flew past Jupiter, it acquired sufficient kinetic energy to carry it completely out of the Solar System.

Departing Jupiter, *Pioneer 10* continued to map the heliosphere (the Sun's giant magnetic bubble, or field, drawn out from it by the action of the solar wind). Then, on June 13, 1983, *Pioneer 10* crossed the orbit of Neptune, which at the time was (and until 1999 will be) the planet farthest out from the Sun, due to the eccentricity in Pluto's orbit, which now takes it inside that of Neptune. (See fig. 1.) This historic date marked the first passage of a human-made object beyond the known boundary of the Solar System—an event that can only occur once in all human history! Beyond the boundary of the Solar System, *Pioneer 10* continues to measure the extent of the heliosphere as it flies through interstellar space. Along with its sistership, *Pioneer 11*, this spacecraft is also helping scientists search for a postulated massive, dark companion to our Sun, called Nemesis.

The *Pioneer 11* spacecraft was launched on April 6, 1973, and swept by Jupiter at an encounter distance of only 42,000 kilometers on December 4, 1973. It provided additional detailed data and pictures on Jupiter and its moons, including the first views of Jupiter's polar regions. Then, on September 1, 1979, *Pioneer 11* swept by Saturn, demonstrating a safe flight path for the more sophisticated Voyager spacecraft to follow through the rings. *Pioneer 11* (officially renamed *Pioneer Saturn*) provided the first close-up observations of Saturn, its rings, satellites, magnetic field, radiation belts and atmosphere. It found no solid surface on Saturn but discovered at least one additional satellite and ring. After rushing past Saturn, *Pioneer 11* also headed out of the Solar System toward the distant stars.

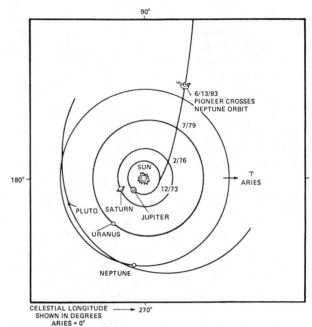

PIONEER 10 BEYOND KNOWN PLANETS

Fig. 1   The *Pioneer 10* trajectory as it became the first spacecraft to leave the Solar System, June 13, 1983. (Drawing courtesy of NASA/Ames Research Center.)

Both Pioneer spacecraft carry a special message to any intelligent alien civilization that might find them wandering through the interstellar void millions of years from now. This message is a drawing-map, engraved on an anodized aluminum plaque. The plaque depicts the location of Earth and the Solar System, a man and a woman, and other points of science and astrophysics that should be decipherable by an intelligent extraterrestrial civilization.

See also: **Jupiter; Nemesis; Pioneer plaque; Saturn**

Pioneer Venus   The *Pioneer Venus* Mission consisted of two separate spacecraft launched to the planet Venus in 1978 by the United States. The *Pioneer Venus* Orbiter was a 553-kilogram spacecraft that contained a 45-kilogram payload of scientific instruments. It was launched on May 20, 1978, and placed into a highly eccentric orbit around Venus on December 4, 1978. The 875-kilogram *Pioneer Venus* Multiprobe spacecraft consisted of a basic bus spacecraft (like the *Pioneer* Orbiter spacecraft), a large probe and three identical small probes. The Multiprobe spacecraft was launched on August 8, 1978, and separated about three weeks before entry into the Venusian atmosphere into four probes and the bus. These five elements entered the Venusian atmosphere at widely separated locations on December 9, 1978, and returned valuable scientific data as they plunged toward the planet's surface. Although the probes were *not* designed to survive landing, one hardy probe did and transmitted data for about one hour after impact.

The *Pioneer* Orbiter spacecraft made an extensive radar

map, covering about 90 percent of the Venusian surface. Using its radar to look through the dense Venusian clouds, the spacecraft revealed that the planet's surface was mostly gentle, rolling plains with two prominent plateaus: Ishtar Terra and Aphrodite Terra. Venus was also found to possess a volcanic structure larger than the Earth's Hawaii-Midway chain and a mountain, called Maxwell Montes, that is higher than Mount Everest.

Pioneer Venus data have revealed two major volcanic areas that vent the planet's internal heat, making Venus the third celestial body in our Solar System now known to have significant volcanic activity. (The Earth and the Jovian satellite Io are the other two celestial bodies.)

The *Pioneer Venus* Orbiter and probe data provided refined information about the Venusian atmosphere. Scientists now know that Venus has an inferno–like surface temperature of about 480 degrees Celsius (900 degrees Fahrenheit) and a surface atmospheric pressure approximately 100 times greater than the atmospheric pressure at sea level on Earth.

The lower atmosphere of Venus consists of approximately 96 percent carbon dioxide, 3 percent nitrogen and 1 percent other gases, including trace amounts of water vapor and sulfur dioxide. The Venusian atmosphere has three distinct regions, and clouds appear to consist of corrosive sulfuric-acid droplets.

The *Pioneer Venus* Mission confirmed that a runaway greenhouse effect is indeed responsible for Venus's infernolike surface temperatures. The highly successful mission also provided supporting data about the absence of an intrinsic magnetic field (a fact observed by earlier spacecraft). *Pioneer Venus* also discovered an excess of noble (also called inert) gases (when compared to the Earth or Mars)—stimulating a possible conflict with contemporary theories on the origin of planetary atmospheres. Future Venus atmospheric-probe missions are needed to verify the Pioneer findings concerning large neon and argon (argon 36 isotope) abundances and the large argon-to-krypton (Ar/Kr) and argon-to-xenon (Ar/Xe) ratios.

See also: **Venus; Venus Atmospheric Probe Mission; Venus Radar Mapper**

planet A nonluminous celestial body that orbits around the Sun or some other star. There are nine such large objects, or "major planets," in our Solar System and numerous "minor planets," or asteroids. The distinction between a planet and a satellite may not always be clear-cut, except for the fact that a satellite orbits around a planet. For example, our Moon is nearly the size of the planet Mercury and is very large in comparison to its parent planet, Earth. In some cases, the Earth and Moon can almost be treated as as "double-planet system," with the same being true for icy Pluto and its large satellite, or moon, Charon.

The largest planet is Jupiter, which has more mass than all the other planets combined. Mercury is the planet nearest the Sun, while (on the average) Pluto is farthest

away. At perihelion (the point in an orbit at which a celestial body is nearest the Sun), Pluto is actually closer to the Sun than Neptune. Saturn is the least dense planet in our Solar System. If we could find some giant cosmic swimming pool, Saturn would float, since it is less dense than water! Seven of the nine planets have satellites, or moons, some of which are larger than the planet Mercury.

See also: **asteroid; Earth; Jupiter; Mars; Mercury; Neptune; Pluto; Saturn; Uranus; Venus**

planetary chauvinism A strongly prejudiced opinion that life has to develop independently on a particular planet. This hypothesis regarding the origin and probability of extraterrestrial life-forms is in direct contrast to the panspermia hypothesis. The panspermia hypothesis states that life, once developed on a single planet somewhere in the Galaxy, can spread unintentionally or intentionally (directed panspermia) via spore migration through interplanetary and interstellar space. Those who support the panspermia hypothesis accept the idea that life on a particular planet, like Earth, might actually have been "seeded" there instead of arising independently from some primeval "chemical soup." The planetary hypothesis, however—namely, that life independently arises on a particular planet—is the more popular current speculation concerning alien life-forms within the Galaxy.

There is another version of planetary chauvinism that holds that life can only originate on a "suitable" planet. This position implies that our own search for extraterrestrial life, possibly intelligent, must be concentrated on star systems that have such suitable planets. Suitable planets are those that lie in the habitable zone—or continuously habitable zone (CHZ), as it is sometimes called—around their parent stars.

Finally, it is also possible that an advanced alien race had its origins on some distant extrasolar planet but then migrated off-planet to occupy selected regions of interplanetary or even interstellar space, tapping energy and material resources of entire star systems. Such an extraterrestrial civilization would no longer be regarded as "planetary" in nature. Perhaps, then, our search for intelligent life in the Universe must also focus on the depths of interstellar space for the telltale signs of cosmic arks, feats of astroengineering and the passage of giant interstellar spaceships.

See also: **extraterrestrial life chauvinisms**

planetary engineering Planetary engineering or terraforming, as it is sometimes called, is the large-scale modification or manipulation of the environment of a planet to make it more suitable for human habitation. In the case of Mars, for example, human settlers would probably seek to make its atmosphere more dense and breathable by adding more oxygen. Early "Martians" would most likely also attempt to alter the planet's harsh temperatures and modify them to fit a more terrestrial thermal pattern. Venus represents an even larger challenge to the planetary engineer. Its current atmospheric pressure would have to be significantly reduced, its inferno-like surface temperatures greatly diminished, the excessive amounts of carbon dioxide in its atmosphere reduced, and perhaps, the biggest task of all, its rotation rate would have to be increased to shorten the length of the solar day.

It should now be obvious that when we discuss planetary engineering projects, we are speaking of truly large, long-term projects. Typical time estimates for the total terraforming of a planet like Mars or Venus range from centuries to a few millennia. However, we can also develop ecologically suitable enclaves or niches, especially on the Moon or Mars. Such localized planetary modification efforts could probably be accomplished within a few decades of project initiation.

Just what are the "tools" of planetary engineering? The planetary pioneers in the next century will need at least the following if they are to convert presently inhospitable worlds into new ecospheres that permit human habitation with little or no personal life support equipment: first, and perhaps the most often overlooked, human ingenuity; second, a thorough knowledge of the physical processes of the particular planet or moon undergoing terraforming (especially the existence and location of environmental pressure points at which small modifications of the local energy or material balance can cause global environmental effects); third, the ability to manipulate large quantities of energy; fourth, the ability to manipulate the surface or material composition of the planet; and fifth, the ability to move large quantities of extraterrestrial materials (for example, small asteroids, comets, water-ice shipments from the Saturnian rings, etc.) to any desired locations within heliocentric space.

One frequently suggested approach to planetary engineering is the use of biological techniques and agents to manipulate alien worlds into more desirable ecospheres. For example, scientists have proposed seeding the Venusian atmosphere with special microorganisms (such as genetically engineered algae) capable of converting excess carbon dioxide into free oxygen and combined carbon. This biological technique would not only provide a potentially more breathable Venusian atmosphere, it would also help to lower the currently intolerable surface temperatures by reducing the runaway greenhouse effect.

Other individuals have suggested the use of special vegetation (such as genetically engineered lichen, small plants or scrubs, etc.) to help modify the polar regions on Mars. The use of specially engineered, survivable plants would reduce the albedo of these frigid regions by darkening the surface, thereby allowing more incident sunlight to be captured. In time, an increased amount of solar energy absorption would elevate global temperatures and cause melting of the long frozen volatiles, including water. This would raise the atmospheric pressure on Mars and possibly cause a greenhouse effect. With the polar caps melted, large quantities of liquid water would be available for transport to other regions of the planet. Perhaps one of the

more interesting Martian projects late in the next century will be to construct a series of large irrigation canals.

Of course, there are other alternatives to help melt the Martian polar caps. The Martian settlers could decide to construct giant mirrors in orbit above the Red Planet. These mirrors would be used to concentrate and focus raw sunlight directly on the polar regions. Other scientists have suggested dismantling one of the Martian moons (Phobos or Deimos) or perhaps a small dark asteroid, and then using its dust to physically darken the polar regions. This action would again lower the albedo and increase the absorption of incident sunlight.

Another approach to terraforming Mars is to use non-biological replicating systems (that is, self-replicating robot systems). These self-replicating machines will probably be able to survive more hostile environmental conditions than genetically engineered microorganisms or plants. To examine the scope and magnitude of this type of planetary engineering effort, we first assume that the Martian crust is mainly silicon dioxide (SiO_2) and then that a general purpose 100-ton, self-replicating system (SRS) "seed machine" can make a replica of itself on Mars in just one year. This SRS unit would initially make other units like itself, using native Martian raw materials. In the next phase of the planetary engineering project, these SRS units would be used to reduce SiO_2 into oxygen that is then released into the Martian atmosphere. In just 36 years from the arrival of the "seed machine," a silicon dioxide reduction capability would be available that could release up to 220,000 tons per second of pure oxygen into the thin atmosphere of the Red Planet. In only 60 years of operation, this array of SRS units would have produced and liberated 4×10^{17} kilograms of oxygen into the Martian environment. Assuming negligible leakage through the Martian exosphere, this is enough "free" oxygen to create a 0.1 bar pressure breathable atmosphere across the entire planet. This pressure level is roughly equivalent to the terrestrial atmosphere at an altitude of 3,000 meters (16,000 feet).

What would be the environmental impact of all these mining operations on Mars? Scientists estimate that the total amount of material that must be excavated to terraform Mars is on the order of 10^{18} kilograms of silicon dioxide. This is enough soil to fill a surface depression 1 kilometer deep and about 600 kilometers in diameter. This is approximately the size of the crater Edom near the Martian equator. The Martians might easily rationalize: "just one small hole for Mars, but a new ecosphere for humankind!"

Asteroids have also played an interesting role in planetary engineering scenarios. People have suggested crashing one or two "small" asteroids into depressed areas on Mars (such as the Hellas Basin) to instantly deepen and enlarge the depression. The goal would be individual or multiple (connected) instant depressions about 10 kilometers deep and 100 kilometers across. These manmade impact craters would be deep enough to trap a more dense atmosphere—

allowing a small ecological enclave or niche to develop. Environmental conditions in such enclaves could range from typical polar conditions to perhaps something almost balmy.

Others have suggested crashing selected asteroids into Venus to help increase its spin. If the asteroid hits the Venusian surface at just the right angle and speed, it could conceivably help speed up the planet's rotation rate—greatly assisting any overall planetary engineering project. Unfortunately, if the asteroid is too small or too slow, it will have little or no effect; while if it is too large or hits too fast, it could possibly shatter the planet! This would be a truly cosmic "oops"!

Another scientist has proposed that several large-yield nuclear devices be used to disintegrate one or more small asteroids that had previously been maneuvered into orbits around Venus. This would create a giant dust and debris cloud that would encircle the planet and reduce the amount of incoming sunlight. This, in turn, would lower surface temperatures on Venus and allow the rocks to cool sufficiently to start absorbing carbon dioxide from the dense Venusian atmosphere.

Finally, other scientists have suggested mining the rings of Saturn for frozen volatiles, especially water-ice and then transporting these back for use on Mars, the Moon or Venus for large-scale planetary engineering projects.

Can you think of anything else planetary engineers might do in the next few centuries to make Mars, Venus, the Moon, or other celestial objects potential "garden spots" in heliocentric space?

planetary exploration For untold millennia, human beings have gazed up at the night sky—wondering, speculating and observing. Ancient astronomers noticed that certain points of light appeared to move relative to the other, "fixed" lights. The ancient Greeks called these particular five celestial objects *planetes*, which meant "wanderers." The powerful Romans named them after their great mythological deities. There was mighty Jupiter (Greek name: Zeus), who was king of the gods and ruler of the heavens. As "lord of the sky," Jupiter hurled terrible thunderbolts that were capable of punishing lesser gods and mortals alike. There was Saturn (Greek name: Cronus, meaning "time"), a Titan and father of Jupiter. The swift Mercury (Greek name: Hermes) was the messenger of the gods and the ancient deity of trade and travel. Venus (Greek name: Aphrodite) was the goddess of love and beauty. And finally, there was Mars (Greek names: Ares), who was the bloodthirsty god of war.

It may be quite difficult for us to realize that just 500 years ago, right around the time Columbus discovered America, the Universe seemed simple, orderly, peaceful and well understood (at least to the people of the time). For in those days, people accepted the Ptolemaic theory of the Universe, which stated that the Earth stood motionless at the center of the Universe, while the Sun, the Moon and the planets circled nearby. The stars were thought of as tiny points of light that were slightly farther away and fixed or embedded on a giant crystal sphere. This model was a quite small, perhaps comfortable and reassuring, Universe to our ancestors. It was a Universe in which humans were assumed to be at the very center of everything!

Then, in only a few centuries following the European renaissance, the scale of the Solar System and of the Universe expanded incredibly. The invention of the telescope and the development of the Copernican theory replaced the Earth with the Sun as the center of our Solar System—that is, humans developed a heliocentric concept for the Solar System. The five wandering lights of the ancient astronomers became individual worlds to study. Through the intervening years, new discoveries in science made possible the study of the distant stars themselves. As telescopes grew larger, we moved the Sun from the center of things to the outer regions of the Milky Way Galaxy—a huge system of over 100 billion stars. Then, as we recently observed the cosmos with even better instruments, the Milky Way itself became for us only one of literally billions of galaxies in an expanding Universe.

But it is really only in the last two decades or so that we have finally overcome the two major physical barriers to our full exploration of the Universe: gravity and the Earth's atmosphere. These barriers fell at the birth of the Space Age, when *Sputnik I* was launched by the Soviet Union on October 4, 1957. Advancements in rocketry during the 1950s have provided us machines to gently break the gravitational grip of Gaea (also spelled Gaia), the Greek goddess of Earth, within whose loving grasp we had previously spent all human history. People and machines could now travel to other worlds, study them at close range and even bring back extraterrestrial samples for subsequent analysis on Earth. Sophisticated robot spacecraft have orbited and landed upon Mars and Venus. Interplanetary spacecraft have flown past Mercury, Jupiter and Saturn, providing spectacular, close-range observations.

In fact, during a single human generation, we have visited all the planets known to the ancients! Instead of fuzzy, wandering lights, these worlds have now become individual, familiar celestial objects that we can see and touch—if not yet in person, at least through "electronic sight" and the other senses provided us by sophisticated orbiter and lander spacecraft. Future historians can only view these amazing pioneering flights to the planets as the "Golden Age of Planetary Exploration." These missions represent some of the most remarkable human achievements of all time!

The marvels of space technology have also allowed us to overcome the blurring and dimming caused by the Earth's atmosphere when we stand on the ground and peer out at the heavens. Once out in space, remote sensing instruments and human eyes can view the Universe directly in all its incredible violence, immensity and variety. Since 1957 we have ventured out into the cosmic ocean for the first time. Life has once again begun to expand its habitat—this time, however, into the extraterrestrial environment of outer space. The last such significant expansion of terrestrial life occurred many hundreds of millions of years ago, when living creatures first ventured out of the oceans of a primeval Earth and occupied the land. Now, we terrans have left footprints on the Moon—and there can be no turning back. It is the Universe, or nothing!

Because we are living through it, it is very difficult for many of us to appreciate the unfolding of a great historic event—the creation of humanity's extraterrestrial civilization. The current exploration of heliocentric (Sun-centered) space may be only the latest episode in a long history of human exploration, seeking and wondering—but the scientific discoveries, the new sights and the sudden explosion of people and machines into outer space have greatly exceeded anything that has gone before in human history. Within the past two decades, we have "discovered" not just one new world, but dozens! Human beings have visited, sampled and returned from an alien world, the Moon. Sophisticated robot laboratories have gathered up and sampled pieces of Martian soil. The sensitive electronic "eyes" of our interplanetary spacecraft have provided spectacular views of celestial objects never before seen by the ancients. And we have even cast our messages of goodwill and friendship into the cosmic void in hopes that an alien intelligence would someday learn that the third planet of a common star in the outer fringes of the Milky Way gave rise to a magnificent life-form called humans!

At this point, it is very instructive, and perhaps even a little amusing, to look back at the prevalent view of the Universe as it existed before the start of the Space Age. Back in 1957, for example, space scientists thought the Sun was a basically stable, steadily burning star. Mercury and Venus were just blurs in the telescope, and we really didn't have any information about their unusual and remarkable surface features. Earth's trapped radiation belts were still undiscovered. The Galilean moons of Jupiter were mere points of light, hardly more visible than they were in 1610 to their discoverer, the famous Italian astronomer Galileo. Only Saturn among the planets was known to have rings. No one knew what the Moon was really made of; and of course, no one had yet touched a lunar rock. Human beings had not yet ventured beyond the Earth's atmosphere into outer space. In fact, in 1957 it was widely doubted that people could even survive or work effectively in space—with its suspected insurmountable dangers of radiation, weightlessness and meteorites. A manned expedition to the Moon was consigned to the realm of science fiction—a feat that would not happen until at least the end of the 20th century... if at all. That's what we thought about our extraterrestrial environment in 1957.

What has happened, of course, in the intervening years is nothing short of mind-boggling. During a single human generation, space scientists and astronomers have had the unique opportunity to compile a remarkable list of accomplishments and discoveries. Please appreciate the truly startling fact that in the last 20 or so years, we have learned more about our extraterrestrial environment than in all the centuries that have gone before. Yet this is just the beginning—just the basis for many more exciting explorations and discoveries to come. We still have many unanswered questions. How did the Universe form? How will it end? How did life originate? Is life unique to the planet Earth, or is there a cosmic prevalence of life? Where are the "little green men"? Are they hiding from us—or are we really alone amid billions and billions of stars? Of course, the answers to these exciting questions lie in our future. What is most interesting to realize, however, is that the answers no longer lie outside our technological grasp.

And just what lies ahead for us in planetary exploration in the next few decades? In 1983 the Solar System Exploration Committee of the NASA Advisory Council identified four goals for the United States planetary-exploration program through the year 2000. These are: (1) to continue the scientific exploration of the Solar System in order to comprehend its origin, evolution and present state; (2) to gain a better understanding of the Earth by comparative studies with other planets; (3) to understand how the appearance of life relates to the chemical and physical history of the Solar System; and finally, (4) to survey the resources available in near-Earth space in order to develop a scientific basis for future utilization of these extraterrestrial resources.

The primary purpose of American planetary exploration is to achieve a deep and thorough understanding of our Solar System. To understand the origins of the Solar System is one of the oldest objectives of human thought. In planetary research we are trying to discover how the basic physical laws operate to produce the world in which we live. This understanding, in turn, will allow us to predict and possibly control those natural phenomena—raising the quality of life for all. Planetary science uses theory, experiment and observation to convert our knowledge of natural laws and phenomena into a more effective understanding of our own "Spaceship Earth."

In many ways, the Voyager flybys of Saturn can be regarded as the culmination of the first period of planetary exploration. As *Voyager 1* sped toward Saturn in November 1980, Carl Sagan eloquently placed the Voyager mission in a cosmic perspective. Sagan said, "We are at the end of the first extraordinary stage of planetary exploration, where all the wandering lights known to the ancients are about to be visited and scrutinized by these wonderful, sophisticated robots we send out to explore the Solar System. I believe we are at a moment that will be remembered for tens, hundreds, perhaps even thousands of years." Of course, humanity has always been characterized by its desire for knowledge. Sagan articulated this particular point when he concluded, "The exploratory instinct which has taken us to the vicinity of Saturn is part of the reason for our success as a species."

After its encounter with Saturn, the *Voyager 1* spacecraft continued on a trajectory that will eventually take it out of the Solar System. *Voyager 2's* postencounter trajectory, on the other hand, will carry it to further encounters—with Uranus in January 1986 and then on to Neptune in August 1989 before it, too, escapes the Solar System.

With these marvelous achievements in exploration, past and pending, we can now consciously recognize that the entire Solar System represents an extended environment for Earth's inhabitants. As we enter the next millennium, we are not conceptually prevented from thinking about the extension of a sphere of human activity that ultimately fills a heliocentric (Sun-centered) environmental niche. Stimulated by even more exciting future programs of planetary exploration and the potential new discoveries they offer, we can confidently address a very simple choice facing us. We can decide to follow our magnificent machines and personally expand into the cosmos, growing, learning and discovering—or we can reject things extraterrestrial, reject the call to the stars, and suffer the consequences of planetary stagnation and eventual extinction!

Planetary Image Facility   The Smithsonian Institution, in cooperation with NASA, has established the Planetary Image Facility to provide the scientific community access to the extensive collections of image data gathered by U.S. interplanetary probes. Included in this reference collection are images from the Moon, Mercury, Venus, Mars and its satellites, Jupiter and its satellites, and Saturn and its satellites. (See figs. C8–C12 on color pages D–F.) Photographs of Earth taken from the Gemini through Apollo-Soyuz missions, as well as selected images from Landsat and other terrestrial remote sensing spacecraft, are also available for use by planetary scientists.

The Image Facility is located in the Center for Earth and Planetary Studies of the National Air and Space Museum, Washington, D.C. The overall objectives of this facility are to assist in scientific research (such as comparative planetology) using the results of interplanetary missions and to provide an image reference library for investigators in the field of planetary science.

For further information or to make arrangements to use this facility, contact:

Director, Center for Earth and Planetary Sciences
Regional Planetary Image Facility
National Air and Space Museum
Smithsonian Institution
Washington, D.C. 20560

See also: **images from space; National Air and Space Museum; remote sensing**

planetesimals   Small celestial objects in the Solar System, such as asteroids, comets and moons.

planet fall   The landing of a spacecraft or space vehicle on a planet.

planetoid   An asteroid, or minor planet.

Pluto   Even though it was discovered more than a half-century ago by the Kansan farm boy-astronomer Clyde Tombaugh, Pluto still remains to a great extent an astronomical mystery. (See fig. C21 on color page I.) Orbiting the Sun at the frigid extremes of our Solar System, this planet lies at the resolution limit of ground-based optical telescopes. It will therefore be a most interesting object for study by the Space Telescope.

Astronomers today think that Pluto has about the same radius as the Earth's Moon, but only one-fifth the Moon's mass. Its size is actually dwarfed by the major satellites of Jupiter and Saturn.

Pluto is about 6 billion kilometers (39.7 astronomical units [AU]) away from the Sun; for most of its orbit, it is the outermost of the nine known planets. However, from now until 1999, it is actually orbiting closer to the Sun than Neptune! Pluto's physical discovery in 1930 came as no real astronomical surprise. Its existence had already been predicted by astronomers at the turn of the century as a result of perturbations (disturbances) observed in the orbits of both Uranus and Neptune. Pluto was found after a deliberate search in just about the anticipated location. This discovery was, however, somewhat fortuitous, because Pluto, with its tiny mass, could not be the source of these outer-planet orbital perturbations. In fact, the planetary-perturbation data that stimulated the original search for a planet beyond Neptune were mainly erroneous, the result of inadequate information about Neptune's orbit. However, recent and more precise orbit data for Neptune and Uranus have indicated other unexplained perturbations. Astronomers are not presently sure what the source of these perturbations is. Pluto, because of its tiny mass, is not considered responsible. The speculation that there is yet another undiscovered planet—a "Tenth Planet," or "Planet X"—located somewhere beyond Pluto does not seem to match the observed perturbation data either. Consequently, the question currently remains open.

In 1978 astronomers discovered that Pluto had a companion, or moon, and named the celestial object Charon. As you might remember, Charon was the boatman in Greek and Roman mythology who ferried the dead across the River Styx to the Underworld, ruled by Hades (Pluto in Roman mythos).

Current estimates for the mass and radius of Pluto indicate that it has a very low specific gravity, less than unity. Water has a specific gravity of one. Therefore, if these estimated data are correct, Pluto would actually float in some giant cosmic body of water! This low specific-gravity value suggests that Pluto and possibly its moon, Charon, are made up of frozen volatiles. In fact, these frozen celestial bodies, closely resembling a "double-planet" system, have been called "snowballs" by astronomers. One speculative model for Pluto's interior suggests a thin layer of frozen methane over silicate rock and (water) ice. On the surface of this planet, there may even be outcroppings of rocky materials in addition to a methane frost.

Since Pluto is both very cold and very small, there are only a few noncondensable gases that could possibly make up any atmosphere it might possess. Candidates include methane mixed with neon or argon.

Currently estimated data about Pluto and its moon are presented in tables 1 and 2. These data are highly speculative and subject to considerable variation in the technical literature.

Table 1   Dynamic Properties and Physical Data for Pluto

Radius	1,600 km ± 300 km
Mass	1.4–2.0×10^{22} kg
Mean density	0.55–1.75 g/cm³
Albedo (geometric)	0.25–0.50
Surface temperature	~50 K
Surface atmospheric pressure	<50 mbar
Atmosphere (?)	methane, possibly mixed with neon or argon
Semimajor axis	5.94×10^9 km (39.72 AU)
Eccentricity	0.25235
Orbit inclination	17.139°
Mean orbital velocity	4.74 km/sec
Orbital period	247.69 years
Distance from the Sun	
Aphelion	7.375×10^9 km (49.2 AU) [409.2 light-min]
Perihelion	4.425×10^9 km (29.5 AU) [245.3 light-min]
Mean solar flux (at 30 AU)	1.5 W/m²
Number of natural satellites	1

NOTE: These data are speculative.
SOURCE: NASA.

Table 2   Physical Data and Dynamic Properties for Pluto's Moon, Charon

Radius	600–1,000 km
Mass	1.3×10^{21} kg ± 10%
Semimajor axis (of orbit around Pluto)	19,700 km ± 300 km
Period	6.387 days
Inclination to Pluto's equator	94° ± 3°

NOTE: These data are speculative.
SOURCE: NASA.

Plutonian Of or relating to the planet Pluto; (in science fiction) a native of the planet Pluto.

pressurized habitable environment Any module or enclosure in space in which an astronaut may perform activities in a "shirt-sleeve" environment.

primary body The celestial body about which a satellite, moon or other object orbits or from which it is escaping or toward which it is falling. For example, the primary body of the Earth is the Sun; the primary body of the Moon is the Earth.

primitive atmosphere The atmosphere of a planet or other celestial object as it existed in the early stages of its formation. For example, the primitive atmosphere of the Earth some 3 billion years ago was thought to consist of water vapor, carbon dioxide, methane and ammonia.

principle of mediocrity A general assumption or speculation often used in discussions concerning the nature and probability of extraterrestrial life. It assumes that things are pretty much the same all over—that is, it assumes that there is nothing special about the Earth or our Solar System. By invoking this hypothesis, we are guessing that other parts of the Universe are pretty much as they are here. This philosophical position allows us to then take the things we know about the Earth, the chemical evolution of life that occurred here and the facts we are learning about other objects in our Solar System and extrapolate these to develop concepts of what may be occurring on alien worlds around distant suns.

The simple premise of the principle of mediocrity is very often employed as the fundamental starting point for contemporary speculations about the cosmic prevalence of life. If the Earth is indeed *nothing special*, then perhaps a million worlds in our own Galaxy (which is one of billions of galaxies) not only are suitable for the origin of life but have witnessed its chemical evolution in their primeval oceans and are now (or at least were) habitats for a myriad of interesting living creatures. Some of these living systems may also have arisen to a level of intelligence where they are at this very moment gazing up into the heavens of their own world and wondering if they, too, are alone!

If, on the other hand, the Earth and its delicate biosphere really are something special, then life—especially intelligent life capable of comprehending its own existence and contemplating its role in the cosmic scheme of things—may be a rare, very precious jewel in a vast, lifeless cosmos. In this latter case, the principle of mediocrity would be most inappropriate to use in estimating the probability that extraterrestrial life exists elsewhere in the Universe.

Today, we cannot pass final judgment on the validity of the principle of mediocrity. We must, at an absolute minimum, wait until human and robot explorers have made more detailed investigations of the interesting objects in our own Solar System. Celestial objects of particular interest to exobiologists include the planet Mars and certain moons of the giant outer planets Jupiter and Saturn. Once we have explored these alien worlds in depth, scientists will have a much more accurate technical basis for suggesting that we are either "something special" or "nothing special"—as the principle of mediocrity implies.

Project Cyclops A very large array of dish antennas proposed for use in a detailed search of the radio-frequency spectrum (especially the 18- to 21-centimeter wavelength "waterhole") for interstellar signals from intelligent alien civilizations. The engineering details of this SETI (search for extraterrestrial intelligence) configuration were derived in a special summer institute design study sponsored by NASA at Stanford University in 1971. The stated object of the Project Cyclops study was:

> To assess what would be required in hardware, manpower, time, and funding to mount a realistic effort, using present (or near-term future) state-of-the-art techniques, aimed at detecting the existence of extraterrestrial (extrasolar system) intelligent life.

Named for the one-eyed giants found in Greek mythology, the proposed Cyclops Project would use as its "eye" a large array of individually steerable 100-meter-diameter parabolic dish antennas (see fig. C22 on color page J). These Cyclops antennas would be arranged in a hexagonal matrix, so that each antenna unit was equidistant from all its neighbors. A 292-meter separation distance between antenna dish centers would help avoid shadowing. In the Project Cyclops concept, an array of about 1,000 of these antennas would be used to simultaneously collect and evaluate radio signals falling on them from a target star system. The entire Cyclops array would function like a single giant radio antenna, some 30 to 60 square kilometers in size.

Project Cyclops can be regarded as one of the foundational studies in our contemporary search for extraterrestrial intelligence. Its results—based on the pioneering efforts of such individuals as Dr. Frank Drake, Dr. Philip Morrison, Dr. John Billingham and Dr. Bernard Oliver—have established the technical framework for subsequent SETI activities. Project Cyclops also reaffirmed the interstellar microwave window, the 18- to 21-centimeter-wavelength "waterhole," as perhaps the most suitable part of the electromagnetic spectrum for interstellar civilizations to communicate with each other and for us here on Earth as we begin our cosmic search for signals from intelligent alien beings.

See also: **Drake equation; extraterrestrial civilizations; Fermi paradox; search for extraterrestrial intelligence**

Project Daedalus The name given to the most extensive study of interstellar space exploration yet undertaken. From 1973 to 1978 a team of scientists and engineers under the auspices of the British Interplanetary Society studied the feasibility of performing a simple interstellar mission using only current technology and reasonable extrapolations of imaginable near-term capabilities.

In Roman mythology Daedalus was the grand architect of King Minos's labyrinth for the Minotaur on the island of Crete. But Daedalus also showed the Greek hero Theseus, who slew the Minotaur, how to escape from the labyrinth. An enraged King Minos imprisoned both Daedalus and his son Icarus. Undaunted, Daedalus (a brilliant engineer) fashioned two pairs of wings out of wax, wood and leather. Before their aerial escape from a prison tower, Daedalus cautioned his son not to fly too high, so that the Sun would not melt the wax and cause the wings to disassemble. They made good their escape from King Minos's Crete, but while over the sea, Icarus, an impetuous teen-ager, ignored his father's warnings and soared high into the air. Daedalus (who reached Sicily safely) watched his young son, wings collapsed, tumble to his death in the sea below.

The proposed Daedalus spaceship structure, communications systems and much of the payload were designed entirely within the parameters of 20th-century technology. Other components, such as the advanced machine intelligence flight controller and on-board computers for in-flight repair, required artificial-intelligence capabilities expected to be available in the mid-21st century. The propulsion system, perhaps the most challenging aspect of any interstellar mission, was designed as a nuclear-powered, pulsed-fusion rocket engine that burned an exotic thermonuclear fuel mixture of deuterium and helium 3 (a rare isotope of helium). This pulsed-fusion system was believed capable of propelling the robot interstellar probe to velocities in excess of 12 percent of the speed of light (that is, > 0.12 c). The best source of helium 3 was considered to be the planet Jupiter, and one of the major technologies that had to be developed for Project Daedalus was an ability to mine the Jovian atmosphere for helium 3. This mining operation might be achieved by using "aerostat" extraction facilities (floating balloon-type factories).

The Project Daedalus team suggested that this ambitious interstellar flyby (one-way) mission might possibly be undertaken at the end of the 21st century—when the successful development of humankind's extraterrestrial civilization had generated the necessary wealth, technology base and exploratory zeal. The target selected for this first interstellar probe was Barnard's star, a red dwarf (spectral type M 5) about 5.9 light-years away in the constellation Ophiuchus.

The Daedalus spaceship would be assembled in cislunar space (partially fueled with deuterium from Earth) and then ferried to an orbit around Jupiter, where it could be fully fueled with the helium 3 propellant that had been mined out of the Jovian atmosphere. These thermonuclear fuels would then be prepared as pellets, or "targets," for use in the ship's two-stage pulsed-fusion power plant. Once fueled and readied for its epic interstellar voyage, somewhere around the orbit of Callisto, the ship's mighty pulsed-fusion first-stage engine would come alive. This first-stage pulsed-fusion unit would continue to operate for about two years. At first-stage shutdown, the vessel would be traveling at about 7 percent of the speed of light (0.07 c).

The expended first-stage engine and fuel tanks would be jettisoned in interstellar space, and the second-stage pulsed-fusion engine would ignite. The second stage would also operate in the pulsed-fusion mode for about 2 years. Then, it, too, would fall silent, and the giant robot spacecraft, with its cargo of sophisticated remote sensing equipment and nuclear (fission)-powered probe ships, would be traveling at about 12 percent of the speed of light (0.12 c). It would take the Daedalus spaceship about 47 years of coasting (after second-stage shutdown) to encounter Barnard's star.

When the Daedalus interstellar probe was about 3 light-years away from its objective (about 25 years mission elapsed time), smart computers on board would initiate long-range optical and radio astronomy observations. A special effort would be made to locate and identify any extrasolar planets that might exist in the Barnardian system.

Of course, traveling at 12 percent of the speed of light, Daedalus would only have a very brief passage through the target star system. This would amount to a few days of "close-range" observation of Barnard's star itself and only "minutes" of observation of any planets or other interesting objects by the robot mothership.

However, several years before the Daedalus mothership passed through the Barnardian system, it would launch its complement of nuclear-powered probes (also traveling at 12 percent of the speed of light initially). These probe ships, individually targeted to objects of potential interest by computers on board the robot mothership, would "fly ahead" and act as data-gathering scouts. A complement of 18 of these scout craft or small robotic probes was considered appropriate in the Project Daedalus study.

Then, as the main Daedalus spaceship flashed through the Barnardian system, it would gather data from its own on-board instruments as well as information telemetered to it by the numerous probes. Then, over the next day or so, it would transmit all these mission data (billions and billions of data bits) back toward our Solar System, where team scientists would patiently wait the approximately six years it takes for these information-laden electromagnetic waves, traveling at light speed, to cross the interstellar void.

Its mission completed, the Daedalus mothership—without its probes—would continue on a one-way journey into the darkness of the interstellar void, to be discovered perhaps millennia from then by an advanced alien race, which might puzzle over humankind's first attempt at the direct exploration of another star system.

The main conclusions that can be drawn from the Project Daedalus study might be summarized as follows:

(1) exploration missions to other star systems are, in principle, technically feasible; (2) a great deal could be learned about the origin, extent and physics of the Galaxy, as well as the formation and evolution of stellar and planetary systems, by missions of this type; (3) the prerequisite interplanetary and initial interstellar space system technologies necessary to successfully conduct this class of mission also contribute significantly to humankind's search for extraterrestrial intelligence (for example, smart robot probes and interstellar communications); (4) a long-range societal commitment on the order of a century would be required to achieve such a project; and (5) the prospects for interstellar flight by human beings do *not* appear very promising using current or foreseeable 21st-century technologies.

The Project Daedalus study also identified three key technology advances that would be needed to make even a robot interstellar mission possible. These are: (1) the development of controlled nuclear fusion, especially the use of the deuterium/helium 3 thermonuclear reaction; (2) advanced machine intelligence; and (3) the ability to extract helium 3 in large quantities from the Jovian atmosphere.

While the choice of Barnard's star as the target for the first interstellar mission was somewhat arbitrary, if future human generations can build such an interstellar robot spaceship and successfully explore the Barnardian system, then with modest technology improvements, all star systems within 10 to 12 light-years of Earth become potential targets for a more ambitious program of (robotic) interstellar exploration.

See also: **Barnard's star; fusion; interstellar contact; robotics in space; search for extraterrestrial intelligence; space nuclear propulsion; starship**

Project Orion (I)   A design study sponsored by NASA's Office of Space Science in 1976 to examine techniques and instruments that could be used in a search for extrasolar planets. Are planets common throughout our Galaxy? Are they formed regularly from a nebula of gas and dust surrounding protostars (developing stars)? (This is called the nebula hypothesis of planetary formation.) Or are they created only rarely, through such unusual cosmic events as the close passage of a rogue star, a stellar collision or a supernova event? The answers to these and many other intriguing questions about the cosmic prevalence of extrasolar planets will help us better understand the origin of our own Solar System and will assist us in planning the search for extraterrestrial intelligence.

At present, most astronomers support the nebula hypothesis of planetary formation. They assume that planetary systems are the rule rather than the exception. This implies that planets occur as a natural, if not inevitable, part of the stellar-formation process. A careful, systematic investigation of how frequently such extrasolar planets occur would provide observational support for modern theories of stellar formation.

In addition, a precise hunt for extrasolar planets helps solve a most puzzling cosmic mystery: "Are we alone?" If planets are necessary for the origin of life and the eventual rise of intelligent creatures (as occurred here on Earth), then we really must know how often extrasolar planets occur in order to develop credible estimates of the probability of encountering intelligent alien beings. For example, if planets—especially Earthlike planets around Sun-like stars—are really rare, then the chances for finding intelligent alien creatures around distant suns would indeed be very remote. If, on the other hand, suitable planets occur as part of the formation process of most stars, then there may indeed be many life-bearing worlds in our Galaxy—some spawning intelligent planetary civilizations like our own!

Unfortunately, the detection of extrasolar planets is a difficult task, since obtaining the extremely precise astrometric data that would be required is beyond the capability of today's technology. Present observational techniques and ground-based instruments are only marginal in their ability to conduct such an effort. NASA's Project Orion examined design concepts for an advanced ground-based astrometric system that could support a technically realistic search effort. The Space Telescope, scheduled for deployment and orbital operation in the late 1980s, may also provide additional help in our search for alien worlds around neighboring stars.

Scientific and technical advancement at the turn of the century made it possible to measure the separation between two photographically recorded star images with accuracies of a few thousandths of a millimeter. As a result, astronomers were able to measure stellar angular separations as small as a few hundredths of an arc-second. (An arc-second is an angle equal to 1/3,600 degree.)

Astrometry is that branch of astronomy that deals with the precise measurement of the position and motion of stars. Massive, but unseen, or "dark," stellar companions will cause a star to wobble from its predicted course. The traditional technique in searching for extrasolar planets is consequently to perform precise astrometric studies. If astronomers detect a slight wobble in a star's predicted motion, the presence of planetary companions could be implied.

However, the first unseen object to be detected outside the Solar System was not a planet, but rather a new type of star called a white dwarf. For a long time, astronomers had been puzzled by the tiny variations they observed in the motion of the star Sirius. Instead of traveling in a straight line, it was seen to wobble from one side to the other of its predicted trajectory. Its motion was much like the wobble pattern described above. Astronomers have now established that this tiny wobble (about 4 arc-seconds) in Sirius's projected trajectory is being caused by a dark stellar companion (a white dwarf star) that has about the same mass as our Sun.

Although severely limited by the measurement capabilities of existing ground-based astrometric telescopes, astronomers over the last few decades have attempted to search for evidence of extrasolar planets by carefully looking for a telltale wobble or perturbation (disturbance) in a particular star's motion. Unfortunately, these efforts currently remain scientifically inconclusive. For example, a great deal of interest has recently centered around Barnard's star, a red dwarf about 5.9 light-years away. Certain wobble data have suggested that this star could possess one or more Jovian-mass planets. But because these data lie at the very threshold of current measurement accuracies, such conclusions are not regarded as "technically firm." More accurate wobble data are needed before scientists can confidently accept the possible fact that one of our stellar neighbors is accompanied by massive planetary companions.

Three techniques for detecting extrasolar planets were considered in the Project Orion study. These are: (1) very accurate astrometric searches for perturbations or wobbles in a star's motion, (2) infrared searches for intrinsic thermal radiation from extrasolar planets and (3) optical searches for reflected stellar (visible) radiation.

The first technique, using precise astrometric observations, is often referred to as "indirect" detection of extrasolar planets. This is because the presence of a planetary companion around an alien sun is deduced from observations of the parent star, and not from observations of the extrasolar planet itself.

The techniques for detecting extrasolar planets by looking for telltale planetary radiations are called "direct" detection techniques, because detection arises from a direct observation of the planet. Planetary companions to stars can be considered as sources of electromagnetic radiation. We can characterize such planetary radiation as either thermal or reflected optical. A planet will radiate at some temperature that is determined by its overall energy balance. A planetary energy balance is simply the sum of the incoming energy (radiant energy from the parent star) plus internal planetary energy sources (such as radioactive decay) minus outgoing energy (energy radiated away by the planet). This planetary radiation is generally most pronounced in the infrared portion of the electromagnetic spectrum and is often called "intrinsic thermal radiation."

A planet might also reflect radiation from its parent star. This reflected radiation is usually strongest at those wavelengths where the star emits the majority of its radiant energy—typically, in the visible portion of the spectrum. The amount of reflected visible radiation is a function of both the size of the extrasolar planet and the nature of the reflecting medium (that is, the composition of the planet's atmosphere, its surface conditions, and so on).

The main effort of Project Orion was to develop a design concept for an improved ground-based astrometric telescope. This telescope needed the following features: (1) reflecting optics, (2) large aperture and (3) "electric" (automated) detection of interference fringes from stars. Project Orion's imaging stellar interferometer (ISI) design incorporated as many technical advantages as possible,

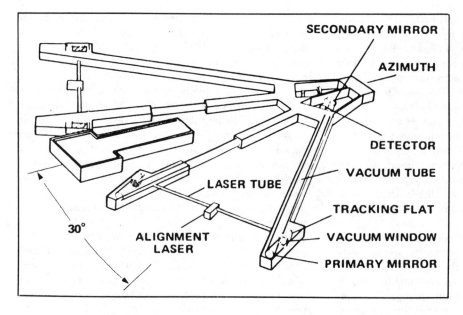

SECONDARY MIRROR

AZIMUTH

DETECTOR

VACUUM TUBE

TRACKING FLAT

VACUUM WINDOW

PRIMARY MIRROR

LASER TUBE

ALIGNMENT LASER

30°

Fig. 1 An artist's drawing of the proposed Orion Imaging Stellar Interferometer. (Courtesy of NASA.)

while avoiding most of the problems found in other astrometric systems (see fig. 1). It is essentially a long-baseline interferometer that simultaneously "images" the white-light fringes of many stars.

An interferometer splits an incoming beam of electromagnetic radiation (for example, a beam of light) into two or more components that are then reunited after each has traveled a different pathlength. As a result, an interference pattern is formed. If the incoming radiation is in the visible portion of the spectrum, light and dark rings or bands are formed. Stellar interferometers are used by optical astronomers to measure very small angles, such as the "apparent" diameter of a single star or the angular separation of a binary star system. Astronomers use the word *apparent* to describe a measured or observed property. Apparent values must often be corrected to obtain the actual, or true, value. For example, astronomers must compensate for the effects of the Earth's atmosphere.

Each of the three extrasolar planet-detection techniques considered during Project Orion appeared to be feasible. In the next few years, improvements in both ground-based astrometric telescopes and space-based astronomy should greatly assist astronomers in their search for positive evidence of planetary companions around alien suns.

The search for extrasolar planets involves many different scientific disciplines. Knowledge about how frequently planets occur in our Galaxy and their distribution as a function of star type would help scientists confirm current theories about stellar evolution, including hypotheses about the origin of our Solar System. Data on the frequency of extrasolar planets will also help us in our efforts to search for extraterrestrial intelligence.

See also: **Drake equation; electromagnetic spectrum; extrasolar planets; search for extraterrestrial intelligence; Space Telescope; stars**

Project Orion (II) *Project Orion* was also the name given to a nuclear-fission pulsed rocket concept studied in the early 1960s. A manned, interplanetary spaceship would be propelled by exploding a series of nuclear-fission devices behind it. A giant pusher plate, mounted on large shock absorbers, would receive the energy pulse from each successive nuclear detonation, and the spaceship configuration would be propelled forward by Newton's action-reaction principle.

In theory, this concept is capable of achieving specific-impulse values ranging from 2,000 to 6,000 seconds, depending on the size of the pusher plate. Specific impulse is a performance index for rocket propellants. It is defined as the thrust produced by the propellant divided by the mass flow rate. As a point of comparison, the best chemical rockets have specific-impulse values ranging from 450 to 500 seconds.

A manned Orion spaceship would move rapidly throughout interplanetary space at a steady acceleration of perhaps 0.5 g (one-half the acceleration of gravity on the Earth's surface). Typically, a 1- to 10-kiloton fission device would be exploded every second or so close behind the giant pusher plate. A kiloton is the energy of a nuclear explosion that is equivalent to the detonation of 1,000 tons of TNT (trinitrotoluene). A few thousand such detonations would be needed to propel a crew of 20 astronauts to Mars or the moons of Jupiter.

Work by the United States on this nuclear-fission pulse rocket concept came to an end in the mid-1960s, as a result of the Limited Test Ban treaty of 1963. This treaty prohibited the signatory nations from testing nuclear devices in the Earth's atmosphere, underwater or in outer space.

Advanced versions of the original Orion concept have also emerged. In these new spaceship concepts, externally detonated nuclear-fission devices have been replaced by many small, controlled, thermonuclear-fusion explosions taking place inside a specially constructed thrust chamber. These mini-thermonuclear explosions would occur in an inertial confinement fusion (ICF) process in which many powerful laser, electron or ion beams simultaneously impinge on a tiny fusion fuel pellet. Each miniature thermonuclear explosion would have an explosive yield equivalent to a few tons of TNT. The expanding shell of very hot, ionized gas from the thermonuclear explosion would be directed into a thrust-producing exhaust stream. Such pulsed nuclear-fusion spaceships, when developed, open up our entire Solar System to human visitation. For example, Earth-to-Neptune travel would take less than 15 days at a steady, comfortable acceleration of one g. Pulsed-fusion systems also represent a possible propulsion system for interstellar travel.

See also: **fission (nuclear); fusion; Project Daedalus; space nuclear propulsion**

Project Ozma   The first attempt to detect interstellar radio signals from an intelligent extraterrestrial civilization. It was conducted by Dr. Frank Drake at the National Radio Astronomy Observatory (NRAO) in Green Bank, West Virginia, in 1960. Drake derived the name for this effort from the queen of the imaginary land of Oz, since in his own words Oz was "a place very far away, difficult to reach, and populated by exotic beings."

A frequency of 1,420 megahertz (MHz) was selected for this initial search—the frequency of the 21-centimeter interstellar hydrogen line. Since this is a radio frequency at which most emerging technical civilizations would first use narrow-bandwidth, high-sensitivity radio telescopes, scientists reasoned that this would also be the frequency that more advanced alien civilizations would use in trying to signal emerging civilizations across the vast interstellar void.

In 1960 the 29.5-meter (85-foot) Green Bank radio telescope was aimed at two sunlike stars about 11 light-years away, Tau Ceti and Epsilon Eridani. Patiently, Frank Drake and his Project Ozma team listened for intelligent signals. But after over 150 hours of listening, no evidence of strong radio signals from intelligent extraterrestrial civilizations was obtained. Project Ozma is generally considered the first serious attempt to listen for intelligent interstellar radio signals in our search for extraterrestrial intelligence—the birth of modern SETI!

See also: **Drake equation; extraterrestrial civilizations; search for extraterrestrial intelligence**

protein   A complex polymer made up of amino acids that contain the elements carbon, hydrogen, nitrogen, oxygen, sometimes sulfur and occasionally other chemical elements such as iron and phosphorus. Proteins are essential constituents of all living cells. They are synthesized by plants from raw materials but can also be assimilated by animals as separate amino acids.

See also: **amino acid; life in the Universe**

protogalaxy   A galaxy at the early stages of its evolution.

protoplanet   Any of a star's planets as it emerges during the process of accretion (accumulation), or the formative period of the star system.

protostar   A star in the making. Specifically, the stage in a young star's evolution after it has separated from a gas cloud but prior to its collapsing sufficiently to support thermonuclear reactions.

protosun   The Sun as it emerged in the formation of the Solar System.

pulsed laser   A laser that produces its radiation in short pulses, as distinct from a continuous-wave (CW) laser. A pulsed laser emits its burst of coherent (optical) radiation in less than 0.25 second.

quark   One of several postulated particles that are thought to represent the structural components of elementary particles. Some physicists speculate that very condensed matter—as found, for example, in neutron stars—might be composed of a mixture of free quarks.

quasars   Mysterious objects that appear almost like stars but are far more distant than any individual star we can now observe. These unusual objects were first discovered in the 1960s with radio telescopes, and they were called "quasi-stellar radio sources"—or "quasars," for short. Quasars emit tremendous quantities of energy from very small volumes. The most distant quasars observed to date are so far away that they are receding at more than one-half the speed of light.

As bright, concentrated radiation sources, quasars resemble the nuclei of Seyfert galaxies but are far more luminous. The optical brightness of some quasars has been observed to change by a factor of two in about a week, with detectable changes occurring in just one day. Therefore, astrophysicists now speculate that such quasars cannot be much larger than about one light-day across (a distance about twice the dimension of our Solar System), since a light source cannot change brightness significantly in less time than it takes for light itself to travel. The problem facing scientists today is to explain how quasars can generate more energy than is possessed by an entire galaxy and generate this energy in so small a region of space! As we place more sophisticated observatories in space, we will learn more about these very unusual extraterrestrial objects.

See also: **astrophysics; Seyfert galaxy**

quiet sun   A term used by solar physicists to describe the minimum portion of the sunspot cycle for our Sun.

See also: **Sun**

R

radar astronomy   The use of radar by astronomers to study objects in our Solar System, such as the Moon, the planets, asteroids and even planetary ring systems. For example, a powerful radar telescope, like the Arecibo Observatory, can hurl a radar signal through the "opaque" Venusian clouds (some 80 kilometers thick) and then analyze the faint return signal to obtain detailed information for the preparation of high-resolution surface maps. Radar astronomers can precisely measure distances to celestial objects, estimate rotation rates and also develop unique maps of surface features, even when the actual physical surface is obscured from view by thick layers of clouds.

See also: **Arecibo Observatory; Venus Radar Mapper**

radiation laws   The collection of theoretical and empirical laws that describe radiation phenomena—in particular, the four physical laws that, together, fundamentally describe the behavior of blackbody radiation. These are: (1) the Kirchhoff law, which is essentially a thermodynamic relationship between emission and absorption of any given wavelength of radiation at a given temperature; (2) the Planck radiation law, which describes the variation of intensity of a blackbody radiation source at a given temperature as a function of wavelength; (3) the Stefan-Boltzmann law, which relates the time rate of radiant-energy emission from a blackbody to its absolute temperature; and (4) the Wien law, which relates the wavelength of maximum intensity emitted by a blackbody to its absolute temperature.

See also: **blackbody**

radiation pressure   The very slight pressure exerted on an absorbing object by all propagating electromagnetic waves.

See also: **solar sail**

radioactivity   The spontaneous decay or disintegration of an unstable atomic nucleus. This phenomenon is usually accompanied by the emission of ionizing radiation (such as alpha particles, beta particles or gamma rays).

radio astronomy   That branch of astronomy which collects and evaluates radio signals from extraterrestrial sources. Radio astronomy is a relatively young branch of astronomy. It was started in 1931, when the American radio engineer Karl Jansky detected the first extraterrestrial radio signals. Until Jansky's discovery, astronomers had only used the visible portion of the electromagnetic spectrum to view the Universe. The detailed observation of cosmic radio sources, however, is very difficult, because these sources shed so little energy on the Earth.

The radio telescope has been used to discover some extraterrestrial radio sources so unusual that their very existence had not even been imagined or predicted by scientists. One of the strangest of these cosmic radio sources is the pulsar—a collapsed giant star that emits pulsating radio signals as it spins. When the first pulsar was detected in 1967, it created quite a stir in the scientific community. Because of the regularity of its signal, scientists thought they had just detected the first interstellar signals from an intelligent alien civilization!

Another interesting celestial object is the quasar, or quasi-stellar radio source. Discovered in 1964, quasars are now considered to be entire galaxies in which a very small part (perhaps only a few light-days across) releases enormous amounts of energy—amounts equivalent to the total annihilation of millions of stars. Quasars are the most distant known objects in the Universe, some of which are receding from us at one-half the speed of light.

Scattered throughout interstellar space are atoms and molecules of materials such as hydrogen, formaldehyde and methyl alcohol. Some of these interstellar molecules represent the basic ingredients of life as we know it. Radio telescopes can be used to collect valuable information about the abundance and nature of these interstellar molecules, as well as to search for other freely floating chemicals in the interstellar medium. If these potential "seeds of life" are found fairly commonly throughout interstellar space, then their existence might be considered as substantiating the principle of mediocrity; that is, the chemistry needed to initiate life on Earth may be typical of what occurs elsewhere.

See also: **Arecibo Observatory; life in the Universe; quasars; stars**

radio frequency (RF)   In general, a frequency at which electromagnetic radiation is useful for communication purposes; specifically, a frequency above 10,000 hertz (cycles per second) and below 3×10^{12} hertz.

radio galaxy   A galaxy that produces very strong signals at radio wavelengths.

radio telescope   A large, metallic device that collects and focuses radio waves onto a sensitive radio-frequency (RF) receiver.

See also: **Arecibo Observatory; astrophysics; search for extraterrestrial intelligence**

radio waves Electromagnetic waves that have a wavelength between one millimeter and several thousand kilometers; the corresponding frequencies range between 300 gigahertz and a few hertz.

See also: **electromagnetic spectrum**

rectenna A term for the "*rec*eiving an*tenna*" on Earth that would intercept and convert beamed microwave energy from an orbiting Satellite Power System into electricity for use in a terrestrial power grid.

See also: **Satellite Power System**

relativistic cosmology The application of Einstein's general relativity theory to cosmology; especially useful in the initial development of "Big Bang" cosmology models.

See also: **"Big Bang" theory; cosmology; relativity**

relativity The theory of space and time developed by Albert Einstein, which has become one of the foundations of 20th-century physics. Einstein's theory of relativity is often discussed in two general categories: the special theory of relativity, which he first proposed in 1905, and the general theory of relativity, which he presented in 1915.

The special theory of relativity is concerned with the laws of physics as seen by observers moving relative to one another at constant velocity—that is, by observers in nonaccelerating or inertial reference frames. Special relativity has been well demonstrated and verified by many types of experiments and observations.

Einstein proposed two fundamental postulates in formulating special relativity:

• *First postulate of special relativity:* The speed of light (*c*) has the same value for all (inertial-reference-frame) observers, regardless and independent of the motion of the light source or the observers.

• *Second postulate of special relativity:* All physical laws are the same for all observers moving at constant velocity with respect to each other.

The first postulate appears contrary to our everyday "Newtonian mechanics" experience. Yet the principles of special relativity have been more than adequately validated in experiments. Using special relativity, scientists can now predict the space-time behavior of objects traveling at speeds from essentially zero up to those approaching that of light itself. At lower velocities the predictions of special relativity become identical with classical Newtonian mechanics. However, when we deal with objects moving close to the speed of light, we must use relativistic mechanics.

What are some of the consequences of the theory of special relativity?

The first interesting relativistic effect is called *time dilation*. Simply stated—with respect to a stationary observer/clock—time moves more slowly on a moving clock/system. This unusual relationship is described by the equation:

$$\Delta t = (1/\beta) \, \Delta T_p \qquad (1)$$

where Δt is called the time dilation (the apparent slowing down of time on a moving clock relative to a stationary clock/observer).

ΔT_p is the "proper time" interval as measured by an observer/clock on the moving system.

$$\beta \equiv \sqrt{1 - (v^2/c^2)} \qquad (2)$$

where v is the velocity of the object.

c is the velocity of light.

Let's now explore the time-dilation effect with respect to a postulated starship flight from our Solar System. We start with twin brothers, Astro and Cosmo, who are both astronauts and are currently 25 years of age. Astro is selected for a special 40-year-duration starship mission, while Cosmo is selected for the ground control team. This particular starship, the latest in the fleet, is capable of cruising at 99 percent of the speed of light (0.99 c) and can quickly reach this cruising speed. During this mission, Cosmo, the twin who stayed behind on Earth, has aged 40 years. (We are taking the Earth as our fixed or stationary reference frame "relative" to the starship.) But Astro, who has been on board the starship cruising the Galaxy at 99 percent of the speed of light for the last 40 Earth-years, has aged just 5.64 Earth-years! Therefore, when he returns to Earth from the starship mission, he is a little over 30 years old, while his twin brother, Cosmo, is now 65 and retired in Florida. Obviously, starship travel (if we can overcome some extremely challenging technical barriers) also presents some very interesting social problems.

The time-dilation effects associated with near-light speed travel are real, and they have been observed and measured in a variety of modern experiments. All physical processes (chemical reactions, biological processes, nuclear-decay phenomena and so on) appear to slow down when in motion relative to a "fixed" or stationary observer/clock.

Another interesting effect of relativistic travel is *length contraction*. We first define an object's proper length (L_p) as its length measured in a reference frame in which the object is at rest. Then, the length of the object when it is moving (L)—as measured by a stationary observer—is always smaller, or contracted. The relativistic length contraction is given by:

$$L = \beta \, (L_p) \qquad (3)$$

This apparent shortening, or contraction, of a rapidly moving object is seen by an external observer (in a different inertial reference frame) only in the object's direction of motion. In the case of a starship traveling at near light speeds, to observers on Earth this vessel would appear to shorten, or contract, in the direction of flight. If an alien starship was 1 kilometer long (at rest) and entered our Solar System at an encounter velocity of 90 percent of the speed of light (0.9 c), then a terrestrial observer would see a starship that appeared to be about 435 meters long. The aliens on board and all their instruments (including tape measures) would look contracted to external observers but would not appear any shorter to those on board the ship (that is, to observers within the moving reference frame). If this alien starship was "burning rubber" at a velocity of 99 percent of the speed of light (0.99 c), then its apparent contracted length to an observer on Earth would be about 141 meters. If, however, this vessel was a "slow" interstellar freighter that was limping along at only 10 percent of the speed of light (0.1 c), then it would appear about 995 meters long to an observer on Earth.

Special relativity also influences the field of dynamics. Although the rest mass (m_o) of a body is invariant (does not change), its "relative" mass increases as the speed of the object increases with respect to an observer in another fixed or inertial reference frame. An object's relative mass is given by:

$$m = (1/\beta) \, m_o \qquad (4)$$

This simple equation has far-reaching consequences. As an object approaches the speed of light, its mass becomes infinite! Since things can't have infinite masses, physicists conclude that material objects cannot reach the speed of light. This is basically the *speed-of-light barrier*, which appears to limit the speed at which interstellar travel can occur. From the theory of special relativity, scientists now conclude that only a "zero-rest-mass" particle, like a photon, can travel at the speed of light. There is one other major consequence of special relativity that has greatly affected our daily lives—the equivalence of mass and energy from Einstein's very famous formula:

$$E = \Delta m \, c^2 \qquad (5)$$

where E is the energy equivalent of an amount of matter (Δm) that is annihilated or converted completely into pure energy.

c is the speed of light.

This simple, yet powerful, equation explains where all the energy in nuclear fission or nuclear fusion comes from. The complete annihilation of just one gram of matter releases about 9×10^{13} joules of energy.

In 1915 Einstein introduced his general theory of relativity. He used this development to describe the space-time relationships developed in special relativity for cases where there was a strong gravitational influence—such as white dwarf stars, neutron stars and black holes. One of Einstein's conclusions was that gravitation is not really a force between two masses (as Newtonian mechanics indicates) but rather arises as a consequence of the curvature of space and time. In our four-dimensional Universe (X −, Y −, Z − and time), space-time becomes curved in the presence of matter.

The fundamental postulate of general relativity is also called *Einstein's principle of equivalence:* The physical behavior inside a system in free-fall is indistinguishable from the physical behavior inside a system far removed from any gravitating matter (that is, the complete absence of a gravitational field).

Several experiments have been performed to confirm the general theory of relativity. These experiments included observation of the bending of electromagnetic radiation (starlight and radio-wave transmissions from Project Viking on Mars) by the Sun's immense gravitational field and recognizing the subtle perturbations (disturbances) in the orbit (at perihelion—that is, at the point of closest approach to the Sun) of the planet Mercury as caused by the curvature of space-time in the vicinity of the Sun. While scientists do not think that these experiments have conclusively demonstrated the validity of general relativity, more favorable experimental evidence is anticipated in the upcoming years—especially when we start investigating phenomena such as neutron stars and black holes with more powerful space-based observatories.

remote sensing The sensing of an object, event or phenomenon without having the sensor in direct contact with the object being studied. Information transfer from the object to the sensor is accomplished through the use of the electromagnetic spectrum. Modern remote sensing technology uses many different portions of the electromagnetic spectrum, not just the visible portion we see with our eyes. As a result, very different and very interesting "images" are often created by modern remote sensing instruments. For example, figure 1 is a radar image of the Los Angeles, California region taken from an altitude of 800 kilometers by NASA's Seasat satellite in July, 1978. This radar image shows an area of approximately 100 kilometer by 100 kilometer. Downtown Los Angeles is located slightly lower and left of center in this figure. Seasat's radar instrument was called a Synthetic Aperture Radar, or "SAR" for short. This device measured the intensity (brightness) of its radar waves as they were reflected back from the Earth's surface. The bright patch in the city of Burbank (upper left portion of fig. 1) and other bright patches, were caused by reflections from building walls and hillsides that were at or near a 90 degree angle to the radar beams from Seasat. The spacecraft was located over the Pacific Ocean to the west, when the image shown here was made. In this type of remote sensing imagery, streets, highways, airports, flood channels and calm water appear dark. This is because little or no radar beams were reflected back to the spacecraft.

Since 1966 NASA has conducted an Earth Observations Program. This program explores the use of multispectral remote sensing in many application areas, including: the measurement of environmental pressures on the terrestrial

Fig. 1 A synthetic aperture radar (SAR) view of Los Angeles taken at an altitude of 800 km (500 mi) by NASA's SEASAT (July 28, 1978). (Courtesy of NASA/Jet Propulsion Laboratory.)

1974
12 JULY
12.0 BUSHELS/ACRE

SEVERE STRESS

1975
15 JULY
27.1 BUSHELS/ACRE

Fig. 2 LANDSAT remote sensing images showing healthy and stressed crops. (Courtesy of NASA.)

LANDSAT CLASSIFICATION MAP OF TRAVIS COUNTY, TEXAS

GRASSLAND   AGRICULTURE

RIVER

URBAN (AUSTIN)

LAKE

TEXAS OAK

AIRPORT

CEDAR

Fig. 3 The use of LANDSAT imagery for land classification. (Courtesy of NASA.)

biosphere; the search for mineral and petroleum resources; the state of health and yield estimations for agricultural crops; and the monitoring of the marine environment. These studies began with Earth surveys from conventional aircraft. Then, in 1972 NASA's Landsat 1, previously called the Earth Resources Technology Satellite or ERTS, provided scientists with a new experimental tool for collecting valuable data about the terrestrial environment from space. The initial effect of Landsat 1 was the availability of large quantities of remotely-sensed data about the Earth, with regional coverage repeated at 18 day intervals.

For example, figure 2 shows an important Landsat capability—to detect crop disease. The Landsat image on the right shows healthy crops, while the Landsat image on the left shows severely "stressed" crops. Similarly, figure 3 is a Landsat image of Travis County, Texas, that illustrates the application of modern remote sensing platforms in land management. In a false color, multispectral image, each shade represents a particular land-use category, such as urban, forest and farm land. Landsat imagery has also been used in water resource surveys, coastal zone management and in the identification of wildlife areas.

Over the past decade, multispectral (visible and infrared), thermal infrared, and active-microwave remote sensing systems have been used to gather research data and to demonstrate the application of remote sensing techniques in agriculture, land use analysis and planning, hydrology and geology. Spacecraft like the Landsat D-D′ represent the current generation of remote sensing spacecraft. The Earth Observation Spacecraft (EOS), a much larger system, represents the future of spacebased remote sensing platforms.

The Landsat D-D′ class spacecraft represents a new modular design, consisting of NASA's Multimission Modular Spacecraft (MMS) configuration and a mission-unique instrument module. The basic MMS includes the attitude control, communications, data handling and power subsystems, and a propulsion module. The instrument module contains the Multispectral Scanner (MSS), the Thematic Mapper (TM), a wideband communications subsystem, the high-gain and other antennas and a solar array for electric power. The Multispectral Scanner provides data in the same format as previous Landsat spacecraft, while the Thematic Mapper provides improved Earth resources management information because of its greater spectral and spatial resolution capabilities. Some of the primary applications of contemporary Landsat data include: crop acreage and yield assessment, water resources management, mineral and petroleum exploration, land use planning and environmental monitoring.

Remote sensing of the Earth from outer space is one of the major contributions of modern space technology toward improving our overall quality of life. Future remote sensing systems, like the very large Earth Observation Spacecraft, provide even more detailed information to assist us in the careful stewardship of our world as we enter the next millennium. The EOS is one of several large space antenna system designs currently being considered by NASA for development in the next decade or so. Space Shuttle astronauts would deploy this spacecraft on orbit, extending it to its final 120 meters by 60 meters configuration. The mast of the system would be some 116 meters long. The EOS would be designed to observe terrestrial resources. Specific capabilities would include: detailed measurement of soil moisture levels, sea surface conditions, and ice fields and their boundaries.

See also: **electromagnetic spectrum; humanization of space**

rendezvous   The close approach of two spacecraft or space objects in the same orbit so that docking can take place.

ringed world   A planet with a ring or series of rings encircling it. In our Solar System Jupiter, Saturn and Uranus are presently known to possess ring systems. This indicates that ring systems may be a common feature of Jovian type planets. (Only Neptune lacks a known ring system.)

Astronomers speculate that there are three general ways such ring systems are formed around a planet. The first is meteoroidal bombardment of a large body, so that the fragments inside the planet's Roche limit form a ring. The second is condensation of material inside the Roche limit when the planet was forming. This trapped nebular material can neither join the parent planet nor form a moon or satellite. Finally, ring systems can also form when a satellite or other celestial object (for example, comet) comes within the planet's Roche limit and gets torn apart by tidal forces.

The Roche limit, named after the French mathematician Edouard Roche who formulated it in the 19th century, is the critical distance from the center of a massive celestial object or planet within which tidal forces will tear apart a moon or satellite. If we assume that (1) both celestial objects have the same density; and (2) that the moon in question is held together only by gravitational attraction of its matter, then the Roche limit is typically about 1.2 times the diameter of the parent planet or primary celestial body.

See also: **Jupiter; Saturn; Uranus**

RNA   Ribonucleic acid, a nucleic acid usually associated with the control of chemical reactions. Messenger RNAs serve as blueprints or templates for the formation of proteins.

See also: **exobiology; life in the Universe**

robotics in space   Robotics is basically the science and technology of designing, building and programming robots. It is a rapidly growing field that is playing an increasingly important role in humanity's exploration and exploitation of the extraterrestrial environment. For example, robotic devices have been used on lunar and planetary missions (see fig. 1) and on the Space Shuttle (see fig. C23 on color page J). Advanced robotic devices are now being considered for servicing spacecraft, for assembling large space structures and for conducting sophisticated exploration of both the lunar and Martian surfaces. (See fig. C24 on color page J.)

Robotic devices, or robots as they are usually called, are primarily "smart machines" with manipulators that can be programmed to do a variety of manual or human labor tasks automatically. A robot therefore is simply a machine that does mechanical, routine tasks on human command. The expression "robot" is attributed to the Czechoslovakian writer, Karel Capek, who wrote the play *R.U.R. (Rossum's Universal Robots)*. This play first appeared in English in 1923 and is a satire on the mechanization of civilization. The word "robot" is derived from *robata*, a

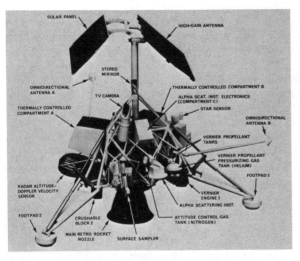

Fig. 1 The *Surveyor VII* robot spacecraft. The *Surveyor* space-craft were landed on the Moon in the 1960s to prepare the way for the manned Apollo expeditions. (Courtesy of NASA/Jet Propulsion Laboratory.)

Czechoslovakian word meaning compulsory labor or servitude.

Since then, robots have become the essential elements found in many exciting science fiction stories. These mechanical "actors" have played the part of both villain and hero. The well-known writer, Isaac Asimov, formulated his three classic "Laws of Robotics" in the 1940s to describe a "machine code of ethics" by which robots could be built and programmed so as not to harm their human masters. Asimov's three laws of robotics are as follows:

Law 1: A robot may not injure a human being, or, through inaction, allow a human being to come to harm.

Law 2: A robot must obey the orders given it by human beings except where such orders would conflict with the First Law.

Law 3: A robot must protect its own existence as long as such protection does not conflict with the First and Second Law.

If we try to build robots that obey such laws, the following design conditions should certainly be satisfied: (a) our robot must not be capable of deliberately injuring a human being; (b) our robot should be able to exercise common sense; (c) our robot must be intelligent; and (d) our robot must also be conscious. For a robot or any smart machine to exercise common sense, intelligence and consciousness, scientists must make great advances in the current field of artificial intelligence.

Actually, today's robots bear very little resemblance to the delightful robots found in such science fiction adventure stories as George Lucas' *Star Wars*. Instead of a beeping R2D2 or an apologetic C3PO, Space Age robots take the form of complex arm-like devices that move objects from one place to another; or they may be very sophisticated, automated surface vehicles that wander

across alien worlds in search of resources (see fig. C24 on color page J).

Before we begin a detailed discussion about the use of robots in space, it will be helpful to review some of the basic concepts and terminology related to robots here on Earth.

A PRIMER ON ROBOTS

A typical robot consists of one or more manipulators (arms), end effectors (hands), a controller, a power supply and possibly an array of sensors to provide information about the environment in which the robot must operate. Because the majority of modern robots are used in industrial applications, their classification is currently based on these industrial functions. Terrestrial robots are frequently divided into the following classes: the non-servo robot (or pick-and-place robot); the servo robot; the programmable robot; the computerized robot; the sensory robot; and the assembly robot.

The non-servo robot is the simplest type of robot. It picks up an object and places it at another location. The robot's freedom of movement is usually limited to two or three directions.

The servo robot represents several categories of industrial robots. This type of robot has servo-mechanisms for the manipulator and end effector to enable it to change direction in mid-air without having to trip or trigger a mechanical limit switch. Five to seven directions of motion are common, depending on the number of joints in the manipulator.

The programmable robot is essentially a servo robot that is driven by a programmable controller. This controller memorizes (stores) a sequence of movements and then repeats these movements and actions continuously. This type of robot is programmed by "walking" the manipulator and end effector through the desired movement.

The computerized robot is simply a servo robot run by a computer. This kind of robot is programmed by instructions fed into the controller electronically. These "smart robots" may even have the ability to improve upon their basic work instructions.

The sensory robot is a computerized robot with one or more artificial senses to observe and record its environment and to feed information back to the controller. The artificial senses most frequently employed are sight (robot or computer vision) and touch.

Finally, the assembly robot is a computerized robot, generally with sensors, that is designed for assembly line and manufacturing tasks, both on Earth and eventually in space.

In industry, robots are mainly designed for manipulation purposes. The actions that can be produced by the end effector or hand include: (1) motion (from point to point; along a desired trajectory; or along a contoured surface); (2) a change of orientation; (3) rotation.

Non-servo robots are capable of point-to-point motions. For each desired motion, the manipulator moves at full speed until the limits of its travel are reached. As a result, non-servo robots are often called "limit sequence," "bang-bang," or "pick-and-place" robots. When non-servo robots reach the end of a particular motion, a mechanical stop or limit switch is tripped, stopping the particular movement.

Servo robots are also capable of point-to-point motions; but their manipulators move with controlled variable velocities and trajectories. Servo robot motions are controlled without the use of stop or limit switches.

Four different types of manipulator arms have been developed to accomplish robot motions. These are the rectangular, cylindrical, spherical and anthropomorphic (articulated or jointed) arm. Each of these manipulator arm designs features two or more degrees of freedom (DOF). The term "degrees of freedom" or DOF refers to each direction a robot's manipulator arm is able to move. For example, a simple straight line or linear movement represents one DOF. If the manipulator arm is to follow a two-dimensional curved path, it needs two degrees of freedom: up and down and right and left. Of course, more complicated motions will require many degrees of freedom. To locate an end effector at any point and to orient this effector in a particular work volume requires six DOF. If the manipulator arm needs to avoid obstacles or other equipment, even more degrees of freedom will be required. For each DOF, one linear or rotary joint is needed. Robot designers sometimes combine two or more of these four basic manipulator arm configurations to increase the versatility of a particular robot's manipulator.

Actuators are used to move a robot's manipulator joints. There are three basic types of actuators currently used in contemporary robots. These are: the pneumatic, the hydraulic and the electrical actuator. Pneumatic actuators employ a pressurized gas to move the manipulator joint. When the gas is propelled by a pump through a tube to a particular joint, it triggers or actuates movement. Pneumatic actuators are inexpensive and simple, but their movement is not precise or "squooshy." Therefore, this kind of actuator is usually found in non-servo or pick-and-place robots. Hydraulic actuators are quite common and capable of producing a large amount of power. The main disadvantages of hydraulic actuators are their accompanying apparatus (pumps and storage tanks) and problems with fluid leaks. Electrical actuators provide smoother movements, can be very accurately controlled and are very reliable. However, these actuators cannot deliver as much power as comparable mass hydraulic actuators. Nevertheless, for modest power actuator functions, electrical actuators are often preferred.

Many industrial robots are fixed in place or move along rails or guideways. Some terrestrial robots are built into wheeled carts, while others use their end effectors to grasp handholds and pull themselves along. Advanced robots use articulated manipulators as legs to achieve a walking motion.

A robot's end effector (hand or gripping device) is generally attached to the end of the manipulator arm. Typical functions of this end effector include: grasping, pushing and pulling, twisting, using tools, performing insertions, and various types of assembly activities. End effectors can be mechanical, vacuum or magnetically operated, can use a snare device or have some other unusual design feature. The final design of an end effector is determined by the shapes of the objects that the robot must grasp. Most end effectors are usually some type of gripping or clamping device.

Robots can be controlled in a wide variety of ways, from simple limit switches which are tripped by the manipulator arm to sophisticated computerized remote sensing systems which provide machine vision, touch and hearing. In the case of computer controlled robots, the motions of the manipulator and end effector are programmed; that is, the robot "memorizes" what it's supposed to do. Sensor devices on the manipulator help to establish the proximity of the end effector to the object to be manipulated and feed information back to the computer controller concerning any modifications needed in the manipulator's trajectory.

CONTEMPORARY SPACE ROBOTS

Robotic systems will play a major role in mankind's exploration and exploitation of the Solar System. Some contemporary applications include: the Space Shuttle's versatile remote manipulator system (RMS); automated servicing of orbiting spacecraft; the construction of very large space structures; and sophisticated interplanetary spacecraft, probes, landers and surface rovers. On the more distant extraterrestrial horizon, one might envision highly automated interstellar probes and perhaps one of the most interesting smart machine devices of all, the self-replicating system (SRS).

The Space Shuttle has been designed primarily to transport a variety of payloads between the Earth and low Earth orbit. The Shuttle Orbiter contains a cargo bay large enough to carry one and one-half school buses! Once the Orbiter vehicle has achieved low Earth orbit, payloads in the cargo bay need to be deployed. For example, spacecraft are removed from the cargo bay and deployed to their higher altitude, operational orbits. In addition, defective spacecraft that are already on orbit must be brought into the cargo bay for on-orbit repair work or for a return to Earth. To handle these spacecraft deployment and retrieval operations as well as to permit the assembly of large structures (such as a permanent space station in orbit), a very versatile robot arm has been installed in the Space Shuttle Orbiter along the port (left) side cargo bay door hinges. This robot arm is called the remote manipulator system (RMS). (See Fig. C23 on color page J.)

The RMS was designed and built by the National Research Council of Canada. It is a highly sophisticated robotic device that is similar to a human arm. The 15-meter-long RMS features a shoulder, wrist and hand—although its hand does not at all look like a human hand. The skeleton of this mechanical arm is made of lightweight graphite composite materials. Covering the skeleton are skin layers consisting of thermal blankets. The

Fig. C18   The Orion Nebula. (Courtesy of NASA.)

Fig. C21   Artist's concept of frozen Pluto and its moon, Charon, looking back across the Solar System at the Sun. (Courtesy of NASA.)

Fig. C19   An artist's concept of the *Voyager 2* spacecraft looking back on Neptune and its moon Triton after an encounter in 1989. (Courtesy of NASA/Jet Propulsion Laboratory.)

Fig. C20   Scientist-astronaut Owen Garriott performing an extra-vehicular activity (EVA) during the *Skylab 3* mission, August 1973. (Courtesy of NASA/Johnson Space Center.)

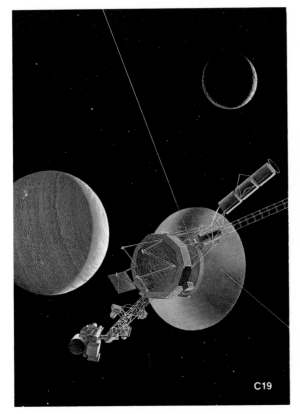

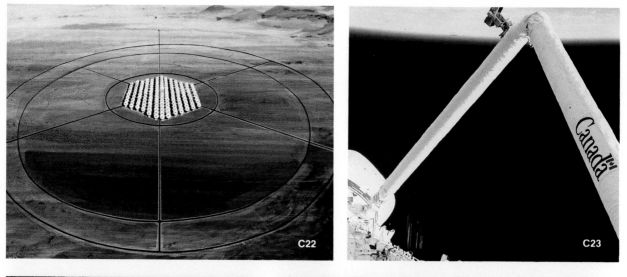

C22

C23

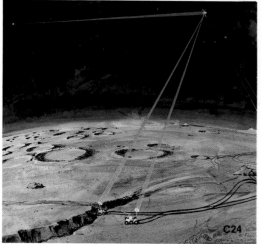

C24

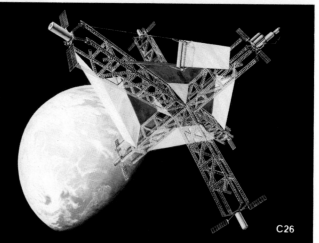

C26

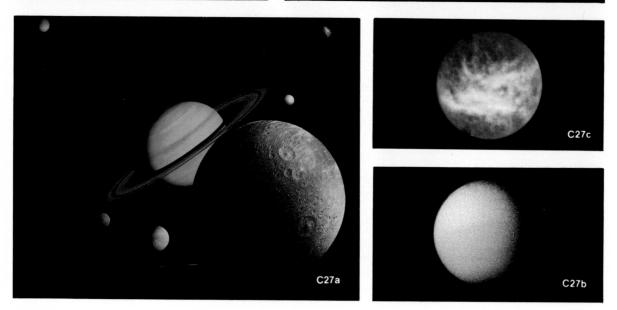

C27a

C27b

C27c

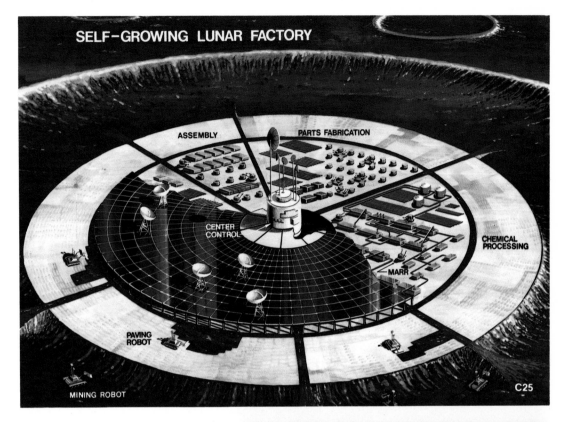

SELF-GROWING LUNAR FACTORY

ASSEMBLY PARTS FABRICATION

CENTER CONTROL

CHEMICAL PROCESSING

MARR

PAVING ROBOT

MINING ROBOT

C25

Fig. C22 An aerial view of a Project Cyclops site on the Earth. Shown in this artist's rendering are the central control and processing building and a hexagonal array of 100-meter antennas— listening for radio signals from intelligent civilizations among the stars. (Courtesy of NASA.)

Fig. C23 The highly versatile robot arm used on the Space Shuttle is called the Remote Manipulator System or RMS for short. It was designed and built in Canada. (Courtesy of NASA.)

Fig. C24 An artist's rendering of a Mars Rover Mission. Here two pair of robot rovers gather scientific data, while communicating with an automated, orbiting "mothership" that then relays mission data back to Earth. (Courtesy of NASA.)

Fig. C25 Self-Growing Lunar Factory. (Courtesy of George von Tiesenhausen; NASA/Marshall Space Flight Center.)

Fig. C26 An artist's concept of a giant SPS unit (microwave transmission concept) in geosynchronous orbit. (Courtesy of NASA/Marshall Space Flight Center.)

Fig. C27a This montage of images of the Saturnian system was created using *Voyager 1* spacecraft photographs taken during its Saturn encounter in November 1980. Shown are Dione in the forefront, Saturn rising behind, Tethys and Mimas fading in the distance to the right, Enceladus and Rhea off Saturn's rings to the left, and Titan, the distant object at the top. (Courtesy of NASA.)

Fig. C27b A Picture of Saturn's moon Rhea assembled from *Voyager 1* spacecraft imagery taken in November 1980. (Courtesy of NASA.)

Fig. C27c Saturn's cloud-enshrouded moon Titan, captured in its true colors by the *Voyager 1* spacecraft in November 1980. (Courtesy of NASA.)

Fig. C31 A torus-shaped space settlement for 10,000 people. (Courtesy of NASA.)

C31

Fig. C28 An artist's concept of a large chemical processing plant in low Earth orbit. This type of facility represents an industrial space base. (Courtesy of NASA/Johnson Space Center.)

Fig. C29 A permanent lunar base with mass driver and solar power array. (Courtesy of NASA.)

Fig. C30 Nuclear-powered geosynchronous information services platform (in transit to its operational orbit). (Courtesy of Los Alamos National Laboratory.)

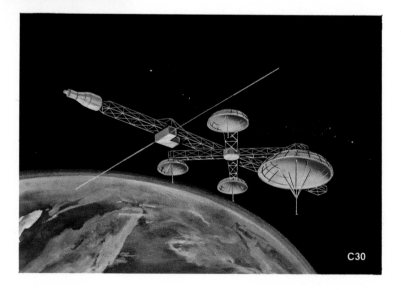

L

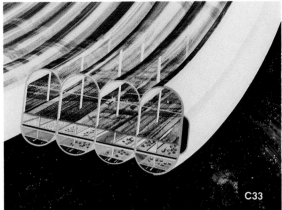

Fig. C32   The residential area of a torus–shaped space settlement. (Courtesy of NASA.)

Fig. C33   A cutaway view of one of the agricultural toruses used with the Bernal Sphere space settlement design. (Courtesy of NASA/Johnson Space Center.)

Fig. C34   Very large, cylindrical space settlement design. Each cylindrically–shaped habitat can house several hundred thousand people. (Courtesy of NASA.)

M

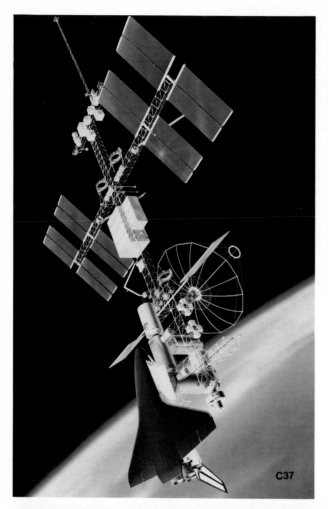

Fig. C35   A close-up view of the *Skylab* Space Station as photographed from the *Skylab 3* Command/Service Module during rendezvous and docking. (Courtesy of NASA/ Johnson Space Center.)

Fig. C36   T-Shaped space station design (unmanned space platform appears to the right of main station.) (Courtesy of NASA.)

Fig. C37   "Power Tower" space station design concept. The facility illustrated is equipped with habitats, supply modules and instruments for viewing the Sun, the Earth and the heavens. (Courtesy of NASA.)

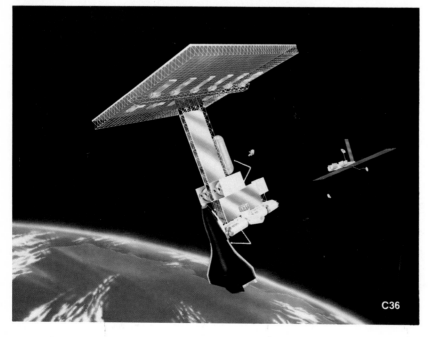

N

Fig. C38a   Postulated halo of neutrinos around the Milky Way Galaxy emitting ultraviolet light. (Courtesy of NASA.)

Fig. C38   This photograph of the Sun, taken on December 19, 1973 by the *Skylab 4* mission, shows one of the most spectacular solar flares ever recorded. The solar flare spans more than 588,000 km across the solar surface. (Courtesy of NASA.)

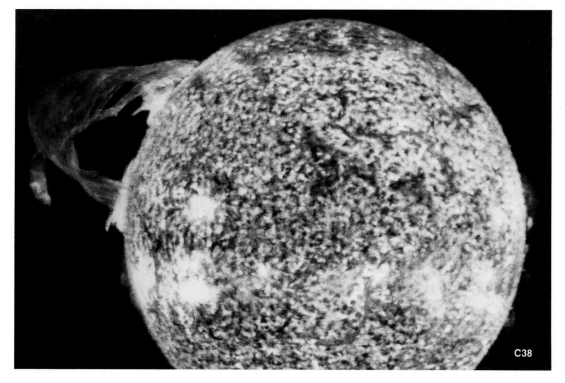

Fig. C39   Artist concept of the greenish-blue ringed world, Uranus, seventh planet out from our Sun. (Courtesy of NASA.)
Fig. C40   The planet Venus as seen from the *Pioneer Venus* spacecraft. (Courtesy of NASA.)
Fig. C41   Artist's concept of *Pioneer Venus Multiprobe* transporter bus entering the Venusian atmosphere. (Courtesy of NASA/Ames.)

muscles driving the joints are electric actuators (motors). Built in sensors act like nerves and sense joint positions and rotation rates.

Beyond the initial similarities to the human arm, the RMS features additional capabilities. To provide all the needed degrees of freedom, the RMS has six independently operating joints. There is a pitch joint (up and down) and a yaw joint (right and left) in the RMS shoulder. The mechanical elbow has a pitch joint, while the RMS wrist has pitch, yaw and roll (rotation) joints—a definite improvement over the human arm. The shoulder joint allows the RMS end effector to move to any point along the surface of a sphere. The elbow joint gives the RMS freedom to reach any point within that sphere, while the wrist joints provide any desired orientations of the end effector.

The RMS also includes two closed-circuit television cameras, one at the wrist and one at the elbow. These cameras allow an astronaut, who is operating the RMS from the Orbiter's aft flight deck, to see critical points along the arm and the target toward which the arm is moving. The television camera at the RMS wrist is also linked to a computer, producing a machine vision that enables the arm to operate in a fully automatic mode. To permit this automatic operation, the target or grapple fixture to be grasped by the RMS must have four white dots. The computer in the machine vision system will sense the physical separation of these white dots. From that it will determine the distance and direction of the RMS end effector from the target, as well as the target's rate of motion.

The RMS end effector represents a radical departure from the human arm analogy we have used so far. Instead of finger-like clamps, the end effector has a three wire snare. Each payload to be grasped by the RMS must have a special grapple fixture. This fixture is a small knobby projection that is mounted to a disc. To grasp the payload, the end effector slides over the fixture and the snare wires are pulled together in the center. Once tightly held by the wires, the disc is pulled into the effector and the payload is ready for deployment or retrieval.

The versatile RMS can be operated in several modes, ranging from fully manual to fully automatic. In the manual mode, an astronaut controls the robot arm with rotational and translational hand controllers. These devices look similar to the controllers on home video games. The astronaut, watching the arm through windows or on closed circuit television, can move the RMS end effector with ease. In the automatic mode, the payload manipulator task to be performed is programmed into the computer and, like a dutiful servant, the RMS carries it out.

Robotics and machine intelligence have also been used in the exploration of the Solar System. The development of advanced integrated circuits, microprocessors and silicon chip technology has made it possible to build systems with compact memories and processors that provide robot explorers with a form of portable machine intelligence. These space exploration robots are able to sense their environments, plan and execute actions, and perform complicated manipulations, frequently to a degree of dexterity normally done by human beings.

By means of telecommunication, human operators can activate and control instruments stationed on alien worlds or flying through the deep regions of interplanetary space. During the 1960s such procedures were developed as part of the planetary exploration program and became known as "teleoperation." Teleoperators are simply machine systems that augment and extend human sight, touch and even thinking abilities to remote places. In this context, the term "robot" is often applied to the remote portion of a teleoperator system if this part of the machine has at least some degree of independent sensing, decision-making or movement capabilities.

System autonomy is necessary in space exploration, because the distances between the planets are so great. Even communicating at the speed of light, it still takes many minutes for instructions from human operators on Earth to reach these robot explorers. This makes direct control of space exploration instruments either a very slow and tedious process, or even technically impossible.

However, a remote machine with independent decision-making can accomplish a great deal on its own without step-by-step instructions from its human supervisors. In fact, less "talking" between human and smart robot can often result in the successful accomplishment of more technical and scientific tasks by the robot—thereby enhancing the overall results of a particular exploration mission.

In such extraterrestrial applications, these robots will take on many forms, although few will look like the popular concepts of androids and robots that are found in science fiction. Their appearance will follow strictly functional lines and will change with the specific requirements of a particular mission. For example, contemporary robot space explorers may take the form of a large, automated telescope in Earth-orbit, or they may be sophisticated interplanetary flyby and orbiter spacecraft. They may be stationary landers, like the *Viking 1* and *2 Landers* placed on Mars in 1976, or they may take the form of wheeled-surface rovers that travel across the plains of an alien world.

One interesting robot that has been designed at NASA's Jet Propulsion Laboratory is a research tool to study the problems of combining visual and manipulatory systems that are needed in future space robots, including planetary rovers. This JPL rover may be regarded as the forerunner of robot explorer vehicles that will be used in planetary sample return missions and for performing chemical and mineralogical studies on alien worlds. The research rover weighs approximately 318 kilograms (700 pounds) and measures 1.5 meters long by 1.3 meters wide (59 inches by 51 inches). It moves by means of a four-legged loopwheel mobility system with jointed or knee-action suspension. This suspension includes interconnected "knee" and "thigh" joints that enable each loopwheel foot to adjust independently to varying heights and depths, while maintaining the rover's boxlike chassis in an approximately level position.

The rover travels at a speed of about one meter per minute and is capable of negotiating 60-centimeter (24 inch) obstacles or depressions, such as rocks and small craters. In addition, the vehicle can climb or descend slopes of up to 30 degrees inclination. Future versions of this type rover will be equipped with a vision system and other sensors to make it capable of "seeing" and assessing its extraterrestrial environment. In addition, the rover will use one or more manipulators and drills to study alien soil and rock samples.

During future planetary missions, the robot rover would move itself to a new position on the planet's surface every day. Each day the vehicle would observe the alien environment and relay appropriate terrain imagery and scientific data back to Earth (see fig. C24 on color page J). Based on the mission's overall technical objectives and on the topographical features encountered on Mars or on one of the major moons of the outer planets, scientists on Earth would issue general commands to the robot explorer vehicle. The vehicle or vehicles would then proceed autonomously throughout the next day performing automated tasks and scientific investigations with no or minimal interaction from humans on Earth.

A prototype robot vision system now under study at NASA's JPL consists of two television cameras that are mounted on a mast above the chassis and tied into an onboard computer system. These television cameras, which operate in stereo, serve as the robot rover's eyes and allow it to view the surrounding alien landscape for obstacles and interesting objects. A laser range finder, also mounted on the chassis, can independently measure the distance between the rover and any obstacles or objects of interest. The rover's computer brain then combines these sensory data, programs the optimum course toward a given destination and sends instructions to parts of the robot vehicle to move along this optimum path. The rover's manipulator arm uses the computer vision system to determine its relationship to any interesting objects it will handle. The rover's on-board computer analyzes data from the cameras and then generates appropriate instructions to the manipulator system. Together, the computer vision cameras, the on-board computer, and the manipulator arm provide the robot rover with an "eye/hand" system. All locomotion, vision and computer systems on the rover will be powered by a dependable radioisotope thermoelectric generator (RTG) system, capable of providing about 200 watts of electricity.

Space industrialization will also depend extensively on automated and robotic machine technology. Industrial activities in space will include the construction of very large antennas and solar arrays, and the construction of permanent habitats, such as space stations and space bases. Robotic cranes, manipulators and teleoperator systems will make these construction jobs practical. In addi-

tion, orbital manufacturing and processing stations will most likely be operated and maintained through highly automated control systems and robotic devices.

Future automated space systems now under consideration include: spacebased manufacturing modules to produce biological, metallic and fluid products for scientific and commercial applications on Earth; systems to provide on board health care to space workers, astronaut crews and even space tourists; lunar rovers to collect and stockpile materials for the construction of lunar bases; and complete lunar base systems to perform mining, processing and manufacturing operations.

THE NEED FOR MACHINE INTELLIGENCE

Tomorrow's advanced robotic space systems promise to take over much of the data processing and information sorting activities that are now performed by human mission controllers here on Earth. In the past, the amount of data made available by space missions has been considerably larger than scientists could comfortably sift through. For example, the Viking missions to Mars returned image data of the Red Planet that were transferred onto approximately 75,000 reels of magnetic tape. Smart robot spacecraft, with onboard computers capable of deciding what information gathered by a spacecraft or surface rover is worth relaying back to Earth and what information should be stored or discarded, would greatly relieve the current extraterrestrial data glut.

Robots with advanced machine intelligence, capable of making these kinds of decisions, would have a large number of pattern classification templates or "world models" stored in their computer memories. These templates would represent the characteristics of objects or features of interest in a particular mission. The robot explorers would compare the patterns or objects they see with those stored in their memories and discard any unnecessary or unusable data. As soon as something unusual appeared, the smart machine explorer would examine this object or event more closely. The robot explorer would then dutifully alert its human controllers on Earth and report the unusual findings. Through these automated selection and data filtering operations, the smart system would free human experts for more demanding and judgmental intellectual tasks.

The advanced machine intelligence (or artificial intelligence) requirements for general purpose robotic space exploration systems can be summarized mainly in terms of two fundamental tasks: (1) the smart robot must be capable of learning about new environments; and (2) it must be able to formulate hypotheses about these new environments. Hypothesis formation and learning represent the key problems in the successful development of machine intelligence. Deep interplanetary and interstellar robotic space systems will need a machine intelligence system capable of autonomously conducting intense studies of alien objects. The machine intelligence levels supporting these missions must be capable of producing scientific knowledge concerning previously unknown objects. Since the production of scientific knowledge is a high-level intelligence capability, the machine intelligence requirements for "smart" autonomous space missions are often called "advanced-intelligence machine intelligence" or simply "advanced machine intelligence."

For a really autonomous deep-space exploration system to undertake knowing and learning tasks, it must have the ability to mechanically or artificially formulate hypotheses, using all three of the logical patterns of inference: analytic, inductive and abductive. Analytic inference is needed by the robot explorer system to process raw data and to identify, describe, predict and explain extraterrestrial events and processes in terms of existing knowledge structures. Inductive inference is needed so that the robot explorer can formulate quantitative generalizations and abstract the common features of events and processes occurring on alien worlds. Such logic activities amount to the creation of new knowledge structures. Finally, abductive inference is needed by the smart robotic explorer system to formulate hypotheses about new scientific laws, theories, concepts, models, etc. The formulation of this type of hypothesis is really the key to the ability to create a full range of new knowledge structures. These new knowledge structures, in turn, are needed if we are to successfully explore and investigate alien worlds.

Although the three patterns of inference just described are distinct and independent, they can be ranked by order of difficulty or complexity. Analytic inference is at the low end of the new knowledge creation scale. An automated system that performs only this type of logic could probably successfully undertake only extraterrestrial reconnaissance missions. A machine capable of performing both analytic and inductive inference could most likely successfully perform space missions combining reconnaissance and exploration. This assumes, however, that the celestial object being visited is represented well enough by the world models with which the smart robot has been preprogrammed. However, if the target alien world cannot be well represented by such fundamental world models, then automated exploration missions will also require an ability to perform abductive inference. This logic pattern is the most difficult to perform and lies at the heart of knowledge creation. An automated space system capable of abductive reasoning could successfully undertake missions combining reconnaissance, exploration and intensive study. Figure 2 summarizes the adaptive machine intelligence required for advance robotic space exploration systems.

TITAN DEMONSTRATION MISSION

Scientists and engineers have recently examined a concept for an advanced robotic space mission that could eventually lead to a deep space exploration system. This autonomous robot explorer would incorporate advanced machine intelligence technology capable of performing NASA's three general phases of extraterrestrial object investigation: reconnaissance, exploration and intensive study.

A general purpose robotic explorer spacecraft to Saturn's interesting moon, Titan, was proposed as a possible mission to demonstrate this advanced technology. Titan was chosen as the target because it lies far enough from the Earth to prevent effective ground-based investigation or even simple teleoperator control of robot explorer vehicles. Yet it still lies close enough for effective system monitoring and even human intervention during the operation of this advanced robot system. The knowledge gained from this initial automated mission would be applied to other deep-space exploration missions within our Solar System and would also serve as the basis for the design and development of the first generation of interstellar robot probes.

Tables 1 through 3 provide some of the technical features of this conceptual Titan Demonstration Mission. As described in these tables, the mission would involve a full complement of sophisticated robotic devices. Perhaps the single most important technology driver for this type of automated mission is advanced machine intelligence, especially a sophisticated machine intelligence (MI) system capable of learning about new environments and generating scientific hypotheses using analytic, inductive and abductive reasoning. Numerous other supporting technologies are also needed to perform such automated space exploration missions in the next century. These technolo-

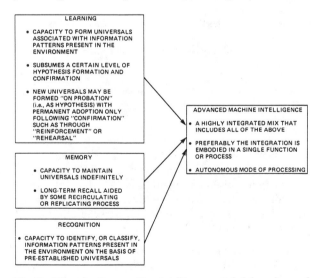

Fig. 2   The adaptive machine intelligence needed for advanced interplanetary and interstellar space exploration. (Drawing courtesy of NASA.)

Table 1   The General Features of a Proposed Titan Demonstration Mission

Status: Opportunity Mission (not in current NASA plans)
Lifetime: 10 years; includes 5 years at Titan
Launch/transfer vehicle: Shuttle/400 kW Nuclear Electric Propulsion (NEP)
Operational location: Titan, Saturn's largest satellite
Total mass: 13,000–17,000 kg
Total power: About 400 kW

SOURCE: NASA.

Table 2   Candidate Space Systems for the Proposed Titan Demonstration Mission

Spacecraft type	Typical number	Operational location	Mass, kg	Power, kW
Nuclear electric propulsion	1	Earth to Titan orbit	10,000[a]	400
Main orbiting spacecraft	1	Circular polar Titan orbit at 600 km altitude	1,200	—[b]
Lander/Rover	2	Surface	1,800	1
Subsatellites	~3	One at a Lagrange point; others on 100 km tethers from NEP	300	0.3
Atmospheric probe	~6	Through Titan atmosphere to surface	200	0.1
Powered air vehicle	1	Atmosphere	1,000	10
Emplaced science	~6	Surface	50	0.1

[a]Does not include propellant.
[b]Uses NEP power.
SOURCE: NASA.

Table 3   Possible Accomplishments of Titan Demonstration Mission

Spacecraft type	Possible accomplishments
Nuclear electric propulsion	Spiral escape from low Earth orbit; interplanetary transfer to Saturn; rendezvous with Titan; and spiral capture into 600 km circular polar orbit.
Main orbiting spacecraft	Automated mission operations during interplanetary and Titan phases: this includes interfacing with one supporting other spacecraft before deployments; deploying other spacecraft; communicating with other spacecraft and with Earth; studying Titan's atmosphere and surface using remote sensing techniques at both global characterization and intensive study levels; and selecting landing sites.
Lander/Rover	Lands at preselected site, avoids hazards; intensive study of Titan's surface; selects, collects and analyzes samples for composition, life, etc., explores several geologic regions.
Subsatellite	Lagrange point satellite monitors environment near Titan and is continuous communications relay; tethered satellite measure magnetosphere and upper atmosphere properties.
Atmospheric probe	Determines surface engineering properties and atmospheric structure at several locations/times.
Powered air vehicle	Intensive study of Titan's atmosphere; aerial surveys of surface; transport of surface samples or surface systems.
Emplaced science package	Deployed by long-range rover to form meteorological and seismological network. (Alternatives are penetrators or extended lifetime probes.)

SOURCE: NASA.

gies include: low thrust propulsion systems; general purpose surface exploration vehicles that can function on both solid and liquid surfaces; smart sensors and sensor networks that can reconfigure themselves; flexible, highly adaptive, general purpose robotic manipulators; and distributed intelligence/database systems.

Once we have developed the sophisticated robotic devices needed for the detailed investigation of the outer regions of our own Solar System, the next step becomes quite obvious. Sometime in the 21st century, humankind will build and launch its first automated explorer to a nearby star system. This interstellar probe will be capable of searching for extrasolar planets around alien suns, targeting any suitable planets for detailed investigation, and then initiating the search for extraterrestrial life. Light-years away, terrestrial scientists will patiently wait for its faint, distant radio signals by which the robot starship describes any new worlds it has encountered and sheds light on the greatest cosmic mystery of all: does life exist elsewhere in the depths of space?

ROLE OF ROBOTICS IN ADVANCED SPACE MANUFACTURING ACTIVITIES

Two of the most important products that will be manufactured in space in the next century are robots and teleoperator systems. The ultimate goals for advanced space manufacturing facilities cannot be achieved without a large expansion of the automation equipment initially provided from Earth. Eventually, space robots and teleoperators must be manufactured in space, drawing from the working experience gathered during the use of the first generation of space industrial robots. These second and third generation robotic devices must be far more versatile and fault-tolerant than the first generation devices created on Earth and shipped to extraterrestrial locations as "seed" or starter machines. The most critical technologies needed for the manufacture of second- and third-generation robots and teleoperator systems appear to be space-adaptive sensors and computer vision. Enhanced decision-making capabilities and self-preservation features must also be provided for use in space robots and teleoperators.

Once we develop the ability to make robots in space, another step in robotics becomes possible—that of the self-replicating system (SRS). A single self-replicating system or SRS unit is a machine system that contains all the elements required to maintain itself, to manufacture desired products, and even (as the name implies) to reproduce itself! In fact, an SRS unit would behave much like a biological cell.

THE THEORY AND OPERATION
OF SELF-REPLICATING SYSTEMS

The brilliant Hungarian-American mathematician, John von Neumann, was the first person to seriously consider the problem of self-replicating systems. He became interested in the study of automatic replication as part of his wide ranging interests in complicated machines. His work during the World War II Manhattan Project led him into automatic computing and he became fascinated with the idea of large, complex electronic computing machines. In fact, he invented the scheme used today in the great majority of general purpose computers—the von Neumann concept of serial processing stored-program—which is also referred to as the "von Neumann machine." Following this pioneering work, von Neumann decided to tackle the larger problem of developing a self-replicating machine. Von Neumann actually conceived of several types of self-replicating systems which he called the "kinematic machine," the "cellular machine," the "neuron-type machine," the "continuous machine" and the "probabilistic machine." Unfortunately, before his death in 1957, he was only able to develop a very informal description of the kinematic machine.

The kinematic machine is the one we hear about most often in discussing von Neumann type self-replicating systems. For this type of SRS, the brilliant mathematician had envisioned a machine residing in a "sea of spare parts." Von Neumann's kinematic machine would have a memory tape that instructed it to go through certain mechanical procedures. Using manipulator arms and its ability to move around, this SRS device would gather and assemble parts. The stored computer program would instruct the machine to reach out and pick up a certain part, and then go through an identification and evaluation

routine to determine whether the part selected was or was not the one called for by the master tape. (Note: in von Neumann's day microcomputers, floppy disks and bubble memory devices did not exist.) If the component picked up by the manipulator arm did not meet the selection criteria, it was tossed back into the "sea of parts." The process would continue until the required part was found and then an assembly operation would be performed. In this way, von Neumann's kinematic SRS would eventually make a complete replica of itself—without, however, really understanding what it was doing. When the duplicate was physically completed, the parent machine would make a copy of its own memory tape on the (initially) blank tape in its offspring. The last instruction on the parent machine's tape would be to activate the tape of its mechanical progeny. The offspring kinematic SRS could then start searching the "sea of parts" for components to build yet another generation of SRS units.

In dealing with his self-replicating machine concepts, von Neumann concluded that they should include the following characteristics and capabilities: (1) logical universality; (2) construction capability; (3) constructional universality; and (4) self reproduction. Logical universality is simply the device's ability to function as a general-purpose computer. To be able to self replicate, a machine must be capable of manipulating information, energy and materials. This is what is meant by "construction capability." The closely related term "constructional universality" is a characteristic which implies the machine's ability to manufacture any of the finite sized machines that can be built from a finite number of different parts that are available from an indefinitely large supply. The characteristic of self-reproduction means that the original machine, given a sufficient number of component parts (of which it is made) and sufficient instructions, can make additional replicas or copies of itself. One characteristic of SRS devices that von Neumann did not address, but has been addressed by subsequent investigators, is the concept of evolution. If we have a series of machines making machines, making machines etc., can successive generations of machines learn to make themselves better?

One conceptual problem associated with the kinematic SRS concept is that the device lives in a "sea of parts," that is, the SRS unit inhabits a universe that provides it precisely what it needs to duplicate itself. This, of course, brings up the issue of closure, a major problem in thinking about self-replicating machines that will be discussed shortly.

From von Neumann's work and the more recent work of other investigators, we arrive at five broad classes of SRS behavior:

(1) *Production*—The generation of useful output from useful input. In the production process, the unit machine remains unchanged. Production is a simple behavior demonstrated by all working machines, including SRS devices.

(2) *Replication*—The complete manufacture of a physi-

cal copy of the original machine unit by the machine unit itself.

(3) *Growth*—An increase in the mass of the original machine unit by its own actions, while still retaining the integrity of its original design. For example, the machine might add an additional set of storage compartments in which to keep a larger supply of parts or constituent materials.

(4) *Evolution*—An increase in the complexity of the unit machine's function or structure. This is accomplished by additions or deletions to existing sub-systems, or by changing the characteristics of these sub-systems.

(5) *Repair*—Any operation performed by a unit machine on itself that helps reconstruct, reconfigure or replace existing sub-systems, but does not change the SRS unit population, the original unit mass or its functional complexity.

In theory, replicating systems can be designed to exhibit any or all of these machine behaviors. When such machines are actually built, however, a particular SRS unit will most likely emphasize just one or several kinds of machine behavior, even if it were capable of exhibiting all of them. For example, the fully autonomous, general-purpose self-replicating lunar factory, proposed by Georg von Tiesenhausen and Wesley A. Darbo of the Marshall Space Flight Center in 1980, is an SRS design that is intended for unit replication. There are four major sub-systems that make up this proposed SRS unit. First, a materials processing sub-system gathers raw materials from its extraterrestrial environment and prepares industrial feedstock. Next, a parts production sub-system uses this feedstock to manufacture other parts or entire machines. At this point, the conceptual SRS unit has two basic outputs. Parts may flow to the universal constructor sub-system where they are used to make a new SRS unit (this is replication); or else, parts may flow to a production facility sub-system where they are made into commercially useful products. This self-replicating lunar factory has other secondary sub-systems, such as a materials depot, parts depot, product depot, power supply, and command and control center (see fig. C25 on color page K). The universal constructor (UC) manufactures complete SRS units that are exact replicas of the original SRS unit. Each replica can then make more replicas of itself until a preselected SRS unit population is achieved. The universal constructor would retain overall command and control (C&C) responsibilities for its own SRS unit as well as for its mechanical progeny—until, at least, the C&C functions themselves have been duplicated and transferred to the new units. To avoid cases of uncontrollable exponential growth of such SRS units in some planetary resource environment, the human masters of these devices may reserve the final step of C&C function transfer to themselves or so design the SRS units such that the final C&C transfer function from machine to machine can be overridden by external human commands.

EXTRATERRESTRIAL IMPACT
OF SELF-REPLICATING SYSTEMS

The issue of closure (total self-sufficiency) is one of the fundamental problems in designing self-replicating systems. In an arbitrary SRS unit there are three basic requirements necessary to achieve closure: (1) matter closure; (2) energy closure; and (3) information closure. In the case of matter closure we ask: Can the SRS unit manipulate matter in all the ways needed for complete self-construction. If not, the SRS unit has not achieved matter or material closure. Similarly, we ask whether the SRS unit can generate a sufficient amount of energy and in the proper form to power the processes needed for self-construction. Again, if the answer is no, then the SRS unit has not achieved energy closure. Finally, we must ask: Does the SRS unit successfully command and control all the processes necessary for complete self-construction? If not, information closure has not been achieved.

If the machine device is only partially self-replicating, then we say that only partial closure of the system has occurred. In this case, some essential matter, energy or information must be provided from external sources, or else the machine system would fail to reproduce itself.

Just what are the applications of self-replicating systems? The early development of self-replicating system technology for use on Earth and in space will trigger an era of superautomation that will transform most terrestrial industries and lay the foundation for efficient spacebased industries. One interesting machine proposed by physicist Theodore Taylor is called the "Santa Claus" machine. In his concept of an SRS unit, a fully automatic mining, refining and manufacturing facility gathers scoopfuls of terrestrial or extraterrestrial materials. It then processes these raw materials by means of a giant mass spectrograph that has huge super-conducting magnets. The material is converted into an ionized atomic beam and sorted into stockpiles of basic elements, atom by atom. Then, to manufacture any item, the Santa Claus machine selects the necessary materials from its stockpile, vaporizes them and injects them into a mold that changes the materials into the desired item. Instructions for manufacturing, including directions on adapting new processes and replication, are stored in a giant computer within the Santa Claus machine. If the product demands become excessive, the Santa Claus machine would simply reproduce itself.

SRS units might be used in very large space construction projects (such as lunar mining operations) to facilitate and accelerate the exploitation of extraterrestrial resources and to make possible feats of planetary engineering. For example, we could deploy a seed SRS unit on Mars as a prelude to permanent human habitation. This machine would use local Martian resources to automatically manufacture *in situ* a large number of robot explorer vehicles. This armada of vehicles would be disbursed over the surface of the Red Planet searching for the minerals and frozen volatiles needed in the establishment of a Martian civilization. In perhaps just a few years, a population of some

1,000 to 10,000 smart machines would scurry across the planet, completely exploring its entire surface and preparing the way for permanent human settlements.

Replicating systems would also make possible large-scale interplanetary mining operations. Extraterrestrial materials could be discovered, mapped and mined, using teams of surface and subsurface prospector robots that were manufactured in large quantities in an SRS factory complex. Raw materials would be mined by hundreds of machines and then sent wherever they were needed in heliocentric space. Some of the raw materials might even be refined in transit, with the waste slag being used as the reaction mass for an advanced propulsion system.

Atmospheric mining stations could be set up at many interesting and profitable locations throughout the Solar System. For example, Jupiter and Saturn could have their atmospheres mined for hydrogen, helium (including the very valuable isotope, helium-3) and hydrocarbons using "aerostats." Cloud-enshrouded Venus might be mined for carbon dioxide; Europa for water; and Titan for hydrocarbons. Large quantities of useful volatiles might be obtained by intercepting and mining comets with fleets of robot spacecraft. Similar mechanized space armadas might mine water-ice from Saturn's ring system. All of these smart robot devices would be mass produced by seed SRS units. Finally, extensive mining operations in the main asteroid belt would yield large quantities of heavy metals. Using extraterrestrial materials, these replicating machines could, in principle, manufacture huge mining or processing plants or even ground-to-orbit or interplanetary vehicles. This large-scale manipulation of the extraterrestrial environment would occur in a very short period of time, perhaps within one or two decades of the initial introduction of replicating machine technology.

From the viewpoint of our extraterrestrial civilization, perhaps the most exciting consequence of the self-replicating system is that it would provide a technological pathway for organizing potentially infinite quantities of matter. Large reservoirs of extraterrestrial matter might be gathered and organized to create an ever widening human habitat throughout the Solar System. Self-replicating space stations, space settlements and domed cities on certain alien worlds of our Solar System would permit a diversity of environmental niches never before experienced in the history of the human race.

The SRS unit would provide such a large amplification of matter manipulating capability that it is possible even now to start seriously considering terraforming or planetary engineering strategies for the Moon, Mars, Venus and certain other alien worlds. SRS technology also appears to be our key to exploration and to human habitat expansion beyond the very confines of the Solar System. Although such interstellar missions may today appear highly speculative, and indeed they certainly require technologies that exceed contemporary or even projected levels in many areas, a consideration of possible interstellar or even intergalactic applications is actually quite an exciting and useful mental exercise. It illustrates immediately the fantastic power and virtually limitless potential of the SRS concept.

It appears likely that before humans move out across the interstellar void, smart robot probes will be sent ahead as scouts. Interstellar distances are so large and search volumes so vast, that self-replicating probes represent a highly desirable, if not totally essential, approach to surveying other star systems for suitable extrasolar planets and for extraterrestrial life. One recent study on galactic exploration suggests that search patterns beyond the 100 nearest stars would most likely be optimized by the use of SRS probes. In fact, reproductive probes might permit the direct reconnaissance of the nearest one million stars in about 10,000 years and the entire Galaxy in less than one million years—starting with a total investment by the human race of just one self-replicating interstellar robot spacecraft!

Of course, the problems in keeping track of, controlling and assimilating all the data sent back to the home star system by an exponentially growing number of robot probes is simply staggering. We might avoid some of these problems by sending only very smart machines capable of greatly distilling the information gathered and transmitting only the most significant quantities of data, suitably abstracted, back to Earth. We might also set up some kind of command and control hierarchy, in which each robot probe only communicates with its parent. Thus, a chain of "ancestral repeater stations" would be used to control the flow of messages and exploration reports. Imagine the exciting chain reaction that might occur as one or two of the leading probes encountered an intelligent alien race. If the alien race proved hostile, an extraterrestrial alarm would be issued, taking light years to ripple across interstellar space, repeater station by repeater station, until humans at "Central Robot Probe Control" received notification that: "Robot 24-76-AX-JA-2 was laser-blasted by hostile aliens in the vicinity of the Rigel system." Would we retaliate and send more sophisticated, possibly predator robot probes to that area of the Galaxy—or would we elect to place warning beacons all around the area, signalling any other robot probes to swing clear of the alien hazard?

In time giant space arks, representing an advanced level of synthesis between human crew and robot "crew," will depart from the Solar System and plunge into the interstellar medium. Upon reaching another star system that contained suitable planetary resources, the space ark itself could undergo replication. The human passengers (perhaps several generations of humans beyond the initial crew that departed the Solar System) would then redistribute themselves between the parent space ark and offspring space arks. Consequently, the original space ark would serve as an extraterrestrial refuge for humanity and any terrestrial lifeforms included on the giant, mobile habitat. This dispersal of humanity to a variety of ecological niches among the stars would ensure that not even disaster on a stellar scale, such as our Sun going supernova, could threaten the complete destruction of man and all his accomplishments. These self-replicating space arks would allow their human crews to literally green the Galaxy with life as we know it.

Finally, self-replicating systems could also be used by advanced civilizations to perform gigantic feats of astro-engineering. The harnessing of the total radiant energy output of a star, through the robot-assisted construction of a Dyson sphere, is an example of such large-scale astro-engineering projects that might be undertaken.

CONTROL OF SELF-REPLICATING SYSTEMS

What happens if a self-replicating system gets out of control? Before we seed the Solar System or the Galaxy with even a single SRS unit, we should also know how to pull its plug, if things get out of control. Some people have already raised a very obvious and legitimate concern about SRS technology: Do these smart machines represent a long-range threat to human life? In particular, will machines with advanced levels of artificial intelligence become our adversaries, whether they can replicate or not? Similarly, even in the absence of very advanced levels of machine intelligence, the self-replicating system might represent a threat just through uncontrollable growth. These questions can no longer remain only in the realm of science fiction. We must start examining the implications of developing advanced machine intelligences and self-replicating systems *before* we bring them into existence—or perhaps someday find ourselves in some mortal conflict over scarce resources with our own intelligent machine creations.

Of course, we need smart machines to help us improve life on Earth and to develop our extraterrestrial civilization. So we should proceed with their development, but include safeguards to avoid undesirable future situations in which the machines turn on their human masters and eventually enslave or exterminate us. Asimov's "Three Laws of Robotics" (discussed earlier in this entry) appears to represent a good starting point in developing safe smart machines.

However, any machine sophisticated enough to survive and reproduce in largely unstructured environments, would probably also be capable of performing a certain amount of automatic or self reprogramming. This type of SRS unit might eventually be able to program itself around any rules of behavior that were stored in its memory banks by its human creators. In learning about its environment, the smart SRS unit might decide to modify its behavior patterns to better suit its own needs. Let's use Asimov's First Law of Robotics ("A robot may not injure a human being, or through inaction, allow a human being to come to harm") as a reference. Well, if our very smart SRS really "enjoys" being a machine and making (and perhaps improving) other machines, then when faced with a situation in which it must save a human master's life at the cost of its own, the smart machine may decide to shut down instead of performing the life-saving task it was preprogrammed to do. Thus, while it didn't harm the

endangered human being, it also didn't help the person out of danger either. Science fiction contains many stories about robots, androids and even computers turning on their human builders. (Perhaps there is a legitimate warning being expressed.)

An SRS population in space might be controlled by one or all of the following techniques. First, the human builders could implant machine-genetic instructions that contained a hidden or secret cutoff command. This cutoff command would be activated after the SRS units had undergone a predetermined number of replications. For example, after each machine replica is made, one generation command could be deleted—until, at last, the entire replication process is terminated upon construction of the last, predetermined replica. Second, a special, predetermined signal from Earth on some emergency frequency channel might be used to cut the power of the main bus for individual, selected groups or all SRS units at any time.

For low-mass SRS units (perhaps the 100 to 10,000 kilogram class) population control might prove more difficult because of the shorter replication times in comparison to larger SRS factory units. To keep these mechanical "critters" in line, we might use a predator robot. These predator robots would be programmed to attack and destroy only the SRS-type unit, whose population needed control. Of course, we could also develop a universal destructor (UD). This machine would be capable of taking apart any other machine encountered and it would stockpile the victim machine's parts and store any information found in the victim's memory banks. Predator species are used today in wildlife refuges on Earth to keep animal populations in balance. Similarly, perhaps, we could use a linear supply of non-replicating machine predators to control an exponentiating population of SRS "prey" units. We might also design the initial SRS units to be sensitive to population density. They could then become infertile, stop their operations or even (like lemmings on Earth) report to a central facility for disassembly, when they sensed overcrowding or overpopulation. Unfortunately, if these smart SRS units reflect their human creators too exactly, even without preprogramming overcrowding might force such machines to compete among themselves for dwindling supplies of extraterrestrial resources—dueling, mechanical cannibalism, and even robotic warfare might result! Hopefully, however, we will create these machines to reflect the best of the human mind and spirit, and with their help sweep through the Galaxy spreading life, intelligence and organization in a golden age of interstellar development.

In the very long term, there appear to be two general pathways for mankind: either we are a very important biological stage in the overall evolutionary scheme of the Universe; or else we are an evolutionary dead end. If we limit ourselves to just the fragile biosphere of "Spaceship Earth," a natural disaster or our own technological fool-hardiness will almost certainly terminate our existence, perhaps in just centuries or maybe in a few millennia.

Even excluding these unpleasant consequences, without an extraterrestrial frontier, our planetary society will simply stagnate while other civilizations flourish and populate the Galaxy.

Replicating system technology provides use with very interesting options for continued evolution beyond the boundaries of the planet Earth. We might decide to create autonomous, interstellar self-replicating probes (which in some sense are our "offspring") and send these across the interstellar void on missions of exploration. Or, we could develop a closely knit (symbiotic) human-machine system, a highly automated interstellar ark, that is capable of crossing interstellar regions and then replicating itself when it encounters star systems with suitable planets.

According to some scientists, any intelligent civilization that wants to explore a portion of the Galaxy more than 100 light years from their parent star system, would probably find it more efficient to use self-replicating robot probes. This galactic exploration strategy would produce the largest amount of data about other star systems for a given period of exploration. For example, it has been estimated that the entire Galaxy could be explored in about one million years, assuming the replicating interstellar probes could achieve speeds of one-tenth the speed of light. If many advanced alien civilizations follow this approach, then the most probable interstellar probe we are likely to encounter would be of the self-replicating type.

One very large advantage of using interstellar robot probes versus interstellar (radio) beacons in the search for extraterrestrial intelligence (SETI) is the fact that these probes could also serve as cosmic safety deposit boxes, carrying the cultural treasures of civilizations through the Galaxy long after the parent civilization has perished. The gold-anodized records we included on the *Voyager* spacecraft and the plaques placed on the *Pioneer 10* and *11* spacecraft are our first attempts at achieving a small degree of cultural immortality in the cosmos. Starfaring self-replicating machines should be able to keep themselves running for a long time. It has been estimated by certain scientists that there may exist at present only 10 percent of all the alien civilizations that have ever lived in the Galaxy (the remaining 90 percent having perished). If this is true, then statistically at least, nine out of every 10 robot star probes within the Galaxy are actually the only surviving emissaries from long-dead civilizations.

If we ever encounter such alien probes and are able to decipher their contents, we may eventually learn about some incredibly ancient alien societies. Those now extinct societies may, in turn, lead us to many others. In a sense, by encountering and successfully interrogating an alien robot star probe, we may actually be treated to a delightful edition of the proverbial *Encyclopedia Galactica*—a literal compendium of the technical, cultural and social heritage of thousands of extraterrestrial civilizations within the Galaxy (most of which are now extinct).

This raises a number of fundamental ethical questions about the use of interstellar self-replicating probes. Is it morally right, or even equitable, for a self-replicating machine to enter an alien star system and convert a portion of that star system's mass and energy to satisfy its own purposes? Does an intelligent race legally "own" its parent star, home planet and any materials residing on other celestial objects within its star system? Does it make a difference whether the star system is inhabited by intelligent beings, or is there some lower threshold of galactic intelligence below which starfaring races may ethically (on their own value scale) invade such a star system and appropriate the resources needed to continue on their mission? If an alien probe enters a star system, by what criteria does it judge any indigenous lifeforms to be "intelligent" so as not to severely disturb existing ecospheres in their interstellar scavenger hunt for resources? And, of course, the really important question: "Now that humanity has developed space technology, are we above (or below) the cosmic appropriations threshold?"

In summary, the self-replicating system is a very powerful tool with ramifications on a cosmic scale. With properly developed and controlled SRS technologies, mankind could set in motion a chain reaction that spreads life, organization and consciousness across the Galaxy in an expansion wave limited in speed only by the speed of light itself. With these smart machines as our close partners in interstellar exploration, we could literally green the Galaxy in about one million years!

Roentgen Satellite (ROSAT)   A cooperative program between the United States (NASA) and the Federal Republic of Germany. The main objective of the Roentgen Satellite will be to conduct a complete sky survey for electromagnetic radiations in the 0.041 to 2.0 kiloelectron volt (keV) range. The spacecraft's scientific payload includes a large X-ray Telescope (XRT) and an Extreme Ultraviolet (EUV) Wide Field Camera. The X-ray Telescope has been designed to study x-ray sources with high resolution in the 0.1 to 2.0 kiloelectron volt (keV) regime; while the Wide Field Camera will examine the currently uncharted extreme ultraviolet (EUV) regions of the sky.

This spacecraft will have a mass of approximately 2,300 kilograms and will be placed into its 475-kilometer altitude, 57 degree inclination orbit around the Earth by the Space Shuttle.

See also: **astrophysics**

rogue star   A wandering star that passes close to a solar system, disrupting the celestial objects in that system and triggering cosmic catastrophes on life-bearing planets. Close passage of a rogue star could result in the stimulation of massive comet showers, giant tidal surges or the disruption of minor planet (asteroid) orbits. The impact of comets or asteroids on a life-bearing planet could, in turn, trigger the mass extinction of many species, since the planetary biosphere would be violently disturbed in a very short period of time. Other names for a rogue star include "death star" and "interstellar vagrant."

See also: **Nemesis**

S

Satellite Power System (SPS)   A very large space structure constructed in Earth orbit that takes advantage of the nearly continuous availability of sunlight to provide useful energy to a terrestrial power grid (see fig. C26 on color page J). The original SPS concept, called the "Solar Power Satellite," was first presented in 1968 by Dr. Peter Glaser. The basic concept behind the SPS is quite simple. Each SPS unit is placed in geosynchronous orbit above the Earth's equator, where it experiences sunlight over 99 percent of the time. These large orbiting space structures then gather the incoming sunlight for use on Earth in one of three general ways: microwave transmission, laser transmission or mirror transmission.

In the fundamental microwave transmission SPS concept, solar radiation would be collected by the orbiting SPS and converted into radiofrequency (RF) or microwave energy. This microwave energy would then be precisely beamed to a receiving antenna on Earth, where it is converted into electricity.

In the laser transmission SPS concept, solar radiation is converted into infrared laser radiation, which is then beamed down to special receiving facilities on the Earth for the production of electricity.

Finally, in the mirror transmission SPS concept, very large (kilometer dimensions) orbiting mirrors would be used to reflect raw sunlight directly to terrestrial energy conversion facilities 24 hours a day.

In the microwave transmission SPS concept described above, incoming sunlight is converted into electrical power on the giant space platform by either photovoltaic (solar cell) or heat engine (thermodynamic cycle-turbogenerator) techniques. This electric power is then converted at high efficiency into microwave energy. The microwave energy, in turn, is focused into a beam and aimed precisely at special ground stations. The ground station receiving antenna (also called a rectenna) reconverts the microwave energy into electricity for distribution in a terrestrial power grid.

Because of its potential to relieve long-term national and global energy shortages, the Satellite Power System concept has been studied extensively. Some of these recent studies have suggested that the SPS units be constructed using extraterrestrial materials from the Moon and Earth-approaching asteroids with manufacturing and construction activities being accomplished by space workers who would reside in large, permanent space settlements in cislunar space. Other SPS studies examined the development and construction of SPS units using only terrestrial materials, which were placed in low Earth orbit (LEO) by a fleet of special heavy-lift launch vehicles (HLLV). In this scenario, construction work could be accomplished in LEO, perhaps at the site of a permanent space station or space construction base. Assembled SPS sections would then be ferried to geosynchronous Earth orbit (GEO) by a fleet of orbital transfer vehicles (OTVs). At GEO, a crew of space workers would complete final assembly and prepare the SPS unit for operation.

It is perhaps too early to fully validate the SPS concept or to totally dismiss it. What can be stated at this time, however, is that the controlled beaming to Earth of solar energy (either as raw concentrated sunlight or as converted microwaves or laser radiation) appears to represent a major space industrialization pathway. This particular commercial avenue could be a very powerful stimulus toward the creation of our extraterrestrial civilization. However, the SPS concept also involves potential impacts on the terrestrial environment. Some of these environmental impacts are comparable in type and magnitude to those arising from other large-scale terrestrial energy technologies; while other impacts are unique to the SPS concept. Some of these SPS-unique environmental and health impacts are: potential adverse effects on the Earth's upper atmosphere from launch vehicle effluents and from energy beaming (that is, microwave heating of the ionosphere); potential hazards to terrestrial life forms from nonionizing radiation (microwave or infrared laser); electromagnetic interference with other spacecraft, terrestrial communications and astronomy; and the potential hazards to space workers, especially exposure to ionizing radiation doses well beyond currently accepted industrial criteria. These issues will have to be favorably resolved, if the SPS concept is to emerge as a major pathway in humanity's creative use of the resources of outer space.

In 1980 the U.S. Department of Energy (DOE) and NASA defined an SPS reference system to serve as the basis for conducting initial environmental, societal and comparative assessments; alternative concept trade-off studies; and supporting critical technology investigations. This SPS reference system is, of course, not an optimum or necessarily the preferred system design. It does, however, represent one potentially plausible approach for achieving SPS concept goals.

This reference SPS system configuration and its main technical characteristics are summarized in Table 1. The proposed configuration would provide 5 gigawatts (GW) of electric power at the terrestrial grid interface. In the reference scenario, 60 such SPS units would be placed in geostationary orbit and thus provide some 300 gigawatts of electric power for use on Earth. It has been optimistically estimated that only about six months would be required to construct each SPS unit.

Table 1 SPS Reference System Characteristics

System Characteristics	
General capability (utility interface)	Number of units: 60
300 GW - total	Design life: 30 years
5 GW - single unit	Deployment rate: 2 units/year
Satellite	
Overall dimensions: 10 × 5 × 0.5 km	Satellite mass: 35–50 × 10⁶ kg
Structural material: graphite composite	Geostationary orbit: 35,800 km
Energy Conversion System	
Photovoltaic solar cells: silicon or gallium aluminum arsenide	
Power Transmission and Reception	
D.C.–R.F. conversion: klystron	Rectenna construction time: ≃2 years
Transmission antenna diameter: 1 km	Rectenna peak power density: 23 mW/cm²
Frequency: 2.45 GHz	Power density at rectenna edge: 1 mW/cm²
Rectenna dimensions (at 35° latitude)	Power density at exclusion edge: 0.1 mW/cm²
Active area: 10 × 13 km	Active, retrodirective array control system with pilot beam
Including exclusion area: 12 × 15.8 km	reference
Space Transportation System	
Earth-to-LEO - Cargo: vertical takeoff, winged two-stage (425 metric ton payload)	
Personnel: modified Shuttle	
LEO-to-GEO - Cargo: electric orbital transfer vehicle	
Personnel: two-stage liquid oxygen/liquid hydrogen	
Space Construction	
Construction staging base - LEO: 480 km	Construction crew: 600
Final construction - GEO: 35,800 km	System maintenance crew: 240
Satellite construction time: 6 months	

SOURCE: NASA/Department of Energy.

The SPS reference system would have a solar array with dimensions of 10 by 5 by 0.5 kilometers, or a rectangular surface area of about 50 square kilometers. This is a truly large space structure! The array mass would be between 35 and 50 million kilograms, depending on the materials used for photovoltaics and structures.

Of the energy conversion techniques originally studied, two photovoltaic options were selected for the reference SPS design. In one option, silicon (Si) solar cells were used, having a basic cell efficiency of 17.3 percent. In the other option, gallium-aluminum-arsenide (GaAlAs) solar cells were used. The GaAlAs cells had a basic cell efficiency of 20 percent and the array used concentrators to focus sunlight on them.

Although silicon solar cell technology is currently more advanced than the gallium aluminum arsenide technology, the GaAlAs cell option has the potential for providing both a lighter mass system and solar cells capable of self-annealing some of the space radiation damage they will accumulate.

A space construction crew of about 600 persons would be needed to assemble the reference SPS unit on orbit. In one construction scenario, heavy lift launch vehicles are used to transport building materials from the Earth's surface to LEO. Each heavy lift launch vehicle would have a payload capability of about 400 to 500 metric tons. About 200 to 400 HLLV flights would be needed to build two SPS units in space each year.

A modified version of the Space Shuttle would also be

required to transport workers between the surface of the Earth and low Earth orbit (LEO). This personnel launch vehicle (PLV) would accommodate 75 persons per trip. Approximately 30–40 trips per year would be made between Earth and LEO to support the construction of two SPS units.

Cargo orbit transfer vehicles (COTVs) and personnel orbit transfer vehicles (POTVs) would be used to transport materials and space workers between LEO and GEO. The payload capacity of COTVs would typically be in the 4,000-metric-ton range, while the POTVs would have a 400-metric-ton payload capacity, including human cargo. Again, to construct two SPS units, it is estimated that approximately 20 to 30 COTV flights between LEO and GEO would be required each year, and some 10 to 20 POTV flights.

The solar energy collected and converted to electricity by each reference SPS unit would be converted to microwave energy and then beamed to Earth at a transmission frequency of 2.45 gigahertz (GHz). The maximum power density at the ground-based rectenna (receiving antenna) would be 23 milliwatts per square centimeter. This microwave beam intensity would drop off to one milliwatt per square centimeter at the edge of the rectenna and 0.1 milliwatt per square centimeter at the edge of the ground site exclusion area. These receiving antenna sites would be shaped like an ellipse with a major axis of approximately 13 kilometers and a minor axis of 10 kilometers. An additional distance of 0.7 kilometers beyond the antenna edge is included in each direction to accommodate support buildings and electrical conversion equipment. The reference rectenna design involves a series of steel mesh (75–80 percent optical transparency) panels perpendicular to the incident microwave beam. It has been suggested that the land under these microwave receiving antennas could be used for agriculture.

Saturn Saturn is the sixth planet from the Sun and to many the most beautiful celestial object in our Solar System. To the naked eye the planet is yellowish in color. Because of its great distance from the Earth, Saturn appears the least bright and moves the most slowly through the Zodiac. This planet is named after the elder god and powerful titan of Roman mythology (Cronus in Greek mythology), who ruled supreme until he was dethroned by his son Jupiter (Zeus to the early Greeks) (see fig 1).

Composed mainly of hydrogen and helium, Saturn is so light that it would float on water if there were some cosmic ocean large enough to hold it. The planet takes about 29.5 Earth years to complete a single orbit around the Sun. But a Saturnian day is only approximately 10 hours and 39 minutes long.

The first telescopic observations of the planet were made by Galileo Galilei in 1610. The existence of its magnificent ring system was not known until the Dutch astronomer Christian Huygens, using a better resolution telescope, properly identified it in 1655. Actually, Galileo had seen the Saturnian rings but mistook them for large moons on either side of the planet. Huygens is also credited with the discovery of Saturn's largest moon, Titan, which also occurred in 1655.

Astronomers had very little information about Saturn, its rings and constellation of moons, until the *Pioneer 11* spacecraft (September 1, 1979), *Voyager 1* spacecraft (November 12, 1980) and *Voyager 2* spacecraft (August 26, 1981) encountered the planet. These robot spacecraft encounters revolutionized our understanding of Saturn and have provided the bulk of our current information about this interesting giant planet, its beautiful ring system and its large complement of moons (20 identified at present) (see fig. C27 on color page J).

Saturn is a giant planet, second in size only to mighty Jupiter. Like Jupiter, it has a stellar-type composition, rapid rotation, strong magnetic field and intrinsic internal heat source. Saturn has a diameter of approximately 120,660 kilometers at its equator, but 10 percent less at the poles because of its rapid rotation. Saturn has a mass of 5.68×10^{26} kilograms, which is about 95 times the mass of the Earth; yet its average density is only 0.71 g/cm³—the lowest of any planet—indicating that much of Saturn is in a gaseous state. Table 1 lists contemporary data for the planet.

Planetary scientists believe that Saturn is composed of "star stuff," the same mixture of elements that formed the Sun and the initial solar nebula and from which the

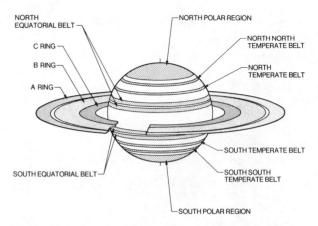

Fig. 1 Major features of Saturn. (Courtesy of NASA.)

planetary system formed some 4.6 billion years ago. Hydrogen is its most abundant element, thought to make up about 80 percent of Saturn's mass; the remaining mass is made up mostly of helium. All the other elements together constitute only a little more than two percent of Saturn's composition, with the most abundant of these being oxygen (1.0 percent), carbon (0.4 percent), iron (0.2 percent), neon (0.2 percent), nitrogen (0.1 percent) and

Table 1   Physical and Dynamic Properties for the Planet Saturn

Radius (equatorial)	60,330 km at top of clouds
Mass	5.684×10^{26} kg
Density (mean)	0.71 g/cm³
Acceleration due to gravity (at "surface")	12.9 m/sec²
Escape velocity	39.4 km/sec
Magnetic field intensity (at equator)	0.2 gauss
Albedo	0.45
Internal heat flux	2.4 ± 0.8 watts/m²
Ratio of total planetary emission to absorbed sunlight	2.2 ± 0.7
Temperature (effective)	94.4 ± 3 kelvins (K)
Number of natural satellites	20+
Semimajor axis	1427×10^6 km (9.539 AU) [79.33 light-min]
perihelion distance	1350×10^6 km (9.015 AU) [74.98 light-min]
aphelion distance	1510×10^6 km (10.066 AU) [83.72 light-min]
Eccentricity	0.056
Inclination of orbit	2.49 degrees
Mean orbital velocity (linear)	9.7 km/sec
Sidereal year	29.46 years
Sidereal day	10 hr 39.4 min
Saturn-to-Earth distance	
Apogee	1571×10^6 km (10.50 AU) [87.33 light-min]
Perigee	1264×10^6 km (8.448 AU) [70.26 light-min]

SOURCE: NASA.

silicon (0.1 percent). Some of these elements combine in the Saturnian atmosphere to form simple compounds like water, methane and ammonia.

Theoretical models of the deep interior of Saturn suggest that most of the silicon, iron and other heavy elements should be concentrated in a core, together with much of the water, methane and ammonia. These materials exist at extremely high temperatures and pressures and are probably in the form of a dense liquid. Outside the core and extending as much as halfway to the surface region is a zone of liquid metallic hydrogen, capped by a deep hydrogen atmosphere, which is gaseous in the upper extremities and gradually becomes a liquid as the pressure increases (see fig. 2). Saturn's magnetic field is thought to originate in the metallic hydrogen region.

One of the most interesting properties of Saturn's interior is the heat that is being released there that then slowly diffuses to the planet's surface. Measurements made by the Pioneer and Voyager spacecraft have confirmed that Sat-

urn was significantly warmer than would be expected from solar heating alone. This internal heat source appears to have a value of about 10^{17} watts. Scientists first thought that all of this radiated energy was primordial in origin, as occurs on Jupiter. But the Saturnian heat source is greater than that on Jupiter, even though the planet is smaller. One popular current theory suggests that the heat is caused by a separation of helium from hydrogen that occurs in the planet's interior. If some of the helium can form drops and "rain down" to deeper levels, this process would release enough thermal energy to account for the observed internal heat source strength.

In addition to hydrogen and helium, the visible portion of Saturn's atmosphere contains ammonia (NH_3) and methane (CH_4), and trace quantities of phosphene (PH_3), ethane (C_2H_6), acetylene (C_2H_2), methylacetylene (C_3H_4) and propane (C_3H_8). Although we cannot see below the ammonia clouds, we can make some intelligent guesses about the conditions in the middle and lower parts of Saturn's atmosphere. The heat escaping from the planet's interior forces the gas to circulate, with bubbles of warm hydrogen rising and colder gas sinking. This convective mixing, in turn, creates a nearly constant increase of temperature in proportion to depth at a rate of about 0.85 degrees Kelvin per kilometer. Saturn, of course, has no solid surface—just gas that keeps increasing in density and temperature down to a depth at which the physical conditions change the hydrogen from a gas to a liquid. None of the anticipated major clouds on Saturn, even at a considerable depth, are colored. The subdued colors that we see appear to represent the effects of trace chemical components of currently unknown composition.

SATURN'S MAGNIFICENT RING SYSTEM

Although Jupiter and Uranus both have ring systems, Saturn's magnificent ring system with its billions and billions of icy particles whirling around the planet in orderly fashion is uniquely beautiful—a natural wonder of the Solar System. These rings consist mainly of water-ice particles, ranging in size from tiny dustlike crystals to giant boulders. Figure 3 presents the results of a computerized model of the number and size of ice particles populating a typical 3-meter by 3-meter (room size) section of Saturn's outermost main ring, the A-ring. This computerized population model is based on radio signal measurements taken during the *Voyager 1* encounter. The particle sizes shown range from marbles (two centimeters in diameter) to beachball (70 centimeters) dimensions (see fig 4).

The main rings stretch from about 7,000 kilometers above Saturn's atmosphere out to the F Ring—a total span of 74,000 kilometers (see table 2). Within this vast region, the ring particles are generally organized into ringlets, each typically less than 100 kilometers wide. A few wide gaps, essentially empty of icy particles, provide a convenient separation into major ring areas. For example, between the B and A rings lies the Cassini Division, a complex region about 5,000 kilometers across, that con-

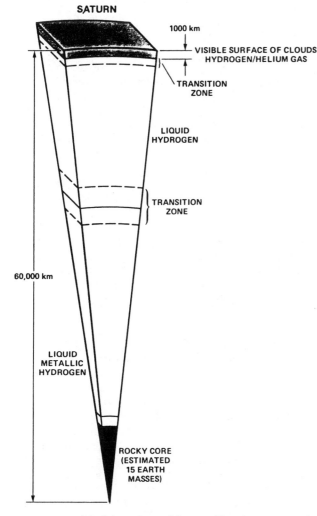

Fig. 2   A model of the interior of Saturn. (Drawing courtesy of NASA.)

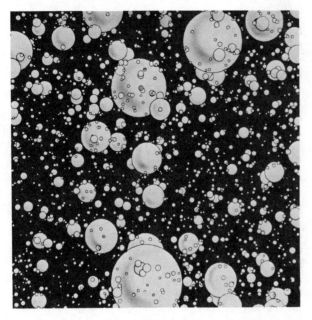

Fig. 3   A computer–generated model of Saturn's A Ring particle population (based on *Voyager 1* radio signal attenuation and scattering data). (Courtesy of NASA.)

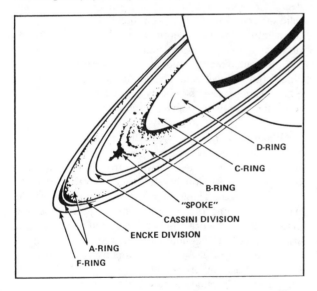

Fig. 4   Nomenclature for Saturn's complex ring structure. (Courtesy of NASA.)

tains several major gaps and hundreds of individual ring features. There are also several gaps in the A Ring, the most prominent of which is called the Keeler Gap (or the Encke Gap).

As a result of the Voyager missions, scientists now know that there are particles of many different sizes in Saturn's ring system, ranging from submicrometer specks to boulders 10 meters and more across. Different parts of the rings are dominated by different particle sizes. For example, the F Ring is made up mostly of small particles, while

Table 2   The Rings of Saturn

Feature	Distance from Center of Saturn (km)	(R_s)	Period (hr)
Cloud tops	60,330	1.00	10.66
D Ring inner edge	67,000	1.11	4.91
C Ring inner edge	73,200	1.21	5.61
B Ring inner edge	92,200	1.53	7.93
B Ring outer edge	117,500	1.95	11.41
Cassini Division (middle)	119,000	1.98	11.75
A Ring inner edge	121,000	2.01	11.93
Keeler (Encke) Gap	133,500	2.21	13.82
A Ring outer edge	136,200	2.26	14.24
F Ring	140,600	2.33	14.94
G Ring	170,000	2.8	19.9
E Ring (middle)	230,000	3.8	31.3

SOURCE: NASA.

in the Cassini division many particles are tens of meters in diameter and submicrometer size particles are almost entirely absent.

THE SATELLITES OF SATURN

Before the first spacecraft encounter, Saturn was believed to have 11 satellites. Saturn is now known to have at least 20 satellites. Three of these were discovered by *Voyager 1*; three from an analysis of *Voyager 2* imagery; and three others by recent ground-based observations (see tables 3 and 4). The moons of Saturn form a diverse and remarkable constellation of celestial objects. The largest satellite, Titan, is in a class by itself. The six other major satellites have much in common, all being of intermediate size (400 to 1,500 kilometers in diameter) and consisting mainly of water-ice. The 13 smaller moons appear to include both captured asteroids and fragments from inter-satellite collisions. Moving outward from the planet, these six icy moons are: Mimas, Enceladus, Tethys, Dione, Rhea and Iapetus. All are in orbits that are nearly circular and located within the equatorial plane of Saturn. The inner five of these icy moons occupy adjacent orbits, while Iapetus is much farther out, beyond Titan and Hyperion, in a highly inclined orbit about 3.5 million kilometers from Saturn. In terms of size, these six satellites can be conveniently divided into three groups of two each: Mimas and Enceladus (400 to 500 kilometers in diameter); Tethys and Dione (about 1,000 kilometers in diameter); and Rhea and Iapetus (about 1,500 kilometers diameter).

Titan is the largest and most interesting of Saturn's satellites. It is the second largest moon in the Solar System and the only one known to have a dense atmosphere. The atmospheric chemistry presently taking place on Titan may be similar to those processes that occurred in the Earth's atmosphere several billion years ago. For this reason, Titan is one of the most interesting objects in the Solar System to exobiologists and planetary scientists.

Larger in size than the planet Mercury, Titan has a density that appears to be about twice that of water-ice. Scientists believe therefore that it may be composed of

Table 3   The Major Satellites of Saturn

Name	Distance from Saturn (R_S)[a]	Orbital Period (hr)	Diameter (km)	Mass (10^{23} g)	Density (g/cm³)	Albedo
1980S28	2.28	14.4	30	—	—	0.4
1980S27	2.31	14.7	100	—	—	0.6
1980S26	2.35	15.1	90	—	—	0.6
1980S3	2.51	16.7	120	—	—	0.5
1980S1	2.51	16.7	200	—	—	0.5
Mimas	3.08	22.6	392	0.45	1.4	0.7
Enceladus	3.95	32.9	500	0.84	1.2	1.0
Tethys	4.88	45.3	1060	7.6	1.2	0.8
1980S13	4.88	45.3	30	—	—	0.6
1980S25	4.88	45.3	25	—	—	0.8
Dione	6.26	65.7	1120	10.5	1.4	0.6
1980S6	6.26	65.7	35	—	—	0.5
Rhea	8.74	108	1530	25	1.3	0.6
Titan	20.3	383	5150	1346	1.9	0.2
Hyperion	24.6	511	300	—	—	0.3
Iapetus	59.0	1904	1460	19	1.2	0.5, 0.05
Phoebe	215	13211	220	—	—	0.06

[a]where R_S = 60,330 kilometers

SOURCE: NASA.

Table 4   The Smaller Saturnian Satellites

Name	Informal Name	Distance From Saturn (R_S)*	Orbital Period (hr)	Dimensions (km)
1980S28	A Ring Shepherd	2.276	14.45	20 × 40 × ?
1980S27	F Ring Shepherd	2.310	14.71	140 × 100 × 80
1980S26	F Ring Shepherd	2.349	15.09	110 × 90 × 70
1980S3	Co-Orbital	2.510	16.66	140 × 120 × 100
1980S1	Co-Orbital	2.511	16.67	220 × 200 × 160
1980S13	Tethys Trojan (leading)	4.884	45.31	34 × 28 × 26
1980S25	Tethys Trojan (trailing)	4.884	45.31	34 × 22 × 22
1981S10	—	(orbits between Tethys and Dione)	—	~15
1980S6	Dione Trojan (leading)	6.256	65.69	36 × 32 × 30
1981S7	Dione Trojan (trailing)	6.256	(65)	~20
1981S9	—	(orbits between Dione and Rhea)	—	~15
Hyperion	—	24.55	511	410 × 260 × 220
Phoebe	—	215	13211	220 × 220 × 220

*where R_S = 60,330 kilometers

SOURCE: NASA.

nearly equal amounts of rock and ice. Titan's surface is hidden from the normal view of spacecraft cameras by a dense, optically thick photochemical haze whose main layer is about 300 kilometers above the moon's surface. Several distinct, detached layers of haze have also been observed above the optically opaque haze layer. Titan's southern hemisphere appears slightly brighter than the northern one, possibly because of seasonal effects.

The atmospheric pressure near Titan's surface is about 1.6 bars, some 60 percent greater than on the surface of the Earth. The Titanian atmosphere is mostly nitrogen, which also happens to be the major constituent of the Earth's atmosphere. The existence of carbon-nitrogen compounds on Titan is possible because of the great abundance of both nitrogen and hydrocarbons. Titan is unique in providing in quantity all of the building blocks for complex organic compounds. Scientists now believe that large, complex organic molecules are forming continuously in Titan's atmosphere. In the process, hydrogen is released and escapes rapidly to space. As a result, the chemical processes involving hydrogen release are not reversible, resulting in the accumulation of hydrocarbon products. Eventually, these heavy organic molecules drift down to the moon's surface, while atmospheric methane is renewed by slow diffusion upward from the surface and the lower atmosphere. Thick clouds of frozen methane probably exist in Titan's troposphere, and may even extend down to its surface.

What does the surface of Titan look like? There must be large quantities of methane on the surface, enough perhaps to form methane rivers or even a methane sea. The temperature on the surface is about 91 degrees Kelvin, which is close enough to the temperature at which methane can exist as a liquid or solid. Some people have therefore suggested that Titan is like the Earth, with methane playing the role that water plays on Earth. This analogy, if correct, leads to visions of oceans of methane near Titan's equator and frozen methane ice caps in the moon's polar regions. Titan's surface also experiences a constant rain of organic compounds from the upper atmosphere, perhaps creating up to a 100-meter-thick layer of tar-like materials.

Most of the sunlight striking Titan is absorbed in the high layers of photochemical haze and smog that occur about 200 kilometers above the satellite's surface. Only a small percentage of sunlight reaches the methane clouds in the troposphere and even less gets to the surface, which is probably quite dark and gloomy.

In any event, after the Voyager missions, we now know that Titan is a fascinating celestial object—in many ways more like the Earth than any other object in our Solar System. Its unique atmospheric conditions cause the production of large quantities of organic materials and their accumulation on the surface. Titan's methane, through photochemistry, is converted to ethane, acetylene, ethylene and (when combined with nitrogen) hydrogen cyanide. The last is an especially important molecule, since it is a building block of amino acids. However, Titan's low temperature most likely inhibits the development of more complex organic compounds.

To the exobiologist, Titan is perhaps the most interesting alien world in our Solar System. It represents a celestial laboratory in which the processes of prebiotic organic chemistry—that ultimately gave rise to life on Earth some four billion years ago—may be studied today!

Mimas, Enceladus, Tethys, Dione and Rhea are approximately spherical in shape and appear to be composed of mostly water-ice. Enceladus reflects almost 100 percent of the sunlight that strikes it. All five satellites represent a size moon not previously explored.

Mimas, Tethys, Dione and Rhea are all cratered. Enceladus appears to have the most active surface of any satellite in the Saturnian system (except perhaps for Titan, whose surface has not yet been observed). At least five types of surface terrain features have been identified on Enceladus. Because it reflects so much sunlight, Enceladus has an observed surface temperature of only 72 degrees Kelvin.

Images of Mimas reveal a huge impact crater. This crater, named Arthur, is about 130 kilometers wide—one-third the diameter of Mimas. Arthur is 10 kilometers deep, with a central mountain almost as large as Mount Everest on Earth.

Voyager 2 photographs of Tethys reveal an even larger impact crater, nearly one-third the diameter of Tethys and larger than the moon, Mimas. In contrast to Mimas' Arthur, the floor of the large crater on Tethys has returned to approximately the original shape of the surface, most likely a result of Tethys' larger gravity and the relative fluidity of its water-ice. A gigantic fracture covers three-fourths of Tethys' circumference. The fissure is about the size scientists would predict if this moon was once fluid and its crust hardened before the interior. This canyon has been named Ithaca Chasma. The surface temperature on Tethys is about 86 degrees Kelvin.

Hyperion shows no evidence of internal activity. Its irregular shape and evidence of bombardment by meteoritic material suggest that it is the oldest surface in the Saturnian system.

Iapetus has long been known to have large differences in surface brightness. Brightness of the surface material on the trailing side has been measured at 50 percent, while the leading side material reflects only five percent of the sunlight that strikes it. Most of the dark material is distributed in a pattern directly centered on the leading surface, causing some speculation that dark material in orbit around Saturn was swept up by Iapetus. The trailing face of Iapetus, however, has several craters with dark floors. This implies that the dark material may have originated in the moon's interior. It is possible that the dark material on the leading hemisphere was exposed by the ablation (erosion) of a thin, overlying bright surface covering.

Phoebe was imaged by the *Voyager 2* after the spacecraft had passed Saturn. Phoebe orbits in a retrograde direction (the opposite direction of the other moons of Saturn) in a plane much closer to the ecliptic than Saturn's equatorial plane. Phoebe is roughly circular in shape and reflects about six percent of the incident sunlight. This moon is also quite red in color. Scientists believe that Phoebe may be a captured asteroid with its composition unmodified since its formation in the outer Solar System. If this is true, then it represents the first such "minor planet" photographed at close enough range to show shape and surface brightness.

Both Dione and Rhea have bright, wispy streaks that stand out against an already bright surface. The streaks are most likely the results of ice that evolved from the interior along fractures in the crust.

Saturn has many small satellites, ranging in size from Hyperion (about 300 kilometers across) down to the A Ring Shepherd which is only about 30 kilometers in diameter. Objects 1980S26, 1980S27 and 1980S28 are called "shepherd satellites" because of their assumed role in helping to keep particles in Saturn's ring system within their proper orbits.

The innermost satellite, 1980S28, orbits near the outer edge of the A-Ring and is about 40 by 20 kilometers in size. It was discovered through *Voyager 1* imagery.

As we go out from Saturn, the next satellite is 1980S27, which shepherds the inner edge of the F-Ring. It is about 140 by 100 by 80 kilometers in size. Next is 1980S26, outer shepherd of the F-Ring. It measures 110 by 90 by 70 kilometers. Both of these shepherd satellites were also discovered using *Voyager 1* photographs.

Next are 1980S1 and 1980S3, called the "co-orbitals," because they share approximately the same orbit (namely

91,100 kilometers) above Saturn's clouds. As they near each other, these satellites trade orbits (the outer orbit is about 50 kilometers farther from Saturn than the inner one). The object 1980S1 measures 220 by 200 by 160 kilometers, while 1980S3 measures 140 by 120 by 100 kilometers. Both of these satellites were discovered from ground-based (terrestrial) observations of Saturn.

Two newly discovered satellites, 1980S6 and 1981S7, share the orbit of Dione. Because these moons are about 60 degrees ahead and 60 degrees behind their larger celestial companion, they are called the Dione Trojans. These moons are about 33 kilometers and 20 kilometers in diameter, respectively. 1980S6 was discovered through ground-based photography, while 1981S7 was discovered using Voyager mission imagery.

Two other small satellites, 1980S13 and 1980S25, are called the Tethys Trojans because they circle Saturn in the same orbit as Tethys, about 60 degrees ahead of and behind that celestial body. 1980S13 is called the leading Trojan, while 1980S25 is the trailing Trojan. These objects were discovered in 1981, using ground-based observations made in 1980, and are comparable in size to 1980S6.

Saturn has two other known satellites, 1981S10 and 1981S9. They are both about 15 kilometers in diameter. 1981S10 circles Saturn between the orbits of Tethys and Dione, while 1981S9 orbits between Dione and Rhea. Both of these moons were discovered through Voyager imagery.

THE MAGNETOSPHERE

The size of Saturn's magnetosphere is determined by external pressure of the solar wind. Several distinct regions have been identified within Saturn's magnetosphere. About 400,000 kilometers inside there is a torus of $H+$ and $O+$ ions, probably originating from water-ice that sputtered off the surfaces of Dione and Tethys. (These ions are positively charged atoms that have lost one electron.) Strong plasma-wave emissions appear to be associated with the inner torus.

At the exterior regions of the inner torus some ions have been accelerated to high velocities. Outside the inner torus is a thick sheet of plasma that extends out to about 1 million kilometers. The source for material in the outer plasma sheet is probably Saturn's ionosphere, Titan's atmosphere and the neutral hydrogen torus that surrounds Titan between 0.5 and 1.5 million kilometers.

See also: *Pioneers 10, 11*; **Voyager**

Saturnian   of or relating to the planet Saturn; (in science fiction) a native of the planet Saturn.

Saturn Orbiter   The Saturn Orbiter spacecraft will provide an extended study of this interesting planet and its complex assembly of satellites, rings and field phenomena. The general goals of the orbiter mission are: (1) to determine the structure and dynamic behavior of Saturn's atmosphere, rings and magnetosphere; and (2) to determine the composition of the surfaces (such as minerals and ices)

and geological history of the Saturnian satellites.

The Saturn Orbiter spacecraft will be launched using the Space Shuttle/Centaur configuration. It will reach Saturn after a flight of about three and one half years. Upon reaching Saturn, the spacecraft will be propulsively decelerated, also taking advantage of Titan's gravity field to slow down. (Titan is Saturn's largest satellite.) Orbit changes over the duration of the mission will be made possible by encounters with Titan. These encounters will also permit detailed mapping and investigation of Titan itself.

For about two years this spacecraft will experience close fly bys or encounters with many of the Saturnian moons. In addition, a great variety of measurements will be made on Saturn's rings from a complete range of viewing angles. Finally, in-place measurements will be performed on Saturn's magnetosphere for a large range of geometries with respect to the incident solar wind. This mission will help unravel some of the current scientific mysteries surrounding one of the wonders of our Solar System and its interesting constellation of moons and beautiful rings.

Saturn probe   Unlike the terrestrial planets, the giant outer planets offer space scientists an opportunity to address important questions about their internal structures and bulk compositions through detailed studies of their atmospheres. The main purpose of the Saturn probe is to determine the chemical composition and physical condition of that planet's atmosphere. In addition, the transport of energy within the Saturnian atmosphere will also be studied. This is important in developing an overall understanding of the internal structure and evolution of the planet, Saturn.

The fundamental probe mission includes a Galileo Project-like atmospheric probe and a carrier spacecraft. Launch opportunities for this mission occur approximately every 13 months, with the trip time to Saturn taking about three and one half years. A *Mariner Mark II* modular spacecraft is being considered by NASA for the role of carrier spacecraft in this mission, which is planned for 1995–2000.

See also: **Galileo Project; Saturn**

Schwarzschild radius   The "event horizon" or boundary of no return of a black hole. Anything crossing this boundary can never leave the black hole.

See also: **black holes**

science fiction   A form of fiction in which technical developments and scientific discoveries represent an important part of the plot or story background. Frequently, science fiction involves the prediction of future possibilites based on new scientific discoveries or technical breakthroughs. Some of the most popular science fiction predictions waiting to happen are: interstellar travel, contact with extraterrestrial civilizations, the development of exotic propulsion or communication devices that would permit us to break the speed-of-light barrier, travel forward or backward in time, and very smart machines and robots.

According to the popular writer, Isaac Asimov, one very important aspect of science fiction is not just its ability to predict a particular technical breakthrough, but rather its ability to predict change itself through technology. Change plays a very important role in our modern life. As we enter the next millennium, people responsible for societal planning must not only consider how things are now, but how they will (or at least might be) in the upcoming decades. Gifted science fiction writers, like Jules Verne, H.G. Wells, Isaac Asimov and Arthur C. Clarke, are also skilled technical prophets who help many people peek at tomorrow before it arrives.

For example, the famous French writer, Jules Verne, wrote *De la terre à la lune (From the Earth to the Moon)* in 1865. This science fiction account of a manned voyage to the Moon originated from a Floridian launch site near a place Verne called "Tampa Town." A little over 100 years later, directly across the state from the modern city of Tampa, the once isolated regions of the east central Florida coast shook to the mighty roar of a *Saturn V* rocket. The crew of the *Apollo 11* mission had embarked from Earth and man was to walk for the first time on the lunar surface!

search for extraterrestrial intelligence (SETI) The search for extraterrestrial intelligence, or SETI, is basically a manifestation of our natural curiosity and desire to explore. These instincts are some of the oldest and most fundamental aspects of our human nature. It is highly probable that many ancient peoples looked up into the night sky and wondered about the existence of other worlds and other beings. However, it is only within the last few decades, with the arrival of the Age of Space, that mankind could do more than simply speculate about extraterrestrial life, including other life forms that had achieved intelligence and developed technology. The classic paper by Giuseppe Cocconi and Philip Morrison entitled "Searching for Interstellar Communications," (*Nature*, 1959) is often regarded as the start of modern SETI. The entire subject of extraterrestrial intelligence (ETI) has left the realm of science fiction and is now regarded as a scientifically respectable (though currently speculative) field of endeavor.

Well, just where do we look for "little green men" or how do we listen for their signals? Scientists have wrestled with both these questions over the last two decades. In an effort to guess the number of technically advanced alien civilizations that might now exist in our Galaxy, Frank Drake and other scientists developed a mathematical relationship called the Drake Equation. Based on the Drake Equation, pessimistic estimates on the number of communicative extraterrestrial civilizations are very low (10 to perhaps 100); while more optimistic speculations indicate a Galaxy with perhaps 10,000 to 100,000 advanced civilizations!

The major aim of modern SETI programs is to listen for evidence of microwave signals generated by intelligent extraterrestrial civilizations. This search includes radio astronomy mapping of a major portion of the sky and the study of manmade (artificial) radio frequency interference for use in future SETI projects.

The current understanding of stellar formation leads scientists to think that planets are normal and frequent companions of most stars. As interstellar clouds of dust and gas condense to form stars, they appear to leave behind clumps of material that form into planets. The Milky Way Galaxy contains at least 100 billion to 200 billion stars.

Present theories on the origin and chemical evolution of life indicate that life is probably not unique to Earth but may be common and widespread throughout our Galaxy. Scientists further believe that life on alien worlds could have developed intelligence, curiosity and the technology necessary to build the devices needed to transmit and receive electromagnetic signals across the interstellar void.

If this is true, then some scientists also believe that alien civilizations might be searching at this very moment for intelligent companions. There may even be some type of galactic community in which interstellar communications are shared by many different extraterrestrial civilizations. However, to date, none of our efforts here on Earth to detect and identify radio wave signals from alien civilizations have been successful. Since Dr. Frank Drake's initial SETI activities under Project Ozma in 1960, there have been over 35 major attempts around the world to listen for signals from intelligent alien sources. These pioneering SETI efforts are summarized in table 1. Unfortunately, none of these searches yielded signals that could be positively identified as originating from alien civilizations among the stars.

Until now only very narrow portions of the electromagnetic spectrum have been examined for artificial signals (generated by intelligent alien civilizations). Man-made radio and television signals, the kind radio astronomers reject as clutter and interference, are actually similar to the signals SETI researchers are hunting for.

The sky is full of radio waves. In addition to the electromagnetic signals we generate as part of our technical civilization (radio, TV, radar, etc.) the sky also contains natural radio wave emissions from such celestial objects as our Sun, the planet Jupiter, quasars, radio galaxies and pulsars. Even interstellar space is characterized by a constant, detectable radio-noise spectrum.

However, SETI scientists are looking for radio wave signals that are considerably different from known natural extraterrestrial (ET) radio sources. Typically, a natural ET radio signal occupies a wide bandwidth, perhaps a kilohertz (1,000 cycles per second) or more. The radio wave signals that might be generated by intelligent alien races should not exhibit such wide bandwidths. For example, man-made radio emissions usually have strong carrier components that occupy less than one hertz (1 cycle per second). No natural ET radio sources have been found to date that broadcast on such narrow frequencies; and none may actually exist.

In conducting their search for intelligent radio signals from ET civilizations, SETI scientists must consider four general parameters or dimensions: (1) the location of the transmitting source; (2) the frequency range within which the source is transmitting; (3) the modulation or method used to impart information to the carrier electromagnetic signal; and (4) the signal power that can be detected by the receiving antenna.

At the best sensitivity now available, a SETI effort might be able to detect directive transmitters like the largest radio telescope/transmitter we now possess (the Arecibo facility) at a distance of more than 1,000 light years. In past efforts, many SETI observers had assumed that alien transmissions might be associated with star systems like or at least similar to our Sun. Stars of much greater luminosity than the Sun were considered too short-lived to permit the chemical evolution of life to occur; while stars very much less luminous than our Sun appear to have unfavorable ecospheres made inhospitable, in part, by violent coronal activities. SETI observers have also avoided stars that have departed the main sequence because even an advanced alien race would be hard pressed to survive when their parent star became a red giant or violently exploded as a supernova.

SETI observers have developed a variety of strategies and search scenarios (recall again the Drake equation). However, for many such efforts an all-sky survey for intense artificial sources, complemented by detailed investigations of nearby star systems for weaker signals, appears to be a favorable strategy to adopt. This is especially true when you consider the current lack of knowledge on this fascinating subject. As a rule of thumb, SETI scientists suggest that because we really don't know about any other planetary civilization but our own, it is most wise to make as few assumptions as possible in developing search strategies and signal detection schemes. For example, even if the Galaxy is bursting with intelligent civilizations (a speculation subject to extensive debate), what part of the electromagnetic spectrum would these civilizations use in communicating with each other? And would mature civilizations use the same type of transmissions to communicate with both advanced civilizations and emerging civilizations? Do you use the same language when you talk to adults and when you talk to little children?

Consequently, the signal broadcast frequency has been the subject of wide and varied speculation among SETI scientists. Some strongly believe that the region from 1.4 to 1.7 gigahertz represents an excellent prospect for detecting intelligent alien signals. It happens that this region lies between the natural radio wave emissions of hydrogen (H) and the radical hydroxyl (OH). Some SETI scientists have rather romantically called this region of the electromagnetic spectrum the "water hole." It is one of the most favored SETI observation regions because of the important role water plays in life on Earth and the fact that the region is also one of relative radio quiet, making any artificial signals within it fairly easy to detect and identify. Although its choice at present is quite pleasing from a terrestrial viewpoint, the water hole region is only a tiny fraction of the overall electromagnetic spectrum. Human philosophy may not, however, be relevant in other parts of the Galaxy. From a point of view of physics, the frequency band that is the most efficient for electromagnetic communications on an interstellar scale is the microwave "window" which lies between 1 and 100 gigahertz. If we conduct our SETI activities using radio antennas in space or on the far side of the Moon, we might consider this entire bandwidth in the search strategy. If we are limited to using only ground-based radio antenna facilities this window narrows to about one-tenth its size in space because of the effects of the Earth's atmosphere. However, even on Earth the lower end of this microwave window (the water hole) is quite free of natural radio noise.

How would an alien race modulate the signal so that it contained useful information? And then, what information would they send? Source modulation is another area of considerable speculation among SETI observers. If we were to detect a narrow-bandwidth carrier signal with no modulation (that is, no information content) of any kind, SETI scientists would most likely assume it is some type of previously unknown natural ET radio signal. If any alien race is sending a message, they will most likely include some information in it. But what method would an alien race use to modulate their signal? SETI scientists have suggested two basic possibilities: The alien society could use a strong pulsed signal; or perhaps a strong carrier component of narrow bandwidth continuously transmitted. Contemporary SETI equipment here on Earth can detect both types of signals.

SETI observations may be performed using radio telescopes on Earth (see fig. 1), in space or even on the far side of the Moon. Each location has distinct advantages and disadvantages. For example, until a full extraterrestrial civilization matures (one with both lunar settlements and permanent space stations), the construction and operation of large radio antennas (dedicated to SETI activities) on either the lunar surface or in orbit will be prohibitively expensive—negating the technical advantages of using either location. However, once the technical infrastructure has been developed in cislunar space (perhaps in the next century) then extraterrestrial SETI observations from either cislunar space or the Moon's far side surface will become very attractive alternatives. But for the next decade or so, SETI scientists must remain content with using ground-based radio telescopes.

One of the signs of maturity of a planetary civilization is its concern for the cosmic question: "Are we alone?" Since the start of the Space Age, we have taken this question from the back of cereal boxes and placed the intriguing issue in the arena of legitimate scientific curiosity. The detection of but a single, clearly identifiable signal from

Fig. 1   An artist's rendering of a large ground-based array of giant radio telescope antennas listening for radio emissions from alien races who might exist in other star systems. (Courtesy of NASA.)

an extraterrestrial alien civilization will have immeasurable impact on our own civilization and on our cosmic perspective. SETI represents an exciting human activity that we can engage in over the next century. A serious SETI program not only helps us mature as a planetary civilization but has the potential of positively answering one of the greatest mysteries of all.

If an alien signal is ever detected and decoded, then we would face another challenging question: Do we answer? How would you respond to a little green man's radio

message? For the present time, SETI scientists are content to passively listen for intelligent signals coming to us across the interstellar void.

See also: **Arecibo Observatory; consequences of extraterrestrial contact; Drake equation; extraterrestrial civilizations; Fermi paradox; interstellar contact; life in the Universe; Project Cyclops; Project Ozma; water hole; What do you say to a little green man?**

Table 1   Summary of Known SETI Observation Efforts (1960–1983)

SETI Observer/Project	Location	Date	Search Frequency (MHz)[a]	Celestial Targets
Drake Project Ozma	U.S.A.	1960	1420–1420.4	2 stars: Tau Ceti; Epsilon Eridani
Kardashev and Sholomitskii	U.S.S.R.	1963	920	quasar
Kellermann	Australia	1966	many between 350 and 5,000	one galaxy
Troitskii, Gershtein, Starodubtsev, Rakhlin	U.S.S.R.	1968; 1969	926–928; 1421–1423	12 stars
Troitskii, Bondar, Starodubtsev	U.S.S.R.	1970 and on	1863; 927; 600	all sky search for sporadic radio pulses
Slysh	France	1970–72	1667	10 nearest stars
Verschuur/Project Ozpa	U.S.A.	1971; 1972	1419.8–1421; 1410–1430	9 stars
Kardashev and Steinberg	U.S.S.R.	1972	40–500	omnidirectional
Palmer and Zuckerman/ Project Ozma II	U.S.A.	1972–1976	1413–1425; 1420.1–1420.7	674 stars
Kardashev and Gindilis	U.S.S.R.	1972 and on	1337–1863	all sky search for sporadic radio pulses

SETI Observer/Project	Location	Date	Search Frequency (MHz)[a]	Celestial Targets
Dixon, Ehman, Kraus, Raub	U.S.A.	1973 and on	1420.4	all sky search
Bridle and Feldman/ "Qui Appelle?"	Canada	1974–1976	22,235.08	70 stars
Wishnia	(Earth orbit) Copernicus spacecraft	1974	ultraviolet portion of spectrum	3 stars
Shvartsman/Mania	U.S.S.R.	1974 and on	5500	21 unusual objects
Drake and Sagan	U.S.A.	1975; 1976	1420; 1667; 2380	4 galaxies
Israel and deRuiter	Netherlands	1975–1979	1415	50 star fields
Bowyer et al U.C. Berkeley/ Serendip	U.S.A.	1976 and on	1410–1430; 1653–1673	all sky survey
Black, Clark, Cuzzi, Tarter	U.S.A.	1976	8522–8523	4 stars
Black, Clark, Cuzzi, Tarter	U.S.A.	1977	1665–1667	200 stars
Drake and Stull	U.S.A.	1977	1664–1668	6 stars
Wielebinski and Seiradakis	Federal Republic of Germany	1977 and on	1420	3 stars
Horowitz	U.S.A.	1978	1420	185 stars
Cohen, Dickey, and Malkan	U.S.A. and Australia	1978	1665–1667; 1612.231; 22,235.08	25 globular clusters
Knowles and Sullivan	U.S.A.	1978	130–500 (spot)	2 stars
Cole and Ekers	Australia	1979	5000	nearby F, G, and K stars
Freitas and Valdes	U.S.A.	1979	5500 Å (photographs)	searched the Lagrangian libration points L_4 and L_5 in Earth-Moon system for alien probes
JPL and U.C. Berkeley/ Serendip II	U.S.A.	1979 and on	S and X band	observed apparent positions of NASA spacecraft
Clark, Duquet, Lesyna, and Tarter	U.S.A.	1979–1981	1420.4	200 stars
Witteborn	U.S.A.	1980	8.5–13.5	20 stars
Lord and O'Dea	U.S.A.	1981	115,000	north Galactic rotation axis
Israel and Tarter	Netherlands	1981	1420	85 star fields
Biraud and Tarter	France	1981 and on	1665–1667	300 stars
Shostak and Tarter	Netherlands	1981	1420.4	Galactic center
Horowitz, Teague, Chen, Backus, and Linscott/ "Suitcase SETI"	U.S.A.	1982	—	250 stars

SETI Observer/Project	Location	Date	Search Frequency (MHz)[a]	Celestial Targets
Vallee and Simard-Normandin	Canada	1982	—	Galactic center meridian
Horowitz	U.S.A.	1983 and on	—	survey of sky

[a] MHz unless otherwise specified

Data in this table based on information provided by NASA/Ames and the following articles: "The Search for Extraterrestrial Intelligence". *The Planetary Report*, Volume III, No 2. March/April 1983. "Searching for Extraterrestrials". *Astronomy*, October, 1982. *The Search for Extraterrestrial Intelligence*. NASA SP-419, 1977.

Selenian Of or relating to the Earth's Moon. Once a permanent lunar base is established, a native of the Moon.

self-replicating space station An orbiting facility that has achieved the very significant and important capacity to reproduce parts and components for replacement and refurbishment, or for the modular construction of new space stations that are then placed in other orbital locations. If raw materials are provided from Earth, this replication process is semi-autonomous; while if essentially only extraterrestrial materials and energy resources are used, then the replication process is regarded as autonomous of the terrestrial biosphere.

Ultimately, the development of our extraterrestrial civilization will depend on an ability to replicate or "clone" space habitats or facilities using just extraterrestrial resources—both materials and energy (see fig. C25 on color page K).

How extensive can this replication process be? For truly far-reaching thinkers and planners, the Dyson sphere represents a Malthusian "upper limit" of this replication process within our Solar System. It is a swarm of habitats that encircle our parent star at about one astronomical unit distance and that intercept essentially all of its radiant energy output. At that point in human development, portions of civilization will have chosen to leave the Solar System itself and migrate to the stars—replicating human civilization on an even grander scale in the Galaxy.

See also: **space settlement; space station**

Seyfert galaxy A galaxy with a very bright central nucleus; named after the American astronomer, Carl Seyfert, who first observed them in 1943. The bright nuclei of Seyfert galaxies appear to contain hot gases that are in rapid motion.

See also: **galaxy**

"shimanagashi" syndrome Will terrestrial immigrants to extraterrestrial communities suffer from the "shimanagashi" syndrome? During Japan's feudal period, political offenders were often exiled on small islands. This form of punishment was called "shimanagashi." Today, in many modern prisons one can find segregation or isolation units in which inmates who are considered troublemakers are confined for a period of time. Similarly, but to a lesser degree, mainlanders who spend a few years on an isolated island feel a strange sense of isolation—even though the

island (such as Hawaii) may have large cities and many modern conveniences. These mainland visitors start feeling left out and even intellectually crippled despite the fact that life might be physically very comfortable there. Early extraterrestrial communities will also be relatively small and physically isolated from the Earth. However, electronic communications, including the transmission of books, journals and contemporary literature, could avoid or minimize such feelings of isolation. As the actual number of extraterrestrial settlements grows, physical travel between them could also reduce the sense of physical isolation.

Shuttle Infrared Telescope Facility (SIRTF) A cryogenically-cooled (temperature less than 20 degrees Kelvin), one-meter (aperture diameter) class infrared (IR) telescope that will be operated from the Space Shuttle/Spacelab as an observatory for infrared astronomy. Over the two to 200 micrometer portion of the electromagnetic spectrum this planned telescope will be 100 to 1,000 times more sensitive than existing infrared facilities. It will be capable of accommodating many types of focal plane instruments, such as photometric, spectroscopic and polarimetric, that can be refurbished or changed between flights.

The SIRTF will be used to study a wide range of astrophysical phenomena including: the cold regions of space where interstellar dust and gas are condensing into stars; small objects in the outer regions of our Solar System, such as planetary satellites, cometary nuclei and asteroids, to help determine their composition and evolutionary history; and infrared emitting galactic objects such as quasars.

Individual SIRTF missions, lasting typically 14 days, will be flown on the Space Shuttle/Spacelab at about one-year intervals and the telescope will be refurbished between each flight. An orbital altitude between 300 and 400 kilometers is currently planned for a typical SIRTF mission with an inclination of 28.5° or 57°. The SIRTF will be supported on three Spacelab pallets and will use Spacelab mission support equipment. As presently planned, this exciting and powerful infrared instrument will have a useful lifetime of 10 years or more. The mission is planned for the 1990s.

See also: **astrophysics; infrared astronomy; Space Transportation System**

Simmons principle A line of reasoning, credited originally to John Simmons of the United Kingdom, that has been popular in discussions that include very pessimistic or very optimistic assumptions on the possibility of the existence of extraterrestrial life. The Simmons principle applied, for example, to something like the Fermi paradox (that is, "Where are they?") or the search for extraterrestrial intelligence (SETI) might be summarized as follows:

> If we can show that something is impossible using very optimistic assumptions, or that something is possible using very pessimistic assumptions, then the effort has some value. All other results that fall in between these two extremes should be considered subjective and, therefore, of little value.

See also: **Drake equation; Fermi paradox; search for extraterrestrial intelligence**

solar cell Solar cells are proven direct energy conversion devices that have been used for more than two decades to provide electric power for spacecraft. In a direct energy conversion (DEC) device electricity is produced directly from the primary energy source without the need for thermodynamic power conversion cycles involving the heat engine principle and the circulation of a working fluid. A solar cell or "photovoltaic system" turns sunlight directly into electricity. They have no moving parts to wear out and produce no noise, fumes or other polluting waste products.

The foundations of our understanding of how the solar cell works were laid early in this century by such great scientists as Max Planck and Albert Einstein. At the turn of the century, for example, Planck theorized that there was a direct relationship between energy and frequency of electromagnetic radiation. Planck also postulated that energy is emitted and absorbed in individual, tiny packets or units called "quanta" (singular: quantum). Then, in 1917 Albert Einstein provided support for Planck's Quantum Theory, when he developed his Nobel prize-winning explanation of the photoelectric effect. Einstein said that the photoelectric effect could be explained only if we assumed that light is composed of individual packets of energy, called "photons." In effect, Einstein's photons are simply quanta of light (electromagnetic radiation in the visible band).

Scientific evidence shows that when photons strike certain materials, the photons ionize the materials—that is, they dislodge electrons from atoms of the material. In addition, the photons impart their energy to the dislodged electrons. As a result, these energized electrons, called upper energy or conduction electrons, produce an electric current. Semiconductors are a class of solid materials that are most susceptible to such ionization by photons. These semiconductor materials include: silicon (Si), germanium (Ge), gallium (Ga), arsenide, cadmium sulfide and cadmium telluride. A solar cell can be made from any one of these materials. However, because silicon solar cells are very widely used, we will describe the scientific principles of solar cell operation in terms of the silicon solar cell.

By itself a piece of silicon is not a solar cell. When photons (light) strike a piece of silicon, they dislodge electrons from atoms, but these electrons wander about randomly through the silicon. They do not collect into an orderly stream that can become a useful electric current. To create this orderly stream, we must first "dope" the silicon. We dope the semiconductor material by introducing atoms into the substance that provide more or fewer electrons than needed to make an ideal atomic structure. For example, silicon is frequently doped with trace amounts of phosphorus (P) and boron (B).

But how has doping changed the silicon? Chemically, there are still millions and millions of silicon atoms for each boron or phosphorus atom present in the solar cell. However, the boron and phosphorus modify the silicon in such a way that when the cell is illuminated with light, the freed electrons now generally flow in one direction and become a useful electric current.

The phosphorus-doped layer is called n-type (for negative) semiconductor material, because of its extra negatively charged electrons. The other layer is called p-type (for positive) semiconductor material, because of its positively charged holes. The zone or plane between the n-type and p-type layers is called the junction. After silicon is doped, electrons move from the n-type material across the junction to the p-type material; while the holes do just the opposite, moving from the p-type material to the n-type material. (A hole is a place in a bound pair of atoms where an electron is missing.) This movement is due to a phenomenon called diffusion—that is, the spontaneous movement of charges away from regions of high concentration.

Electrons moving from the n-type to the p-type layer of semiconductor material build up a localized negative electric charge adjacent to the junction in the p-type layer. Similarly, holes crossing from the p-type material to the n-type material layer build up a positive charge adjacent to the junction in the n-type layer. This creates a built-in electric field that eventually stops charge diffusion across the junction.

When sunlight strikes a solar cell, photons displace electrons from atoms, imparting energy to these electrons. Scientists say that each electron is being raised to the upper energy or conduction band, leaving a hole in the valence band. The internal electric field drives these conduction electrons through the n-type material toward the metallic contact at the top of the solar cell. The internal electric field also drives holes created by the dislodged conduction electrons toward the bottom of the p-type material layer which is also at the bottom of the cell (see fig. 1). The conduction electrons flow from the metallic contact through the outside circuit creating electricity, which can be used to power spacecraft equipment and systems. These electrons continue through the external or outside circuit and reach the metal contact on the bottom of the solar cell. Here, they reenter the solar cell and unite with holes clustered around the bottom of the cell. At this

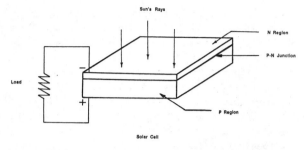

Fig. 1   The basic solar cell. (Courtesy of NASA.)

point, the electrons have returned to their places as valence electrons in the semiconductor material atoms—that is, they have dropped back to the valence band.

As electrons leave holes in the illuminated upper layer (n-type) of the solar cell and accumulate in the lower (p-type) layer, diffusion occurs. Holes move downward from the upper layer material and electrons move upward from the lower layer. When electrons rise to the solar-illuminated area of the cell, they are raised to the conduction or upper energy band. These electrons are, subsequently, swept by the internal electric field to the current collectors at the top of the solar cell. This direct energy conversion cycle can continue indefinitely.

The solar cell has been and will continue to be used extensively on board spacecraft and space platforms to provide electric power. However, many future missions, especially those operating far away from the Sun, or in periods of extensive darkness or shadowing, or requiring very large quantities of power (such as megawatts electric) from compact, mobile systems, may require the use of alternate forms of space power, including space nuclear power.

See also: **Satellite Power System; space nuclear power**

solar constant   The total amount of the Sun's radiant energy that normally crosses to a unit area at the top of the Earth's atmosphere (that is, at one astronomical unit from the Sun). The currently adopted value of the solar constant is 1371 ± 5 watts per square meter. The spectral distribution of the Sun's radiant energy resembles that of a blackbody radiator with an effective temperature of 5800 degrees Kelvin. This means that the majority of the Sun's radiant energy lies in the visible portion of the electromagnetic spectrum, with a peak value near 0.450 micrometer. (A micrometer is one millionth of a meter).

solar cosmic rays   Atomic particles ejected from the Sun in solar flare events. Solar cosmic rays consist of protons (bare hydrogen nuclei), alpha particles (bare helium nuclei) and electrons. They are generally lower in energy than galactic cosmic rays. The stream of solar cosmic rays usually reaches and surrounds the Earth within minutes after a solar flare and reaches peak intensity within hours. Then, within a day or two, the solar

cosmic ray population declines. The effects of solar cosmic rays may be excluded with moderate amounts of shielding (for example, tens of milligrams of shielding material per square centimeter), leaving galactic cosmic rays as the dominant cosmic ray hazard in space.

See also: **cosmic rays; galactic cosmic rays**

solar mass   The mass of our Sun, 1.99×10^{30} kilograms; it is commonly used as a unit in comparing stellar masses.

See also: **stars; Sun**

solar nebula   The cloud of dust and gas from which a star forms (condenses).

See also: **stars**

solar sail   A method of space transportation that uses solar radiation pressure to gently push a giant gossamer structure and its payload through interplanetary space (see fig. 1). As presently envisioned, the solar sail would use a large quantity of very thin (typically 0.0025 millimeter or 0.1 millimeter thick or less) reflective material to produce a net reaction force by reflecting incident sunlight. Because solar radiation pressure is very weak and decreases as the square of the distance from the Sun, enormous sails—perhaps 100,000 to 200,000 square meters—would be needed to achieve useful accelerations and payload transport.

The main advantage of the solar sail would be its long duration operation as an interplanetary transportation system. Unlike rocket propulsion systems which must expel their onboard supply of propellants to generate thrust, solar sails have operating times only limited by the effective lifetimes in space of the sail materials. The solar

Fig. 1   The Space Shuttle is dwarfed by a giant solar sail that is being prepared for a comet intercept mission (artist's concept). (Courtesy of NASA.)

photons that do the "pushing" constantly pour in from the Sun and are essentially "free." This makes the concept of solar sailing particularly interesting for cases where we must ship large amounts of non-priority payloads through interplanetary space—as for example, a shipment of special robotic exploration vehicles from the Earth to Mars.

However, since the large reflective solar sail cannot generate a force opposite to the direction of the incident solar radiation flux, its maneuverability is limited. This lack of maneuverability along with long transit times represents the major disadvantages of the solar sail as a space transportation system.

solar system When used in the lower case, the term refers to any star and its gravitationally-bound collection of planets, asteroids and comets.

Solar System When used with capital letters, our Sun and the collection of celestial objects that are bound to it gravitationally. These celestial objects include: the nine known major planets, over 40 moons, more than 2,000 minor planets or asteroids and a very large number of comets (see fig. 1). Except for the comets, all of these celestial objects orbit around the Sun in the same direction, and their orbits lie close to the plane defined by the Earth's own orbit and the Sun's equator.

The nine major planets can be divided into two general categories: (1) the terrestrial or Earth-like planets, consisting of Mercury, Venus, Earth and Mars; and (2) the outer or Jovian planets, consisting of the gaseous giants Jupiter, Saturn, Uranus and Neptune. Pluto is currently regarded as a "frozen snowball" and along with its moon, Charon, appears to be in a class by itself. As a group, the terrestrial planets are dense, solid bodies with relatively shallow or no atmospheres. In contrast, the Jovian planets are believed to contain modest sized rock cores, surrounded by layers of frozen hydrogen, liquid hydrogen and gaseous hydrogen. Their atmospheres also contain other gases such as helium, methane and ammonia.

See also: **asteroid; comet; Earth; Jupiter; Mars; Mercury; Neptune; Pluto; Saturn; Sun; Uranus; Venus**

solar wind An electrically neutral stream of atomic particles consisting mainly of protons and electrons, that continuously flows outward from the Sun into the interplanetary medium. The solar wind interacts in a variety of complex ways with a planetary magnetic field, shaping something scientists call the magnetosphere.

See also: **Earth's trapped radiation belts; magnetosphere**

solipsism syndrome A psychological disorder that could happen to the inhabitants of space bases or space settlements. It is basically a state of mind in which a person feels that everything is a dream and not real. The whole of life becomes a long dream from which the individual can never awaken. A person with this syndrome feels very lonely and detached, and eventually becomes apathetic and indifferent. This syndrome might easily be caused in a space habitat environment where everything is artificial or man-made. To avoid or alleviate the tendency toward solipsism syndrome in space habitats, we would use large geometry interior designs (that is, have something beyond the obvious horizon); place some things beyond the control or reach of the inhabitants' manipulation (that is, an occasional rainy day weather pattern variation or small animals that have freedom of movement); and provide growing things like vegetation, animals and children.

See also: **hazards to space workers; space settlement**

South Atlantic Anomaly A region of the Earth's trapped radiation particle zone that dips close to the planet in the southern Atlantic Ocean southeast of the Brazilian coast. This region represents the most important source of (ionizing) radiation for space travelers and workers in low Earth orbit (LEO).

See also: **Earth's trapped radiation belts; hazards to space workers**

space base A large, permanently inhabited space facility located in orbit around a celestial body or on its surface. The space base would have a much larger crew than a space station with from 10 up to perhaps 1,000 occupants.

Orbiting space bases can be built in modular fashion,

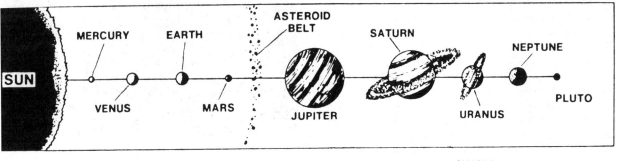

Fig. 1   The major components of our Solar System. (Drawing courtesy of NASA.)

using space station hardware as the building blocks; while bases on alien worlds could be expanded from an initial "seed complex," using extraterrestrial materials to the greatest extent possible. For example, a lunar base could use lunar regolith (soil) for shielding material to protect the occupants from both micrometeoroid hits and ionizing radiation.

Space bases in low Earth orbit would operate nominally at about 500-kilometer (270-mile) altitude at typical inclinations of either 28.5, 55 or 97 degrees. Such bases would be built up from initial space station complexes and would serve the following objectives: (1) support a long-term human presence in space for a much larger population of astronauts; (2) support a wide variety of operational Earth observation activities (such as climatology, crop forecasting, mineral prospecting, marine resource utilization, etc.); (3) conduct detailed scientific investigations of near-Earth space and other celestial objects within the Solar System; (4) develop and demonstrate the technology and operational capability for spacebased manufacturing and materials processing; (5) develop and demonstrate the engineering techniques needed for on-orbit construction and assembly of very large space structures; (6) service operational orbital transfer vehicles; and (7) prepare interplanetary payloads for their missions (such as the vehicle for a manned Mars mission sometime in the early 21st century).

A space base serving as an orbiting chemical processing plant is shown in figure C28 on color page L. In this facility, lunar material that has been shipped from the lunar mining base (see fig. C29 on color page L) is chemically separated to provide the elemental chemicals serving as the feedstock for a variety of space-based industries. For example, anorthosite, a common lunar mineral, can provide aluminum, silicon, oxygen and other useful elements. This conceptual plant could process 30,000 metric tons of anorthosite (90 percent pure) each year. The large hexagonal object appearing in the above figure is a 30 megawatt solar power array that measures 0.5 kilometer long on each side and has a total mass of 120 metric tons. A dual habitat for the construction and operating crews is shown beneath the huge solar array. About 20 to 40 persons would continuously operate this plant.

Geostationary Earth orbit (GEO) is the preferred location for a number of information transfer, Earth observation and scientific sensor systems. GEO construction and maintenance bases might range in size from a small modular eight person "work shack" to a larger, more extensive 50-person construction base.

Our return to the Moon will be marked by the development of lunar surface bases. These bases might include: (1) a 6-10 person temporary surface base that provides life support for up to six months; (2) a permanent 10-20 person science and engineering technology base (see fig. C25 on color page K); and (3) a larger, permanent complex of laboratories, greenhouses, habitats and pilot factories for 100 to 1,000 lunar inhabitants. These lunar surface bases

will demonstrate the technologies necessary for permanent habitation of the Moon and for full exploitation of its resources.

Manned expeditions to Mars in the next century will be complemented by the establishment of surface bases—at first temporary and then permanent. The permanent bases on the Red Planet will be our foothold in heliocentric space and serve as the trail head or staging point for manned missions to the main asteroid belt and the giant outer planets beyond.

See also: **lunar bases and settlements; Mars; space settlement; space station**

space construction   Large structures in space, such as modular space stations, global communication and information services platforms, and satellite power systems (SPSs), will all require on-orbit assembly operations by space construction workers. Figure 1 shows astronauts equipped with the manned maneuvering unit (MMU) in the process of constructing and aligning a large space structure. The initial fabrication of a proposed satellite power system structural member in low Earth orbit (LEO) is illustrated in figure 2.

Space construction requires protection of the work force and some materials from the hard vacuum, intense sunlight and natural radiation environment encountered above the Earth's protective atmosphere. Outer space, however, is also an environment that, in many ways, is ideal for the construction process. First, because of the absence of significant gravitational forces (that is the microgravity experienced by the free-fall condition of orbiting objects), the structural loads are quite small, even minute. Structural members may, therefore, be much lighter than terrestrial structures of the same span and stiffness. Second, the absence of gravitational forces greatly facilitates the movement of material and equipment. On Earth, the movement of material during a

Fig. 1   Space workers in manned maneuvering units performing on–orbit assembly operations. (Courtesy of NASA.)

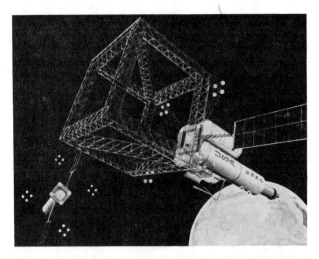

Fig. 2   Fabrication of a large structure in space. (Courtesy of NASA.)

construction operation absorbs a large portion of the total work effort expended by construction personnel and their machines. Third, the absence of an atmosphere, with its accompanying wind loads, inclement weather and unpredictable change, permits space work to be accurately planned and readily executed without environmental interruptions (except, perhaps due to solar flares which would increase the radiation hazard).

In order to minimize transportation costs for large space construction projects made with terrestrial materials, the construction materials shipped from Earth should be packaged in a very dense form. The desire to make very low density, lightweight systems from high density materials leads space technologists and engineers to consider automated, on-orbit fabrication techniques. Such techniques will allow the materials payloads to be densely packaged for launch in the form of tightly wound rolls and spools. Since repetitive operations are more easily automated, regular uniform cross-section structural members are being considered for building large space structures. For example, figure 3 shows the basic satellite power system structural elements.

A process similar to the terrestrial roll-forming of light sheet metal members has been adapted to produce these structural members at a rapid pace using an automated beam builder. This machine consists of roll-forming, heating and cooling, and ultrasonic welding components that function to "produce" a finished beam. These basic structural members may be produced from rolls of aluminum strip stock or from a graphite fiber/thermoplastic-impregnated roll. The latter space construction material has the advantage of having a low coefficient of thermal expansion. The basic structural element, a triangular cross-section shape, may be assembled into primary-structure triangular trusses. These members are then assembled into a trussed box structure.

The Space Shuttle will be used to explore a variety of space construction techniques, including erection, auto-

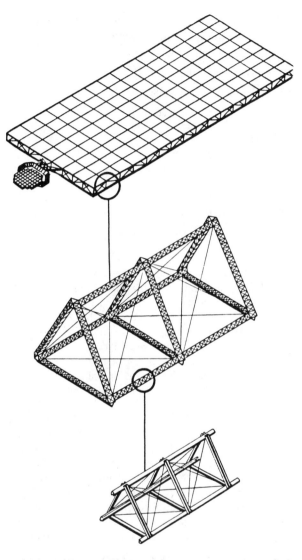

Fig. 3   Basic structural elements for space construction projects, such as a satellite power system. (Courtesy of NASA.)

mated fabrication methods and deployment. As a matter of fact, automated fabrication is believed to be a key requirement for viable space construction activities. Table 1 identifies some typical space construction hardware and supporting systems. It should be quite obvious from these listings that working in space will require the close interaction of astronaut and very smart machine. Such smart machines will in turn also trigger a robotics technology explosion on Earth.

The manned maneuvering unit (MMU) is a self-contained, life supporting backpack with gaseous nitrogen propelled jet thrusters that enable a space worker to leave the Shuttle on an extravehicular activity (EVA) mission and travel back and forth to various construction locations. The automated beam builder is a machine designed for fabricating "building-block" structural beams in space. Combined with a space structure fabrication sys-

Table 1   Typical Space Construction Equipment and Supporting Systems

- Manned Maneuvering Unit (MMU)

- Automated Beam Builder

- Space Structure Fabrication System

- Remote Astronaut Work Stations
    "closed cherry picker"
    "open cherry picker"
    free–flyer work station

- Manned Orbital Transfer Vehicle (MOTV)

- Cargo Orbital Transfer Vehicle (COTV)

- Personnel Launch Vehicle (PLV)

- Heavy Lift Launch Vehicle (HLLV)

SOURCE: Author.

tem, the beam builder allows space workers to manufacture and assemble intricate structures in low Earth orbit using the Space Shuttle as an early construction camp. Eventually, permanent space stations or bases in LEO will be used to serve as construction sites for even more complicated space-assembled and manufactured systems. Remote astronaut work stations can be mounted on the end of the Shuttle Orbiter's remote manipulator system. These "open cherry pickers" would have a convenient tool and parts bin, a swing-away control and display panel, and lights for general and point illumination. The closed version of this cherry picker would involve a pressurized manned remote work station that contains life support equipment and controls and displays for operating dexterous manipulators. The design shown in figure 4 provides cabin space for three space workers, a "shirt-sleeve" work environment while on the job and additional protection (over a spacesuit) against the space radiation environment. The free-flyer work station is another variant of the

basic cherry picker. This enclosed astronaut work station has jet thrustors and propellant tanks, giving the space worker sufficient mobility to cover many widely separated locations of a large space construction project. Twin robot arms provide the "shirt-sleeved" astronaut greater reach and strength for handling large structural members.

The cargo orbital transfer vehicle (COTV) and the manned orbital transfer vehicle (MOTV) are used to move materials and personnel between various orbital locations—for example, from low Earth orbit to geosynchronous orbit. COTV concepts include a chemically fueled, high thrust system with a short trip time and a high degree of reusability; a nuclear rocket COTV featuring a propulsion system capable of high thrust and high specific impulse; and an electric propulsion COTV (either solar or nuclear powered) for non-time critical transfer of massive quantities of cargo or for the gentle movement of very delicate structures. One MOTV design involves a reusable chemical propellant (probably liquid oxygen/liquid hydrogen, abbreviated LOX/LH$_2$) vehicle for the rapid, reliable transfer of space workers between orbital locations. Typical configurations can accommodate up to 160 passengers and 100 metric tons of supplies. The personnel launch vehicle (PLV) carries space workers and their personal equipment from the Earth's surface to low Earth orbit. Concepts for the PLV presently include a modification of the Space Shuttle's cargo bay to carry between 50 and 100 passengers. The heavy lift launch vehicle (HLLV) is essentially a cargo transport vehicle designed to haul massive quantities of cargo from a terrestrial spaceport to a space station at LEO. Space construction based on terrestrial raw materials will depend on making these freight hauling costs as low as possible—or else we will have to also explore the use of extraterrestrial materials as may be found on the Moon or Earth-approaching asteroids.

Fig. 4   A three–person work capsule (closed cherry picker) with robot manipulator arms. (Courtesy of the Boeing Company.)

Two particular orbital locations have been studied for primary space construction activities: geostationary orbit and low Earth orbit. Construction in geostationary orbit (GEO) offers the advantage of continuous sunlight, thereby decreasing the differential thermal effects on a large space structure during construction and reducing the need for artificial illumination of the construction site. Remember, a construction site in low Earth orbit would travel around the Earth experiencing night and day cycles every hundred minutes or so. Additionally, at GEO the construction process can be designed to produce the large structure in its final operational form. This avoids the requirement to transport a large space system from LEO to its operational location at GEO.

Construction in LEO, however, does provide the advantage of operating in close proximity to the Earth. Space construction workers and their facilities at LEO are afforded some protection from the radiation environment of space by the trapped radiation belts (also called the Van Allen belts). Gravitational attraction varies with altitude above the Earth, so that very large structures built in LEO must overcome relatively large gravity gradient torques during construction and transport to GEO. To lessen these gravity-induced torques, the structure could be built in LEO in modules, with the final assembly occurring at GEO.

Construction activities in outer space impose a need for protecting the space worker from the environmental hazards encountered there. Provisions for a life-supporting atmosphere and protection from solar thermal and ultraviolet radiation are well understood engineering problems. Protection against the space radiation environment (such as energetic particles) requires that additional knowledge be gathered on the anticipated dynamic range of these phenomena at various potential space construction locations, so that engineering measures to shield space workers can be most effective from an overall protection and cost point of view. Radiation exposure standards for space workers must also be established to determine the necessary protective measures that will be designed into work stations, habitats and orbital transfer vehicles.

See also: **hazards to space workers; large space structures; robotics in space; Satellite Power System; space industrialization**

spacecraft charging In orbit or in deep space, spacecraft and space vehicles can develop an electric potential up to tens of thousands of volts relative to the ambient extraterrestrial plasma (the solar wind). Large potential differences (called "differential charging") can also occur on the space vehicle. One of the consequences is electrical discharge or arcing, a phenomenon that can damage space vehicle surface structures and electronic systems. Many factors contribute to this complex problem including the spacecraft configuration, the materials from which the spacecraft is made, whether the spacecraft is operating in sunlight or shadow, the altitude at which the spacecraft

is performing its mission, and environmental conditions such as the flux of high energy solar particles and the level of magnetic storm activity.

Wherever possible, spacecraft designers use conducting surfaces and provide adequate grounding techniques. These design procedures can significantly reduce differential charging, which is generally a more serious problem than the development of a high spacecraft-to-space (plasma) electric potential.

See also: **hazards to space workers**

space debris Space junk or derelict man-made space objects in orbit around the Earth. Space debris represents a hazard to astronauts, spacecraft and large space facilities like a space station. Man-made debris in space differs from natural solid matter occasionally encountered in the form of meteoroids because the space debris is in permanent Earth orbit during its lifetime and is not transient through regions of interest (like meteoroid showers). At present the very high velocity collision hazard is real but not yet extremely severe. However, as the level of space activity increases—for example, with more frequent Space Shuttle launches into low Earth orbit and with the construction of permanent space stations—this hazard will continue to grow throughout cislunar space.

Current space debris in Earth orbit is mainly composed of explosion fragments from upper stage rocket motors that have blown up on orbit. But this space junk also includes malfunctioning or derelict spacecraft, spent upper-stage propulsion systems, a variety of space mission hardware (including payload shrouds, separation bolts and clamps), and even tools and equipment that were left behind by astronauts and cosmonauts during extravehicular activities.

The only natural mechanism for removing space debris from Earth orbit is atmospheric drag, assisted by the gentle tugging effects of the Earth's gravitational field. Unfortunately, this cleanup process takes a long time, especially for tiny pieces of junk or garbage found at higher altitude orbits.

Space technologists and engineers must consider the space debris problem when they design future space systems. They must make them as litter-free in design as possible and should also design the systems for retrieval or removal at the termination of their useful operations. We may also see the start of an exciting new extraterrestrial career field—space refuse collector! If the space debris problem in the 21st century gets too much out of hand, robotic or human-operated "extraterrestrial garbage trucks" might be used to sweep orbital regions and to clear cislunar trajectories of such collision hazards.

See also: **hazards to space workers**

space industrialization A new wave in man's sociotechnical development in which the special environmental conditions and properties of outer space are used for the economic and social benefit of people on Earth. Some interesting properties of cislunar space include hard vacuum, microgravity, low vibration levels (in orbiting spacecraft), a wide-view angle of Earth and the heavens and complete isolation from the terrestrial biosphere.

Recent studies have attempted to look some 50 years into the future and to match anticipated human needs with growing space opportunities. These space industrial opportunities can be conveniently broken down into four general categories: (1) information services; (2) products; (3) energy; (4) human activities (see table 1).

In a real sense, the information service area of space industrialization already exists. Space platforms are now providing valuable communication, navigation, meteorological, and environmental services to people around the globe. Further expansion of such services involves the use of larger platforms in orbit (especially at geosynchronous Earth orbit or GEO) and much higher power levels. Recent aerospace industry evaluations indicate that greatly expanded information transmission services from space represent some of the most beneficial industrialization activities that could be accomplished in the next decade or so.

A multifunction platform of major capability is needed at GEO—one capable of continued reliable operation. A baseline geosynchronous platform with a 500 kilowatt electric nuclear power supply is shown in figure C30 on color page L. This giant platform would provide five new nationwide information services: direct broadcast television (five nationwide channels, 16 hours per day); pocket telephones (45,000 private channels linked to the current telephone system); national information services (using pocket telephone hardware); electronic teleconferencing (150 two-way video, voice and facsimile channels); and electronic mail (40 million pages transferred among 800 sorting centers overnight). The exciting age of information services through space platforms is just beginning.

Space-manufactured products will include both organic and inorganic items. In the field of energy, space industrialization supports terrestrial energy conservation programs and the search for and exploitation of new energy resources for mankind (for example, nuclear fusion research or the development of satellite power systems). Finally, the area of human activities includes new career opportunities and the establishment of challenging physical and psychological frontiers for the human race.

Through the process of space industrialization, humanity will learn to use outer space, its unique properties and vast potential material and energy resources to create wealth and improve the quality of life on Earth.

See also: **extraterrestrial resources; humanization of space; materials processing in space; Satellite Power System**

Table 1   Major Areas of Space Industrialization

INFORMATION SERVICES

Information Transmission
   [education, medical aid, electronic mail, news services, teleconferences, telemonitoring and teleoperation, time, navigation, search & rescue, . . .]

Data Acquisition
   [Earth resources, crop and forest management, water resources, weather and climate, ocean resources, mineral resources, environmental monitoring, land use surveys, . . .]

PRODUCTS

Organic
   [biochemicals: isozymes, urokinase, . . .]

Inorganic
   [large single crystals, high-strength fibers, perfect glasses, new alloys, high-strength magnets, . . .]

ENERGY

Power From Space
   [nuclear or solar]

Nuclear Fusion Research In Space

Illumination From Space

HUMAN ACTIVITIES

Medical and Genetic Research

Orbiting Scientific Laboratories

Spacebased Education [i.e. "The University of Space"]

Space Therapeutics
   [e.g. "Micro-gravity" hospital, sanitarium . . .]

Space Tourism

Entertainment and the Arts

SOURCE: Author, NASA and Rockwell International.

space launch vehicles How do we get people and cargo off the planet Earth and into the extraterrestrial environment of outer space? The answer—use rocket-propelled space launch vehicles. Whatever space mission is undertaken, the vehicle carrying the payload must be propelled into space by rocket power. All unmanned rockets presently used by the United States have more than one stage and are usually called launch vehicles, or expendable launch vehicles, since they are used only once. The manned Space Shuttle is a unique, reusable aerospace vehicle and is in a class by itself. An aerospace vehicle is capable of operating both within a planetary atmosphere and in outer space (see fig. 1).

The extraterrestrial destination and the payload size and weight determine what rocket capabilities are needed for a particular space mission. For example, a modest-size, low-weight remote sensing spacecraft designed to operate in near-Earth orbit might be flown on board NASA's smallest launch vehicle, the Scout. On the other hand, sending a manned Apollo spacecraft on an expedition to the Moon required the payload launch capabilities of the massive Saturn V vehicle. The powerful Titan-Centaur launch vehicle/upper stage combination sent large, sophisticated robot explorers (like *Viking* and *Voyager*) to visit other worlds within our Solar System. Atlas-Agena launch vehicles have sent several spacecraft to impact on our nearest celestial neighbor, the Moon; while Atlas-Centaur and Delta vehicles have placed over 200 spacecraft into orbit, supporting a wide variety of space technology applications.

At present, NASA's John F. Kennedy Space Center (KSC), America's Spaceport, conducts operational launches of the Delta and Atlas-Centaur vehicles and of the reusable Space Shuttle. The expendable Delta and Atlas-Centaur vehicles (now being phased out of the flight inventory) are launched from NASA pads located on Cape Canaveral Air Force Station, which is adjacent to the

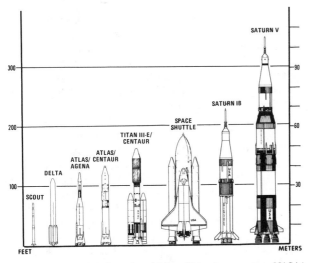

Fig. 1   NASA space launch vehicles. (Drawing courtesy NASA/Kennedy Space Center.)

Kennedy Space Center. The Space Shuttle lifts off from Complex 39 at the Kennedy Space Center.

DELTA

The Delta launch vehicle has been called "the workhorse of the space program." For many years this vehicle has dependably satisfied the requirement for launching intermediate-size payloads. It has successfully transported over 150 scientific, weather, communications and technology applications satellites into outer space. These spacecraft included: the Tiros, Nimbus, and Itos meteorological satellites; the Landsat Earth resources technology spacecraft; the early Intelsat international communications satellites; and a great number of Explorer scientific spacecraft.

First launched in May 1960, the Delta was continuously upgraded over the years. Its most recent configuration involved a vehicle height of 35.4 meters (116 feet). Its first stage is augmented by nine Caster IV strap-on solid propellant motors, six of which ignite at lift off, and three after the initial six burn out some 58 seconds into the flight. The average first-stage thrust with the main engines and the six solid propellant motors burning is about 3.2 million newtons (718,000 lb-force). The *Delta* vehicle has liquid-fueled first and second stages and a solid propellant third stage. A new third stage, called the Payload Assist Module (PAM), has been used in recent launches. The PAM is also used to boost Shuttle carried payloads to higher orbits. (NASA is retiring the Delta launch vehicle and replacing its functions with the reusable Space Shuttle.)

ATLAS-CENTAUR

The Atlas-Centaur configuration has been NASA's standard launch vehicle for high-energy missions, particularly to geostationary Earth orbit and for interplanetary missions. The Centaur was the United States' first high-energy, liquid hydrogen/liquid oxygen propelled launch vehicle stage. It became operational in 1966 with the launch of *Surveyor 1*, the first U.S. spacecraft to soft-land on the Moon.

Since 1966, both the Atlas booster and the Centaur second stage have undergone many improvements. At present, the combined stages can place 4,530 kilograms (10,000 pounds) in low Earth orbit; 1,880 kilograms (4,150 pounds) in a geosynchronous transfer orbit; and 900 kilograms (2,000 pounds) on an interplanetary trajectory. On the pad, an Atlas/Centaur vehicle stands approximately 39.9 meters (131 feet) tall. At lift off, the Atlas booster develops over 1.9 million newtons (431,000 lb-force) of thrust. The Centaur second stage develops 133,450 newtons (33,000 lb-force) of thrust in a vacuum.

Atlas/Centaur vehicles have launched: Orbiting Astronomical Observatories; Applications Technology Satellites; Intelsat IV, IV-A, and V communications satellites; *Mariner* Mars orbiters; the *Mariner 10* spacecraft which encountered both Venus and Mercury; the *Pioneer 10* and *11* spacecraft that accomplished the first encounters of the gaseous giant planets, Jupiter and Saturn, and are now heading out into interstellar space; and the *Pioneer* Venus mission whose orbiter radar-mapped the cloud-covered planet and whose multiprobe spacecraft sent probes hurling through the dense Venusian atmosphere. (After a long and distinguished career, the Atlas/Centaur vehicle combination is also being retired by NASA and will be replaced by the Space Shuttle.)

SCOUT

The Scout vehicle was developed in the late 1950s to launch small payloads into Earth orbit. It became operational in 1960 and has undergone systematic upgrading since 1976. The standard Scout vehicle is a solid propellant, four-stage booster system that is approximately 23 meters (75 feet) in length. It has a launch weight of 21,600 kilograms (46,620 pounds) and a lift-off thrust of 588,240 newtons (132,240 lb-force).

Recent Scout improvements include an up-rated third-stage motor that increases its payload capability. With this up-rated third stage, the Scout can place 211 kilograms (465 pounds) in low Earth orbit. The up-rated third stage has also been given an improved guidance system. The Scout's fourth stage is spin stabilized and an optional fifth stage has been added to provide highly elliptical orbits. Scout vehicles are launched from NASA's Wallops Flight Facility (off the coast of Virginia), from the Western Test Range (Vandenberg AFB, California), and from the San Marco Range (a man-made island at 3 degrees south latitude off the eastern coast of Africa near Kenya). The San Marco Range is operated by the Italian government in cooperation with NASA. Over 100 Scout vehicles have been launched.

ATLAS/AGENA

The Atlas/Agena was a multipurpose, two stage, liquid propellant rocket. It was used to place unmanned spacecraft in Earth orbit or to inject them into the proper trajectories for interplanetary or deep-space missions. The versatile Atlas/Agena combination was used on the early Mariner missions to Mars and Venus; the Ranger photographic missions to the Moon; the Orbiting Astronomical Observatory (OAO), and the early Applications Technology Satellites (ATS). The Agena upper stage was also used as the rendezvous target vehicle for the Gemini spacecraft in 1965-1966. It also launched the Lunar Orbiter spacecraft that circled the Moon taking photographs in preparation for the Apollo landings. The Atlas/Agena vehicle stood 36.6 meters (120 feet) high, and developed a total thrust at lift off of approximately 1.7 million newtons (388,000 lb-force). It was last used in 1968 to launch an Orbiting Geophysical Observatory (OGO).

SATURN V

The Saturn V, America's most powerful staged rocket, was assigned for and successfully accomplished the ambitious task of sending astronauts to the Moon. The first Saturn V vehicle (*Apollo 4*) was launched on November 9, 1967. *Apollo 8*, the first manned flight of the Saturn V, was also the first flight by human beings to the Moon. *Apollo 8* was launched in December 1968 and successfully

orbited the Moon, but did not land. *Apollo 11*, launched by a Saturn V vehicle on July 16, 1969, achieved the first lunar landing—an event often considered the birth of our extraterrestrial civilization!

The Saturn V flew its last manned mission on December 7, 1972, when it sent the *Apollo 17* crew on the final expedition to the Moon under the Apollo Program. The Saturn V was last used as a launch vehicle on May 14, 1973, when it lifted the unmanned *Skylab* space station into low Earth orbit. This first U.S. space station was then occupied by three separate crews for a total of 171 days in 1973-74.

All three stages of the Saturn V used liquid oxygen as the oxidizer. The giant first stage burned kerosene with the oxygen, while the fuel for the upper two stages was liquid hydrogen. The mammoth Saturn V, with the Apollo spacecraft and its small emergency escape rocket on top, stood 111 meters (363 feet) tall, and developed 34.5 million newtons (7.75 million lb-force) of thrust at lift off.

The Saturn 1B vehicle was originally used to launch Apollo spacecraft into Earth orbit as training flights for the manned expeditions to the Moon. The first launch of a Saturn 1B with an unmanned Apollo spacecraft took place in February 1966. A Saturn 1B vehicle launched its first manned Apollo flight (*Apollo 7*) on October 11, 1968. After the Apollo Program, the Saturn 1B vehicle was called upon to place three separate *Skylab* crews aboard the orbiting *Skylab* space station. These launches occurred in 1973. Then, in 1975, the Saturn 1B was used to launch the American crew for the Apollo-Soyuz Test Project—a joint space docking mission with the Soviet Union. The Saturn 1B was 69 meters (223 feet) tall (with an Apollo spacecraft) and developed 7.1 million newtons (1.6 million lb-force) of thrust at lift off.

TITAN III-E/CENTAUR

The Titan III-E/Centaur, first launched in 1974, had an overall height of 48.8 meters (160 feet). It was designed to take advantage of the best features of three well-demonstrated rocket propulsion systems. This vehicle gave the United States an extremely powerful and versatile vehicle for launching large spacecraft on interplanetary missions. The Titan III-E/Centaur was the launch vehicle for the two Viking spacecraft to Mars and the two Voyager spacecraft that encountered both Jupiter and Saturn. This launch vehicle combination also sent two Helios spacecraft toward the Sun. All of these exciting interplanetary missions provided remarkable new information about our Solar System.

The Titan III-E booster was a two-stage liquid-fueled rocket with two large solid propellant rockets attached. At lift off, these solid rockets provided 10.7 million newtons (2.4 million lb-force) of thrust. The Centaur stage (still in use today) produced 133,440 newtons (30,000 lb-force) of thrust from two main engines and burned for up to seven and one-half minutes. The Centaur stage can be restarted several times in space, a design feature that permits more flexibility in launch times.

SPACE SHUTTLE

On April 12, 1981 the first Space Shuttle vehicle lifted off from Launch Complex 39, Pad A, at the Kennedy Space Center—an event that heralded the beginning of a new era in space travel for the world. After a two-day test flight that verified the Orbiter's ability to function in space, the Columbia landed at Edwards Air Force Base in California—becoming the first spaceship to land on Earth. This initial Shuttle mission was also the first time astronauts had been launched on a new space launch vehicle on its maiden flight. The Shuttle vehicle performed extremely well—indicating it was ready to assume its role as the major United States launch vehicle for the 1980s and beyond.

Preparations to launch the Space Shuttle at the Kennedy Space Center required extensive modifications to some existing facilities, such as the Vehicle Assembly Building and the pads at Complex 39; but only two completely new structures were required. One was a 4,600-meter (15,000-foot) runway, and the other was a highly specialized "hangar" called the Orbiter Processing Facility (OPF).

The Space Shuttle flight vehicle consists of a reusable delta-winged space plane called the Orbiter; two solid propellant rocket boosters that are recovered, refurbished and reused; and a giant (currently expendable) External Tank that contains the cryogenic propellants for the Orbiter's main rocket engines. The External Tank contains separate compartments to house liquid hydrogen (fuel) and liquid oxygen (oxidizer).

At launch, the Orbiter's three liquid-fueled rockets (drawing their cryogenic propellants from the huge External Tank) and the two Solid Rocket Boosters (SRBs) burn simultaneously. Together, they generate 30 million newtons (6.7 million lb-force) of thrust under maximum performance conditions. As the Shuttle flight vehicle reaches an altitude of about 50 kilometers (31 miles), the spent solid rockets are detached and parachute into the ocean for recovery and reuse. Meanwhile, the Orbiter and the External Tank continue toward Earth orbit. When the Orbiter's main engines cut off (an event called MECO) just before orbital velocity is achieved, the giant External Tank is jettisoned and tumbles back to Earth impacting in a remote ocean area. Now using its orbital maneuvering engines, the Orbiter with its astronaut crew and payload accelerates into low Earth orbit. There, the Orbiter vehicle and its crew conduct orbital operations for generally between two and eight days.

When the Orbiter and its crew have completed their mission, the aerospace vehicle is prepared for re-entry and descends through the Earth's atmosphere on its return journey to the Earth's surface. Like a glider, it soars through the atmosphere, making an unpowered landing. Touchdown speed is usually above 335 kilometers per hour (210 miles per hour).

The assembled Shuttle vehicle is approximately 56 meters (184 feet) long and 23.3 meters (76 feet) high. The Orbiter measures about 37 meters (122 feet) long and has a wingspan of approximately 24 meters (78 feet). Its payload bay is 18.3 meters (60 feet) long and 4.6 meters (15 feet) wide.

After the Space Shuttle successfully completed its fourth orbital test flight in July 1982, it was declared an operational aerospace vehicle. It is now the major launch vehicle for the United States. It will also carry many international civilian and government payloads into space. One very interesting payload is *Spacelab*, a manned laboratory designed and built by the European Space Agency (ESA) in cooperation with NASA. The first *Spacelab* mission took place from November 28 to December 8, 1983 on board the STS-7 Shuttle flight.

Most of the spacecraft carried into low Earth orbit by the Space Shuttle will be sent to higher operating altitudes by means of orbital transfer vehicles (OTVs) or upper stages. After separating from the Orbiter in low Earth orbit, the OTV will ignite to provide the thrust necessary to transport the spacecraft to a higher orbit or to place the payload on an interplanetary trajectory.

The Space Shuttle is launched into an east-west (low inclination) orbit from the Kennedy Space Center in Florida. Higher inclination (that is, polar) orbits will be provided by Shuttle flights from Vandenberg AFB in California.

The development of the world's first reusable aerospace vehicle is a major step in the creation of our extraterrestrial civilization. The Space Shuttle represents the first in what will be a long line of spaceships capable of carrying human beings and cargo across the "vertical frontier."

See also: **orbital-transfer vehicle; spaceport; Space Transportation System**

space law Space law is basically the code of international law that governs the use and/or control of outer space by different nations on Earth. There are four major international agreements, conventions or treaties that currently govern space activities. These are (1) *Treaty on Principles Governing the Activities of States in the Exploration and Use of Outer Space, Including the Moon and Other Celestial Bodies* (1967), which is also called the "Outer Space Treaty"; (2) *Agreement on the Rescue of Astronauts, the Return of Astronauts and the Return of Objects Launched Into Outer Space* (1968); (3) *Convention on International Liability for Damage Caused by Space Objects* (1972); and (4) *Convention on Registration of Objects Launched into Outer Space* (1975).

The *United Nations Moon Treaty*, or the "Moon Treaty," has been under discussion since late 1971 when the General Assembly adopted resolution 2779, in which it took note of a draft treaty submitted by the Soviet Union (USSR) and requested the Committee on the Peaceful Uses of Outer Space (COPUOS) and its legal subcommittee to consider the question of the elaboration of a draft international treaty concerning the Moon. This Moon Treaty is based to a considerable extent on the 1967 Outer Space Treaty. The Moon Treaty is considered to represent a meaningful advance in international law dealing with outer space. It contains obligations of both immediate and long-term application to such matters as the safeguarding of human life on celestial bodies, the promotion of scientific investigations, the exchange of information about and derived from activities on celestial bodies, and the enhancement of opportunities and conditions for the evaluation, research, and exploitation of the natural resources of celestial bodies. By consensus, the United Nations General Assembly opened the Moon Treaty for signature on December 5, 1979.

With the exception of the 1979 Moon Treaty, the United States has signed and ratified each of the international space agreements discussed above.

Few human undertakings have stimulated so great a degree of legal scrutiny on an international level as has the development of modern space technology. Perhaps it is because space activities involve technologies that do not respect national (terrestrial) boundaries and therefore place new stresses on traditional legal principles. In fact, these traditional legal principles, which are based on the rights and powers of territorial sovereignty, are often in conflict with the most efficient application of new space systems. In order to resolve such complicated and complex Space Age legal issues, both the technologically-developed and developing nations of the Earth have been forced to rely even more on international cooperation.

COPUOS has been and continues to be the main architect of international space law. It was established by resolution of the U.N. General Assembly in 1958 to study the problems associated with the arrival of the Space Age. COPUOS is made up of two subcommittees: one of which studies the technical and scientific aspects of space activities and the other of which studies the legal aspects of space activities. At present COPUOS is conducting international negotiations in the following general areas: (1) remote sensing; (2) direct broadcast satellites; (3) nuclear power sources in space; (4) delimitation of outer space (i.e. where does space begin and national "air space" end from a legal point of view); and (5) military activities in space.

space nuclear power Through the cooperative efforts of the U.S. Department of Energy (DOE), formerly called the Atomic Energy Commission, and NASA, the United States has successfully used nuclear energy in its space program to provide electrical power for many missions, including science stations on the Moon, extensive exploration missions to Jupiter, Saturn and beyond, and even in the search for extraterrestrial life on Mars.

For example, when the *Apollo 12* astronauts departed from the lunar surface on their return trip to Earth (November 1969), they left behind a nuclear-powered science station that sent information back to terrestrial scientists for several years. That system, as well as similar stations left on the Moon by the *Apollo 14* through *Apollo 17* missions operated on electrical power supplied by radioisotope thermoelectric generators (RTGs). In fact, since 1961 nuclear power systems have helped assure the success of many space missions, including the *Pioneer 10* and *11* missions to Jupiter and Saturn, the *Viking 1* and *2* landers on Mars, and the spectacular *Voyager 1* and *2* missions to Jupiter, Saturn and beyond. It should also be realized that these magnificent space exploration missions would not have been possible without the use of nuclear energy (see table 1).

Energy supplies that are reliable, transportable and abundant represent a very important technology in the development of our extraterrestrial civilization. Space nuclear power systems will play an ever expanding role in the more ambitious space exploration and resource exploitation missions of the next few decades. For example, the movement of massive payloads from low Earth orbit (LEO) to high Earth orbit (HEO) or a lunar destination, the operation of very large space platforms throughout cislunar space and the successful startup and expansion of the initial lunar bases can all benefit from the creative application of advanced space nuclear power system technologies. Even more progressive space activities, such as asteroid movement and mining, planetary engineering and climate control, and human expeditions to Mars and the planets beyond, all require compact energy systems at the megawatt and gigawatt levels.

Space nuclear power supplies offer several distinct advantages over the more traditional solar and chemical space power systems. These advantages include: compact size, modest mass requirements, very long operating lifetimes, the ability to operate in extremely hostile environments (for example, intense trapped radiation belts, the surface of Mars, the moons of the outer planets and even interstellar space), and the ability to operate independent of distance from the Sun or orientation to the Sun. It appears that as the energy requirements of our initial extraterrestrial civilization efforts approach hundreds of kilowatts to megawatts, nuclear energy systems represent the only realistic technological option in the next few decades.

Space nuclear power systems use the thermal energy or heat released by nuclear processes. These processes include the spontaneous but predictable decay of radioisotopes, the controlled splitting or fissioning of heavy atomic nuclei (such as uranium-235) in a sustained neutron chain reaction, and the joining together or fusing of light atomic nuclei (such as deuterium and tritium) in a controlled thermonuclear reaction. This "nuclear" heat can then be applied directly or converted by a variety of engineering techniques into electric power. Until we successfully achieve controlled thermonuclear fusion capabilities, nuclear energy applications (both terrestrial and extraterrestrial) will be based on the use of radioisotope decay and nuclear fission reactors.

Figure 1 illustrates a basic radioisotope thermoelectric generator. This RTG consists of two main functional components: the thermoelectric converter and the nuclear heat

Table 1   Summary of Space Nuclear Power Systems Launched by the United States (1961–1985)

Power Source	Spacecraft	Mission Type	Launch Date	Status
SNAP–3A	Transit 4A	Navigational	June 29, 1961	Successfully achieved orbit
SNAP–3A	Transit 4B	Navigational	November 15, 1961	Successfully achieved orbit
SNAP–9A	Transit–5BN–1	Navigational	September 28, 1963	Successfully achieved orbit
SNAP–9A	Transit–5BN–2	Navigational	December 5, 1963	Successfully achieved orbit
SNAP–9A	Transit–5BN–3	Navigational	April 21, 1964	Mission aborted: burned up on reentry
SNAP–10A (reactor)	Snapshot	Experimental	April 3, 1965	Successfully achieved orbit
SNAP–19B2	Nimbus–B–1	Meteorological	May 18, 1968	Misson aborted: heat source retrieved
SNAP–19B3	Nimbus III	Meteorological	April 14, 1969	Successfully achieved orbit
SNAP–27	Apollo 12	Lunar	November 14, 1969	Successfully placed on lunar surface
SNAP–27	Apollo 13	Lunar	April 11, 1970	Mission aborted on way to Moon. Heat source returned to South Pacific Ocean
SNAP–27	Apollo 14	Lunar	January 31, 1971	Successfully placed on lunar surface
SNAP–27	Apollo 15	Lunar	July 26, 1971	Successfully placed on lunar surface
SNAP–19	Pioneer 10	Planetary	March 2, 1972	Successfully operated to Jupiter and beyond; in interstellar space
SNAP–27	Apollo 16	Lunar	April 16, 1972	Successfully placed on lunar surface
Transit–RTG	"Transit" (Triad–01–1X)	Navigational	September 2, 1972	Successfully achieved orbit
SNAP–27	Apollo 17	Lunar	December 7, 1972	Successfully placed on lunar surface
SNAP–19	Pioneer 11	Planetary	April 5, 1973	Successfully operated to Jupiter, Saturn and beyond
SNAP–19	Viking 1	Mars	August 20, 1975	Successfully landed on Mars
SNAP–19	Viking 2	Mars	September 9, 1975	Successfully landed on Mars
MHW	LES 8/9	Communications	March 14, 1976	Successfully achieved orbit
MHW	Voyager 2	Planetary	August 20, 1977	Successfully operated to Jupiter, Saturn and beyond
MHW	Voyager 1	Planetary	September 5, 1977	Successfully operated to Jupiter, Saturn and beyond

SOURCE: NASA, Dept. of Energy.

source. The isotope, plutonium-238, has been used as the heat source in all U.S. space missions involving radio-isotope power supplies. Plutonium-238 has a half-life of 87.7 years and therefore supports a long operational life. (The half-life is the time required for one half the number of unstable nuclei present at a given time to undergo radioactive decay.) In the nuclear decay process, plutonium-238 emits mainly alpha radiation, that has very low penetrating power. Consequently, only light-weight shielding is required to protect the spacecraft from its nuclear radiation. A thermoelectric converter uses the "thermocouple principle" to directly convert a portion of the nuclear heat into electricity.

We can also use a nuclear reactor to provide space nuclear power. The United States flew one space nuclear reactor, called SNAP-10A, in 1965. The objective of this program was to develop a space nuclear reactor power unit capable of producing a minimum of 500 watts-electric for a period of one year while operating in the extraterrestrial environment. SNAP-10A was the first and only space reactor flight tested in Earth orbit by the United States. It was a small zirconium hydride (ZrH) thermal reactor fueled by uranium-235. The SNAP-10A orbital test was successful, although the mission was pre-maturely terminated by the failure of an electronic component outside the reactor.

Work is presently going on to define and develop a 100 kilowatt-electric class space nuclear reactor for applications in the 1990s and beyond. Figure 2 shows a typical reactor "core" design for this advanced type of space

Fig. 1   Typical radioisotope thermoelectric generator (RTG) configuration for a space nuclear power application (the device is a modified SNAP 19 system). (Courtesy of NASA.)

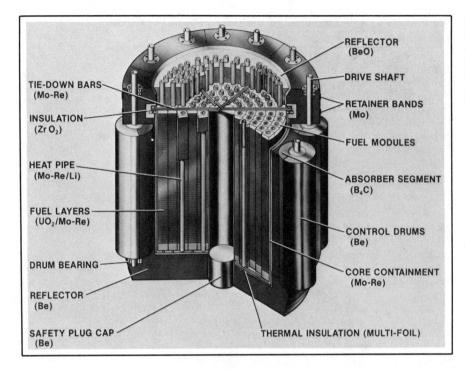

TIE-DOWN BARS
(Mo-Re)

INSULATION
(Zr O₂)

HEAT PIPE
(Mo-Re/Li)

FUEL LAYERS
(UO₂/Mo-Re)

DRUM BEARING

REFLECTOR
(Be)

SAFETY PLUG CAP
(Be)

REFLECTOR
(BeO)

DRIVE SHAFT

RETAINER BANDS
(Mo)

FUEL MODULES

ABSORBER SEGMENT
(B₄C)

CONTROL DRUMS
(Be)

CORE CONTAINMENT
(Mo-Re)

THERMAL INSULATION (MULTI-FOIL)

Fig. 2 The core of a typical advanced–design heat pipe space reactor. (The heat pipes transport the thermal energy out of the core and bring it to energy conversion devices). (Courtesy of Los Alamos National Laboratory.)

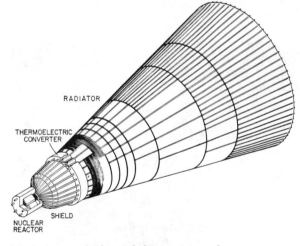

RADIATOR

THERMOELECTRIC
CONVERTER

SHIELD

NUCLEAR
REACTOR

Fig. 3 A typical advanced–design space nuclear power system (100 kilowatt-electric power level). (Drawing courtesy of the Los Alamos National Laboratory and the U.S. Department of Energy.)

power supply; while figure 3 shows the major components of a typical advanced space nuclear power system. It is interesting to note that the reactor itself is generally only the size of a waste-paper basket! This type of system would be capable of continuous operation for many years at a constant power level of 100 kilowatt-electric or better. Such space nuclear power systems would also be made compatible for launch by the Space Shuttle.

Since the United States first used nuclear power in space, great emphasis has been placed on the safety of people and the protection of the terrestrial environment. A continuing major object in any new space nuclear power

program is to avoid undue risks. In the case of radioisotope power supplies, this means designing the system to contain the nuclear isotope fuel under all normal operating and potential accident conditions. For advanced space nuclear reactor systems, this means launching the reactor in a "cold" (non-operating) configuration and starting up the reactor only after a safe, stable Earth orbit or interplanetary trajectory has been achieved.

See also: **fission (nuclear)**; **fusion; nuclear energy; space nuclear propulsion**

space nuclear propulsion Nuclear fission reactors can be used in two basic ways to propel a space vehicle: (1) to generate electric power for an electric propulsion unit; and (2) as a thermal energy or heat source to raise a propellant (working material) to extremely high temperatures for subsequent expulsion out a nozzle. In the second application, the system is often called a nuclear rocket (see fig. 1).

In a nuclear rocket, chemical combustion is not required. Instead, a single propellant, usually hydrogen, is heated by the energy released in the nuclear fission process which occurs in a controlled manner in the reactor's core. Conventional rockets, in which chemical fuels are burned, have severe limitations in the specific impulse a given propellant combination can produce. These limitations are imposed by the relatively high molecular weight of the chemical combustion products. At attainable combustion chamber temperatures the best chemical rockets are limited to specific impulse values of about 4,300 meters per second (440 seconds). Nuclear rocket systems using fission reactions, fusion reactions and even possibly matter-antimatter annihilation reactions (the "photon rocket")

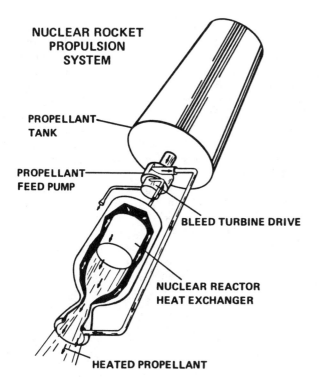

NUCLEAR ROCKET PROPULSION SYSTEM

PROPELLANT TANK

PROPELLANT FEED PUMP

BLEED TURBINE DRIVE

NUCLEAR REACTOR HEAT EXCHANGER

HEATED PROPELLANT

Fig. 1   Nuclear rocket propulsion system. (Drawing courtesy of the U.S. Department of Energy.)

have much greater propulsion performance capabilities.

Engineering developments will be needed in the 21st century to permit the use of advanced fission reactor systems, such as the gaseous core reactor rocket or even fusion powered systems. However, the solid-core nuclear reactor rocket is within a flight-test demonstration of engineering reality. In this nuclear rocket concept, hydrogen propellant is heated to extremely high temperatures while passing through flow passageways within the solid fuel elements of a compact nuclear reactor system that uses uranium-235 as the fuel. The high temperature gaseous hydrogen then expands through a nozzle to produce propulsive thrust. From the mid–1950s until the early 1970s, the United States conducted a nuclear rocket program called Project Rover. The primary objective of Project Rover was to develop a nuclear rocket for a manned mission to Mars. Unfortunately, despite the technical successes of this nuclear rocket program, overall space program emphasis changed and the nuclear rocket and the Mars mission planning were discontinued in 1973.

Engineers who are involved with the development of our extraterrestrial civilization recognize one basic fact: the secret of space travel and of extending our human civilization throughout heliocentric space is energy—large amounts of portable energy for power and propulsion systems. The compact energy advantages of nuclear fission reactors (and eventually controlled fusion systems) will ultimately enable us to develop the powerful, reusable interplanetary spaceships we need to sweep across the

Solar System. In the upcoming decades, nuclear rockets—based initially perhaps on Rover technology and then evolving into more advanced propulsion system technologies—will carry extraterrestrial settlers and their equipment to the surface of the Red Planet, to the minor planets of the main asteroid belt and to the exciting constellations of moons surrounding the gaseous giants, Jupiter and Saturn.

See also: **fission** (nuclear); **fusion; nuclear-electric propulsion system; nuclear energy; Project Daedalus; Project Orion (II); space nuclear power; starship**

Space Operations Center (SOC)   A concept recently investigated by NASA for a manned space station in low Earth orbit (LEO). The Space Operations Center (SOC) would provide a habitable operational base in LEO to perform missions requiring an extended stay in orbit with frequent or continuous astronaut involvement. It would be a self-contained orbital facility built up of several Shuttle-launched modules. With resupply, on-orbit refurbishment and on-orbit maintenance, the SOC would be capable of continuous operation for an indefinite period. In normal operation, the station would be continuously inhabited, although unmanned operation is also possible.

See also: **large space structures; space construction; space industrialization; space station**

space platform   An unmanned, free-flying orbital platform that is dedicated to a specific mission, such as space industrialization activities (materials processing in space) or scientific research (space telescope). It would orbit near a permanently manned space station and would be serviced by either the Space Shuttle or the space station.

See also: **space station**

spaceport   A spaceport is a planet's doorway to outer space. At a spaceport we find the unique facilities and equipment required for the assembly, testing, launching and (in the Shuttle Era) the landing of spaceships.

The John F. Kennedy Space Center (KSC), located on the east central coast of Florida is America's spaceport and the major NASA launch organization for manned and unmanned space missions. As the lead center within NASA for the development of launch procedures, technology and facilities, the Kennedy Space Center launches unmanned interplanetary spacecraft and scientific, weather and communications satellites. KSC also serves as the launch and landing facility for the reusable Space Shuttle, an aerospace vehicle that has opened up a new era in space transportation.

The operations performed at the Kennedy Space Center include: assembly of space vehicles; preflight preparation of space vehicles and their payloads; test and checkout of launch vehicles, spacecraft and facilities; coordination of tracking and data acquisition requirements; countdown and launch operations; and landing operations and refurbishment of the Space Shuttle for future missions.

Supporting this primary mission are a great variety of technical and administrative functions. These include design engineering, safety, quality assurance, documentation, supply, maintenance, computer operations and communications.

Launch Complex 39 at KSC served as the launch site for the epic *Apollo* expeditions to the Moon; the *Skylab* mission (first U.S. space station); and the American astronaut crew that participated in the joint U.S.-U.S.S.R. *Apollo-Soyuz* Test Project. The facilities at Complex 39 were then modified to accommodate the Space Shuttle, the major U.S. space launch vehicle for the 1980s and 1990s.

NASA was created on October 1st, 1958. This was 12 months after the Soviet Union launched *Sputnik 1*, the first artificial satellite placed around the Earth; and nine months after the launch of the first American satellite, *Explorer 1*. The history of NASA reflects the complex task of initiating and implementing a national space program involving various government agencies, industry and the scientific community. The early focus of NASA's launch operations centered on Cape Canaveral—the site selected by the military following World War II for the testing of long-range guided missiles. This world famous piece of land jutting out into the Atlantic Ocean was selected because of the chain of islands that stretched southeastward to Ascension Island which could accommodate tracking stations to measure vehicle trajectories and performance. In 1947, Congress authorized and approved the construction of the Atlantic Missile Range, now called the Eastern Test Range. As part of the Eastern Space and Missile Center (ESMC) of the Air Force Systems Command, the Eastern Test Range operates and maintains the largest missile proving ground in the world. Cape Canaveral Air Force Station (CCAFS) is Station One of the range which extends over 16,000 kilometers into the Indian Ocean. Its mission is to provide facilities and support services for launching missiles and rockets and to gather useful information from these flights. The Eastern Test Range supports NASA launches from both Cape Canaveral and Launch Complex 39 at KSC. The Army and Navy have used the range facilities in the development of rocket-powered weapons systems. NASA also maintains facilities at the Western Test Range, Vandenberg Air Force Base, California, for launching spacecraft into high inclination (polar) orbits.

As the civilian space program got underway, Cape Canaveral became the headquarters of NASA's Launch Operations Center, later named the John F. Kennedy Space Center. In late 1964, the NASA Kennedy Space Center was relocated to a facility on Merritt Island, adjacent to Cape Canaveral Air Force Station. The site, selected in 1961, occupies some 34,000 hectares of land and water; approximately 22,600 additional hectares owned by the State of Florida are also under the control of NASA KSC. Facilities were installed to accommodate the enormously powerful Saturn launch vehicles used to carry

men to the Moon and later to support the *Skylab* and *Apollo-Soyuz* manned missions. Beginning in 1976, these facilities were modified and new ones built to support the launch and landing of the Space Shuttle. All but the operational areas of the spaceport are designated as a national wildlife refuge, much of which is open to the public. In 1975, 16,600 hectares of the spaceport were designated as part of the Canaveral National Seashore.

Complex 39 is considered one of America's major engineering accomplishments. Here manned space vehicles are assembled and checked out in the protective environment of an assembly building, then moved to the launch site for final preparation and launch. Upon return from space, the Shuttle Orbiter is refurbished and prepared for new missions. Originally designed to support the Apollo Lunar Landing Program, the facilities at Complex 39 have now been modified for Space Shuttle operations.

The Complex 39 Space Shuttle facilities consist of:

(1) A 160-meter (525-foot) tall Vehicle Assembly Building (VAB) for assembly and checkout of Space Shuttle flight vehicles.

(2) An Orbiter Processing Facility (OPF), where the Space Shuttle Orbiter is stripped of ordnance and fuel residue, inspected, tested and refurbished as necessary. Here large payloads are inserted horizontally into the Orbiter's cargo bay.

(3) A Shuttle Landing Facility with a 4,572-meter (15,000 foot) long, 91.4-meter (300 foot) wide landing runway for Shuttle Orbiters. This runway is one of the largest in the world, and is located northwest of the VAB.

(4) A Launch Control Center with two firing rooms for control and monitoring of Shuttle vehicle preparation and launch operations.

(5) A Launch Processing System that performs most Shuttle checkout and launch functions automatically.

(6) Two Mobile Launcher Platforms upon which the Shuttle is erected, checked out and then moved to Pad 39.

(7) Two Crawler-Transporters which carry the Mobile Launcher Platforms with the assembled Space Shuttle to the launch pad.

(8) A Crawlerway, as wide as a turnpike, over which the Transporter slowly moves on its way to Pad 39.

(9) Two pads, roughly octagonal in shape, with a Fixed Service Structure to connect the Shuttle vehicle with ground support equipment, and a Rotating Service Structure to install smaller payloads at the pad when the Orbiter is in the vertical (launch) position.

The Industrial Area at KSC is the nerve center of the Spaceport. It is located 8 kilometers south of Launch Complex 39. Here administrators, scientists, engineers and technicians plan and accomplish many of the detailed operations associated with prelaunch processing and testing of space vehicles. The largest structure in the Industrial Area is the Operations and Checkout (O&C) Building. It contains offices, laboratories, astronaut quarters and spacecraft assembly areas. Facilities once used for the assembly and testing of Apollo spacecraft are now used for

Fig. 1   The European Space Agency's *Spacelab* undergoing pre-flight testing in the Operations and Checkout Building at Kennedy Space Center (October 1982). (Courtesy of Kennedy Space Center.)

Spacelab, the manned laboratory designed and built by the European Space Agency, which is carried into space inside the Orbiter's huge cargo bay. *Spacelab* and its many varying configurations and experiments make up the majority of the payloads handled in the O&C Building (see fig. 1).

The Central Instrumentation Facility is the heart of the spaceport's instrumentation and processing operations. It provides instrumentation to receive, monitor, process, display and record information received from space vehicles during testing, launch and landing.

Hazardous checkout operations are conducted in a remote part of the Industrial Area, about 13 kilometers south of the Vehicle Assembly Building. Clusters of small orbital maneuvering system (OMS) engines using hypergolic propellants (chemicals that ignite upon contact with each other) are processed and stored in this area.

With an operational Space Shuttle and the development of a permanent space station in low Earth orbit, the Kennedy Space Center will serve as one of the world's most important "ports." In the upcoming decades, goods and travelers will routinely flow through KSC on their way to help build our extraterrestrial civilization. Unique space-manufactured products and interesting things extraterrestrial found on alien worlds will flow back into the main terrestrial civilization through this spaceport in a manner very analogous to the way the wealth of the American continent (the "new world") flowed back through major port cities in Europe (the "old world") in the 16th, 17th and 18th centuries.

See also: **NASA facilities; space launch vehicles; Space Transportation System**

space settlement   A large extraterrestrial habitat where from 1,000 to perhaps 100,000 people would live, work

and play, while supporting space industrialization activity, such as the operation of a large space manufacturing complex or the construction of satellite power systems. Figure C31 on color page K illustrates a torus-shaped space settlement for about 10,000 people. Its inhabitants, all members of a space manufacturing complex workforce, would return after work to homes on the inner surface of the large torus, which is nearly 1.6 kilometers in circumference (see fig. C32 on color page M). It would rotate to provide the inhabitants with a gravity level similar to that experienced on the surface of the Earth. This habitat would be shielded against cosmic rays and solar flare radiation by a non-rotating shell of material that could be built up from accumulated slag or waste materials from lunar or asteroid mining operations. Outside the shielded area agricultural crops would be grown taking advantage of the intense, continuous stream of sunlight available in space. Docking areas and microgravity industrial zones are located at each end of the settlement; so are the large flat surfaces necessary to radiate waste heat away from the facility to outer space.

Another possible design is a spherical space settlement, called the Bernal sphere. This giant spherical habitat would be approximately two kilometers in circumference. Up to 10,000 people would live in residences along the inner surface of the large sphere (see fig. 1 on page 15). Rotation of the settlement at about 1.9 revolutions per minute (RPM) would provide Earth-like gravity levels at the sphere's equator, but there would be essentially microgravity conditions at the poles. In the settlement's "polar regions" human-powered flying machines could be used and the space settlers would be able to enjoy a variety of microgravity recreational pursuits. Because of the short distances between things in the equatorial residential zone, passenger vehicles would not be necessary. Instead, the space settlers would travel on foot or perhaps by bicycle. The climb from the residential equatorial area up to the sphere's poles would take about 20 minutes and would lead the hiker past small villages, each at progressively lower levels of artificial gravity. A corridor at the axis would permit residents to float safely in microgravity out to exterior facilities, such as observatories, docking ports, industrial and agricultural areas. Ringed areas above and below the main sphere in this type of space settlement would be the external agricultural toruses (see fig. C33 on color page M).

Figure C34 on color page M depicts a very large set of twin 32 kilometer-long, 6.4 kilometer-diameter, cylindrical space settlements. These huge space settlements would be able to house several hundred thousand people. Each cylinder rotates around its main axis once every 114 seconds to create an Earth-like level of artificial gravity. The teacup-shaped containers ringing each cylinder are agricultural stations. Each cylinder is capped by a space industrial facility and a power station. Large movable rectangular mirrors on the sides of each cylinder (hinged at one end to the cylinder) would direct sunlight into the

Table 1   Suggested Physiological Design Criteria for a Space Settlement

Pseudogravity	0.95 ± 0.5 g
Rotation rate	$\leqslant 1$ rpm
Radiation exposure for the general population	$\leqslant 0.5$ rem/yr
Magnetic field intensity	$\leqslant 100 \ \mu$T
Temperature	$23° \pm 8°$ C
Atmospheric composition pO_2	22.7 ± 9 kPa (170 ± 70 mm Hg)
p(Inert gas; most likely N_2)	26.7 kPa $< pN_2 < 78.9$ kPa ($200 < pN_2 < 590$ mm Hg)
pCO_2	<0.4 kPa (<3 mm Hg)
pH_2O	1.00 ± 0.33 kPa (7.5 ± 2.5 mm Hg)

SOURCE: NASA.

Table 2   Suggested Quantitative Environmental Design Criteria for a Space Settlement

Population: men, women, children	10,000
Community and residential, projected area per person, m^2	47
Agriculture, projected area per person, m^2	20
Community and residential, volume per person, m^3	823
Agriculture, volume per person, m^3	915

SOURCE: NASA.

Table 3   Suggested Qualitative Environmental Design Criteria for a Space Settlement

Long lines of sight
Larger overhead clearance
Noncontrollable unpredictable parts of the environment; for example, plants, animals, children, weather
External views of large natural objects
Parts of interior out of sight of others
Natural light
Contact with the external environment
Availability of privacy
Good internal communications
Capability of physically isolating segments of the habitat from each other
Modular construction
    of the habitat
    of the structures within the habitat
Flexible internal organization
Details of interior design left to inhabitants

SOURCE: NASA.

habitat's interior, control the day-night cycles and even regulate the settlement's "seasons." A random number generator somewhere in the mirror's controller loop could be used to provide weather variations that are unpredictable but within certain previously established limits. This type of controlled randomness might be very necessary in overcoming some of the psychological problems that could arise from living in a totally "artificial" or man-made world.

The basic space settlement design will have to satisfy the essentials for life such as: air, food, water and, for extended stays in space, some type of artificial gravity. The space settlement design must ensure not only physiological safety and comfort but must also satisfy the psychological and esthetic needs of the inhabitants. Table 1 describes some of the physiological design criteria that can be applied to a space settlement.

Human beings living in space must have an adequate diet. Food in a large settlement should be nutritious, sufficiently abundant and even attractive. The settlers can get their food supplies from the Earth or the Moon (permanent lunar settlements in the next century will practice space farming and should be able to export food products to markets within cislunar space); or else, the settlers may elect to grow their own. Recent space station and space settlement studies indicate that when more than 10,000 person-days of food are needed each year in cislunar space, agriculture in space becomes economically competitive with food import from Earth. It also appears that for a large space settlement, a modified type of terrestrial agriculture, based on plants and meat-bearing animals, should solve both nutritional requirements and the need for dietary variety.

Photosynthetic agriculture can also be used to help regenerate the space settlement's atmosphere by converting carbon dioxide and generating oxygen. Space agricultural activities might even provide a source of pure water from the condensation of humidity produced by transpiration.

The design of a space settlement must not exert damaging psychological stresses on its inhabitants. A sense of

isolation (the Shimanagashi syndrome) or a sense of artificiality (the Solipsism syndrome) must be avoided through variety, diversity and flexibility of interior designs. Table 2 lists some suggested quantitative environmental design criteria for a large space settlement, while table 3 provides some qualitative design criteria.

The space settlement must also have a form of government or political organization that permits its inhabitants to enjoy a comfortable life-style under conditions that are crowded and physically isolated from other human communities. Since early space settlements may very likely be "company towns" dedicated to some particular space industrial activity, their organizations should also support a fairly high level of productivity and should maintain the physical security of the habitat.

Without proper organization and internal security this type of isolated community could easily become the victim of despots and self-elected demigods. If things really got out of hand in the settlement, a "space marshal" might have to be sent from Earth (arriving, of course, on the noon space tug) to restore law and order on the extraterrestrial frontier. One might even imagine a 21st century "spacer" (ET analog to a western) entitled: *Lasers At the Airlock Canal.*

Full social, political and economic autonomy from Earth does not appear possible for space settlements with populations under about 500,000. For example, a community of from 10,000 to 50,000 would be hard pressed to even support a large university or medical center. Therefore, many of the services and benefits of a full civilization (such as large universities) will still be supplied from Earth—at least until the lunar civilization attains a social critical mass of about half a million selenians.

At that point, some very interesting things will happen. For example, teen-age space settlers and their parents will have the opportunity to examine two sets of college catalogs: one from Earth and one from the Moon. And, as has never occurred in generations past, they will have the distinct opportunity of evaluating the pros and cons for advanced education in institutions on two different worlds! Of course, our young space settler might elect to stay at home and take teleconferenced courses through the Community College of Cislunar Space.

Space settlements, whatever their final design, population or political structure, will emerge as a major part of our extraterrestrial civilization in the next century and beyond. We cannot fully appreciate today the impact that (almost) self-sufficient pockets of humanity sprinkled throughout cislunar space will have on the technical, political, economic and social structure of 21st-century living. In time, as these settlements grow and replicate themselves, we will witness the rise of extraterrestrial city-states throughout cislunar and then heliocentric space.

Sometime in the next century, terrans who had maintained very little interest in things above the atmosphere, will suddenly awaken to find: "We have met the extra-terrestrials and they are us!" Can you imagine the social impact of certain interplanetary relationships! "Your place or mine?" will now become a question of astronomical proportions.

The term "space settlement" is also used to describe permanent habitats for over about 1,000 people on the surface of another world, such as the Moon or Mars.

See also: **Bernal sphere; Dyson sphere; lunar bases and settlements; Mars; space industrialization; space station**

space station   A space station is an orbiting space system that is designed to accommodate long-term human habitation in space. The concept of people living in artificial habitats in outer space appeared in 19th-century science fiction literature in stories such as Edward Everett Hale's "Brick Moon" (1869) and Jules Verne's "Off on a Comet" (1878).

At the turn of the century, Konstantin Tsiolkovsky provided the theoretical underpinnings for this concept with his truly visionary writings about the use of orbiting stations as a springboard for exploring the Cosmos. Tsiolkovsky, the father of Soviet astronautics, provided a more technical introduction to the space station concept in his 1895 work: *Dreams of Earth and Heaven, Nature and Man.* He greatly expanded on the idea of a space station in his 1903 work entitled: "The Rocket into Cosmic Space." In this technical classic, Tsiolkovsky described all the essential ingredients needed for a manned space station

including the use of solar energy, the use of rotation to provide artificial gravity and the use of a closed ecological system complete with "space greenhouse."

Throughout the first half of the 20th century the space station concept continued to technically evolve. For example, the German scientist, Hermann Oberth, described the potential applications of a space station in his classic work: *The Rocket into Interplanetary Space* (1923). These suggested applications included the use of a space station as an astronomical observatory, an Earth-monitoring facility, and a scientific research platform. In 1929 an Austrian named Potočnik (pen name Hermann Noordung) introduced the concept of a rotating, wheel-shaped space station. Potočnik called his design "Wohnrad" or "Living Wheel." Another Austrian, Guido von Pirquet, wrote many technical papers on space flight, including the use of a space station as a refueling node for space tugs. In the late 1920s and early 1930s, von Pirquet also suggested the use of multiple space stations at different locations in cislunar space. After World War II Dr. Wernher von Braun popularized the concept of a wheel-shaped space station in the United States.

The idea of much larger man-made space habitats and mini-planets accompanied the birth of the space station concept. In 1918 Dr. Robert Goddard, the father of American rocketry, introduced the concept of nuclear-powered space arks that could carry an entire civilization from one solar system to another. (Possibly to avoid criticism, Goddard's manuscript, entitled "The Ultimate Migration," describing this space ark was not published until after his death.) The British scientist and writer, J. D. Bernal, described man-made planets and large self-contained worlds in his 1929 work: *The World, the Flesh and the Devil.* The brilliant writer and science prophet Arthur C. Clarke used very large space stations and habitats in his 1952 novel: *Islands In the Sky.* The space engineer and visionary, Dr. Krafft Ehricke, suggested an entire line of evolutionary space stations and space habitats in his writings during the 1960s and 1970s. His uniquely far-reaching concepts included Astropolis (an orbiting city-state) and the Androcell (a miniature man-made planet). And most recently, several groups of space technology thinkers under the initial auspices of NASA's Ames Research Center conceived of several large space settlement designs, whose populations were committed to space manufacturing and the construction of Satellite Power Systems. The efforts of scientists like G. K. O'Neill and Brian O'Leary have helped stimulate much contemporary interest in living and working in outer space. Many of these larger space settlement designs might actually evolve from the modularized space station designs of the next two decades.

SKYLAB—FIRST U.S. SPACE STATION

Even before the Apollo program had successfully landed men on the Moon, NASA engineers and scientists were considering the next giant step in the U.S. manned space flight program. That next step became the simultaneous development of two complementary space technology capabilities. One was a safe, reliable transportation system that could provide routine access to space. The other was an orbital space station where human beings could live and work in space. This space station would serve as a base camp from which other, more advanced space technology developments could be initiated. This long-range strategy set the stage for two of the most significant American space activities carried out in the 1970s and 1980s: *Skylab* and the Space Shuttle.

On May 14, 1973 the United States launched its first space station, called *Skylab*. It was launched by a Saturn V booster from the Apollo program (see fig. C35 on color page N and fig. 1). *Skylab* demonstrated that people could function in space for periods up to 12 weeks and, with proper exercise, could return to Earth with no ill effects.

In particular, the flight of *Skylab* proved that human beings could operate very effectively in a prolonged microgravity environment and that it was not essential to provide artificial gravity for people to live and work in space (at least for periods up to about six months). The *Skylab* astronauts accomplished a wide range of emergency repairs on station equipment, including freeing a stuck solar panel array (a task which saved the entire mission), replacing rate gyros, and repairing a malfunctioning antenna. On two separate occasions the crews installed portable sun shields to replace the originals lost when *Skylab* was launched. These on-orbit activities clearly demonstrated the unique and valuable role people have in space. Table 1 summarizes the Skylab missions.

The Skylab program also provided a realization that extravehicular activity (EVA) can be considered normal and even routine. The *Skylab* astronauts performed more than 82 hours of EVA. Skylab demonstrated the effectiveness of an orbiting scientific laboratory. *Skylab*'s eight different solar telescopes provided a quantum leap in our understanding of the Sun. The *Skylab* astronauts also performed many pioneering experiments in materials

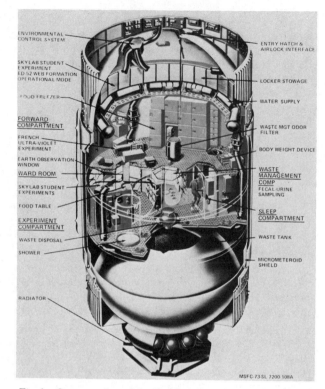

Fig. 1   Cutaway view of the *Skylab* Orbital Workshop. (Courtesy of Johnson Space Center.)

processing, the multispectral remote sensing of Earth and in the life sciences. These results now form the technical basis for more advanced activities on a permanent space station.

Unfortunately, *Skylab* was not designed for a permanent presence in space. The system was not designed to be routinely serviced on orbit (although the *Skylab* crews were able to perform certain repair functions). *Skylab* was not equipped to maintain its own orbit—a design deficiency that eventually caused its fiery demise on July 11, 1979 over the Indian Ocean and portions of western

Table 1   Skylab Program Summary

Mission	Dates	Astronaut Crew	Mission Duration	Remarks
SKYLAB 1	Launched: May 14, 1973	Unmanned	May 14, 1973 to July 11, 1979	100-ton space station visited by three crews. Descended into Earth's atmosphere over Indian Ocean and Australia after 34,981st orbit.
SKYLAB 2	May 25–June 22, 1973	Charles Conrad Jr. Paul J. Weitz Joseph P. Kerwin	28 days, 50 min.	Repaired *Skylab*; 404 revolutions; 392 experiment hours; 3 EVAs total 5 hrs, 34 min.
SKYLAB 3	July 28–September 25, 1973	Alan L. Bean Jack R. Lousma Owen K. Garriott	59 days, 11 hrs, 9 min.	Performed station maintenance; 858 revolutions; 1,081 experiment hours; 3 EVAs total 13 hrs, 42 min.
SKYLAB 4	November 16, 1973–February 8, 1974	Gerald P. Carr William R. Pouge Edward G. Gibson	84 days, 1 hr, 16 min.	Observed Comet Kohoutek; 1,214 revolutions; 1,563 experiment hours; 4 EVAs total 22 hrs, 25 min.

SOURCE: NASA.

Australia. Finally, it was not designed for evolutionary growth and therefore was subject to rapid technological obsolescence. Future space station designs will take these shortcomings into account, and effectively use the highly successful Skylab program experience to develop a permanent, evolutionary and modular U.S. space station.

SOVIET SPACE STATION ACTIVITIES

While the United States was conducting the *Skylab* program, the Soviet Union was embarking on an even more ambitious space station endeavor. The Soviet Union launched its first space station, *Salyut-1*, in April 1971. In the years since that launch, additional Salyut stations have been placed in orbit. The Soviet program reflects an apparently aggressive and expanding national commitment to manned space flight. The two most recent Salyuts (*Salyut-6* and *Salyut-7*) represent a second generation of Soviet space station design. *Salyut-6* demonstrated the ability to refuel the station using the Progress logistics spacecraft. *Salyut-6* cosmonauts lived on orbit for more than 200 days. Many international space travelers, representing nations such as Czechoslovakia, Poland, the German Democratic Republic, Bulgaria, Hungary, Vietnam, Cuba, Mongolia, Romania, France and India were hosted on orbit by *Salyut* cosmonauts.

In a statement provided to the Office of Technology Assessment, Congress of the United States, in December 1983, Dr. Balayan, Vice Chairman of the Intercosmos Council of the USSR Academy of Sciences, said:

> The creation of the Salyut orbital stations is an important stage in the development of Soviet cosmonautics, intended to increase the length of both manned and unmanned flights.

The late Leonid Brezhnev, then president of the Soviet Union, made these remarks in 1978, which appear to echo the visions of Konstantin Tsiolkovsky, the founder of Soviet astronautics:

> Mankind will not forever remain on Earth, but in the pursuit of light and space will first timidly emerge from the bounds of the atmosphere, and then advance until he has conquered the whole of circumsolar space. We believe that permanently manned space stations with interchangeable crews will be mankind's pathway to the Universe.

The Salyut program represents a long-term, methodical development of space capabilities that the Soviet Union considers worthwhile and in their overall national interest. This effort appears to be well financed and has involved many high-energy launches. In the next few years, U.S. space experts anticipate seeing flight tests of a Soviet space shuttle, the test of an extremely powerful Soviet space booster and the appearance of a small, highly maneuverable space plane capable of re-entry through the Earth's atmosphere. In fact, some U.S. space experts now speculate that a Salyut station may provide the core element of a future space base needed to make successful manned missions to Mars possible.

CURRENT U.S. SPACE STATION PROGRAM

An effort is now underway by the United States to achieve the next major step in its space program: the development and deployment of a permanently manned space station. NASA has received firm direction from the U.S. President for a strong, national space program. In his State of the Union Message on January 25, 1984, President Ronald Reagan said:

> We can follow our dreams to distant stars, living and working in space for peaceful economic and scientific gain. Tonight I am directing NASA to develop a permanently manned space station and to do it within a decade.
>
> A space station will permit quantum leaps in our research in science, communications and in metals and life-saving medicines which can be manufactured in space. We want our friends to help us meet these challenges and share in the benefits. NASA will invite other countries to participate so we can strengthen peace, build prosperity and expand freedom for all who share our goals.

The principal functions of a permanent U.S. space station are listed in table 2. The basic need for a space station is associated with providing a permanent facility in orbit that is rich in power, has ample "people-hours" available for astronauts, scientists and technicians, and provides adequate facilities to assemble, deploy and operate very large spacecraft—spacecraft that would be too big to be carried intact in the Space Shuttle Orbiter's cargo bay.

A major purpose of the space station will be the establishment of a laboratory in orbit to take advantage of the unique attributes of the space environment, especially microgravity. The space station will serve as a base for orbital transfer vehicle and spacecraft deployment, servicing and retrieval. In the upcoming decades, the space station will permit the creation of very large spacecraft, orbiting antenna arrays, space power generation systems, large interplanetary vehicles (including a possible manned expedition to Mars) and large factories and materials processing platforms. All of these very interesting space systems will be assembled on orbit at the space station from component parts and sub-systems delivered to orbit by the Space Shuttle.

The contemporary U.S. space station concept includes a system of permanent, manned and unmanned elements in orbit that communicate with support facilities on Earth.

Table 2   Functions of a Space Station

• On–orbit laboratory
—science and applications
—technology and advanced
  development
	• Communications and data
	  processing node

	• Manufacturing facility
• Permanent observatory(s)	• Assembly facility
• Transportation node	• Storage depot
• Servicing facility —free flyers —platforms	

SOURCE: NASA.

This basic concept also includes one or more free-flying, unmanned platforms that are dedicated to specific scientific or commercial activities. These unmanned space platforms would orbit near the manned space station element. The Space Shuttle Orbiter would carry the different space station components and elements into orbit, assist in the assembly of the station and periodically return to the space station with new supplies, fresh crews and new equipment. Transportation between the manned space station and various free-flying space platforms would be accomplished by means of a co-orbiting, unmanned space tug called an Orbital Maneuvering Vehicle (OMV). The OMV would be based at the manned portion of the space station. An Orbital Transfer Vehicle (OTV) would be used to send spacecraft to higher altitude Earth orbits or on interplanetary trajectories from the space station.

Table 3 summarizes some of the initial Space Station capabilities now being considered in the design. Although the final space station design has not yet been selected, it will most likely be an evolutionary, modular design that allows for both expansion and change to meet the future space program needs. Figures C36 and C37 on color page N provide a look at two of the space station configurations and elements that are now being considered by space engineers and design teams.

The extraterrestrial frontier will soon be open to many people, not just those few uniquely qualified individuals who possess "the right stuff." When the space station finally evolves from its initial design concepts in the mid-1980s to an almost routinely operational space facility in the year 2000, we will have started to build our extraterrestrial civilization. With factories on orbit providing unique and special products, with permanent laboratories routinely yielding new knowledge in a wide variety of

disciplines, with sophisticated interplanetary payloads and manned expeditions being assembled and launched on orbit to a number of exciting destinations, the permanent space station will formally identify our transition from a planetary to an interplanetary or Solar System civilization. This can happen only once in the history of a planet—and it will occur within the next few years!

See also: **extraterrestrial civilizations; people in space; space settlement; Space Transportation System**

space suit   Outer space is a very hostile environment. If astronauts are to survive there, they must take part of the Earth's environment with them. Air to breathe, acceptable ambient pressures and moderate temperatures have to be contained in a shell surrounding the space traveler. This can be accomplished by providing a very large enclosed structure or habitat; or else on an individual basis by encasing the astronaut in a protective flexible capsule called the space suit.

Space suits used on previous missions from the Mercury program up through the Apollo/Soyuz test project have provided effective protection for American astronauts. However, they have also been handicapped by certain design problems. These suits were custom-fitted garments. In some suit models, more than 70 different measurements had to be taken of the astronaut in order to manufacture the suit to the proper fit. As a result, a space suit could be worn by only one astronaut on only one mission. These early space suits were stiff and even simple motions such as grasping objects quickly drained an astronaut's strength. Even donning the suit was an exhausting process lasting, at times, more than an hour and requiring the help of an assistant.

A new space suit has been developed for Shuttle era astronauts that provides many improvements in comfort, convenience and mobility over the previous models. This suit, which is worn outside the Orbiter during extravehicular activity (EVA), is modular and features many interchangeable parts. Torso, pants, arms and gloves come in several different sizes and can be assembled for each mission in the proper combination to suit individual male and female astronauts. The Shuttle space suit is called the Extravehicular Mobility Unit (EMU) and consists of three main parts: liner, pressure vessel and primary life support system (PLSS). These components are supplemented by a drink bag, communications set, helmet and visor assembly.

Containment of body wastes is a significant problem in space suit design. In the EMU, the primary life support system (PLSS) handles odors, carbon dioxide and the containment of gases in the suit's atmosphere. The PLSS is a two-part system consisting of a backpack unit and a control and display unit located on the suit chest. A separate unit is required for urine relief. Two different urine relief systems have been designed to accommodate both male and female astronauts. Because of the short time

Table 3   Intitial Space Station Capabilities

	Intitial	Future
Base at 28.5°		
Crew	6–8	12–18
Power	75kw	160kw
Attached payloads	Some	More
Servicing capability	Initial (close by)	Mature
Smart front-end TMS	Available	Available
Data system	300 MBPS	300 MBPS
Utility module	One	More
Laboratory module	One	More
Logistics module	Two	More
Living quarters module	One	More
Multi-berthing adapter	One	Under study
Platforms		
28.5°/15kw	One	Several
Polar/15kw	One	One
Space-Based OTV	No	Yes
Manned Polar Station	No	Under study

SOURCE: NASA.

Fig. 1 This patch is an insignia for the manned maneuvering unit (MMU), first flown on Shuttle flight 41-B. Astronaut Bruce McCandless II using the MMU became the first Earthling to fly untethered in space. (Courtesy of NASA.)

durations for extravehicular activities, fecal containment is considered unnecessary.

The Manned Maneuvering Unit (MMU) is a special tool that enables a space-suited astronaut to leave a spacecraft or space station and perform untethered extravehicular activities. The value of the MMU was shown in April 1984 during the Solar Maximum Mission Satellite repair when, for the first time, astronauts repaired a damaged spacecraft on orbit and restored it to operation (see fig. 1). The MMU device has been called "the world's smallest reusable spacecraft."

Space Telescope (ST) The Space Telescope (ST), or Hubble Space Telescope as it is also called, is a large, unmanned, multipurpose optical telescope that will be placed in Earth orbit by the Space Shuttle in 1986 (see fig. 1). This telescope has been named for the American astronomer, Edwin P. Hubble, who revolutionized our knowledge of the size, structure and makeup of the Universe through his pioneering observations in the first half of the 20th century. With this very innovative astronomical instrument, astronomers and space scientists will be able to observe the Universe with a clarity and to distances never before obtained. The Space Telescope will be used to investigate a wide variety of extraterrestrial problems and puzzles, especially those arising in the areas of extragalactic astronomy and observational cosmology.

From ancient times, men and women have gazed up into the night skies in wonder and awe. To them the distant stars, the wandering planets and the spectacular passages of comets and meteorites were mysterious, perhaps even terrifying. Yet some individuals began to observe that there were patterns and cycles in the movement of these celestial objects—and the science of astronomy was born. In various primitive civilizations this growing science was often interwoven with religion and with at-

tempts to predict the future. But without adequate instruments these ancient astronomers could not determine the exact nature of the tiny dots of light they were observing in the night skies. In fact, it was not until 1610 that we took our first major technical step in opening up the Universe to human observation. In that year, the famous Italian scientist, Galileo Galilei, aimed his first telescope toward the heavens and became the first human to clearly see what was really there! What a feeling of excitement he must have had as he gazed at the craters on the Moon, at the large red spot on Jupiter and the four major Jovian moons (now called the Galilean satellites). He even observed "projections" coming from the planet Saturn, which were later understood to be rings. Galileo's early observations not only mark the birth of modern observational astronomy, but also represent our first enhanced viewing of the extraterrestrial environment.

More than three and one-half centuries later, we took another major technical step in observing and exploring the extraterrestrial environment. The birth of the Space Age in 1957 enabled us to send people and instruments above the Earth's atmosphere. For the first time we have viewed the Cosmos directly, unimpeded by the filtering effects of the terrestrial atmosphere. In just a few short years, thousands of discoveries have been made. For example, the new science of X-ray astronomy has revolutionized our concept of the Universe. We have discovered a Universe of incredible violence, immensity and variety. Astronauts and robot explorers have stood on alien worlds. In the last three decades man and machine together have entered the extraterrestrial realm itself. The last two barriers to our meeting the Universe, gravity and the terrestrial atmosphere, have now fallen!

With the launch of the Hubble Space Telescope, we will witness another such major technical step in our unfolding

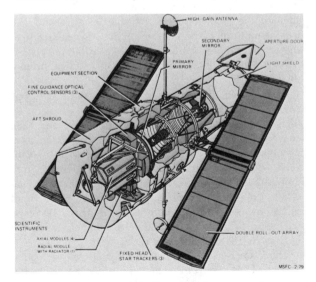

Fig. 1 Cutaway view of the Space Telescope—a multipurpose optical observatory launched and serviced by the Space Shuttle. (Courtesy of NASA.)

of things extraterrestrial. The Space Telescope will exceed the capabilities of all previous optically-instrumented astronomical spacecraft. Space scientists and astronomers predict that the Space Telescope will gather new astronomical data that will cause another unprecedented leap of our knowledge of the Universe. In fact, the difference between the observational capabilities of the orbiting Space Telescope and Earth-based optical telescopes can be compared to the difference between Galileo's first telescope and the main tool of the ancient astronomers, the human eye. Shuttle Era deployment and operation of the Space Telescope represents nothing short of an extraterrestrial renaissance.

The Hubble Space Telescope will be able to look far out into space and back into time. It will create extraterrestrial imagery of unprecedented clarity. We will be able to view star systems, galaxies, quasars, pulsars and exploding galaxies as never before. How can the Space Telescope provide such a quantum leap in observational capability? The answer is quite simple. The Space Telescope will observe the Universe from a 600 kilometer (373 mile) high orbit with a large 2.4 meter (94.5 inch) diameter mirror. By comparison, large optical telescopes on Earth, such as the 5 meter (200 inch) telescope at Mount Palomar, California, are handicapped in the ultimate quality of their imagery by the blurring and dimming effects of the Earth's atmosphere. Observing the Universe through the Earth's atmosphere, even with the most advanced optical instruments, involves an insurmountable barrier—the atmosphere itself. In optical astronomy, excellent conditions for observing occur only a relatively few nights a year. In addition to visible light, only a few very narrow-band portions of the electromagnetic spectrum (infrared and radio) can penetrate through the Earth's atmosphere. Consequently, a ground-based astronomer is ultimately limited by the very air that gives him or her life. You can simulate the problem faced by terrestrial astronomers in looking up through this envelope of gases by looking up at the surface world from the bottom of a swimming pool or pond. How clearly can you see someone in a boat or on a diving board?

Floating silently in outer space, the Space Telescope won't experience this atmospheric distortion problem. Its imagery will be sent to Earth electronically and then converted to clear images that can be used by astronomers and space scientists. It will expand our view of the Universe 350 times over what we can now observe from ground-based telescopes and will permit us to look at celestial objects 50 times fainter and seven to 10 times farther away than ever before seen. The Space Telescope will serve as the primary optical tool for observing the heavens for the next several decades.

The Space Telescope will make observations over a large portion of the spectrum including ultraviolet, visible and infrared measurements. For example, on-board instrumentation will give scientists on Earth the opportunity to collect spectral data needed to study the physical composition of celestial objects as well as their chemical and atomic structures.

The Space Telescope will be used to observe our own Solar System as well as to study objects deep in space. For example, this orbiting facility will be used to routinely view our planetary neighbors and will provide a level of detail that can be bettered only by sophisticated interplanetary probes or orbiting spacecraft. Planetary scientists and astronomers will make regular, detailed examinations of cloud enshrouded Venus, study the seasonal variation of the Martian polar caps, and view mighty Jupiter and its dynamic atmosphere and miniature solar system of moons. The Space Telescope will also be used to study the rings and moons of Saturn and the dynamics of the Saturnian atmosphere. This facility will also let us look at the most distant planetary members of our Solar System: Uranus, Neptune and Pluto, providing valuable data that simply cannot be obtained from ground-based observations.

Why should we still be interested in planetary observations? Well, such observations are very useful in helping scientists understand terrestrial weather, resources and planetary processes. These observations will also be helpful in determining the feasibility of using extraterrestrial resources as part of our expansion into outer space. Finally, the Space Telescope will not only provide information about our own Solar System, but it will also be used to search for clues concerning the existence of planets in other solar systems (extrasolar planets). Although the Space Telescope may not be able to observe any extrasolar planets directly—even those that might be orbiting around even our nearest stellar neighbors such as Alpha Centauri—it should nevertheless be able to observe perturbations in the motions of nearby stars. These perturbations will, in turn, indicate the presence of large orbiting objects, such as planets. There are 37 known stars within 15 light years of our Solar System and 10 Sun-like stars within 10 light years. The Space Telescope will be used to look at these particular stars and at some 100 to 500 other stellar neighbors that might possess planetary systems. The payoff, of course, is truly exciting. The confirmation of the existence of extrasolar planets would greatly facilitate our ability to estimate the probability that extraterrestrial life, possibly even intelligent life, might exist elsewhere in the Universe. Some of the very speculative estimates for planet formation now used in the Drake equation might then be replaced by more credible, observationally-determined numbers.

Astronomers are also interested in studying the many gas clouds in our Milky Way Galaxy and in nearby galaxies. Stars in various stages of development can be examined with the Space Telescope. It can provide a composite history of stellar life cycles from gas condensation birth, through thermonuclear ignition, to one of two contrasting finales. At death, some stars simply exhaust their nuclear fuel and fade quietly into oblivion in the darkness of interstellar space. Other stars put out one last dazzling burst of energy and destroy themselves in a spectacular cataclysmic explosion, called a supernova (plural: supernovae). Although supernovae are rare celestial events, the Space Telescope can help scientists study this dramatic portion of the stellar life cycle by providing data on the gas cloud remnants of stars that have already gone supernova.

With the telescope's powerful instruments, astronomers and cosmologists will be able to compare the rate of recession of nearby galaxies with that of distant galaxies. Such data should help answer the key question of whether the Universe is actually open or closed. It is not currently known whether the Universe will continue to expand forever (open Universe model) or whether it will eventually stop expanding and fall back on itself in an awesome final singularity due to gravitational collapse (closed Universe model).

Some astronomers speculate that the Space Telescope will enable us to "look back in time," viewing much of the Universe as it may have been just after the Big Bang. When we view distant galaxies, we are actually seeing them as they were in the distant past since their light has traveled millions, possibly billions of light years to reach us. With the Space Telescope and its ability to see objects 50 times fainter and seven times farther away than any ground-based observatory, these distant primeval galaxies will be seen as they were being formed, shortly after the beginning of time.

The Space Telescope is designed to work much like a ground-based observatory. The main concept is to collect more light than can be gathered by the human eye, concentrate this light on a focal plane, and then record the image formed on the focal plane.

The Space Telescope will have a launch mass of approximately 11,000 kilograms (25,500 pounds); it will be 13.1 meters (43.5 feet) long and have a diameter of 4.27 meters (14 feet). Two large solar panels are attached to the exterior of the spacecraft. Once on orbit, these panels will be unfurled and will have the dimensions 2.3 by 11.8 meters (7.8 by 39.4 feet). The Space Telescope's power supply system consists of these solar arrays, batteries and power conditioning equipment. It will supply a minimum of 2,400 watts to the spacecraft components two years after the beginning of the mission. This arrangement is sufficient to recharge the spacecraft's six nickel-cadmium batteries after it completes the night-side part of its 90-minute Earth orbit.

Space Telescope data are transmitted and received on Earth by means of the Tracking and Data Relay Satellites. The Space Telescope Operations Control Center (STOCC) at NASA's Goddard Space Flight Center combines two major elements of Space Telescope operations at one facility. These two operations are the Payload Operations Control Center (POCC) and the Science Support Center (SSC). The POCC has the responsibility for conducting Space Telescope mission operations, while the SSC provides the primary interface between the POCC and the scientific users at the Space Telescope Science Institute, located at Johns Hopkins University in Baltimore.

The Space Telescope consists of an Optical Telescope Assembly (OTA), a Support Systems Module (SSM), and Scientific Instruments (SI). In space, all these components function together as one unit to return imagery and scientific data to experiments on Earth.

The very heart of the spacecraft is the Optical Telescope Assembly where the reflecting Cassegrain telescope is located (see fig. 2). This type of telescope features a system where light from a celestial object travels through the aperture, down the assembly past the small secondary mirror, and then strikes the large (2.4 meter) primary mirror. The light is then reflected back a distance of 4.6 meters to the secondary mirror (0.3 meter) where it is narrowed and intensified into a small diameter beam. This beam travels through a 60 centimeter hole in the primary mirror to the focal plane just behind it. The focal plane is located 1.5 meters (4.9 feet) behind the front surface of the primary mirror.

At the focal plane, the light originally captured by the big mirror is now converted into a focused image. Portions of this image enter the apertures of the scientific instruments and are transmitted to Earth as data. Images and other scientific data are converted into electronic signals that are then transmitted to Earth by means of high-gain antennas at a rate up to one megabit (one million bits) per second. After being received on Earth, these data can be reconstructed into images and spectrograms.

The five scientific instruments and three fine guidance sensors are located just behind the primary mirror at the focal plane. These instruments are housed in modular units so that each module can be removed on orbit by astronauts and replaced with another similar unit or more advanced instrument package.

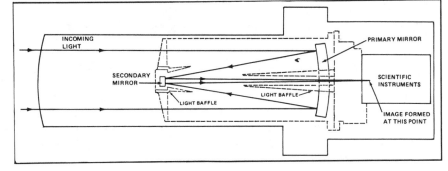

Fig. 2 Stellar light path through the Space Telescope. Light from a celestial object of interest is projected from the primary mirror to the secondary mirror, and is then directed to a focus inside the scientific instruments at the rear. (Courtesy of NASA.)

The Support Systems Module (SSM) provides the Telescope Assembly and Scientific Instruments with the communications, pointing and control, power and other support functions needed for successful operation on orbit. The aperture door is located on the forward portion of the Support Systems Module; it also serves as a light shield. The Space Telescope must be shielded from the Sun, Earth and Moon, so that the very sensitive scientific instruments will not be damaged by a flood of light. As a precaution, while the Space Telescope is in operation, it will not be aimed to within 50 degrees of the Sun. This procedure also eliminates observations of the planet Mercury. Internal baffles in the SSM door prevent scattered light from degrading the focal plane image. The central baffle is located just in front of the primary mirror, while a secondary baffle fits just behind the secondary mirror (see fig. 2).

The Space Telescope pointing control system employs six rate gyros and two fine guidance optical sensors to provide roll, pitch and yaw data. It is designed to keep the orbiting observatory locked on a celestial object for extended periods with an accuracy of less than 0.007 arc-second. This accuracy is equivalent to locking on an object that has the diameter of a dime, when the coin is located in Boston and you are making the observations from Washington, D.C.

The Space Telescope's five scientific instruments (four U.S. and one European) are located behind the primary mirror at the focal plane where they can pick up light from the telescope. These scientific instruments are: the Wide Field/Planetary Camera; the Faint Object Spectrograph; the High Resolution Spectrograph; the High Speed Photometer; and the Faint Object Camera (provided by the European Space Agency). In addition, the Fine Guidance Sensors, because of their capability to accurately locate stars, are frequently considered equivalent to a sixth scientific instrument.

The Wide Field/Planetary Camera can operate in two modes. It has a wide field capability that permits examination of large areas of space. It is also capable of taking high resolution images of objects in our Solar System (except Mercury). These images once transmitted to Earth and reconstructed will be far better than those produced by ground-based observatories. For example, Space Telescope images of Jupiter will be comparable to images taken of the planet by the two Voyager spacecraft in 1979. However, resolution of the planets beyond Saturn is not anticipated to be as high quality as the data returned by eventual fly-by spacecraft. The Wide Field/Planetary Camera will return data in support of the following areas: a determination of cosmic distance scales; cosmic evolution; the comparison of near and far galaxies; stellar population studies; energy distribution in stars and in compact celestial objects; star formation and supernovae; planetary atmosphere observations (within our Solar System); the search for extrasolar planets; and the observation of comets.

The Faint Object Spectrograph is another versatile Space Telescope instrument that can gather the spectra of extremely faint astronomical objects in the ultraviolet and visible portions of the electromagnetic spectrum. The study of such spectra tells astronomers about the nature of the celestial object being viewed, including whether it is hot or cold, dense or rarified, and even something about its chemical composition. Analysis of a spectrum also provides scientists information on the distance away the object is and its velocity. The Faint Object Spectrograph is designed to pick up the image of a star, interstellar cloud, galaxy or other object that appears on the Space Telescope's focal plane as a point of light. Then through an intricate system of optical mirrors and gratings that function like prisms, this light beam is spread out across the spectrum, from the ultraviolet portion through the visible wavelengths. This instrument will be used to support the following activities: observation of the centers (called nuclei) of active galaxies; defining the amounts and types of chemicals found in other galaxies; providing information on the physical properties of quasars; studying the mysterious cosmic jets that appear in visible spectrum images of quasars; studying quasars that are believed to be at the nuclei of certain galaxies; and finally, studying comets before they become chemically changed (active) by close passage to our Sun.

The Space Telescope's High Resolution Spectrograph will use the full resolving capability of the telescope to view much dimmer objects than have ever been viewed by previous orbiting astronomical instruments. It will be looking at only the ultraviolet portion of the electromagnetic spectrum, a region that cannot be seen by ground-based observatories due to the filtering effects of the Earth's atmosphere. This interesting region of the spectrum provides astronomers with very detailed chemical composition information on celestial objects. This particular instrument will examine supernovae, active galaxies, bright quasars and even selected phenomena in our own Solar System. The High Resolution Spectrograph will be used to support investigations concerning the physical composition of exploding galaxies, quasars, and other dense celestial objects; studies involving the loss of mass from one star to another in a binary star system (such as when a black hole devours its binary companion); measurements of the total amount of matter expelled in stellar explosions; investigations concerning the physical composition of interstellar gas clouds; examination of the various stages of stellar evolution; a definition of the atmospheric structure of planets within our own Solar System; and finally, measurement of the chemical elements found in comets.

The High Speed Photometer is designed to provide accurate observations of the total amount of light from a celestial object, record any fluctuations in brightness on a time scale down to microseconds and detail any fine structure related to the light source. These measurements will be made over a wide region of the electromagnetic spectrum, including the ultraviolet. The High Speed Photometer will be used in the following ways: to make precise observations of rapidly pulsing compact celestial objects, exploding variable stars and binary stars; to examine the optical properties of zodiacal light (sunlight reflected from interplanetary dust); to calibrate faint stellar objects; and to examine optical transients from stellar objects, including compact stars and supernovae.

The Faint Object Camera (constructed by the European Space Agency) takes advantage of the spatial resolution capability of the Space Telescope to capture images of very faint objects in the Universe. It is anticipated that this instrument will be able to detect stars as faint as the 28th magnitude and should easily observe stars of magnitude 24, which is the best a large ground-based observatory can currently do. In astronomy, magnitude is a designation of an object's brightness—the lower the number, the brighter the object. For example, the dimmest stars you can see with the naked eye are magnitude 6 stars; while the brightest stars (except for the Sun) are at magnitude 1. It is interesting to note that in this rating scheme the planet Venus is magnitude − 4 (minus four). This instrument will be used in the following applications: to observe extragalactic supergiant stars; to study variable brightness stars; to gather data on globular clusters; to examine binary star systems; to search for extrasolar planets; to establish stellar masses; to perform detailed investigations concerning shock fronts and condensing gas clouds in interstellar space; and to search for direct evidence that quasars might be at the center of faint galaxies.

Information provided by the Space Telescope's Fine Guidance Sensors will be used in the following ways: to provide stability to the Space Telescope; to calibrate the positions of nearby and distant stars and galaxies; to provide new information on the unseen companions of binary star systems; to provide more accurate positional information on the moons of the outer planets; and to establish better positional reference systems on compact stars. These sensors will provide precise measurements on the location of celestial objects (the science of astrometry) and will assist in answering the question of whether we live in an open or closed Universe.

The Space Telescope is a powerful, free-flying spacecraft designed to be placed in orbit and maintained by the Space Shuttle. As necessary, it will be serviced by astronauts who will replace instruments, batteries or other modules while on orbit. When significant repairs are needed, the entire facility will be returned to Earth by the Shuttle for a major refurbishment. Astronomy has always been a science that has helped stimulate human curiosity and imagination. The Space Telescope is a truly remarkable tool that will open up many new aspects of our extraterrestrial environment in the exciting years ahead.

Space Transportation System (STS) The Space Shuttle is the prime element of the U.S. Space Transportation System for space research and applications (see fig. 1). Carrying payloads of up to 29,000 kilograms (65,000 pounds), the Space Shuttle is replacing most expendable launch vehicles that have been used by the United States to launch deep-space missions into their initial low Earth orbit. The Shuttle also represents the world's first space system capable of returning payloads from orbit on a routine basis.

Shuttle crews have retrieved satellites in Earth orbit, have repaired and redeployed them and have brought them back home to Earth for refurbishment and reuse. The Shuttle has also been used to carry out missions in which scientists and technicians performed experiments in orbit.

The Shuttle is a truly versatile aerospace vehicle. It takes off like a rocket, maneuvers in orbit like a spacecraft, and then lands back on Earth like an airplane. The Shuttle is designed to carry heavy payloads into low Earth orbit. While other (expendable) launch vehicles have also done this, they could only be used just once. Unlike these expendable launch vehicles, each Space Shuttle Orbiter can be reused many times. Each Shuttle mission reveals a new capability and marks the start of an entirely new era in space transportation.

The exceptional versatility of this manned aerospace vehicle permits the checkout and repair of automated spacecraft while they are on orbit. Those spacecraft that cannot be repaired in space can be returned to Earth for refurbishment and reuse. This completely changes the design philosophy and procedures that have gone into the development of the first few generations of Earth satellites. In the future, revolutionary engineering designs

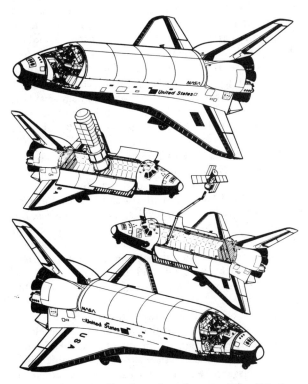

Fig. 1 The Space Shuttle—prime element of the U.S. Space Transportation System. (Courtesy of NASA.)

(perhaps taking advantage of on-orbit assembly) will usher in an entirely new generation of Earth-monitoring spacecraft. The fields of environmental protection and monitoring, energy resource development and conservation, meteorology, oceanography, agriculture, forestry and range management (just to name a few) will all be directly influenced by the next generation of "super-spacecraft" made possible by an operational Space Transportation System. Data from these spacecraft have the potential of greatly enriching the lives of all the peoples of the Earth!

Interplanetary spacecraft are delivered to low Earth orbit by the Shuttle and then placed on their final interplanetary trajectory by an upper propulsive stage called an orbital transfer vehicle.

Unmanned space platforms, such as the Space Telescope (which will greatly expand our view of the Universe) and the Long Duration Exposure Facility (which supports a variety of space environment exposure research projects) can be placed in orbit, deployed, retrieved and returned to Earth by the Space Shuttle.

As the operation of the Space Transportation System matures, Orbiter vehicles will be refurbished and made ready for another journey "across the vertical frontier" in a matter of weeks after landing.

Frequent Shuttle flights provide excellent opportunities to view interesting, but transient, astronomical, geophysical or meteorological phenomena. Unusual information gathered by skilled scientists who have personally made the observations from space will greatly contribute to our overall understanding of many natural phenomena and will also allow timely response to urgent crises such as oil spills, giant forest fires and violent volcanic eruptions. These "people-in-the-loop" observations of Earth from the Shuttle will set the pace for similar but more permanent expert investigations using the Space Station.

The Shuttle is the "mothership" for a complete scientific laboratory called *Spacelab*. Developed by the European Space Agency (ESA), *Spacelab* is designed to take full advantage of the space environment found in low Earth orbit in support of innovative scientific and engineering investigations. Highly skilled scientists can take advantage of microgravity, hard vacuum and synoptic view of the Earth and the heavens in performing pioneering experiments in such fields as astrophysics, planetary science, life science and medicine, materials science and processing in space, and solar physics. *Spacelab* will remain attached in the Orbiter's cargo bay throughout its mission. Upon return to Earth, *Spacelab* will be removed from the Orbiter and prepared for its next mission. It is a reusable space laboratory.

The Space Shuttle now brings within reach extraterrestrial projects and programs that many people considered impossible or impractical only a few years ago. For example, the Shuttle can carry the modular units and components of a large space station and assemble this permanent facility in a piece-by-piece, flight-by-flight fashion. The size of on-orbit facilities or spacecraft is no longer limited by the volume or payload capacity of individual launch vehicles.

The first orbital flight of the Space Shuttle *Columbia* in April 1981 has initiated a new age of space transportation and exploitation. Creative use of this versatile aerospace vehicle will trigger a terrestrial renaissance and establish the foundation of our extraterrestrial civilization. Through the Space Shuttle, and subsequent generations of reusable aerospace vehicles, outer space has the potential of becoming humanity's pathway into the next millennium.

See also: **orbital-transfer vehicle; space launch vehicles; space station**

Starlab A unique facility for the pursuit of a wide variety of astrophysical problems involving the optical and ultraviolet portion of the electromagnetic spectrum. It is now being developed for placement on a free-flying space platform as a cooperative venture between Australia, Canada and the United States. Australia is responsible for the scientific instrument package, Canada for the telescope, and the United States (NASA) for facilities and the space platform. The projected launch date is about 1990.

Starlab will have two specific capabilities: first, very high resolution and large bandwidth imagery over a large field of view; and second, high efficiency and high spatial resolution spectroscopy for extended or point astronomical sources. This orbiting observatory will complement other space-based astrophysical research tools, including the Space Telescope.

The current *Starlab* design consists of a modified Richey-Chretien telescope with a one-meter primary mirror and a full baffled field to prevent stray light from entering and causing unwanted background illumination. The telescope assembly will be attached to a pointing system and the complete package will be mounted on a free-flying space platform for flights up to six months' duration over a period of 20 years. The Shuttle launched and serviced *Starlab* will have an initial orbit of 435 kilometers altitude and 28.5° inclination (or possibly 57° inclination). Refurbishment and instrument upgrades will be performed at regular time intervals.

See also: **astrophysics; Space Telescope**

Starprobe A proposed NASA spacecraft that would represent mankind's first direct exploratory venture to the vicinity of our Sun. Heavily shielded to protect against the Sun's searing heat, the *Starprobe* spacecraft will be placed in a heliocentric orbit that comes within 1.6 million kilometers of the Sun for a truly "close encounter of the solar kind." Instruments on board *Starprobe* will study first hand the physical conditions in the Sun's corona and will be able to sense the structure inside the Sun, providing scientists on Earth their first good look at what really happens inside a star. The proposed spacecraft would have a mass of approximately 1500 kilograms, a lifetime of five years and be powered by radioisotope thermoelectric generators.

stars A star is essentially a self-luminous ball of very hot gas that generates energy through thermonuclear fusion reactions that take place in its core.

Stars may be classified as either "normal" or "abnormal." Normal stars, like our Sun, shine steadily (see fig. C38 on color page O) These stars exhibit a variety of colors: red, orange, yellow, blue and white. Most stars are smaller than the Sun and many stars even resemble it. However, there are a few stars that are also much larger than our Sun. In addition, astronomers have observed several types of abnormal stars including: giants, dwarfs and a variety of variable stars.

Most stars can be put into one of seven general spectral types called O, B, A, F, G, K and M (see table 1). This classification is a sequence established in order of decreasing surface temperature. Perhaps one easy way to remember this classification scheme is to use the mnemonic:

Oh Be A Fun Girl (or Guy), Kiss Me!

Our parent star the Sun is approximately 1.4 million kilometers (865,000 miles) in diameter and has an effective surface temperature of about 5,800 degrees Kelvin. The Sun, like other stars, is a giant nuclear furnace, in which the temperature, pressure and density are sufficient to cause light nuclei to join together or "fuse." For example, deep inside the solar interior, the hydrogen which makes up 90 percent of the Sun's mass is fused into helium atoms, releasing large amounts of energy that eventually works its way to the surface and is then radiated throughout the Solar System. The Sun is currently in a state of balance or equilibrium between two competing forces: gravity (which wants to pull all its mass inward) and the radiation pressure and hot gas pressure resulting from the thermonuclear reactions (which push outward).

Many stars in the Galaxy appear to have companions, with which they are gravitationally bound in binary, triple or even larger systems. Compared to other stars throughout the Galaxy, our Sun is slightly unusual. It does not have a known stellar companion. (However, the existence of a very distant, massive, dark companion called Nemesis has recently been postulated by some astrophysicists in an attempt to explain an apparent "cosmic catastrophe cycle" that occurs here on Earth.)

A STAR IS BORN

Astrophysicists have discovered what appears to be the life cycle of the stars. Stars originate by the condensation of enormous clouds of cosmic dust and hydrogen gas, called nebulae (see fig. C18 on color page I). Gravity is the dominant force behind the birth of a star. According to Newton's Universal Law of Gravitation, all bodies attract each other in proportion to their masses and distance apart. The dust and gas particles found in these huge interstellar clouds attract each other and gradually draw closer together. Eventually, enough of these particles join together to form a central clump that is sufficiently massive to bind all the other parts of the cloud by gravitation. At this point, the edges of the cloud start to collapse inward, separating it from the remaining dust and gas in the region.

Initially, the cloud contracts rapidly, because the thermal energy release related to contraction is easily radiated outward. However, when the cloud grows smaller and more dense, the heat released at the center cannot immediately escape to the outer surface. This causes a rapid rise in internal temperature, slowing down but not stopping the relentless gravitational contraction.

The actual birth of a star occurs when its interior becomes so dense and its temperature so high, that thermonuclear fusion occurs. The heat released in thermonuclear fusion reactions is greater than that released through gravitational contraction, and fusion becomes the star's primary energy producing mechanism. Gases heated by nuclear fusion at the cloud's center begin to rise, counterbalancing the inward pull of gravity on the outer layers. The star stops collapsing and reaches a state of equilibrium between these outward and inward forces. At this point, our star has become what astronomers and astrophysicists call a "main sequence star." Like our Sun, it will then remain in this state of equilibrium for billions of years, until all the hydrogen fuel in its core has been converted into helium.

Table 1   Stellar Spectral Classes

Type	Description	Typical Surface Temperatures (K)	Remarks
O	very hot, large blue stars [*hottest]	28,000–40,000	ultraviolet stars; very short lifetimes (3–6 million years)
B	large, hot blue stars	11,000–28,000	example: Rigel
A	blue-white, white stars	7,500–11,000	Vega, Sirius, Altair
F	white stars	6,000– 7,500	Canopis, Polaris
G	yellow stars	5,000– 6,000	the Sun
K	orange-red stars	3,500– 5,000	Arcturus, Aldebaran
M	red stars [*coolest]	<3,500	Antares, Betelgeuse

SOURCE: NASA.

Are stars being born today in our Galaxy? Scientists studying dense interstellar clouds with orbiting infrared telescopes have discovered an abundance of glowing objects hidden from optical telescopes by the intervening dust and gas. Since collapsing clouds radiate in the infrared portion of the electromagnetic spectrum until nuclear fusion processes begin, these infrared sources are considered to be stars in the birthing process.

MAIN SEQUENCE STARS

How long a star remains on the main sequence, burning hydrogen for its fuel, depends mostly on its mass. Our Sun has an estimated main sequence lifetime of about 10 billion years, of which approximately five billion years have now passed. Larger stars burn their fuels faster and at much higher temperatures. These stars, therefore, have short main sequence lifetimes, sometimes as little as one million years. In comparison, the "red dwarf stars," which typically have less than one-tenth the mass of our Sun, burn up so slowly that trillions of years must elapse before their hydrogen supply is exhausted. When a star has used up its hydrogen fuel, it leaves the "normal" state or departs the main sequence. This happens when the core of the star has been converted from hydrogen to helium by the thermonuclear reactions that have taken place.

RED GIANTS AND SUPERGIANTS

When the hydrogen fuel in the core of a main sequence star has been consumed, the core starts to collapse. At the same time, the hydrogen fusion process moves outward from the core into the surrounding, outer regions. There, the process of converting hydrogen into helium continues, releasing radiant energy. But as this burning process moves into the outer regions, the star's atmosphere expands greatly and it becomes a "red giant." The term giant is quite appropriate. If we put a red giant where our Sun is now, the innermost planet Mercury could be engulfed by it; similarly, if we put a larger "red supergiant" there, this supergiant would extend out past the orbit of Mars!

As the star's nuclear evolution continues, it might become a "variable star," pulsating in size and brightness over periods of several months to years. The visual brightness of such an "abnormal" star might now change by a factor of 100, while its total energy output varies by only a factor of two or three.

As an abnormal star grows, its contracting core may become so hot that it ignites and burns nuclear fuels other than hydrogen, beginning with the helium created in millions to perhaps billions of years of main sequence burning. The subsequent behavior of such a star is complex, but in general it can be characterized as a continuing series of gravitational contractions and new nuclear reaction ignitions. Each new series of fusion reactions produces a succession of heavier elements, in addition to releasing large quantities of energy. For example, the burning of helium produces carbon, the burning of carbon produces oxygen, etc.

Finally, when nuclear burning no longer releases enough radiant energy to support the giant star, it collapses and its dense central core becomes either a compact white dwarf or a tiny neutron star. This collapse may also trigger an explosion of the star's outer layers, which displays itself as a supernova. In exceptional cases with very massive stars, the core (or perhaps even the entire star) might even become a black hole.

WHITE DWARFS, NEUTRON STARS, AND BLACK HOLES

When a star like our Sun has burned all the nuclear fuels available, it collapses under its own gravity until the collective resistance of the electrons within it finally stops the contraction process. The "dead star" has become a "white dwarf" and may now be about the size of the Earth. Its atoms are packed so tightly together that a fragment the size of a sugar cube would have a mass of thousands of kilograms! The white dwarf then cools for perhaps several billion years, going from white, to yellow, to red, and finally becomes a cold, dark sphere called a "black dwarf." (Note that the white dwarf is not experiencing nuclear burning; rather its light comes from a thin gaseous atmosphere that gradually dissipates its heat to space.) Astrophysicists estimate that there may be some 10 billion white dwarf stars in the Milky Way Galaxy alone, many of which have now become black dwarfs. This fate appears to awaiting our own Sun and most other stars in the Galaxy.

However, when a star with a mass of about 1.5 to 3 times the mass of our Sun undergoes collapse, it will contract even further and ends up as a "neutron star," with a diameter of perhaps only 20 kilometers. In neutron stars, intense gravitational forces drive electrons into atomic nuclei, forcing them to combine with protons and transforming this combination into neutrons. Atomic nuclei are, therefore, obliterated in this process and only the collective resistance of neutrons to compression halts the collapse. At this point, the star's matter is so dense that each cubic centimeter has a mass of several billion tons!

But for stars that end their life having more than a few solar masses, even the resistance of neutrons is not enough to stop the unyielding gravitational collapse. In death such massive stars ultimately become "black holes." A black hole is an incredibly dense point mass or singularity that is surrounded by a literal "black region" in which gravitational attraction is so strong that nothing, not even light itself, can escape.

NOVAE, SUPERNOVAE AND PULSARS

Today, many physicists and astronomers relate the astronomical phenomena called supernovae and pulsars with neutron stars and their evolution. The final collapse of a giant star to the neutron stage may give rise to the physical conditions that cause its outer portions to explode, creating a "supernova." This type of cosmic explosion releases so much energy that its debris products will temporarily outshine all the hundreds of millions of ordinary stars in a galaxy.

A regular "nova" (the Latin word for "new," the plural of which is *novae*) that occurs more frequently is far less violent and spectacular. One common class, called "recurrent novae," is due to the nuclear ignition of gas being drawn from a companion star to the surface of a white dwarf. Such binary star systems are quite common and sometimes the stars will have orbits that regularly bring them close enough for one to draw off gas from the other.

When a supernova occurs at the end of a massive star's life, the violent explosion fills vast regions of space with matter that may radiate for hundreds or even thousands of years. The debris created by a supernova explosion will eventually cool into dust and gas, become part of a giant interstellar cloud and perhaps once again be condensed into a star or a planet. Most of the heavier elements found on Earth are thought to have originated in supernovae, since the normal thermonuclear fusion processes cannot produce such heavy elements. The violent power of a supernova explosion can, however, combine lighter elements into the heaviest ones found in nature (such as lead, thorium and uranium). Consequently, both our Sun and its planets were most likely enriched by infusions of material hurled into the interstellar void by ancient supernova explosions. That's right, you are made out of stardust!

Pulsars, first detected by radio astronomers in 1967, are sources of very accurately spaced bursts or pulses of radio signals. These radio wave signals are so regular, in fact, that the scientists who made the first detections were initially startled into thinking that they might have intercepted a radio signal from an intelligent alien civilization.

The pulsar, named because its radio wave signature regularly turns on and off or pulses, is considered to be a rapidly spinning neutron star. One pulsar is located in the center of the Crab Nebula, where a giant cloud of gas is still glowing from a supernova explosion that occurred in the year 1054 A.D.—a spectacular celestial event observed and recorded by ancient Chinese astronomers. The discovery of this pulsar has led scientists to make a great synthesis of our modern understanding of both pulsars and supernovae.

In a supernova explosion, a massive star is literally destroyed in an instant, but the explosive debris lingers and briefly outshines everything in a galaxy. In addition to scattering material all over interstellar space, supernova explosions leave behind a dense collapsed core made of neutrons. This neutron star, with an intense magnetic field, spins many times a second, emitting beams of radio waves, X-rays and other radiations. These radiations are possibly focused by the pulsar's powerful magnetic field and sweep through space much like a revolving lighthouse beacon. The neutron star, the end product of a violent supernova explosion, has become a pulsar!

Astrophysicists must now develop new theories to explain how pulsars can create intense radio waves, visible light, X-rays and gamma rays, all at the same time. Orbiting X-ray observatories, for example, have detected X-ray pulsars, such as Hercules X-1 in 1971. These X-ray pulsars are believed to be caused by a neutron star pulling gaseous matter from a normal companion star in a binary star system. As gas is sucked away from the normal com-

panion to the surface of the neutron star, the gravitational attraction of the neutron star heats up the gas to millions of degrees Kelvin and causes it to emit X-rays.

The advent of the Space Age and the use of orbiting observatories to view the Universe as never before possible has greatly increased our knowledge about the many different types of stellar phenomena that make up the Universe. Most exciting of all, perhaps, is the fact that this process of astrophysical discovery has really only just begun!

See also: **astrophysics; black holes; fusion; Space Telescope; Sun**

starship A starship is a space vehicle capable of traveling the great distances between star systems. Even the closest stars in our Galaxy are often light-years apart. By convention, the word "starship" is used here to describe interstellar spaceships capable of carrying intelligent beings to other star systems; while robot interstellar spaceships are called interstellar probes.

What are the performance requirements for a starship? First, and perhaps most important, the vessel should be capable of traveling at a significant fraction of the speed of light (c). Ten percent of the speed of light (0.1c) is often considered as the lowest acceptable speed for a starship while cruising speeds of 0.9c and beyond are considered highly desirable. This "optic velocity" cruising capability is necessary to keep interstellar voyages to reasonable lengths of time, both for the home civilization and for the starship crew.

Consider, for example, a trip to the nearest star system Alpha Centauri—a triple star system about 4.23 light-years away. At a cruising speed of 0.1c, it would take about 43 years just to get there and another 43 years to return. The time dilation effects of travel at relativistic speeds would not help too much either, since a ship's clock would register the passage of about 42.8 years versus a terrestrial ground elapse time of 43 years. In other words, the crew would age about 43 years during the journey to Alpha Centauri. If we started with 20-year-old crewmembers departing from the Solar System in the year 2100 at a constant cruising speed of 0.1c, they would be approximately 63 years old when they reached the Alpha Centauri star system some 43 years later in 2143! The return journey would be even more dismal. Any surviving crewmembers would be 106 years old when the ship returned to the Solar System in the year 2186. Most if not all the crew would probably have died of old age or boredom. And that's for just a journey to the nearest star!

A starship should also provide a comfortable living environment for the crew and passengers (in the case of an interstellar ark). Living in a relatively small, isolated and confined habitat for a few decades to perhaps a few centuries can certainly overstress even the most psychologically adaptable individuals and their progeny. One common technique used in science fiction to avoid this crew stress problem is to have all or most of the crew placed in

Table 1   Characteristics of Possible Starship Propulsion Systems

PULSED NUCLEAR FISSION SYSTEM (Project Orion)

• Principle Of Operation: Series of nuclear fission explosions are detonated at regular time intervals behind the vehicle; special giant pusher plate absorbs and reflects pulse of radiation from each atomic blast; system moves forward in series of pulses.

• Performance Characteristics: Very low efficiency in converting propellant (explosive device) mass into pure energy for propulsion; limited to number of nuclear explosives that can be carried on board; radiation hazards to crew (needs heavy shielding); probably limited to a maximum speed of about 0.01 to 0.10 the speed of light.

• Potential Applications: Most useful for interplanetary transport (especially for rapid movement to far reaches of Solar System); not suitable for a starship; very limited application for an interstellar robot probe; possible use for a very slow, huge interstellar ark (several centuries flight time). INTERPLANETARY VERSION COULD BE BUILT IN A DECADE OR SO; LIMITED INTERSTELLAR VERSION BY END OF 21st CENTURY

PULSED NUCLEAR FUSION SYSTEM (Project Daedalus)

• Principle of Operation: Thermonuclear burn of tiny deuterium/helium-3 pellets in special chamber (using laser or particle beam inertially confined fusion techniques); very energetic fusion reaction products exit chamber to produce forward thrust.

• Performance Characteristics: Uses energetic single step fusion reaction; thermonuclear propellant carried onboard vessel; maximum speed of about 0.12c considered possible.

• Potential Application: Not suitable for starship; possible use for robot interstellar probe (fly-by) mission or slow interstellar ark (centuries flight time). LIMITED SYSTEM MIGHT BE BUILT BY END OF 21st CENTURY (INTERSTELLAR PROBE).

INTERSTELLAR RAMJET

• Principle of Operation: First proposed by R. Bussard; after vehicle has an initial acceleration to near-light speed, its giant scoop (thousands of square kilometers in area) collects interstellar hydrogen which then fuels a proton-proton thermonuclear cycle or perhaps the carbon-cycle (both of which are found in stars); thermonuclear reaction products exit vehicle and provide forward thrust.

• Performance Characteristics: In principle, not limited by amount of propellant that can be carried; however, construction of light-mass giant scoop is major technical difficulty; in concept, cruising speeds of from 0.1c up to 0.9c might be obtained.

• Potential Applications: Starship; interstellar robot probe; giant space ark; WOULD REQUIRE MANY MAJOR TECHNOLOGICAL BREAKTHROUGHS—SEVERAL CENTURIES AWAY, IF EVER.

PHOTON ROCKET

• Principle of Operation: Uses matter and antimatter as propellant; equal amounts are combined and annihilate each other releasing an equivalent amount of energy in form of hard nuclear (gamma) radiation; these gamma rays are collected and emitted in a collimated beam out the back of vessel, providing a forward thrust.

• Performance Characteristics: The best (theoretical) propulsion system our understanding of physics will permit; cruising speeds from 0.1c to 0.99c.

• Potential Applications: Starship; interstellar probes (including self-replicating machines); large space arks; MANY MAJOR TECHNOLOGICAL BARRIERS MUST BE OVERCOME—CENTURIES AWAY, IF EVER.

some form of "suspended animation" while the vehicle travels through the interstellar void, tended by a ship's company of smart robots.

Any properly designed starship must also provide an adequate amount of radiation protection for the crew, passengers and sensitive electronic equipment. Interstellar space is permeated with galactic cosmic rays. Nuclear radiation leakage from an advanced thermonuclear fusion engine or a matter/antimatter engine (photon rocket) must also be prevented from entering the crew compartment. In addition, the crew will have to be protected from nuclear radiation showers produced when a starship's hull, traveling at near light speed, slams into interstellar molecules, dust or gas. For example, a single proton (which we can assume is "stationary") being hit by a starship moving at 90 percent of the speed of light (0.9c) would appear to those on board like a one billion electron volt (GeV) proton being accelerated at them. Imagine traveling for years at the beam output end of a very high energy particle accelerator! Without proper deflectors or shielding, survival in the crew compartment from such radiation doses is doubtful.

To truly function as a starship, the vessel must be able to cruise at will, light-years from its home star system. The starship must also be able to accelerate to significant fractions of the speed of light; cruise at these near optic velocities; and then decelerate to explore a new star system or to investigate a derelict alien spaceship found adrift in the depths of interstellar space.

We will not discuss the obvious difficulties of navigating through interstellar space at near light velocities. It will be sufficient just to mention that when you "look" forward at near light speeds everything is "blueshifted"; while when you look aft things appear "redshifted." The starship and its crew must be able to find their way from one location in the Galaxy to another—on their own.

What appears to be the major engineering technology needed to make the starship a real part of our extraterrestrial civilization is an effective propulsion system. Interstellar class propulsion technology is the key to the Galaxy for any emerging civilization that has mastered space flight within and to the limits of its own solar system. Despite the tremendous engineering difficulties associated with the development of a starship propulsion system, several concepts have been proposed. These include: the pulsed nuclear fission engine (Project Orion concept), the pulsed nuclear fusion concept (Project Daedalus study), the interstellar nuclear ramjet, and the photon rocket. These systems are briefly described in table 1 along with their potential advantages and disadvantages.

Unfortunately, based on our current understanding of the laws of physics, all known phenomena and mechanisms that might be used to power a starship are either not energetic enough or simply entirely out of the reach of today's technology and even the technology levels anticipated for several tomorrows. Perhaps major breakthroughs will occur in our understanding of the physical laws of the Universe—breakthroughs that provide insight into more intense energy sources or ways around the speed of light barrier now imposed by the theory of relativity. But until such new insights occur (if ever), human travel to another star system on board a starship must remain in the realm of future dreams.

See also: **interstellar contact; Project Daedalus: Project Orion (II); relativity; space nuclear propulsion**

star wars   Interstellar warfare between two or more advanced extraterrestrial civilizations. The question still remains open as to whether intelligent creatures can develop the high-technology tools needed for interstellar travel without destroying themselves and their home planet(s) in the process. If alien creatures learn to live with their advanced technologies, it is highly probable that when they expand out into other star systems it will be a peaceful expansion. The alternative, unfortunately, is a barbaric struggle for domination of a particular star system followed, perhaps (if there are any survivors), by a belligerent expansion of the winning alien faction across interstellar space.

In this latter situation, contact with another intelligent species would almost certainly result in some form of interstellar warfare. If both civilizations have similar levels of technology, a distant planetary system around a mutually prized star might be the scene of violent conflict on an astronomical scale. A peacefully expanding race colliding with a belligerent race might withdraw, might elect to defend itself, or might be quickly dispatched (due to a lack of weapons).

If a belligerent alien race encounters intelligent creatures with greatly inferior technologies, rapid annihilation, mass destruction or enslavement of the inferior civilization can be anticipated. If, however, the technology gap is not as great (say, for example, the star system inhabitants have already developed nuclear technologies and interplanetary travel), then the alien invaders might encounter severe resistance and conquest of the star system would be achieved only after heavy invader losses—a Pyrrhic victory on an interstellar scale.

While speculative, of course, these scenarios for interstellar conflict have numerous analogs in terrestrial history. Are intelligent creatures really the same everywhere in the Universe?

Star Wars   In science fiction, *Star Wars* is the first of George Lucas' famous extraterrestrial adventure movies in which Luke Skywalker, Princess Leia Organa and a charming collection of assorted heroes battle the evil imperial forces of an ancient galactic empire, led by Lord Darth Vader and a host of heavies.

stellar magnitude   The relative luminance or brightness of a celestial body. In this sytem of measuring relative brightness, the smaller the number indicating magnitude, the more luminous the celestial object. Zero magnitude is the brightest star, and naked eye visual detection limits usually extend down to magnitude 6 (although on a perfectly black background the limit for a single luminous point approaches the 8th magnitude). The difference between successive magnitudes (for example, the 3rd and 4th magnitude) is 2.512 (the fifth root of 100), while the difference over five magnitudes (that is, 1st to 6th) is approximately 100. The planets, the Moon and the Sun are all brighter than magnitude zero and are therefore assigned negative magnitudes. For example, at their brightest the planet Venus has a relative visual magnitude of about -4.4 and the full Moon -12.7.

See also: **stars**

stellar wind   The continuous stream of atomic particles and matter ejected by many types of stars, including our Sun. In this Solar System, the stellar wind is called the "solar wind."

See also: **solar wind; stars; Sun**

Sun   The Sun is our parent star and the massive, luminous celestial object about which all other bodies in our Solar System revolve (see fig. C38 on color page O). It provides the light and warmth upon which all terrestrial life depends. Its gravitational field determines the movement of the planets and other celestial bodies. The Sun is a main sequence star of spectral type G-2. In any other place in the Galaxy, it would hardly appear worthy of special study by alien astronomers and astrophysicists; but its central position in our Solar System makes it a unique object for research. Like all main sequence stars, the Sun derives its abundant energy output from thermonuclear fusion reactions involving the conversion of hydrogen to helium and heavier nuclei. Photons associated with these exothermic (energy-releasing) fusion reactions diffuse outward from the Sun's core, until after several thousand years, they reach the convective envelope. Another by-product of the thermonuclear fusion reactions is a flux of neutrinos which freely escape from the Sun.

At the center of the Sun is the core, where energy is released in thermonuclear reactions. Surrounding the core are concentric shells called the radiative zone, the convective envelope (which occurs at approximately 0.8 of the Sun's radius), the photosphere (which is the layer from which visible radiation emerges), the chromosphere and, finally, the corona (which is the Sun's outer atmosphere).

Energy is transported outward through the convective envelope by convective (mixing) motions that are organized into cells. The Sun's lower or inner atmosphere, the photosphere, is the region from which energy is radiated directly into space. Solar radiation approximates a Planck distribution (blackbody source) with an effective temperature of 5,800 Kelvin. This means that the bulk of the emitted solar energy lies between 150 nanometers and 10 micrometers wavelength, with a maximum at 450 nanometers (that is, it peaks in the visible portion of the electromagnetic (EM) spectrum.)

The chromosphere, which extends for a few thousand kilometers above the photosphere, has a maximum temperature of approximately 10,000 Kelvin. The corona, which extends several solar radii above the chromosphere, has temperatures of approximately 2 million degrees Kelvin. These regions emit electromagnetic radiation in the ultraviolet (UV), extreme ultraviolet (EUV) and X-ray portions of the spectrum. This shorter wavelength EM radiation, though representing a relatively small portion of the Sun's total energy output, still plays a dominant role in forming planetary ionospheres and in photochemistry reactions occurring in planetary atmospheres.

Since the Sun's outer atmosphere is heated, it expands into the surrounding interplanetary medium. This continuous outward flow of plasma is called the solar wind. It consists of protons, electrons, and alpha particles, as well as small quantities of heavier ions. Typical particle velocities in the solar wind fall between 300 and 400 kilometers per second, but these velocities may get as high as 1,000 kilometers per second.

Although the total energy output of the Sun is remarkably steady, its surface displays many types of irregularities. These include sunspots, faculae, plages (bright areas), filaments, prominences and flares. All are believed to be ultimately the result of interactions between ionized gases in the solar atmosphere and the Sun's magnetic field. Most solar activity follows the sunspot cycle. The appearances of individual sunspots on the solar disk cannot be predicted, but they statistically follow a generally regular periodic pattern. The solar magnetic cycle is believed to be the basic underlying cause of all the observed fluctuations over long (years) and short (days, minutes and even seconds) periods. This solar cycle was first observed in the systematic rise and fall of the number of visible sunspots. The apparent cyclic nature of solar activity with an eleven year fundamental period can be seen in the sunspot number record that extends back to the beginning of the 18th century.

Table 1 provides a summary of the physical properties of the Sun. The solar constant represents the total amount of EM radiation crossing a unit area in space perpendicular to the solar flux. At one astronomical unit (AU), the currently adopted value of the solar constant is 1371 ± watts/m².

The Space Age has allowed us to more fully investigate the Sun, our nearest star. Scientists, using space-based

Table 1   Physical Properties of the Sun

Distance from the Earth	1.496×10^8 km (1 AU)
Mass	1.991×10^{30} kg
Radius	6.960×10^5 km
Surface Area	6.09×10^{18} m²
Luminosity	3.9×10^{26} watts
Equivalent Black Body Temperature	5780 K
Rotation Period (varies with latitude zones)	(approx) 27 days
General Magnetic Field	1 gauss (local fields can be much greater)
Magnetic Period	Overall reversal approx every 11 years (22-year period)
Tilt of Solar Equator to Ecliptic	7 degrees
Solar Constant	1371 ± 5 W/m² (at 1 AU)
Radiation Output per Unit Surface Area	6.4×10^7 W/m²

SOURCE: NASA.

instruments, have now been able to observe the Sun's radiation output across the entire electromagnetic spectrum and not just the portion that filters down through the Earth's atmosphere. They have also been able to make direct measurements of solar particle emissions. For example, the *Skylab* missions (1972–1974) led to a significant improvement in our understanding of the solar corona and the solar wind.

Since the Sun is a churning ball of very hot gases, it does not rotate uniformly at all latitudes as does a ball of solid matter. Observations indicate that the equatorial regions rotate around the Sun's axis every 27 days, while the areas around the poles may take as long as 30 days.

The intense heat of the Sun strips electrons from their atoms, producing a sea of positive and negative charges that scientists call a plasma. As this hot plasma rises towards the Sun's surface, it drags a magnetic field with it, bending its lines of force. The magnetic field becomes more and more compressed as the wrapping continues and sometimes one of them will pop through the Sun's surface. This occurs because the increased magnetic pressure causes the balancing gas pressure to decrease, and it becomes buoyant. This buoyant low-density blob is a sunspot.

Sunspots were originally observed by Galileo in 1610. They are less bright than the adjacent portions of the Sun's surface, because they are not as hot. A typical sunspot temperature might be 4,500 Kelvin compared to the photosphere's temperature of 5,800 Kelvin. Sunspots appear to be made up of gases boiling up from the Sun's interior. A small sunspot may be about the size of the Earth, while larger ones could hold several hundred or even thousands of earth-sized planets. Extra-bright solar regions, called plages, often overlie sunspots. The number and size of sunspots appear to rise and fall through an eleven year cycle. The greatest number occur in years when the Sun's magnetic field is the most severely twisted. Solar physicists think that sunspot migration causes the Sun's magnetic field to reverse its direction. It then takes another 22 years for the Sun's magnetic field to return to its original configuration.

A solar flare is the sudden release of tremendous energy and material from the Sun (see fig. C38 on color page O). A flare may last minutes or hours and it usually occurs in complex magnetic regions near sunspots. Exactly how or why enormous amounts of energy are liberated in solar flares is still unknown, but scientists think the process is associated with electrical currents generated by changing magnetic fields. The maximum number of solar flares appears to accompany the increased activity of the sunspot cycle. As a flare erupts, it discharges a large quantity of material outward from the Sun. This violent eruption also sends shock waves through the solar wind.

Space-based solar observatories, including *Skylab* and Solar Maximum Mission data, have indicated that prominences (condensed streams of ionized hydrogen atoms) appear to spring from sunspot groups. Their looping shape suggests that these prominences are controlled by strong magnetic fields. About 100 times as dense as the solar corona, prominences can rise at speeds of hundreds of kilometers per second. Sometimes the upper end of a prominence curves back to the Sun's surface forming a "bridge" of hot glowing gas hundreds of thousands of kilometers long. On other occasions, the material in the prominence jets out and becomes part of the solar wind.

High energy particles are released into heliocentric space by solar events, such as large solar flares. Because of their close association with large flares, these bursts of energetic particles are relatively infrequent. For example, only 34 such events occurred with particle energies greater than 30 million electron volts in solar cycle 19 (1954 to 1965), while for cycle 20 (1965 to 1976) there were fewer than 20 energetic particle events. Typical proton fluxes at one astronomical unit are 10 to 100 protons/(cm²-steradian-s) with energies above 10 million electron volts. This proton flux falls to below 10 protons/(cm²-ster-s) for particles above 30 million electron volts. The total number of protons associated with single energetic particle events is 10^7 to 10^8 protons/cm² on the average, but particularly intense bursts may deposit 10^{10} protons/cm² at energies greater than 10 million electron volts. Solar flares represent a real hazard to space workers in cislunar space.

About five billion years from now, the Sun will have used up all the hydrogen fuel in its core and converted this hydrogen into helium. It will also have expanded and cooled. The hydrogen in the shell around the core will then begin thermonuclear burning. In the core itself, a major event called helium flash will occur. This is the initiation of a new thermonuclear reaction in which helium begins fusing and creates carbon (from three helium atoms) and oxygen (from one carbon atom and one helium atom). The expansion and cooling of the Sun's exterior surface will be accelerated. Our parent star will leave the main sequence and become a red giant. During this expansion, the Sun will probably grow large enough to engulf the Earth—boiling off all water and incinerating the land. This double shell burning of hydrogen and helium will then continue until thermal instabilities develop. These instabilities will cause the Sun to pulsate and eventually it will eject its outer shell of gases into space. The remaining core will contract until it is about the size of the Earth, forming an incredibly dense white dwarf star. This white dwarf will continue to cool itself by emitting ultraviolet radiation for many billions of years. By that time, the starfaring descendents of the human race will have spread throughout the Galaxy. Among these terrestrial progeny, some astrophysicists will look back to this portion of the Galaxy and observe the death of the star system that gave rise (perhaps uniquely) to intelligent life in the Galaxy!

See also: **fusion; stars**

sunlike stars   Yellow, main sequence stars with 5,000 to 6,000 degree Kelvin surface temperatures; spectral type G stars.

See also: **stars; Sun**

superior planets   Planets that have orbits around the Sun that lie outside the Earth's orbit. These planets include: Mars, Jupiter, Saturn, Uranus, Neptune, and Pluto.

synchrotron radiation   Electromagnetic radiation emitted by charged particles, especially electrons, moving at relativistic velocities in a magnetic field. For example, when a very energetic electron spirals around a magnetic field line, it emits characteristic synchrotron radiation in a narrow cone in its direction of motion. Synchrotron radiation displays a high degree of polarization. Space scientists now realize that synchrotron radiation is one of the most significant ways that astrophysical objects, such as supernova remnants and radio galaxies, emit radio waves. A supernova remnant is the expanding shell of gas ejected by a supernova explosion.

See also: **astrophysics; radio galaxy**

tachyon   A hypothetical faster-than-light particle. If it exists, advanced alien civilizations may use it in some way to achieve more rapid interstellar communications.

teleportation   A concept used in science fiction to describe the instantaneous movement of material objects to other locations in the Universe.

See also: **science fiction**

Tenth Planet   Astronomers speculate that the currently estimated mass of Pluto is too small to account for recently observed perturbations in the orbits of Uranus and Neptune. Although the source of these perturbations is not presently known, one hypothesis that has been offered is that there is a massive object circling our Sun far beyond the orbit of Pluto. This theorized "Tenth Planet" (also informally referred to as "Planet X") has never been detected, and today the Tenth Planet hypothesis itself is not widely accepted as an adequate explanation of the perturbation data.

See also: **Pluto**

terran   Of or relating to the planet Earth; a native of the planet Earth.

See also: **terrestrial**

terrestrial   Of or pertaining to the Earth; an inhabitant of the planet Earth.

terrestrial planets   In addition to the Earth itself, the terrestrial (or inner) planets include Mercury, Venus and Mars. These planets are similar in their general properties and characteristics to the Earth; that is, they are small, relatively highly dense bodies, composed of metals and silicates with shallow (or no) atmospheres as compared to the gaseous outer planets.

See also: **Earth; Mars; Mercury; Venus**

theorem of detailed cosmic reversibility   A premise developed by Francis Crick and Leslie Orgel in support of their directed panspermia hypothesis. This theorem states that if we can now contaminate another world in our Solar System with terrestrial microorganisms, then it is also reasonable to assume that an intelligent alien civilization could have developed the advanced technologies needed to "infect" or seed the early prebiotic Earth with spores, microorganisms or bacteria.

See also: **extraterrestrial contamination; panspermia**

thermonuclear   A reaction in which very high temperatures (millions of degrees Kelvin) are needed to bring about the fusion or joining of two light atomic nuclei, such as the hydrogen isotopes [deuterium (D) and tritium (T)], to form a heavier nucleus. The thermonuclear fusion reaction is accompanied by the release of large quantities of energy.

See also: **fusion**

Titan Probe   The atmosphere of Titan, the largest satellite of Saturn, is uniquely interesting to scientists from the standpoint of the chemical evolution of life—that is, from the standpoint of organic chemical evolution. Please note, however, that the term "organic" as used here carries the traditional chemical science meaning, rather than implying biological origin. In chemistry, organic refers to molecules based on carbon [except for a few extremely simple carbon-based molecules such as carbon dioxide (CO_2) and carbon monoxide (CO)]. Of great interest to exobiologists is the fact that the organic chemistry now taking place on Titan provides the only planetary-scale laboratory for the study of processes that could have been important in the primitive (pre-life) terrestrial atmosphere.

A Titan Fly-by/Probe mission now being considered by NASA would have the following major goals: (1) a determination of the structure and chemical composition of Titan's atmosphere; (2) a determination of energy flow patterns within Titan's atmosphere; and finally (3) a characterization (at least locally) of the surface of Titan.

This mission involves a Galileo Project-like probe that can be either carried to Titan on a fly-by carrier spacecraft, or else, flown to this interesting celestial body on a Saturn orbiter spacecraft. As the carrier spacecraft approaches Saturn, the probe would be deployed into Titan's atmosphere. Probe release from this carrier spacecraft would happen about 20 days before the spacecraft's closest approach to Titan. The probe descent into Titan's atmosphere would last approximately one hour. At present, scientists think that Titan's atmosphere is composed of the following major constituents: nitrogen (N_2) [82%–99%]; argon (Ar) [0–12%]; and methane (CH_4) [1%–6%]. The carrier spacecraft would (radar) map a portion of Titan's surface during its closest approach to the cloud enshrouded moon. Following the encounter with Titan, a small propulsive maneuver would adjust the trajectory of the carrier spacecraft, allowing it to encounter Iapetus, another Saturnian moon.

The Titan Probe mission is a critical step in our study of life in the Universe. It will attempt to answer such intriguing questions as: What does the surface of Titan really look like? Are there oceans of ethane (C_2H_6) and methane (CH_4)? What complex organic molecules are present on Titan, and what can they tell us about the origin of life on Earth?

See also: **exobiology; Galileo Project; life in the Universe; planetary exploration; Saturn**

Tunguska event   A violent explosion that occurred in a remote part of Siberia (Soviet Union) in June 1908. It is currently believed that this event was caused by the entrance of a cometary nucleus into the Earth's atmosphere. However, a few of the original investigators speculated that this destructive event was caused by the explosion of an alien spacecraft. (No firm technical evidence has been accumulated to support the latter hypothesis.)

See also: **comet; extraterrestrial catastrophe theory**

U

ultraviolet (UV) astronomy Astronomy based on the ultraviolet (UV) (10 to 400 nanometer wavelength) portion of the electromagnetic (EM) spectrum. Because of the strong absorption of UV radiation by the Earth's atmosphere, ultraviolet astronomy must be performed using high altitude balloons, rocket probes and orbiting observatories. Ultraviolet data are extremely useful in investigating interstellar and intergalactic media.

The Orbiting Astronomical Observatory-3 (OAO-3), also called *Copernicus*, was launched in 1972 and carried the first instrumentation for performing ultraviolet spectroscopy of the stars. Data from Copernicus have led scientists to theorize that low-density cavities or "bubbles" in interstellar space are caused by supernova explosions and are filled with gases that are much hotter than the surrounding interstellar medium. They have further suggested that a series of supernova explosions may cause bubbles or cavities that connect to form tunnels of hot gas, threading through the "cold" clouds and "warm" interstellar medium.

In 1975 white dwarf stars were detected with an extreme ultraviolet (EUV) telescope on board the American spacecraft used in the international Apollo-Soyuz mission. The International Ultraviolet Explorer (IUE) spacecraft detected a corona or ring of 100,000 degree Kelvin gas surrounding our Milky Way in 1979. This corona may be fed with hot materials by "superbubbles" expanding from the galactic plane. Other scientists have speculated that this ultraviolet radiation may be indicative of a halo of neutrinos that surround our Galaxy. (See fig. C38a on color page O.) Up until very recently, scientists thought that neutrinos, like photons, did not have any mass. Several scientists have now speculated that if neutrinos really did have a little mass, those produced during the moments of creation would now form a halo of countless billions of neutrinos around our Galaxy. As these neutrinos experienced decay interactions, ultraviolet radiation would be emitted.

See also: **astrophysics; electromagnetic spectrum; interstellar medium; stars**

ultraviolet (UV) radiation That portion of the electromagnetic spectrum that lies beyond visible (violet) light and is longer in wavelength than X-rays. Generally taken as electromagnetic radiation with wavelengths between 400 nanometers (just past violet light in the visible spectrum) and 10 nanometers (the extreme ultraviolet cutoff and the beginning of X-rays).

See also: **electromagnetic spectrum**

unidentified flying object (UFO) A flying object (apparently) seen in the terrestrial skies by an observer who cannot determine its nature. The vast majority of such "UFO" sightings can, in fact, be explained by known phenomena. However, these phenomena may be beyond the knowledge or experience of the person making the observation. Common phenomena that have given rise to UFO reports include: artificial Earth satellites, aircraft, high altitude weather balloons, certain types of clouds and even the planet Venus.

There are, nonetheless, some reported sightings that cannot be fully explained on the basis of the data available (which may be insufficient or too scientifically unreliable) or on the basis of comparison with known phenomena. It is the investigation of these relatively few UFO sighting cases that has given rise, since the end of World War II, to the "UFO hypothesis." This popular (though technically unfounded) hypothesis speculates that these unidentified flying objects are under the control of extraterrestrial beings who are surveying and visiting the Earth.

Modern interest in UFOs appears to have begun with a sighting report made by a private pilot named Kenneth Arnold. In June 1947 he reported seeing a mysterious formation of shining disks in the daytime near Mount Rainier in the State of Washington. When newspaper reporters heard of his account of "shining saucer-like disks"—the popular term "flying saucer" was born.

In 1948 the United States Air Force began to investigate these UFO reports. Project Sign was the name given by the Air Force to its initial study of UFO phenomena. In the late 1940s Project Sign was replaced by Project Grudge, which in turn became the more familiar Project Blue Book. Under Project Blue Book the Air Force investigated many UFO reports from 1952 to 1969. Then on December 17, 1969 the Secretary of the Air Force announced the termination of Project Blue Book.

The Air Force decision to discontinue UFO investigations was based on the following circumstances: (1) an evaluation of a report prepared by the University of Colorado and entitled, "Scientific Study of Unidentified Flying Objects" (this report is also often called the Condon report after its principal author); (2) a review of this University of Colorado report by the National Academy of Sciences; (3) previous UFO studies; and (4) Air Force experience from nearly two decades of UFO report investigations.

As a result of these investigations and studies and of experience gained from UFO reports since 1948, the conclusions of the Air Force were: (1) no UFO reported, investigated and evaluated by the Air Force has ever given any indication of threatening national security; (2) there has been no evidence submitted to or discovered by the Air Force that sightings categorized as "unidentified" represent technological developments or principles beyond the range of present-day scientific knowledge; and (3) there has been no evidence to indicate that the sightings categorized as "unidentified" are extraterrestrial vehicles.

With the termination of Project Blue Book, the Air

Force regulation establishing and controlling the program for investigating and analyzing UFOs was rescinded. All documentation regarding Project Blue Book investigations was then transferred to the Modern Military Branch, National Archives and Records Service, 8th Street and Pennsylvania Avenue N.W., Washington, D.C. 20408. This material is presently available for public review and analysis. If you wish to review these files personally, you need to simply obtain a researcher's permit from the National Archives and Record Service.

Today, reports of unidentified objects entering United States air space are still of interest to the military as part of its overall defense surveillance program. But beyond that, the Air Force no longer investigates reports of UFO sightings.

Similarly, while NASA is the focal point for answering public inquiries to the White House concerning UFOs, it is not engaged in a research program involving these UFO phenomena or sightings—nor is any other agency of the United States government.

One interesting result that emerged from Project Blue Book is a scheme, developed by Dr. J. Allen Hynek, to classify or categorize UFO sighting reports. Table 1 describes the six levels of classification that have been used. A Type-A UFO report generally involves seeing bright lights in the night sky. These sightings usually turn out to be a planet (typically Venus), a satellite, an airplane or meteors. A Type-B UFO report often involves the daytime observation of shining disks (that is, flying saucers) or cigar-shaped metal objects. This type of sighting usually ends up as a weather balloon, a blimp or lighter-than-air ship or even a deliberate prank or hoax. A Type-C UFO report involves unknown images appearing on a radar screen. These signatures might linger, be tracked for a few moments or simply appear and then quickly disappear—often to the amazement and frustration of the scope operator. These radar visuals often turn out to be something like swarms of insects, flocks of birds, unannounced aircraft, and perhaps the unusual phenomena radar operators call

"angels." (To radar operators, angels are anomalous radar wave propagation phenomena.)

Close encounters of the first kind (visual sighting of a UFO at moderate to close range) represent the Type-D UFO reports. Usually, the observer reports something unusual in the sky that "resembles an alien spacecraft." In the Type-E UFO report, not only does the observer claim to have seen the alien spaceship but also reports the discovery of some physical evidence in the terrestrial biosphere (such as scorched ground, radioactivity, mutilated animals, etc.) that is associated with the alien craft's visit. This type of sighting has been named a close encounter of the second kind. Finally, in the last type of UFO report, a close encounter of the third kind, the observer claims to have seen and sometimes to have been contacted by the alien visitors. Extraterrestrial contact stories range from simple sightings of "ufonauts," to communication with them (usually telepathic), to cases of kidnapping and then release of the terrestrial observer. There are even some reported stories in which a terran was kidnapped and then seduced by an alien visitor—a challenging task of romantic compatibility even for an advanced starfaring species!

Despite numerous stories about such UFO encounters, not a single shred of scientifically credible, indisputable evidence has yet to be acquired! If we were to judge these reports on some arbitrary proof scale, table 2 might be used as a guide for helping us determine what type of data or testimony we will need to convince ourselves that the "little green men" have arrived in their flying saucer. Unfortunately, we do not have any convincing data to support categories 1 to 3 in table 2. Instead, all we have are large quantities of eyewitness accounts of various UFO encounters (category 4 items in table 2). Even the most sincere human testimony changes in time and is often subject to wide variations and contradictions. The scientific method puts very little weight on human testimony in validating a hypothesis.

Even from a more philosophical point of view, it is very difficult to logically accept the UFO hypothesis. While intelligent life may certainly have evolved elsewhere in the Universe, the UFO encounters reported to date hardly

Table 1　UFO Report Classifications

A. NOCTURAL (Nighttime) LIGHT

B. DIURNAL (Daytime) DISK

C. RADAR CONTACT (Radar Visual or "RV")

D. VISUAL SIGHTING OF ALIEN CRAFT AT MODEST TO CLOSE RANGE
   [also called: Close Encounter of the First Kind, CE I]

E. VISUAL SIGHTING OF ALIEN CRAFT PLUS DISCOVERY OF (hard) PHYSICAL EVIDENCE OF CRAFT'S INTERACTION WITH TERRESTRIAL ENVIRONMENT
   [also called: Close Encounter of Second Kind, CE II]

F. VISUAL SIGHTING OF ALIENS THEMSELVES, INCLUDING POSSIBLE PHYSICAL CONTACT
   [also called: Close Encounter of Third Kind, CE III]

SOURCE: Derived from work of Dr. J. Allen Hynek and Project Blue Book.

Table 2　Proposed "Proof Scale" to Establish Existence of UFOs

Highest Value*		
(1)	The alien visitors themselves; or the alien spaceship
(2)	Irrefutable physical evidence of a visit by aliens or the passage of their spaceship
(3)	Indisputable photograph of an alien spacecraft or one of its occupants
(4)	Human eyewitness reports
Lowest Value		

*from a standpoint of the scientific method and validating the UFO hypothesis with "hard" technical data.

SOURCE: Based on work of Dr. J. Allen Hynek and Project Blue Book.

reflect the logical exploration patterns and encounter sequences we might anticipate from an advanced, starfaring alien civilization.

From our current understanding of the laws of physics, interstellar travel appears to be an extremely challenging, if not technically impossible, undertaking. Any alien race that developed the advanced technologies necessary to travel across vast interstellar distances would most certainly be capable of developing sophisticated remote sensing technologies. With these remote sensing technologies they could study the Earth essentially undetected—unless, of course, they wanted to *be* detected. And if they wanted to make contact, they could most surely observe where the Earth's population centers are and land in places where they could communicate with competent authorities. It is insulting not only to their intelligence but to our own human intelligence as well to think that these alien visitors would repeatedly contact only people in remote, isolated areas, scare the dickens out of them and then lift off into the sky. Why not once land in the middle of the Orange Bowl during a football game or near the site of an international meeting of astronomers and astrophysicists! And why only short, momentary intrusions into the terrestrial biosphere? After all the *Viking* Landers we sent to Mars gathered data for years. It's really hard to imagine that an advanced culture would make the tremendous resource investment to send a robot probe or even to come themselves and then only flicker through an encounter with beings on this planet. Are we that uninteresting? If that's the case, then why so many reported visits? From a simple exercise of logic, the UFO hypothesis just doesn't make sense—terrestrial or extraterrestrial!

Hundreds of UFO reports have been made since the late 1940s. Again, why are we so interesting? Are we at a galactic crossroads? Are our outer planets an "interstellar truck stop" where alien starships pull in and refuel? (Some people have already proposed this hypothesis.) Let's play a simplified interstellar traveler game to see if so many reported visits are realistic—even if we are very interesting. First, we'll assume that our Galaxy of over 100 billion stars contains about 100,000 different starfaring alien civilizations which are more or less uniformly dispersed. (This is a very *optimistic* number according to the Drake Equation and scientists who have speculated about the likelihood of Kardashev Type II civilizations.) Then each of these ET civilizations has, in principle, one million other star systems to visit without interfering with any other civilization. Yes, the Galaxy is a big place! What do you think the odds are of two of these civilizations both visiting our Solar System and only casually exploring the planet Earth during the last three decades? The only realistic conclusion that can be drawn is that the UFO reports are not credible indications of extraterrestrial visitations!

See also: **ancient astronauts; Drake Equation; extraterrestrial civilizations; interstellar travel**

Universe Everything that came into being at the moment of the Big Bang, and everything that has evolved from that initial mass or energy; everything that we can (in principle) observe.

Uranian Of or relating to the planet Uranus; (in science fiction) a native of the planet Uranus.

Uranus Uranus, unknown to ancient astronomers, was discovered by Sir William Herschel in 1781. Herschel attempted to name the new planet *Georgium Sidus* (or "George's star") after the reigning King of England, George III. Other 18th century astronomers chose to call the newly found celestial object, "Herschel," in honor of its discoverer. In the end, however, the use of mythological names for the planets won out and by the middle of the 19th century the seventh planet from the Sun was called Uranus after the ancient Greek god of the sky and father of the Titan, Cronos (Saturn in Roman mythology).

Uranus is a cold, gaseous giant, smaller than Saturn and appearing like a greenish disc, even in the largest terrestrial telescopes (see fig. C39 on color page P). This planet has one particularly interesting property—its axis of rotation lies in the plane of its orbit rather than nearly vertical to the orbital plane as occurs with the other planets. Because of this curious situation, Uranus moves around the Sun like a barrel rolling along on its side rather than like a top spinning on its end.

At nearly 3 billion kilometers from the Sun, Uranus is too distant from Earth to permit telescopic imaging of its features. Because of methane in its atmosphere, the planet (like Neptune) appears as a greenish disc or blob. Some of the presently known physical and dynamic properties of Uranus are presented in table 1.

Any solid surface Uranus has is well hidden below its thick atmosphere. The structure now theorized for this planet is that of a rock and metal core surrounded by layers of ice, liquid hydrogen and gaseous hydrogen, in that ascending order. There would be no clear boundary between the liquid and gaseous hydrogen. Instead, as the hydrogen becomes progressively less dense with increasing distance from the planet's core, the elemental substance tends to change from a liquid to a gas. Hydrogen appears to be the major component of the Uranian atmosphere, making up about 85 percent of its total. Helium makes up about 10 percent of the atmosphere, while the remainder consists of other gases such as methane.

One of the most exciting astronomical discoveries in the Solar System in this century, perhaps second only to the discovery of Pluto, was the discovery in 1977 that Uranus (like Saturn and Jupiter) also has rings. This discovery was made by Dr. James Elliot of Cornell University and his associates using a telescope mounted in NASA's Kuiper Airborne Observatory (a C-141 aircraft modified for astronomical studies). The scientists were observing the planet to determine its precise diameter when, quite unexpectedly, the previously unknown rings blocked the light

Table 1   Physical and Dynamic Properties of the Planet Uranus

radius	
(equatorial)	26,145 km
(polar)	25,518 km
mass	8.7×10^{25} kg
"surface" gravity	8.3 m/sec²
mean density (estimated)	1.2 g/cm³
albedo	0.37
temperature	58 K
magnetic field	expected to be small
rings	yes
atmosphere	hydrogen ($\sim 85\%$), helium ($\sim 10\%$), other gases including methane ($\sim 5\%$)
surface features	none yet observed
eccentricity	0.0505
mean orbital velocity	6.8 km/sec
sidereal year (an "uranian year")	84.01 years
inclination of planet's equator to its orbit around Sun	97.88 deg
number of natural satellites	5
period of rotation (estimated?)	11–24 hours (retrograde)
solar flux at average distance of planet from Sun	~ 3.8 W/m²
distance from the Sun	
average	2.88×10^9 km [19.28 AU] (160.1 light-min)
perihelion	2.74×10^9 km [18.3 AU] (152.2 light-min)
aphelion	3.01×10^9 km [20.1 AU] (167.2 light-min)

SOURCE: NASA.

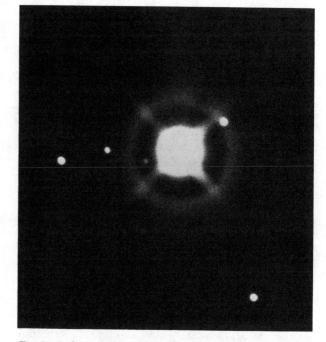

Fig. 1   A photograph of Uranus (overexposed) and its five moons: Miranda, Ariel, Umbriel, Titania and Oberon, in order of increasing distance from the planet. (Courtesy of NASA, photo by the Lunar and Planetary Laboratory, University of Arizona.)

2 presents current data about these moons.

In 1986 the *Voyager 2* spacecraft will encounter Uranus and provide astronomers with their first close-up view of the gaseous giant.

Uranus Probe   Unlike the terrestrial planets, the giant outer planets such as Jupiter, Saturn, Uranus and Neptune provide space scientists an opportunity to investigate key questions concerning their bulk compositions and internal structures through detailed studies of their atmospheres. The main purpose, therefore, of the Uranus Probe (planned for 2000–2010) is to determine the physical composition of that planet's atmosphere. In addition, the transport of energy within the atmosphere of Uranus would also be studied. This is important in developing an overall understanding of the internal structure and evolution of that planet.

from a distant star. The Uranian rings are thin, lie within a small range of orbital distances and have a low albedo. These factors now lead some astronomers to speculate that the rings resulted from the breakup of a moon or other celestial object which came too close to the planet.

From Earth, astronomers have been able to detect five moons around Uranus (see fig. 1). These natural satellites form a very regular system with the four larger ones all orbiting in the planet's equatorial plane. The moons are numbered I through V and are named: Ariel (I), Umbriel (II), Titania (III), Oberon (IV), and Miranda (V). Recent observations have led to the identification of water-ice on the surfaces of Ariel, Umbriel, Titania and Oberon. Table

Table 2   Moons of Uranus (Some of these data are tentative)

Name*	Year Discovered	Mean Distance from Uranus (km)	Sidereal Period (days)	Diameter** (km)	Mass** (kg)	Inclination (degrees)
Miranda (V)	1948	130,000	1.4	320	3.4×10^{19}	3.4
Ariel (I)	1851	192,000	2.5	860	6.7×10^{20}	0
Umbriel (II)	1851	267,000	4.1	900	7.6×10^{20}	0
Titania (III)	1787	438,000	8.7	1040	1.2×10^{21}	0
Oberon (IV)	1787	586,000	13.5	920	8.2×10^{20}	0

*moons presented in order of increasing distance from planet.

**currently estimated values (tentative).

SOURCE: NASA.

The basic Uranus Probe mission includes a Galileo Project-like atmospheric probe and a carrier spacecraft. Launch opportunities for this mission occur approximately every 13 months, with the trip time to Uranus taking about five and one-half years—if a very light carrier spacecraft (like the *Pioneer 10/11* spacecraft) is used. The *Mariner Mark II* modular spacecraft is also being considered by NASA as a candidate carrier spacecraft for this mission.

V

Venus Venus is the second planet out from the Sun. Often called the Evening Star or the Morning Star, it is named after the Roman goddess of love and beauty. Among the planets in our Solar System, it is the only one named after a female deity. Venus is called an inferior planet because it revolves around the Sun within the orbit of the Earth. It maintains an average distance of about 0.723 astronomical unit (AU) from the Sun. The planet appears to observers on Earth as either an evening star or a morning star. In fact, ancient astronomers regarded these two bright "wandering stars" as separate objects and even gave them different names. The early Greeks, for example, called the evening star Hesperos, and the morning star Phosphoros (see fig. C40 on color page P).

In the not too distant past, it was quite popular to think of Venus as literally Earth's twin. People thought that since Venus' diameter, density and gravity were only slightly less than Earth's, it must be similar, especially since it had an obvious atmosphere and was a little nearer the Sun. Visions of a planet with oceans, tropical forests, and even giant reptiles and primitive natives frequently appeared in science fiction stories. Unfortunately, spacecraft probes soon dispelled these romantic fantasies of a prehistoric world that mirrored a younger Earth. Except for a few physical similarities of size and gravity, the Earth and Venus are very different worlds.

Why should Venus be so different from Earth? Today, the environment on Venus differs significantly from the terrestrial biosphere. Its surface is an inferno and its atmosphere is nearly 100 times as dense as that of the Earth. Also, Venus rotates much slower and in retrograde fashion. The surface of Venus is enshrouded by thick clouds. In the ultraviolet portion of the spectrum these clouds exhibit markings that appear to rotate about the planet in a period of four (Earth) days. The predominantly carbon dioxide Venusian atmosphere contains only minute amounts of water vapor. Venus does not possess a

significant magnetic field, so the interaction of the planet with the solar wind is quite different from that of the Earth. Venus also does not have a natural satellite or moon. Table 1 provides physical and dynamic data for our nearest planetary neighbor. (At closest approach, Venus is approximately 42 million kilometers from Earth.)

Despite its closeness to Earth, astronomers using optical telescopes have been unable to unveil any details from the yellowish, brilliant disc of Venus. Then, on May 10, 1961 a radar signal was reflected from the planet. Analysis of the returned echo indicated that it must rotate extremely slowly. Subsequent investigations revealed that Venus rotated about its axis in 243 (Earth) days in the opposite direction (retrograde) to the way the Earth rotates. Because its axial rotation and orbital revolution are in opposite directions and of comparable periods, a local or "solar" day on Venus is actually 116.75 Earth days long. This means that Venus has 58 Earth days of daytime and an equally long nighttime. On Venus, the Sun rises in the west and sets in the east.

Why should Venus rotate so slowly? Most other planets in the Solar System rotate in periods of hours rather than days. The slow rotation of the innermost planet, Mercury, is attributed to tidal effects from the Sun; but Venus is too far from the Sun for such effects to have been significant over the lifetime of the planet. Some scientists speculate that Venus' rotation was slowed down by a grazing collision with an asteroid.

Venus is an almost perfect sphere. Planetary scientists

Table 1   Dynamic and Physical Properties of Venus

Radius (equatorial)	6051.3 km (≈ 0.95 Earth)
Mass	4.88×10^{24} kg (≈ 0.82 Earth)
Density (mean)	5.27 g/cm^3 (Earth $\approx$ 5.41)
Surface Gravity	8.88 m/s^2 (Earth = 9.81)
Escape Velocity at Surface	10.4 km/s (Earth = 11.2)
Albedo (over visible spectrum)	0.80 ± 0.02
Surface Atmospheric Pressure	9616 kPa (Earth = 101 kPa) (1396 psi) (14.7 psi)
Surface Temperature	750 K (approx) [480°C]
Solar Flux	2620 ± 35 W/m^2
Number of Natural Satellites	None
Semi-Major Axis	1.082×10^8 km (0.723 AU)
Perihelion Distance	1.075×10^8 km (0.7184 AU)
Aphelion Distance	1.089×10^8 km (0.7282 AU)
Eccentricity	0.0068
Inclination of Orbit	3.393 degrees
Mean Orbital Velocity	35.01 km/s
Sidereal Year	224.701 days (Earth days)
Sidereal Day (axial rotation period)	243 days (retrograde)
Local (solar) Day	116.75 days
Inclination of Equator to Orbit	2.2 degrees
Earth-to-Venus Distances maximum	2.59×10^8 km (1.73 AU)
minimum	0.42×10^8 km (0.28 AU)

SOURCE: NASA.

hypothesize that its interior is similar to that of Earth—namely, a liquid core, a solid mantle and a solid crust.

Despite the dense atmosphere and clouds enveloping Venus, some sunlight does penetrate to its surface where the solar flux is estimated to be equivalent to that at the Earth's surface on an overcast day in the mid-latitudes. Table 2 describes the composition of the Venusian atmosphere at an altitude of 21.6 kilometers; while figure 1 illustrates the three distinct regions of the Venusian atmosphere. These regions are: the high or upper atmosphere above the clouds, the thick layer of clouds, and the clear atmosphere beneath the clouds. A wind velocity profile is also included in figure 1.

As the nearest planet, Venus has been the target of many probes, fly-bys and orbiter spacecraft from both the United States and the Soviet Union (see table 3). The American *Mariner 2* space probe, launched in 1962, was the first successful interplanetary mission to the mysterious planet. As the spacecraft zoomed by, plunged into the dense Venusian atmosphere and gently landed on its inferno-like surface, all the romantic myths about Earth's twin were laid to rest.

Mariner 2 passed within 35,000 kilometers of Venus on December 14, 1962 and became the first spacecraft to scan another planet. Its instruments made measurements of Venus for 42 minutes. *Mariner 5*, launched by the United States in June 1967, flew much closer to the planet. Passing within 4,000 kilometers of Venus, its instruments measured the planet's (weak) magnetic field and temperatures. On its way to Mercury, *Mariner 10* flew past Venus and provided ultraviolet images that showed cloud circulation patterns in the Venusian atmosphere.

In the spring and summer of 1978, two American spacecraft, jointly called the Pioneer Venus mission, were launched to further help unravel the mystery of Venus. On December 4 the Pioneer Venus Orbiter became the first spacecraft placed in orbit around the planet. Five days later, the five separate components that made up the second spacecraft, called the *Pioneer Venus Multiprobe*, entered the Venusian atmosphere at different locations above the planet. These four independent probes and the main body telemetered data back to Earth on the properties of the Venusian atmosphere as they plunged to the surface (see fig. C41 on color page P).

Radar data from the *Pioneer Venus Orbiter* have helped unveil the mystery surrounding this cloud-enshrouded world. These data revealed an alien world of great mountains, expansive plateaus, enormous rift valleys and shallow basins. Three quite different regions became apparent: an ancient crust at intermediate elevations, relatively smooth lowland plains, and highlands. Some 70 percent of Venus consists of upland rolling plains, on which circular features may possibly be the remains of large impact craters. The lowlands cover about 25 percent of the Venusian surface. One extensive lowland basin, called *Atalanta Planitia*, is about the size of the North Atlantic Ocean basin on Earth. It is interesting to note that except for a few surface features named after they were discovered by Earth-based radar searches, all Venusian surface features are now being given female (deity) names that follow the overall tradition of the planet's mythological name. Finally, there are only two major highland or continental masses on Venus: *Ishtar Terra* and *Aphrodite Terra*. (Ishtar was the Babylonian goddess of love; and Aphrodite the love goddess of the ancient Greeks). *Ishtar Terra* is about the size of Australia or the continental United States and possesses the highest peaks yet discovered on Venus. It consists of three geographical units: *Maxwell Montes*, *Lakshmi Planum* and *Freyja Montes*.

Table 2   Composition of Venusian Atmosphere at 21.6 Kilometers

Gas	Volume Mixing Ratio (Percent)
CO_2	96.4 ± 1.0
N_2	3.41 ± 0.01
H_2O	0.135
O_2	$16.0 \pm 7.4 \times 10^{-4}$
Ar	$67.2 \pm 2.3 \times 10^{-4}$
CO	$19.9 \pm 3.12 \times 10^{-4}$
Ne	$4.31 \pm 4 \times 10^{-4}$
SO_2	$185 (+350 - 155) \times 10^{-4}$

Source: NASA.

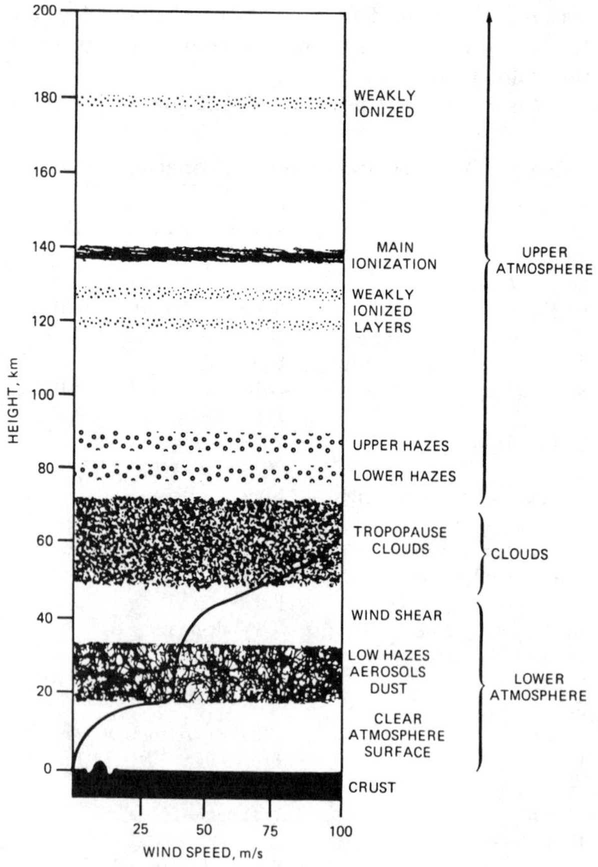

Fig. 1 Three distinct regions of the Venusian atmosphere. (Drawing courtesy of NASA.)

Table 3   Spacecraft Exploration of Venus (1961–1983)

Spacecraft	Country	Launch Date	Comments
Venera 1	USSR	12 Feb 61	Passed Venus at 100,000 km in May 61; but lost radio contact on 27 Feb 61.
Mariner 2	USA	27 Aug 62	1st successful interplanetary probe; passed Venus on 14 Dec 62 at 35,000 km.
Venera 2	USSR	12 Nov 65	Passed Venus on 27 Feb 66 at 24,000 km; but communications failed.
Venera 3	USSR	16 Nov 65	Impacted on Venus on 1 Mar 66; but communications failed earlier.
Venera 4	USSR	12 Jun 67	Probed Venusian atmosphere; fly–by spacecraft and descent module.
Mariner 5	USA	14 Jun 67	Venus fly–by on 19 Oct 67 within 3,900 km.
Venera 5	USSR	5 Jan 69	Descent probe entered atmosphere on 16 May 69.
Venera 6	USSR	10 Jan 69	Descent module entered atmosphere on 17 May 69.
Venera 7	USSR	17 Aug 70	Descent module soft landed on Venus on 15 Dec 70.
Venera 8	USSR	26 Mar 72	Descent module soft landed on Venus on 22 Jul 72.
Mariner 10	USA	3 Nov 73	Fly–by investigation of Venus on 5 Feb 74 at 5800 km; Mariner 10 continued on to Mercury.
Venera 9	USSR	8 Jun 75	Orbiter and descent module arrived at Venus on 22 Oct 75; descent module soft landed and returned picture; orbiter circled planet at 1545 km.
Venera 10	USSR	14 Jun 75	Orbiter and descent module arrived at Venus on 25 Oct 75; descent module soft landed and returned picture; orbiter circled planet at 1665 km.
Pioneer Venus			
Orbiter	USA	20 May 78	Orbited Venus on 4 Dec 78; radar mapping mission.
Multiprobe	USA	8 Aug 78	3 small probes, 1 large probe, and main bus entered atmosphere on 9 Dec 78.
Venera 11	USSR	9 Sep 78	Descent module softlanded; fly–by vehicle passed planet at 35,000 km on 25 Dec 78.
Venera 12	USSR	12 Sep 78	Descent module softlanded; fly–by vehicle passed planet at 35,000 km on 21 Dec 78.
Venera 13	USSR	30 Oct 81	Orbiter and descent module; descent module softlanded on 3 Mar 82 and returned color picture.
Venera 14	USSR	4 Nov 81	Orbiter and descent module; descent module softlanded on 5 Mar 82 and returned color picture.
Venera 15	USSR	2 Jun 83	Orbiter; radar mapping mission
Venera 16	USSR	7 Jun 83	Orbiter; radar mapping mission

SOURCE: NASA.

There are many intriguing questions still to be answered about Venus. What happened to the early Venusian oceans, if there ever were any? Because this planet lies on the inner edge of the Solar System's ecosphere, Venus has been considered an eventual candidate for planetary engineering. If we could reduce its currently intolerable surface temperatures and reverse the ongoing runaway greenhouse effect, Earth's twin might someday (perhaps centuries from now) become the balmy, tropical world envisioned by many science fiction writers. Until then, however, it must remain the excessively hot, desolate and waterless world revealed by our robot spacecraft in the last few decades.

See also: **Pioneer Venus; planetary engineering; Venus Atmospheric Probe Mission; Venus Radar Mapper**

Venus Atmospheric Probe Mission The American Pioneer-Venus mission and the Soviet Venera missions have raised many interesting questions about the atmosphere of Venus. For example, did Venus once have oceans? How did the Venusian atmosphere originate and evolve to its present condition? Space scientists feel that such intriguing questions can only be answered by an atmospheric probe instrumented for *in situ* (in the atmosphere) analysis.

NASA is considering a Venus Atmospheric Probe Mission to determine the composition of the Venusian atmosphere. In particular, it will measure the abundance of the noble gases, helium through xenon. The base-line mission includes a Pioneer-Venus-like probe and carrier spacecraft (called the "probe bus") that can be launched to Venus during any launch window (about every 19 months) by the Space Shuttle/Payload Assist Module (PAM)-A vehicle configuration. This probe will be targeted for a daylight entry into the Venusian atmosphere and will transmit data directly to Earth.

A mission option also being considered is to insert the Venus Atmospheric Probe during the Venus swing-by phase of a fly-by mission to the planet Mercury.

See also: **Venus**

Venusian   Of or relating to the planet Venus; (in science fiction) a native of the planet Venus.

Venus Radar Mapper (VRM)   An approved NASA Solar System exploration mission to the planet Venus. The Venus Radar Mapper has three main objectives: (1) to obtain a near-global map of the planet through the thick clouds that enshroud it by using a synthetic aperture radar (SAR) that has sub-kilometer spatial resolution; (2) to provide global and local topographic (surface) information with a radar altimeter that will have a vertical (height) resolution of about 100 meters and a spatial resolution of less than 50 kilometers; and (3) to extend the global gravity field measurements made by the *Pioneer Venus* spacecraft.

The Venus Radar Mapper will be launched sometime in the late 1980s by a Space Shuttle/Centaur-G upper stage configuration. When the spacecraft arrives at the planet, it will be inserted into a nearly polar, elliptical orbit (approximately 250 by 10,300 kilometers). Radar mapping will be performed for 42 minutes during that part of the mapper's orbit in which it is closest to Venus. Then for the remaining 110 minutes of each orbit, it will transmit the data back to Earth from an on-board recorder. After the Venusian surface has been thoroughly mapped for about 243 days, the VRM's primary mapping mission will be terminated.

Scientists anticipate that the Venus Radar Mapper will make significant contributions to their understanding of the planet's evolution. They expect to learn more about the planet's age, history and surface characteristics. They also hope to discover the age of the Venusian atmosphere and whether water and oceans were ever present. Other important questions that might be answered include: how Venus dissipates its internally generated heat and whether there is a presence or absence of tectonic activity (that is, plate tectonics).

See also: *Pioneer Venus;* **Venus**

Viking Project   The Viking Project was the culmination of a series of American missions to explore the planet Mars. This series of interplanetary missions began in 1964 with *Mariner 4,* and continued with the *Mariner 6* and 7 fly-by missions in 1969, and then the *Mariner 9* orbital mission in 1971 and 1972.

Viking was designed to orbit Mars and to land and operate on the surface of the Red Planet. Two identical spacecraft, each consisting of a lander and an orbiter, were built.

The Orbiters carried the following scientific instruments:

1. A pair of cameras with 1,500-millimeter focal length that performed systematic searches for landing sites; then looked at and mapped almost 100 percent of the Martian surface. Cameras onboard the *Viking 1* and 2 Orbiters took more than 51,000 photographs of Mars.

2. A Mars Atmospheric Water Detector that mapped the Martian atmosphere for water vapor and tracked seasonal changes in the amount of water vapor.

3. An Infrared Thermal Mapper that measured the temperatures of the surface, polar caps and clouds; it also mapped seasonal changes. In addition, although the Viking Orbiter radios were not considered scientific instruments, they were used as such. By measuring the distortion of radio signals as these signals traveled from the Orbiter spacecraft to the Earth, scientists were also able to measure the density of the Martian atmosphere.

The *Viking Landers* carried the following instruments:

1. The Biology Instrument that consisted of three separate experiments designed to detect evidence of microbial life in the Martian soil. There was always a chance that larger life forms could be present on Mars. But exobiologists thought the life forms most likely to be there (if any at all) would be microorganisms.

2. A Gas Chromatograph/Mass Spectrometer (GCMS) that searched the Martian soil for complex organic molecules. These organic molecules could be the precursors or the remains of living organisms.

3. An X-ray Fluorescence Spectrometer that analyzed samples of the Martian soil to determine its elemental composition.

4. A Meteorology Instrument that measured air temperature and wind speed and direction at the landing sites. These instruments returned the first extraterrestrial weather reports in the history of meteorology.

5. A pair of slow-scan cameras that were mounted about one meter apart on the top of each Lander. These cameras provided black-and-white, color and stereo photographs of Mars.

6. A seismometer had been designed to record any "Marsquakes" that might occur on the Red Planet. Such information would have helped planetologists determine the nature of the planet's internal structure. Unfortunately, the seismometer on Lander 1 did not function after landing and the instrument on Lander 2 observed no clear signs of internal (tectonic) activity.

7. An Upper Atmosphere Mass Spectrometer that conducted its primary measurements as each Lander plunged through the Martian atmosphere on its way to the landing site. The *Viking* Lander's first important scientific discovery—the presence of nitrogen in the Martian atmosphere—was made by this instrument.

8. A Retarding Potential Analyzer that measured the Martian ionosphere, again during entry operations.

9. Accelerometers, a Stagnation Pressure Instrument and a Recovery Temperature Instrument that helped determine the structure of the lower Martian atmosphere as the Landers approached the surface.

10. A Surface Sampler Boom that employed its collector head to scoop up small quantities of Martian soil to feed the biology, organic chemistry and inorganic chemistry instruments. It also provided clues to the soil's physical properties. Magnets attached to the sampler, for example, provided information on the soil's iron content.

11. The Lander radios were also used to conduct scientific experiments. Physicists were able to refine their estimates of Mars' orbit by measuring the round trip time for radio signals to travel between Mars and Earth. The great accuracy of these radio wave measurements also allowed scientists to confirm portions of Einstein's General Theory of Relativity.

Both Viking spacecraft were launched from Cape Canaveral, Florida. *Viking 1* was launched on August 20, 1975 and *Viking 2* on September 9, 1975. The Landers were sterilized before launch to prevent contamination of Mars with terrestrial microorganisms. These spacecraft spent nearly a year in transit to the Red Planet. *Viking 1* achieved Mars orbit on June 19, 1976; and *Viking 2* began orbiting Mars on August 7, 1976. *Viking 1* performed the first landing on Mars on July 20, 1976, on the western slope of *Chryse Planitia* (the Plains of Gold) at 22.3 degrees north latitude, 48.0 degrees longitude. The *Viking 2* Lander landed on September 3, 1976 at *Utopia Planitia* located at 47.7 degrees north latitude, 48.0 degrees longitude.

The Viking mission was planned to continue for 90 days after landing. Each Orbiter and Lander, however, operated far beyond its design lifetime. *Viking* Orbiter 1 exceeded four years of active flight operations in orbit around Mars.

The Viking Project's primary mission ended on November 15, 1976, just 11 days before Mars passed behind the Sun (an astronomical event called a "superior conjunction"). After conjunction, in mid-December 1976, telemetry and command operations were reestablished and extended mission operations began.

The *Viking* Orbiter 2 mission ended on July 25, 1978 due to an exhaustion of attitude-control system gas. The Orbiter 1 spacecraft also began to run low on attitude-control system gas, but through careful planning it was possible to continue collecting scientific data (at a reduced level) for another two years. Finally, with its control gas supply exhausted, the Orbiter 1's electrical power was commanded off on August 7, 1980, after 1,489 orbits of Mars.

The last data from the *Viking 2* Lander were received on April 11, 1980. The Lander 1 made its final transmission to Earth on November 11, 1982. After over six months of effort to regain contact with the Lander 1, the Viking mission came to an end on May 21, 1983.

With the single exception of the seismic instruments, the scientific instruments of the Viking Project acquired far more data than ever anticipated. The seismometer on Lander 1 did not function after touchdown, while the seismometer on Lander 2 detected only one event that might have been of seismic origin. Nevertheless, this instrument still provided data on surface wind velocity at Landing Site 2 (supplementing the meteorology experiment) and also indicated that the Red Planet has a very low level of seismicity.

The primary objective of the *Viking* Lander was to determine whether life exists on Mars. The evidence provided by the *Viking* Landers is currently being interpreted by scientists as strongly indicative that life does *not* now exist on Mars!

Three of the Lander instruments were capable of detecting life on Mars. The Lander cameras could have photographed living creatures large enough to be seen with the human eye. These cameras could also have observed growth in organisms such as plants and lichens. Unfortunately, the cameras at both sides observed nothing that could be interpreted as living.

The gas chromatograph/mass spectrometer (GCMS) could have found organic molecules in the soil. (Organic compounds combine carbon, nitrogen, hydrogen and oxygen.) These compounds are present in all living matter on Earth. The GCMS was programmed to search for heavy organic molecules, those large molecules that contain complex combinations of carbon and hydrogen and are either life precursors or the remains of living systems. To the surprise of exobiologists, the GCMS (which easily detects organic matter in the most barren terrestrial soils) found no trace of any organic molecules in the Martian soil samples.

Finally, the Lander Biology Instrument was the primary device used to search for extraterrestrial life. It was a one cubic foot box, loaded with the most sophisticated scientific instrumentation yet built and flown in space. The Biology Instrument actually contained three smaller instruments that examined the Martian soil for evidence of metabolic processes like those used by bacteria, green plants and animals on Earth.

The three biology experiments worked flawlessly on each Lander. All showed unusual activity in the Martian soil—activity that mimicked life—but exobiologists here on Earth needed time to understand the strange behavior of the Red Planet's soil. Today, according to most scientists who helped analyze these data, it appears that the chemical reactions were not caused by living things.

Futhermore, the immediate release of oxygen, when the

Martian soil contacted water vapor in the Biology Instrument, and the lack of organic compounds in the soil, indicate that oxidants are present in both the Martian soil and atmosphere. Oxidants, such as peroxides and superoxides, are oxygen-bearing compounds that break down organic matter and living tissue. Consequently, even if organic compounds evolved on Mars, they would be quickly destroyed.

Evaluation of the Martian atmosphere and soil has revealed that all the elements essential for life (as we know it on Earth)—carbon, hydrogen, nitrogen, oxygen and phosphorus—are also present on the Red Planet. However, exobiologists also consider the presence of liquid water as an absolute requirement for the evolution and continued existence of life. The Viking Project discovered ample evidence of Martian water in two of its three phases—vapor and solid (ice), and even evidence of large quantities of permafrost. But under current environmental conditions on Mars, it is impossible for water to exist as a liquid on the planet's surface.

Therefore, the conditions now known to occur on and just below the surface of the Red Planet do not appear adequate for the existence of living (carbon-based) organisms. However, exobiologists, though disappointed in their first serious search for extraterrestrial life, add that the case for life sometime in the past history of Mars is still open.

While the Gas Chromatograph/Mass Spectrometer found no sign of organic chemistry at either landing site, it did provide a precise and definitive analysis of the composition of the Martian atmosphere. The GCMS, for example, found previously undetected trace elements. The Lander X-ray Fluorescence Spectrometer measured the elemental composition of the Martian soil.

The two Landers continuously monitored weather at the landing sites. The midsummer Martian weather proved repetitious but in other seasons the weather varied and became more interesting. Cyclic variations in Martian weather patterns were observed. Atmospheric temperatures at the southern (*Viking 1*) landing site were as high as − 14 degrees Celsius (+ 7 degrees Fahrenheit) at midday; while the predawn summer temperature was typically − 77 degrees Celsius (− 107 degrees Fahrenheit). In contrast, the diurnal temperatures at the northern (*Viking 2*) landing site during the midwinter dust storm varied as little as 4 degrees Celsius (7.2 degrees Fahrenheit) on some days. The lowest observed predawn temperature was − 120 degrees Celsius (− 184 degrees Fahrenheit), which is about the frost point of carbon dioxide. A thin layer of water frost covered the ground near the *Viking* Lander 2 each Martian winter.

The barometric pressure was observed to vary at each landing site on a semiannual basis. This occurred because carbon dioxide (the major constituent of the Martian atmosphere) freezes out to form an immense polar cap—alternately at each pole. The carbon dioxide forms a great cover of "snow" and then evaporates (or sublimes) again with the advent of Martian "spring" in each hemisphere. When the southern cap was largest, the mean daily pressure observed by Lander 1 was as low as 6.8 millibars; while at other times during the Martian year it was as high as 9.0 millibars. Similarly, the pressures at the Lander 2 site were 7.3 millibars (full northern cap) and 10.8 millibars. (For comparison, the sea-level atmospheric pressure on Earth is about 1,000 millibars or 1 bar.)

Martian surface winds were also typically slower than anticipated. Scientists had expected these winds to reach speeds of hundreds of kilometers per hour. But neither Lander recorded a wind gust in excess of 120 kilometers (74 miles) per hour; and average speeds were considerably lower.

Photographs of Mars from the *Viking* Landers and Orbiters surpassed all expectations in both quantity and quality (see fig. 1 on page 106 and fig. C13 on color page G). The Landers provided over 4,500 images and the Orbiters over 52,000. The Landers provided the first close-up view of the surface of the Red Planet, while the Orbiters mapped almost 100 percent of the Martian surface, including detailed images of many intriguing surface features.

The Infrared Thermal Mapper and the Atmospheric Water Detector on board the Orbiters provided essentially daily data. Through these data it was determined that the residual northern polar ice cap that survives the northern summer is composed of water ice, rather than frozen carbon dioxide (dry ice), as scientists once believed.

The *Viking* Orbiters also provided high-quality photographs of the two Martian satellites, Phobos and Deimos.

Today, after all the Viking robot explorers have fallen silent, we are heir to billions of bits of valuable scientific data about Mars and now possess over 50,000 outstanding photographs from this project alone.

But intriguing questions about the Red Planet still remain. Is there a remote possibility that life exists in some crevice or biological niche on this mysterious world? Did life once evolve there, only to have vanished millions of years ago? And how did climatic conditions change so radically that great floods of water, which apparently raged over the Martian plains, have now vanished, leaving behind the dry, sterile world found by the Viking Project explorers? Only further exploration, including human expeditions, will resolve these intriguing questions.

See also: **Mars**

Voyager Once every 175 years the giant outer planets, Jupiter, Saturn, Uranus, and Neptune, align themselves in such a pattern that a spacecraft launched from Earth to Jupiter at just the right time might be able to visit the other three planets on the same mission, using a technique called "gravity assist." NASA space scientists named this multiple planet encounter mission the Grand Tour, and took advantage of a unique celestial alignment opportunity by launching two sophisticated spacecraft, called *Voyager 1* and 2 (see fig. 1).

Fig. 1 The 825–kilogram *Voyager* spacecraft and its complement of sophisticated scientific instruments. (Courtesy of NASA.)

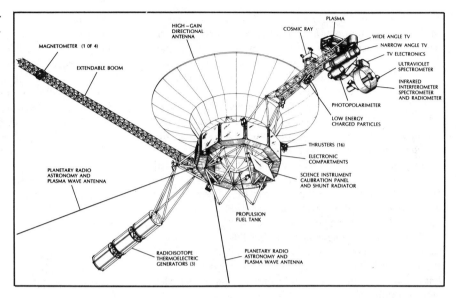

Each Voyager spacecraft had a mass of 825 kilograms and carried a complement of scientific instruments to investigate the outer planets and their many moons. These instruments recorded spectacular close-up images, examined magnetic environments, explored the Saturnian ring system and measured properties of the interplanetary medium. Since both spacecraft will eventually journey beyond the Solar System, a special interstellar message, in the form of a record called "The Sounds of Earth," was also placed on each spacecraft in the hope that perhaps millions of years from now some intelligent alien race will find either Voyager spacecraft floating through the interstellar void. If they are able to decipher the instructions for using this record, they will learn about the terrestrial civilization that built it.

Taking advantage of the 1977 Grand Tour launch window, the *Voyager 2* spacecraft lifted off from Cape Canaveral, Florida on August 20, 1977 on board a Titan-Centaur rocket. (NASA called the first Voyager spacecraft launched *Voyager 2*, because the second Voyager to be launched would eventually overtake it and become *Voyager 1*). *Voyager 1* was launched on September 5, 1977. It followed the same trajectory as its *Voyager 2* twin and overtook its sistership, just after entering the asteroid belt in mid-December 1977.

A year after entering the main asteroid belt, *Voyager 1* started drawing near to Jupiter. Some 50 million kilometers out from the giant planet, the Voyager's cameras began taking pictures which soon surpassed the best photographs ever taken by ground-based telescopes. *Voyager 1*'s closest approach to Jupiter occurred on March 5, 1979; while *Voyager 2* had its closest approach on July 9, 1979. *Voyager 1* completed its Jovian encounter in early April after taking almost 19,000 images and accomplishing many other scientific measurements. The *Voyager 2* spacecraft picked up the extraterrestrial baton in April 1979 and continued observing Jupiter and its moons until August. Together, the two spacecraft took more than 33,000 images of Jupiter and its five major satellites.

Although astronomers had studied Jupiter for several centuries with Earth-based telescopes, they were quite surprised by many findings of the Voyager spacecraft. Scientists now understand that very important physical, geologic and atmospheric processes are occurring in the giant planet, in its magnetosphere and on its many moons.

The discovery of active volcanism on the Galilean moon, Io, was probably the greatest surprise. It was the first time *active* volcanoes had been observed on another world in our Solar System. Voyager imagery provided a dramatic view of Io displaying two simultaneously occurring "extraterrestrial" volcanic eruptions.

Scientists now think that volcanic activity on Io affects the entire Jovian system. Io appears to be the source of matter that pervades the Jovian magnetosphere. (The magnetosphere is the region of space surrounding a planet that is primarily influenced by the planet's strong magnetic field.) Sulfur, oxygen and sodium were detected by the Voyager spacecraft as far away as the outer edge of the Jovian magnetosphere. These materials are apparently associated with Io's many volcanic eruptions and may have been sputtered off the moon's surface by the impact of high energy particles. Sulfur, oxygen and sodium particles were also detected inside Io's orbit where they were accelerated to more than 10 percent of the speed of light.

Close-up views of the other Galilean moons also revealed that Europa had the smootheset surface; Ganymede had experienced some tectonic activity before it froze solid about three billion years ago; and Callisto's crater-pocked surface is as ancient as that of our Moon or Mercury. Ganymede and Callisto are composed of about equal parts of water, ice and rock; Europa has substantial quantities of water; while Io is a waterless world.

The *Voyager 1* and *2* encounters with Saturn occurred nine months apart, in November 1980 and August 1981, respectively. The *Voyager 1* spacecraft is now heading out of the Solar System on an interstellar trajectory; while the *Voyager 2* spacecraft is currently enroute to a January 1986 encounter with Uranus, and then (if all remains well with this intrepid space traveler) an encounter with distant Neptune in August 1989.

The two encounters with Saturn increased our knowledge and altered our understanding of Saturn and some of its moons. Voyager findings indicate that Saturn's rings and moons are for the most part composed of water ice. The exceptions are the giant satellite Titan (which is half rock) and the outermost satellite, Phoebe (which is mostly rock). Voyager images of Phoebe show that it resembles an asteroid and some scientists believe that it may be a captured outer Solar System asteroid.

Voyager measurements also toppled Titan from its position as the Solar System's largest satellite. They found Jupiter's moon, Ganymede, to be slightly larger. However, both of these moons are actually bigger than either Mercury or Pluto.

Voyager instruments confirmed that neither Jupiter nor Saturn have solid surfaces. Their data indicated that while Jupiter's radiant heat may originate from gravitational contraction or the release of thermal energy accumulated during its formation, Saturn's radiant energy comes from the gravitational separation of helium and hydrogen in the planet's interior. Both planets radiate about two and one-half times more thermal energy than they receive from the Sun.

Voyager 2 discovered that nitrogen makes up about 82 percent of Titan's atmosphere. Nitrogen also makes up about 78 percent of the Earth's atmosphere. The other known constituents of Titan's atmosphere are methane, ethane, acetylene, ethylene, hydrogen, cyanide and other organic compounds. Titan's atmosphere is believed to be similar to that of the primeval Earth. Its composition reflects an abundance of water. An orange-colored smog enshrouds the moon and prevents direct observation of its surface. This smog is thought to result from the chemical reaction of solar radiation with methane in Titan's atmosphere. The Voyager spacecraft found that the atmospheric pressure on the surface of Titan is about 1.6 times the sea-level atmospheric pressure on Earth.

Some space scientists and exobiologists now speculate that the smog in Titan's atmosphere could have created a greenhouse effect that over the last few billion years warmed the satellite enough for organic chemicals to evolve toward prelife or prebiotic forms. *Voyager 2* measurements show that Titan's surface temperature is 95 Kelvin (−288 Fahrenheit), too cold for water to liquefy or for significant progress in prelife chemistry (for life as we now know it). Around this temperature, methane could exist as a liquid, vapor or solid—just as water exists in three phases in the Earth's biosphere. On Titan, therefore, methane rain or snow may fall from methane clouds! Methane rivers may flow through icy methane channels and empty into giant methane seas.

The Voyager encounters with the two dominant planets of our Solar System were an outstanding achievement in space exploration. With their prime missions successfully accomplished, *Voyager 1* now continues on a trajectory that will take it out of the Solar System by 1990; while *Voyager 2* will first encounter Uranus (January 1986) and then Neptune (August 1989), before entering interstellar space. Along the way, both spacecraft will search for the physical boundary of our Solar System—their instruments looking for the place where the solar wind fades away and is replaced by the stellar wind and cosmic rays.

See also: **images from space; Jupiter; Neptune; Saturn; Uranus; Voyager record**

Voyager record The *Voyager 1* and *2* spacecraft, launched during the summer of 1977, will eventually leave our Solar System. As these spacecraft wander through the Milky Way Galaxy over the next million or so years, each has the potential of serving as an interstellar ambassador —since each carries a special message from Earth on the chance that an intelligent alien race might eventually find one of the spacecraft floating among the stars. The *Voyager*'s interstellar message is a phonograph record called "The Sounds of Earth." Electronically imprinted on it are words, photographs, music and illustrations that will tell an extraterrestrial civilization about our planet. Included are greetings in over fifty different languages, music from various cultures and periods and a variety of natural terrestrial sounds such as the wind, the surf and different animals. The *Voyager* record also includes a special message from former president Jimmy Carter. Dr. Carl Sagan describes in detail the full content of this "phonograph message to the stars" in his delightful book, *Murmurs Of Earth* (see table 1).

Table 1   A Partial List of the Contents of the Voyager Record, *The Sounds of Earth*

A. SOUNDS OF EARTH

Whale, volcanoes, rain, surf, cricket frogs, birds, hyena, elephant, chimpanzee, wild dog, laughter, fire, tools, Morse code, train whistle, Saturn V rocket liftoff, kiss, baby.

B. MUSIC

Bach: *Brandenberg Concerto #2*, 1st movement; Zaire: "Pygmy Girls" initiation song; Mexico: mariachi band playing: "El Cascabel"; Chuck Berry: "Johnny B. Goode"; Navajo: night chant; Louis Armstrong: "Melancholy Blues"; China (zither) "Flowering Streams"; Mozart: Queen of the Night aria (*Magic Flute*); Beethoven: Symphony #5, 1st movement.

C. GREETINGS (in 55 different languages)

(example) English: "Hello from the children of planet Earth"

D. PICTURES (digital data to be reconstructed into images)

Calibration circle; solar location map; the Sun; Mercury; Mars; Jupiter; Earth; fetus; birth; nursing mother; group of children; sequoia (giant tree); snowflake; seashell; dolphins; eagle; Great Wall of China; Taj Mahal; U.N. Building; Golden Gate Bridge; radio telescope (Arecibo); Titan Centaur launch; astronaut in space.

Each record is made of copper with gold plating and is encased in an aluminum shield that also carries instructions on how to play it (see fig. 1). Look at figure 1, without reading beyond this paragraph. Can you decipher the instructions we've given to alien civilizations? No, it does not mean "Batteries not included!" If you can decipher these instructions, congratulations—you qualify as an extraterrestrial interpreter. If not, please do not feel disappointed; we shall now explore them together.

In the upper left is a drawing of the phonograph record and the stylus carried with it. Written around it in binary notation is the correct time for one rotation of the record, 3.6 seconds. Here, the time units are 0.70 billionths of a second, the time period associated with a fundamental transition of the hydrogen atom. The drawing further indicates that the record should be played from the outside in. Below this drawing is a sideview of the record and stylus, with a binary number giving the time needed to play one side of the record (approximately one hour).

The information provided in the upper–right portion of the instructions is intended to show how pictures (images) are to be constructed from the recorded signals. The upper–right drawing illustrates the typical wave form that occurs at the start of a picture. Picture lines 1, 2 and 3 are given in binary numbers and the duration of one of the picture "lines" is also noted (about 8 milliseconds). The drawing immediately below shows how these lines are to be drawn vertically, with a staggered interlace to give the correct picture rendition. Immediately below this is a

drawing of an entire picture raster, showing that there are 512 vertical lines in a complete picture. Then, immediately below this is a replica of the first picture on the record. This should allow extraterrestrial recipients to verify that they have properly decoded the terrestrial pictures. A circle was selected for this first picture to guarantee that any aliens who find the message use the correct aspect ratio in picture reconstruction.

Finally, the drawing at the bottom of the protective aluminum shield is that of the same pulsar map drawn on the *Pioneer 10* and *11* plaques. (These spacecraft are also headed on interstellar trajectories.) The map shows the location of our Solar System with respect to 14 pulsars, whose precise periods are also given. The small drawing with two circles in the lower-right-hand corner is a representation of the hydrogen atom in its two lowest states, with a connecting line and digit 1. This indicates that the time interval associated with the transition from one state to the other is to be used as the fundamental time scale, both for the times given on the protective aluminum shield and in the decoded pictures.

If you were making up a message to be included on an interstellar phonograph record, what would you like to say to some distant alien civilization?

See also: **interstellar communication; Pioneer plaque; *Voyager***

Vulcan   A planet that some 19th-century astronomers believed existed in an extremely hot orbit between Mercury and the Sun. Named after the Roman god of fire and metalworking craftsmanship, Vulcan's existence was postulated to account for gravitational perturbations observed in the orbit of Mercury. Modern astronomical observations have failed to reveal this celestial object and Einstein's theory of relativity has enabled 20th-century astronomers to account for the observed irregularities in Mercury's orbit. As a result, the planet Vulcan, created out of theoretical necessity by 19th-century astronomers, has now quietly disappeared in contemporary discussions about our Solar System.

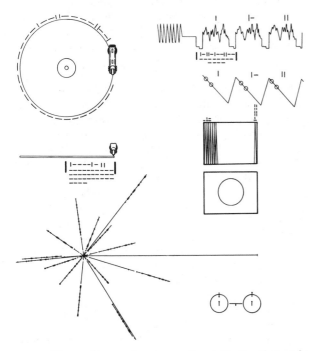

Fig. 1   The set of instructions to any alien civilization that might find the *Voyager 1* or *2* spacecraft, explaining how to operate the *Voyager* record and where the spacecraft and message came from. (Courtesy of NASA.)

water hole   A term used in the search for extraterrestrial intelligence (SETI) to describe a narrow portion of the electromagnetic spectrum that appears especially appropriate for interstellar communications between emerging and advanced civilizations. This band lies in the radio frequency (RF) part of the spectrum between 1420 megahertz frequency (21.1 centimeter wavelength) and 1660 megahertz frequency (18 centimeter wavelength).

Hydrogen (H) is abundant throughout interstellar space. When hydrogen experiences a "spin-flip" transition (due to an atomic collision), it emits a characteristic 1420 megahertz frequency (or 21.1 centimeter wavelength) radio wave. Any intelligent race throughout the Galaxy that has risen to the technological level of radio astronomy will eventually detect these emissions. Similarly, there is another grouping of characteristic spectral emissions centered near the 1660 megahertz frequency (18 centimeter wavelength) that are associated with hydroxyl (OH) radicals.

As we all know from elementary chemistry: H + OH = H_2O. So we have, as suggested by Dr. Bernard Oliver and other SETI investigators, two radio wave emission signposts associated with the dissociation products of water that "beckon all water-based life to search for its kind at the age-old meeting place of all species: the water hole."

Is this high regard for the 1420 to 1660 megahertz frequency band reasonable, or simply a case of terrestrial chauvinism? Well, many exobiologists currently feel that if other life exists in the Universe, it will most likely be carbon-based life and water is essential for carbon-based life as we know it. In addition, for purely technical reasons, if we scan all the decades of the electromagnetic spectrum in search for a suitable frequency at which to send or receive interstellar messages, we would arrive at the narrow microwave region between 1 and 10 gigahertz as the most suitable candidate for conducting interstellar communication. The two characteristic emissions of dissociated water, namely 1420 megahertz for H and 1660 megahertz for OH, are situated right in the middle of this optimum communication band.

Based on this type of reasoning, the water hole has been strongly recommended by scientists currently engaged in SETI projects. They generally feel that this portion of the electromagnetic spectrum represents a logical starting place for us to listen for interstellar signals from other intelligent civilizations.

See also: **search for extraterrestrial intelligence**

weightlessness   The condition of free fall or "zero gravity" found in an orbiting spacecraft. Astronauts experience the sensation of being weightless; and loose objects float about a crew cabin or space station.

See also: **microgravity**

What do you say to a little green man?   Suppose there is an intelligent extraterrestrial species out there, say 75 light years away, that is not only capable of receiving our radio messages, but also wants to communicate back. If we really are successful at CETI, just what will we talk about? Remember, radio waves travel at the speed of light—so it will take 75 years for a message to travel each way in this hypothetical interstellar communication. There are going to be some very long pauses in this

extraterrestrial "phone call." In fact, the first interstellar conversation with all the excitement it will cause on Earth might start something like this:

EARTH: "This is Earth calling. Is anyone there?"

—very long pause (150 years)

THEM: (possible response #1)

"Yes, we are here. Who are you?"

or

THEM: (possible response #2)

"Yes, someone is here. What do you want?"

or finally

THEM: (possible response #3)

"Yes, we're here. Who and *where* are you?"

Well, just how would you respond to any of these replies?

Maybe we could just send out news about ourselves—much like the Arecibo Observatory radio message sent in 1974. But that's almost like a holiday greeting card with a little, nonpersonalized printed newsletter tucked inside. We really have to choose the items selected for transmission carefully to guarantee the durability of the information. For example, if we choose to send news about our latest technological achievements, after the passage of a century and a half or so, we could become very embarrassed by the "backward" society these data appear to represent. Just think of the technical changes that have taken place on Earth in the last fifty years—computers, lasers, the microelectronics revolution, atomic energy, space travel, to name a few. Perhaps we might decide to send information about our current world situation or the welcoming speeches from terrestrial politicians. But we may quickly lose our extraterrestrial audience with such domestic trivia. For even the hottest news items of today will in all likelihood become rather insignificant in the overall context of planetary history.

So what do we say to the "little green men" who patiently wait for messages from Earth? One scientist feels that we could talk about mathematics, physics and astronomy. The long pauses between exchanges might give subsequent generations of humans something to look forward to—as they await the next message from the stars. Another scientist has suggested we send music and, if electronic communication techniques permit, even images of terrestrial works of art. Perhaps the creation and love of beauty might represent a common basis of communication throughout the Galaxy.

Along these lines, the first communication sequence between beings from Earth and an extraterrestrial culture could boil down to something like:

EARTH: "Hello! Is anyone there?"

(—very long pause; many years—)

THEM: "Yes, we're here. Who are we talking to?"

(—very long pause; again many years—)

EARTH: "We're beings from the planet Earth; listen to some of our beautiful music...."

(—very long pause; many years pass—)

THEM: "You've got a wrong number." Click!

Seriously, however, before we attempt to reach out and

touch someone or something across interstellar space, we, as a planetary society, must think about what we'd really say if someone answered. What would you say? And equally important, what would you ask "THEM"?

See also: **interstellar communication**

why humans explore The Age of Space is an era of exploration, discovery and scientific achievement without equal in all human history. Human intelligence has now begun to personally expand beyond our terrestrial biosphere. We have left footprints on the lunar surface. Our sophisticated machines have probed the red sands of Mars and whirled past the gaseous giant planets Jupiter and Saturn, returning spectacular images of these magnificent frozen worlds and many of their moons.

In previous explorations on Earth, a few brave adventurers, driven by a lust for gold, religious zeal or that unquenchable human curiosity for the unknown, went forward into uncharted regions, while the remainder of the sponsoring population anxiously waited for months or even years to learn what they had discovered. Today, through the wonders of modern communications, we have been able to watch the spectacular explorations of the Space Age—essentially as they were occurring. In the last two decades, human beings around the globe watched man first set foot upon the Moon, enjoyed the never before seen beauty of a Martian sunset and witnessed close up Titan's frigid, alien landscape and the unexpected volcanoes of Io.

Not since Galileo first lifted his crude, yet revolutionary, telescope to the night skies have we been able to observe our own star with such an increase in clarity. X-ray and infrared telescopes lifted above the Earth's atmosphere have revealed an invisible, incredibly energetic and violent Universe beyond our Solar System. Through the developments of modern space technology, we have acquired new tools to help us understand not only our cosmic origin but also our cosmic destiny! Who are we? Where did we come from? Where are we going? Are we alone? These are all questions to be asked and hopefully answered (if not immediately in full, at least in part) as we explore and meet the Universe firsthand.

It has been said that men and women are the vehicles of consciousness in the Universe. If this is so, then the exploration of space and the development of our extraterrestrial civilization are a major unfolding of that consciousness beyond the confines of one tiny planetary biosphere. The last such major evolutionary unfolding occurred some 350 million years ago when prehistoric fish, called crossopterygians, first left the seas of an ancient Earth and crawled upon the land. These early "explorers" are considered to be the ancient ancestors of all terrestrial animals with backbones and four limbs. Perhaps the deep human drive to explore will now serve as the very catalyst for some larger evolution of consciousness on a grand, cosmic scale.

From birth, curiosity drives us to explore our surroundings. As a child, we want to see the unseen and to learn what lies beyond our immediate view. Unfortunately, as we get older, this deep and early urge to explore and discover is conditioned by our environment, channeled into an acceptable pattern of behavior and, to a great extent, repressed. Nevertheless, despite intense social pressures to conform, not to take risks, not to be the first into "uncharted waters," a few hardy spirits in each generation have always maintained this deep, early sense of curiosity.

These are the individuals who as adults have always been bothered by blank areas on the map or even in the human mind. They have therefore worked and studied, striven and sometimes even given their lives to fill in these gaps. Over the centuries humankind has greatly benefitted from the restless urges of these curious and brave few. Step by painful step they have taken the human spirit into unknown and previously unreachable places—to the New World, to the South Pole, and even to the Moon. This impulse to explore and this drive to challenge the unknown now carry the minds of men and women inward into the microscopic domain of subatomic particles and outward into the vast realm of the galaxies.

Initially, the pressures of survival, especially the need for food and the escape from severely hostile environmental conditions, provided the exploration stimuli for early man. This was followed by competition for resources. When primitive man became a toolmaker, he or she needed to search further for certain materials. As ancient tribal populations increased due to more efficient use of very early hunting technologies, these larger populations had to expand the size of their hunting grounds and also to search for new, larger habitats. Throughout history, this pattern repeated itself many times. As the society or civilization developed, its resource needs also expanded, and an imaginative, determined few (driven by this deep-seated human urge to explore) helped advance knowledge and helped guide the many along a pathway to a better tomorrow.

Such explorations have almost always been worthwhile. In some cases they have produced tangible riches, such as precious jewels and metals, fertile new lands, favorable trading arrangements or extremely profitable advances in technology. In other cases the rewards of exploration were an increase in knowledge, especially a more thorough understanding of our world, its physical processes and the fundamental laws governing such processes. Such knowledge, even more than physical riches themselves, has helped pave the way for our modern civilization.

On July 2, 1976 (the eve of the landing of the *Viking 1* spacecraft on Mars) a special symposium was held by NASA at the California Institute of Technology in Pasadena. The members of the "Why Man Explores" panel were selected as authorities in classical disciplines related to exploration. This panel included Norman Cousins (who served as moderator), Philip Morrison, James Michener, Jacques Cousteau and Ray Bradbury. Some of the major comments made by these participants are provided here, since these statements on exploration are just as relevant today in the context of the development of our extraterrestrial civilization. A complete transcript of this most interesting and stimulating panel discussion can be found in the NASA Educational Publication (EP 125), entitled *Why Man Explores* (1977).

Comments made by the panel moderator, Mr. Norman Cousins, distinguished editor and writer:

The question, 'why explore?' pertains less to the *Viking 1* expedition in particular than to the nature of the human mind in general. We are here to consider not just the phenomenon of a journey to Mars but the phenomenon of intelligence. The fact that we can conceive of the inconceivable, and comprehend the incomprehensible, is perhaps the highest exercise of the human brain, symbolized so dramatically by the exploration of Mars.

Comments made by Dr. Philip Morrison, distinguished theoretical physicist and scholar-philosopher:

Why do human beings explore? I would answer, as I think the Greeks would answer, "Because it is our nature."

We are beings who construct for ourselves, each separately and singly, and as well together in our collectivities, internal models of all that happens, of all we see, find, feel, guess, and conjecture about our experience in the world.

For me, exploration is filling in the blank margins of that inner model, that no human can escape making.

Comments made by James A. Michener, world-renowned novelist and travel writer:

Exploring is one of his (man's) permanent and attractive characteristics.

We never gain as much from it (the exploration) as the wild enthusiasts promise; we invariably gain more than the frightened old men predict. And regardless of predictions, the exploration must go on because it is in man's nature to explore.

Comments made by Jacques-Yves Cousteau, a man who has dedicated his life to the exploration of the oceans:

When man explores for resources, his motivations are clear...[but] what is the origin of the devouring curiosity that drives men to commit their lives, their health, their reputations, their fortunes, to conquer a bit of knowledge, to stretch our physical, emotional, or intellectual territory...Man's motivation for exploration is but the sophistication of a universal instinctive drive deeply ingrained in all living creatures. Life is growth—individuals and species grow in size, in number, and in territory. The peripheral

manifestation of growing is exploring the outside world.

The exploration drive, pure and natural, is associated with risk, freedom, initiative and lateral thinking. The enemies of the exploration spirit are mainly the sense of security and responsibility; red tape; and exclusive vertical thinking.

Comments by Ray Bradbury, renowned science fiction and futuristic fiction writer:

The Space Age is titanic; it's a whole universe we are talking about.

We are going out into the universe. We go there because we love life.

I don't know how far out into the Galaxy we will make it, but indeed we will.

We close this entry by considering: *Man explores because for us it is the Universe, or nothing!*

wormhole   Some scientists speculate that matter falling into a black hole may actually survive! They suggest that under very special circumstances such matter might be conducted by means of passageways, called "wormholes," to emerge in another place or time in this Universe or in another universe. In forming this concept, scientists are theorizing that black holes can play "relativistic tricks" with space and time.

See also: **black holes; stars**

X-ray   A penetrating form of electromagnetic radiation of very short wavelength (approximately 0.01 to 10 nanometers or 0.1 to 100 angstroms) and high photon energy (approximately 100 electronvolts to some 100 kiloelectronvolts). X-rays are emitted when either the inner orbital electrons of an excited atom return to their normal energy states (these photons are called characteristic X-rays) or when a fast-moving charged particle (generally an electron) loses energy in the form of photons upon being accelerated and deflected by the electric field surrounding the nucleus of a high atomic number element (this process is called "bremsstrahlung" or braking radiation). Unlike gamma rays, X-rays are non-nuclear in origin.

See also: **Advanced X-Ray Astrophysics Facility; astrophysics; X-Ray Timing Explorer**

X-Ray Timing Explorer (XTE) A proposed NASA explorer-class payload for the study of the temporal variability of X-ray emitting objects. The main instrument on the XTE, called a large area proportional counter, will be applied to the detailed observation of individual celestial objects over long exposure times, while an all-sky monitor will record all the remaining strong X-ray emitters simultaneously over longer time scales with decreased sensitivity. In this way, the XTE can sample all time scales from microseconds to years for the entire X-ray source catalog. The XTE will be launched by the Space Shuttle during the late 1980s and placed into a 400 kilometer circular orbit at a 28.5 degree inclination. A two year mission lifetime is anticipated.

See also: **astrophysics**

zero-gravity aircraft An aircraft that flies a special parabolic trajectory to create low-gravity conditions (typically 0.01 g) for short periods of time (10 to 30 seconds), where one g here represents the acceleration due to gravity at the Earth's surface (9.8 m/s²).

zodiac The word zodiac comes from the ancient Greek language and means circle of figures or circle of animals. In astronomy zodiac refers to a band in the sky, extending about nine degrees to each side of the ecliptic. Since the earliest times, the zodiac has been divided into intervals of 30 degrees along the ecliptic, with each of these sections being designated by a "sign of the zodiac." The annual revolution of the Earth around the Sun causes the Sun to appear to enter a different constellation of the zodiac each month. These twelve constellations (or signs) are: Aries (Ram), Taurus (Bull), Gemini (Twins), Cancer (Crab),

Leo (Lion), Virgo (Maiden), Libra (Scales), Scorpius (Scorpion), Sagittarius (Archer), Capricornus (Sea-goat), Aquarius (Water-bearer), and Pisces (Fish). Although the signs of the zodiac originally corresponded in position to the twelve constellations just named, because of the phenomenon of precession, the zodiacal signs do not presently coincide with these constellations. For example, when people today say the Sun enters Aries at the vernal equinox, it has actually shifted forward (from ancient times) and is now actually in the constellation Pisces.

zodiacal light A faint cone of light extending upward from the horizon in the direction of the ecliptic. Zodiacal light is seen from the tropical latitudes for a few hours after sunset or before sunrise. It is due to sunlight being reflected by tiny pieces of rock and interplanetary dust in orbit around the Sun much farther out than the Earth.

Zoo hypothesis One response to the Fermi paradox. It assumes that intelligent, technically very advanced species do exist in the Galaxy, but that we cannot detect or interact with them because they have set the Solar System aside as a perfect zoo or wildlife preserve.

The reasoning followed in establishing this hypothesis goes something like this. Technically advanced beings exert a great deal of control over their environment. For example, human beings have a far greater influence on the biosphere than all the other creatures that co-inhabit the planet with us. Occasionally, we have decided not to exert this influence, but rather to set aside certain regions of the Earth as zoos, wildlife sanctuaries or wilderness areas, where other species can develop naturally with little or no human interaction. In fact, the perfect wildlife sanctuary or zoo is one set up so that the species within are not even aware of the presence or existence of their zookeepers. Thus, in response to the question: "Where are they?," the Zoo hypothesis suggests that we cannot and will never be able to detect "them," because our "extraterrestrial zookeepers" have set aside the Solar System as a perfect zoo or wildlife sanctuary and want us to develop "naturally" without awareness of or interaction with them.

See also: **extraterrestrial civilizations; Fermi paradox; Laboratory hypothesis**

APPENDIX A

While many metric units are used in the aerospace industry, space science and space technology, some appear with great frequency. These are the units for mass, length, temperature and time. Other commonly used units, such as energy (work) and power (work per unit time) are derived units, that is, those based on one or more of the basic units identified previously.

When the metric system was developed in France in the late 18th century, the basic unit for length was obtained from nature. This unit, the meter (m), was defined to be one ten-millionth the distance from the equator to the North Pole along the meridian of longitude nearest Paris, France. Accurate measurement of this distance was difficult in those days and, by chance, the resulting unit was very close to the English yard. Similarly, the liter, a volume measurement equal to one cubic decimeter or 0.001 cubic meter, was also very close in size to the English quart.

In 1960, the General Conference on Weights and Measures adopted "Le Système International d'Unites" (International System of Units or SI). The modernized units proposed and periodically revised by SI are still based on natural standards, but these standards are ones that can be measured with greater precision than the arc distance from the Earth's equator to the North Pole.

Some of the more common metric system units are described in Table A-1, while those frequently encountered in astronomical studies or extraterrestrial pursuits appear in Table A-2. Table A-3 provides the names, symbols and conversion factors for SI units frequently encountered in space activities. Table A-4 provides a listing of acceptable SI unit prefixes. Table A-5 describes traditional notation of powers of ten, and Table A-6 gives convenient metric/English conversions for units frequently encountered in space technology.

Table A-1   Common metric system units (International System of Units—SI)

meter (m): [metre, British] SI unit of length equal to 1,650,763.73 wavelengths in a vacuum of the orange-red line of the spectrum of krypton-86.

kilogram (kg): [kilogramme, British] SI unit of mass defined by a reference 90% platinum–10% iridium metal cylinder, called the international prototype kilogram (orginally the mass of one cubic decimeter (dm^3) of water at its maximum density).

second (s): SI unit of time defined as the duration of 9,192,631,770 periods of radiation that correspond to the transition between two hyperfine energy levels of the ground state of the cesium-133 atom (that is, time defined using an atomic clock). Formerly, a second was defined as 1/86,400 of the mean solar day.

newton (N): SI unit of force or thrust needed to accelerate a one kilogram mass one meter per second squared.
$$1 \text{ N} = 1 \text{ kg} - \text{m/s}^2$$

joule (J): SI unit of energy (all forms: thermal, chemical, mechanical, electrical, etc.) A joule is defined as the energy equivalent of applying a force of one newton through a distance of one meter.
$$1 \text{ J} = 1 \text{ N} - \text{m}$$
The joule has also replaced the calorie as the unit of thermal energy.
$$1 \text{ calorie} = 4.1868 \text{ joules}$$

watt (W): SI unit of power (all forms: electrical, mechanical, thermal, etc.) Power is time rate of using/releasing energy.
$$1 \text{ W} = 1 \text{ J/s}$$
In electrical applications,
$$1 \text{ W} = 1 \text{ ampere} \cdot 1 \text{ volt}$$

kelvin (K): SI unit of (absolute) temperature. A degree kelvin is equal to 1/273.16 of the thermodynamic temperature of the triple point of water. A temperature of zero degrees kelvin is absolute zero. Degrees Celsius (C) are frequently encountered as a metric unit of temperature. In this case, Celsius degrees represent a *relative* temperature scale; on the Celsius scale, 0°C is the freezing point of water and 100°C is the boiling point of water at one atmosphere pressure.

(in relative terms) one degree kelvin = one degree Celsius
(in absolute terms) temperature (degrees kelvin) = temperature (degrees Celsius) + 273.16

Table A-2   Special units for astronomical investigations

astronomical unit (AU): The mean distance from the Earth to the Sun—approximately $1.495\ 979 \times 10^{11}$ meters.

light year (ly): The distance light travels in one year's time—approximately $9.460\ 55 \times 10^{15}$ meters.

parsec (pa): The parallax shift of one second of arc (3.26 light years)—approximately $3.085\ 768 \times 10^{16}$ meters.

speed of light (c): $2.997\ 9 \times 10^8$ meters per second.

SOURCE: NASA.

Table A-3   International System (SI) Units and their conversion factors

Quantity	Name of unit	Symbol	Conversion factor
Distance	meter	m	1 km = 0.621 mile 1 m = 3.28 ft 1 cm = 0.394 in. 1 mm = 0.039 in. 1 μm = 3.9 × 10^{-5} in. = 10^4 Å 1 nm = 10 Å
Mass	kilogram	kg	1 tonne = 1.102 tons 1 kg = 2.20 lb 1 gm = 0.0022 lb = 0.035 oz 1 mg = 2.20 × 10^{-6} lb = 3.5 × 10^{-5} oz
Time	second	sec	1 yr = 3.156 × 10^7 sec 1 day = 8.64 10^4 sec 1 hr = 3600 sec
Temperature	kelvin	K	273 K = 0° C = 32° F 373 K = 100° C = 212° F
Area	square meter	m^2	1 m^2 = 10^4 cm^2 = 10.8 ft^2
Volume	cubic meter	m^3	1 m^3 = 10^6 cm^3 = 35 ft^3
Frequency	hertz	Hz	1 Hz = 1 cycle/sec 1 kHz = 1000 cycles/sec 1 MHz = 10^6 cycles/sec
Density	kilogram per cubic meter	kg/m^3	1 kg/m^3 = 0.001 gm/cm^3 1 gm/cm^3 = density of water
Speed, velocity	meter per second	m/sec	1 m/sec = 3.28 ft/sec 1 km/sec = 2240 mi/hr
Force	newton	N	1 N = 10^5 dynes = 0.224 lbf
Pressure	newton per square meter	N/m^2	1 N/m^2 = 1.45 × 10^{-4} lb/in^2
Energy	joule	J	1 J = 0.239 calorie
Photon energy	electronvolt	eV	1 eV = 1.60 × 10^{-19} J; 1 J = 10^7 erg
Power	watt	W	1 W = 1 J/sec
Atomic mass	atomic mass unit	amu	1 amu = 1.66 × 10^{-27} kg

Customary Units Used With the SI Units

Quantity	Name of unit	Symbol	Conversion factor
Wavelength of light	angstrom	Å	1 Å = 0.1 nm = 10^{-10} m
Acceleration of gravity	g	g	1 g = 9.8 m/sec^2

SOURCE: NASA.

Table A-4   Recommended SI Unit prefixes

Prefix	Abbreviation	Factor by which unit is multiplied
tera	T	10^{12}
giga	G	10^9
mega	M	10^6
kilo	k	10^3
hecto	h	10^2
centi	c	10^{-2}
milli	m	10^{-3}
micro	μ	10^{-6}
nano	n	10^{-9}
pico	p	10^{-12}

SOURCE: NASA.

Table A-5   Powers of ten notation

Increasing	Decreasing
10^2 = 100	10^{-2} = 1/100 = 0.01
10^3 = 1,000	10^{-3} = 1/1000 = 0.001
10^4 = 10,000 etc.	10^{-4} = 1/10,000 = 0.0001, etc.

Examples: | Example:

2 × 10^6 = 2,000,000
2 × 10^{30} = 2 followed by
    30 zeros

5.67 × 10^{-5} = 0.000 056 7

SOURCE: NASA.

Table A-6   Common Metric/English conversion factors (for space technology activities)

	Multiply	*By*	*To Obtain*
Length:	inches	2.54	centimeters
	centimeters	0.393 7	inches
	feet	0.304 8	meters
	meters	3.281	feet
	statute miles	1.609 3	kilometers
	kilometers	0.621 4	statute miles
	kilometers	0.54	nautical miles
	nautical miles	1.852	kilometers
	kilometers	3 281.	feet
	feet	0.000 304 8	kilometers
Weight and Mass	ounces	28.350	grams
	grams	0.035 3	ounces
	pounds	0.453 6	kilograms
	kilograms	2.205	pounds
	tons	0.907 2	metric tons
	metric tons	1.102	tons
Liquid Measure	fluid ounces	0.029 6	liters
	gallons	3.785 4	liters
	liters	0.264 2	gallons
	liters	33.814 0	fluid ounces
Temperature	degrees Fahenheit plus 459.67	0.555 5	kelvins
	degrees Celsius plus 273.16	1.0	kelvins
	kelvins	1.80	degrees Fahrenheit minus 459.67
	kelvins	1.0	degrees Celsius minus 273.16
	degrees Fahrenheit minus 32	0.555 5	degrees Celsius
	degrees Celsius	1.80	degrees Fahrenheit plus 32
Thrust (Force)	pounds force	4.448	newtons
Pressure	newtons	0.225	pounds
	millimeters mercury	133.32	pascals (newtons per square meter)
	pounds per square inch	6.895	kilopascals (1000 pascals)
	pascals	0.007 5	millimeters mercury at 0° C
	kilopascals	0.145 0	pounds per square inch

SOURCE: NASA.

SPECIAL REFERENCE LIST

The following sources were especially useful in the overall development and assembly of this book.

Adelman, Saul J., and Adelman, Benjamin. *Bound for the Stars.* Englewood Cliffs, NJ: Prentice Hall, Inc., 1981.

Angelo, Joseph A., Jr. *The Dictionary Of Space Technology.* New York: Facts On File, 1982.

———, and Buden, David. *Space Nuclear Power.* Malabar, FL: Orbit Book Company, Inc., 1985.

Calder, Nigel. *Spaceships of the Mind.* New York: The Viking Press, 1978.

Crick, Francis. *Life Itself.* New York: Simon and Schuster, 1981.

Goldsmith, Donald. *The Quest for Extraterrestrial Life: A Book Of Readings.* Mill Valley, CA: University Science Books, 1980.

———, and Owen, Tobias. *The Search for Life in the Universe.* Menlo Park, CA: The Benjamin/Cummings Publishing Company, Inc., 1980.

Hoyle, Fred. *The Intelligent Universe.* New York: Holt, Rinehart, and Winston, 1983.

———, and Wickramasinghe, N.C. *Evolution from Space.* New York: Simon and Schuster, Inc., 1981.

Illingworth, Valerie. *The Facts On File Dictionary Of Astronomy.* New York: Facts On File, 1979.

NASA. "Advanced Automation for Space Missions," NASA Conference Publication 2255. 1982.

NASA. "Life In the Universe." NASA Conference Publication 2156. 1981.

NASA. *The Martian Landscape.* NASA SP-425. 1978.

NASA. "A Meeting With The Universe." NASA EP-177. 1981.

NASA. "Pioneer: First To Jupiter, Saturn, and Beyond." NASA SP-446. 1980.

NASA. "Pioneer Venus." NASA SP-461. 1983.

NASA. "Project Orion: A Design Study of a System for Detecting Extrasolar Planets." NASA SP-436. 1980.

NASA. "The Search For Extraterrestrial Intelligence." NASA SP-419. 1977.

NASA. "Space Resources and Space Settlements." NASA SP-428. 1979.

NASA. "Space Settlements: A Design Study." NASA SP-413. 1977.

NASA. *Viking Orbiter Views Of Mars.* NASA SP-441. 1980.

NASA. "Voyages to Saturn." NASA SP-451. 1982.

NASA/Ames Research Center. "Project Cyclops: A Design Study of a System for Detecting Extraterrestrial Intelligent Life." CR 114445. Revised ed. July, 1973.

Nicogossian, Arnauld, and Parker, James. *Space Physiology and Medicine.* NASA SP-447. 1982.

Oberg, James E. *Mission To Mars.* Harrisburg, PA: Stackpole Books, 1982.

———. *New Earths.* New York: The New American Library, 1983.

Rood, Robert T., and Trefil, James S. *Are We Alone?* New York: Charles Scribner's Sons, 1981.

Sagan, Carl. *The Cosmic Connection.* New York: Doubleday & Company, 1973.

———. *Cosmos.* New York: Random House, 1980.

———. *Murmurs Of Earth.* New York: Random House, Inc., 1978.

Vajk, J. Peter. *Doomsday Has Been Cancelled.* Culver City, CA: Peace Press, 1978.

SELECTED READING LIST

1. SPACE SCIENCE & SPACE TECHNOLOGY

Angelo, Joseph A. Jr. *The Dictionary of Space Technology*. New York: Facts On File, 1982.

Asimov, Isaac. *Exploring the Earth and the Cosmos*. New York: Crown Publishers, 1982.

Feldman, A. *Space*. New York: Facts On File, 1980.

French, B.M., and Maran, S.P., eds. *A Meeting with the Universe*. NASA EP-177. 1982.

Illingworth, Valerie. *The Facts On File Dictionary Of Astronomy*. New York: Facts On File, 1979.

Jastrow, R. *Red Giants and White Dwarfs*. New York: Warner Books, 1979.

NASA. *Atoms and Astronomy*. NASA EP-128. 1977.

———. *Chemistry Between the Stars*. NASA EP-127. 1977.

———. *Extragalactic Astronomy*. NASA EP-129. 1977.

———. *High Energy Astronomy Observatory*. NASA EP-167. 1980.

———. *The Supernova*. NASA. EP-126. 1977.

Pasachoff, J.M. *Astronomy: From the Earth to the Universe*. Philadelphia: W.B. Saunders and Co., 1983.

Vajk, J. Peter. *Doomsday Has Been Cancelled*. Culver City, CA: Peace Press, 1978.

2. PEOPLE IN SPACE

Allaway, H. *The Space Shuttle at Work*. NASA SP-432. 1979.

Belwe, L.F., ed. *Skylab, Our First Space Station*. NASA SP-400. 1977.

Brooks, C.G., et al. *Chariots for Apollo: A History of Manned Lunar Spaceflight*. NASA SP-4205. 1979.

Kaplan, M. *Space Shuttle: America's Wings to the Future*. Fallbrook, CA: Aero Publishers, Inc., 1982.

NASA. *Apollo-Soyuz*. NASA EP-109. 1977.

———. *Apollo-Soyuz Test Project*. Vol. 1: *Astronomy, Earth Atmosphere and Gravity Field, Life Sciences, and Materials Processing*. NASA SP-412. 1977.

———. *Skylab EREP Investigations Summary*. NASA SP-399. 1978.

———. *Skylab, Our First Space Station*. NASA SP-400. 1977.

———. *Skylab's Astronomy and Space Sciences*. NASA SP-404. 1979.

———. *Why Man Explores*. NASA EP-123. 1977.

Nicogossian, Arnauld, and Parker, James. *Space Physiology and Medicine*. NASA SP-447. 1982.

3. OUR SOLAR SYSTEM

Beatty, J.K., et al., eds. *The New Solar System*. Cambridge, MA: Sky Publishing Co.; New York: Cambridge Univ. Press, 1981.

Briggs, G.A., and Taylor, F.W. *The Cambridge Photographic Atlas of the Planets*. New York: Cambridge Univ. Press, 1982.

Chapman, C.R. *Planets of Rock and Ice*. New York: Charles Scribner's and Sons, 1982.

Hartmann, W.K. *Moons and Planets*. 2nd ed. Belmont, CA: Wadsworth Publishing Co., Inc., 1983.

Lauber, P. *Journey to the Planets*. New York: Crown Publishers Inc., 1982.

Miller, R., and Hartman, W.K. *The Grand Tour: A Traveler's Guide to the Solar System*. New York: Workman Publishing Co., Inc., 1981.

Taylor, G.J. *Volcanoes in Our Solar System*. New York: Dodd, Mead, and Co., 1983.

Von Braun, W., and Ordway, F.I. *New Worlds: Discoveries from Our Solar System*. Garden City, NY: Doubleday and Co., 1979.

(SUN)

Asimov, Isaac. *The Sun Shines Bright*. 1981 Reprint. New York: Avon Books, 1983.

Eddy, J.A. *A New Sun: The Solar Results from Skylab*. NASA SP-402. 1979.

Gibson, E.G. *The Quite Sun*. NASA SP-303. 1973.

Goodwin, J., et al., eds. *Fire of Life: The Smithsonian Book of the Sun*. Washington, DC: Smithsonian Inst., 1981.

Herman, H.R., and Goldberg, R.A. *Sun, Weather, and Climate*. NASA SP-426. 1979.

(MERCURY & VENUS)

Asimov, Isaac. *Venus, Near Neighbor of the Sun*. New York: Lothrop, Lee & Shepard Books, 1981.

Davies, M.E., et al. *Atlas of Mercury*. NASA SP-423. 1978.

Dunne, J.A., and Burgess, E. *The Voyage of Mariner 10*. NASA SP-424. 1978.

Fimmel, R.O., et al. *Pioneer Venus*. NASA SP-461. 1983.

(EARTH)

NASA. *Skylab Explores the Earth*. NASA SP-380. 1977.

Nicks, O.W., ed. *This Island Earth*. NASA SP-250. 1970.

Short, N.M. *The Landsat Tutorial Workbook*. NASA RP-1078. 1982.

——— et al. *Mission to Earth: Landsat Views the World*. NASA SP-360. 1976.

(MOON)

Collins, Michael. *Flying to the Moon and Other Strange Places*. New York: Farrar, Straus, and Giroux, 1976.

French, B.M. *What's New on the Moon*. NASA EP-131. 1976.

Kosofsky, L.J., and El-Baz, F. *The Moon as Viewed by Lunar Orbiter*. NASA SP-200. 1970.

Masursky, H., et al., eds. *Apollo over the Moon: A View from Orbit*. NASA SP-362. 1978.

Moore, P. *The Moon*. New York: Rand McNally, 1981.

NASA. *Apollo Expeditions to the Moon*. NASA SP-350.

Taylor, G.J. *A Close Look at the Moon*. New York: Dodd, Mead, and Co., 1980.

Zim, H.S. *The New Moon*. New York: Morrow Junior Books. William Morrow and Co., 1980.

(MARS)

Batson, R.M., et al. *Atlas of Mars*. NASA SP-438. 1979.

Carr, M.H., and Evans, N. *Images of Mars: The Viking Extended Mission*. NASA SP-444. 1980.

Cooper, Henry S.F., Jr. *The Search for Life on Mars: The Evolution of an Idea*. New York: Holt, Rinehart, and Winston. 1980.

French, B.M. *Mars: The Viking Discoveries*. NASA EP-146. 1977.

NASA. *The Martian Landscape—Viking Mars Mission and Photographs*. NASA SP-425. 1978.

Oberg, James E. *Mission to Mars: Plans and Concepts for the First Manned Landing*. Harrisburg, PA: Stackpole Publishing Co., 1982.

Spitzer, C.R. *Viking Orbiter Views of Mars.* NASA SP-441. 1980.

(JUPITER)

Fimmel, R.O., et al. *Pioneer: First to Jupiter, Saturn, and Beyond.* NASA SP-446. 1980.

———. *Pioneer Odyssey.* NASA SP-349. 1977.

Morrison, D., and Samz, J. *Voyage to Jupiter.* NASA SP-439. 1980.

(SATURN)

Asimov, Isaac. *Saturn and Beyond.* New York: Lothrop, Lee & Shepard Books, 1979.

Cooper, Henry, S.F., Jr. *Imaging Saturn.* New York: Holt, Rinehart, and Winston, 1982.

Morrison, D. *Voyages to Saturn.* NASA SP-451. 1982.

NASA. *Voyager 1 Encounters Saturn.* NASA JPL 400-100. 1980.

(URANUS, NEPTUNE, PLUTO)

Grosser, M. *The Discovery of Neptune.* New York: Dover Publications, Inc., 1979.

Moore, P., and Tombaugh, C. *Out of the Darkness: The Planet Pluto.* Harrisburg, PA: Stackpole Books, 1980.

Clube, V., and Napier, B. *The Cosmic Serpent.* New York: Universe Books, 1982.

Gehrels, T., ed. *Asteroids.* Tucson, AZ: Univ. of Arizona Press, 1979.

Hutchinson, R. *The Search for Our Beginnings.* New York: Oxford Univ. Press, 1983.

McDonnell, J.A.M., ed. *Cosmic Dust.* New York: John Wiley and Sons, 1978.

Pejovic, B. *Man and Meteorites.* Avening, England: Thames Head, Ltd., 1982.

(COMETS)

Calder, N. *The Comet Is Coming.* New York: Penguin Books, 1981.

Moore, P. *Comets.* New York: Charles Scribner and Sons, 1976.

4. OUR EXTRATERRESTRIAL CIVILIZATION

Adleman, Saul J., and Adelman, Benjamin. *Bound for the Stars.* Englewood Cliffs, NJ: Prentice Hall, Inc., 1981.

Angelo, Joseph A., Jr., and Buden, David. *Space Nuclear Power.* Malabar, FL: Orbit Book Company, Inc., 1985.

Calder, Nigel. *Spaceships of the Mind.* New York: The Viking Press, 1978.

Glenn, J.C., and Robinson, G.S. *Space Trek: The Endless Migration.* New York: Warner Books, 1980.

Heppenheimer, T.A. *Colonies in Space.* New York: Warner Books, 1978.

———. *Toward Distant Suns.* Harrisburg, PA: Stackpole Books, 1979.

Johnson, R.D., and Holbrow, C., eds. *Space Settlements: A Design Study.* NASA SP-413. 1977.

NASA. *Space Resources and Space Settlements.* NASA SP-428. 1979.

Oberg, James E. *Mission To Mars.* Harrisburg, PA: Stackpole Books, 1982.

———. *New Earths.* New York: The New American Library, 1983.

O'Neill, G.K. *The High Frontier: Human Colonies in Space.* New York: William Morrow and Co., 1977.

5. SEARCH FOR EXTRATERRESTRIAL INTELLIGENCE (SETI); LIFE IN THE UNIVERSE

Asimov, Isaac. *Extraterrestrial Civilizations.* New York: Crown Publishers, 1979.

Billingham, J., ed. *Life in the Universe.* NASA CP-2156. 1981; Boston, MA: MIT Press, 1981.

Cocconi, G., and Morrison, P., "Searching for Interstellar Communications." *Nature* 184 (1959): 844.

Cooper, Henry, S.F., Jr. *The Search for Life on Mars: The Evolution of an Idea.* New York: Holt, Rinehart, and Winston, 1980.

Cousins, N. *Why Man Explores.* Symposium Held at Beckman Auditorium, Calif. Inst. of Tech., Pasadena, CA, July 2, 1976. Washington, DC: U.S. Gov't. Printing Off., 1976.

Crick, Francis. *Life Itself.* New York: Simon and Schuster, 1981.

Feinberg, G., and Shapiro, R. *Life beyond Earth: The Intelligent Earthling's Guide to Life in the Universe.* New York: William Morrow and Co., Inc., 1980.

Goldsmith, Donald. *The Quest for Extraterrestrial Life: A Book of Readings.* Mill Valley, CA: University Science Books, 1980.

———, and Owen, Tobias. *The Search for Life in the Universe.* Menlo Park, CA: The Benjamin/Cummings Publishing Company, Inc., 1980.

Hoyle, Fred. *The Intelligent Universe.* New York: Holt, Rinehart, and Winston, 1983.

———, and Wickramasinghe, N.C. *Evolution from Space.* New York: Simon and Schuster, Inc., 1981.

———. *Lifecloud: The Origin of Life in the Universe.* New York: Harper and Row, 1978.

Jastrow, R. *Until the Sun Dies.* New York: W.W. Norton and Co., 1977.

Mallove, E.F., and Forward, R.L. "Bibliography of Interstellar Travel and Communication." *JBIS* 33, no. 6, 1980: 201–48.

———. et al. *A Bibliography on the Search for Extraterrestrial Intelligence.* NASA RP-1021. 1978.

Morrison, P., et al., eds. *The Search for Extraterrestrial Intelligence: SETI.* NASA SP-419. 1977.

NASA. *Life beyond Earth and the Mind of Man.* NASA SP-328. 1973.

———. *Project Orion: A Design Study of a System for Detecting Extrasolar Planets.* NASA SP-436. 1980.

Ridpath, Ian. *Messages from the Stars: Communication and Contact with Extraterrestrial Life.* Harper Colophon Books. New York: Harper & Row, Publishers, 1979.

Rood, Robert T., and Trefil, James S. *Are We Alone?* New York: Charles Scribner's Sons, 1981.

Sagan, Carl, ed. *Communication with Extraterrestrial Intelligence.* Cambridge, MA: M.I.T. Press, 1973.

———. *The Cosmic Connection.* New York: Doubleday & Company, 1973.

———. *Cosmos.* New York: Random House, 1980.

———. *Murmurs Of Earth.* New York: Random House, Inc., 1978.

Shklovskii, I.S., and Sagan, Carl. *Intelligent Life in the Universe.* New York: Holden-Day, 1966.

Sullivan, Walter. *We Are Not Alone.* New York: McGraw-Hill Book Company, 1964.

Sullivan, Walter T., 3rd, et al. "Eavesdropping: The Radio Signature of the Earth." *Science,* 199 (Jan. 27, 1978): 377–88.

Tarter, Jill. "Searching for THEM: Interstellar Communications." *Astronomy,* 10, no. 10 (Oct. 1982).

Tipler, F.J. "Extraterrestrial Intelligent Beings Do Not Exist." *Quarterly Journal of the Royal Astronomical Society* 21 (1980): 267–81.

Troitskii, V.S., et al. "Search for Radio Emissions from Extraterrestrial Civilizations." *Acta Astronautica* 6, no. 1–2. (1979): 81–94.

Wolfe, J.H. et al. "SETI, The Search for Extraterrestrial Intelligence: Plans and Rationale," in *Life in the Universe.* NASA CP-2156; Cambridge, MA: The MIT Press, 1981.

6. UNIDENTIFIED FLYING OBJECTS (UFOs)

Condon, E.U. *Scientific Study of Unidentified Flying Objects.* New York: Bantam Books, 1969.

Hynek, J.A. *The UFO Experience: A Scientific Inquiry.* New York: Ballantine Books, 1975.

Oberg, James E. *UFO's and Outer Space Mysteries: A Sympathetic Skeptic's Report.* Norfolk, VA: Donning Company, 1982.

INDEX OF CROSS REFERENCES